Medicinal Plants with Anti-Ageing Properties

Saikat Gantait • Oksana Sytar
Jayoti Majumder Sarkar • Uma Rani Sinniah
Editors

Medicinal Plants with Anti-Ageing Properties

Advances in Biotechnology and Therapeutic Potential

Springer

Editors
Saikat Gantait
Crop Research Unit (Genetics and Plant Breeding)
Bidhan Chandra Krishi Viswavidyalaya
Mohanpur, West Bengal, India

Jayoti Majumder Sarkar
Department of Floriculture and Landscape Architecture
Bidhan Chandra Krishi Viswavidyalaya
Mohanpur, West Bengal, India

Oksana Sytar
Centre Agronomique et Agroalimentaire de la Province de Hainaut
Agronomic and Agri-food Center of the Province of Hainaut
Rue Paul Pastur, Belgium

Uma Rani Sinniah
Department of Crop Science
Universiti Putra Malaysia
Serdang, Selangor, Malaysia

ISBN 978-981-95-8063-7 ISBN 978-981-95-8064-4 (eBook)
https://doi.org/10.1007/978-981-95-8064-4

This Springer imprint is published by the registered company Springer Nature Singapore Pte Ltd.
The registered company address is: 152 Beach Road, #21-01/04 Gateway East, Singapore 189721, Singapore

Preface

Since ancient times, medicinal plants with naturally synthesizing active secondary metabolites have been used in traditional system of medicine as remedies for a wide range of ailments and to alleviate pain. In most developing nations, the use of conventional or alternative medicines made from medicinal plants for therapeutic purposes has received widespread praise, mainly for the preservation of health and wellness.

On the other hand, global atmospheric temperatures are expected to rise an additional 0.5–0.7 °C between 2020 and 2035, according to recent climate models, resulting in a total increase of 1.5–2 °C by the end of the twenty-first century. Persistent pollution is worsening the harmful effects of global warming. The demand for herbs with anti-aging properties is increasing in both developed and developing countries amid unpredictable weather patterns caused by climate change. Notably, a significant portion of the plant materials harvested and traded comes from wild habitats, with fewer species being cultivated. As a result, several plant species are seriously threatened with extinction due to the expanding trade in medicinal plants. These anti-aging medicinal plants contain highly bioactive components that could be developed into pharmacologically active drugs, making their conservation essential.

In that context, featuring 14 chapters with a comprehensive bibliography, this book highlights the anti-aging properties of relevant medicinal plants and their mechanisms of action, as well as advanced biotechnological techniques essential for mass propagation, including in vitro regeneration, genome editing, omics approaches, nanotechnology applications, cryopreservation, and the production of secondary metabolites through cell culture and bioreactors. Targeted toward scientists, researchers, pharmacologists, drug developers, botanists, and students, this book provides essential insights into recent advancements and prospective directions in biotechnological research concerning medicinal plants associated with anti-aging.

We are highly obliged to Springer Nature Singapore Pte. Ltd. for bringing out this project with successful accomplishment.

Mohanpur, West Bengal, India — Saikat Gantait
Rue Paul Pastur, Belgium — Oksana Sytar
Mohanpur, West Bengal, India — Jayoti Majumder Sarkar
Serdang, Selangor, Malaysia — Uma Rani Sinniah

About the Book

To ensure sustainable production of pharmaceutically significant active metabolites (from natural sources) with anti-aging properties, biotechnological interventions become the need of the hour when global warming and environmental pollution result in increasing UV radiation and, consequently, sunburn and aging of the tissue. In that context, the introductory chapter of this book outlines the concept and importance of medicinal plants available in local, national, and global regions that are rich in bioactive compounds responsible for anti-aging effects, along with their distribution, status, and limitations of conventional propagation and their biotechnological enhancement. The subsequent chapter provides insights into how environmental changes affect the anti-aging properties of metabolites in medicinal plants. There are dedicated chapters that provide insights on the genetic basis of anti-aging activity, the chemistry and physiology of modes of action, as well as the cellular mechanisms of secondary metabolites, along with the role of phytohormones for in vitro mass propagation, somatic embryos, their maturation, and the regeneration of anti-aging medicinal plants. This book also includes salient chapters on enhancement of anti-aging properties and the production of secondary metabolites with the interventions of LEDs, cell suspension, elicitation, bioreactor, auto-polyploidization, induced mutagenesis, and nanotechnology, as well as CRISPR/Cas9-mediated genome editing. Apart from genetic improvements, chapter-wise focus is given on conservation and exchange of anti-aging medicinal plants via cryopreservation and synthetic seed technologies, as well as secondary metabolite extraction techniques and the clinical evidence and significance of these metabolites. Expert scientists from 15 different countries, such as Argentina, Belgium, Botswana, Cameroon, China, Cuba, Czech Republic, Egypt, France, India, Japan, Mexico, South Africa, Türkiye, and Vietnam, have contributed 14 insightful chapters to this book. It is noteworthy to mention that there is a sustainable hunt for possible novel anti-aging compounds from natural resources, especially plants and herbs, and this book will be a potential guideline to facilitate that endeavor.

Contents

Editors and Contributors

About the Editors

Saikat Gantait is an Assistant Professor at the Crop Research Unit (Genetics and Plant Breeding), Bidhan Chandra Krishi Viswavidyalaya (BCKV), India. Following the completion of his PhD in biotechnology from BCKV, he gained advanced research experience on in vitro conservation of plant germplasms at Universiti Putra Malaysia during his post-doctoral tenure. His 20-year research work so far includes tissue and organ culture, induced mutagenesis and polyploidization, encapsulation/synthetic seed, and cryopreservation of medicinal, aromatic, and ornamental plant germplasms. He is Editorial Board Member in multiple journals from Springer Nature and Elsevier. His name has been included in the list of Top 2% Scientists Worldwide (2023, 2024, and 2025) by Stanford University. To date, he has published more than 180 articles/book chapters in peer-reviewed journals/books and edited 5 books.

Oksana Sytar obtained her PhD from Taras Shevchenko National University of Kyiv, Ukraine. She has been awarded several international scholarships to conduct advanced research at the University of Hyderabad (India), Technical University of Berlin (Germany), University of Bretagne Sud (France), and the Slovak University of Agriculture in Nitra (Slovakia). She is a recognized specialist in plant biochemistry, phytoremediation, and the study of secondary metabolites. Her name has been included in the list of Top 2% Scientists Worldwide (2024 and 2025) by Stanford University. Currently, she is engaged in biotechnological research on rare medicinal plants in Belgium.

Jayoti Majumder Sarkar is an Assistant Professor in Floriculture and Landscape Architecture, Bidhan Chandra Krishi Viswavidyalaya (BCKV), India. She received her MSc and PhD degrees from the prestigious Indian Agricultural Research Institute, New Delhi, India, and has been working for more than 15 years to make a measurable impact to the research fields of nutraceuticals, secondary metabolites, edible flowers, and natural pigments. She also served as an Agricultural Research Scientist at ICAR-Directorate of Floricultural Research, Pune, India. Her research works have been published in 44 research/review articles and 16 book chapters in reputed journals/books, and she has edited/authored 4 books. She guided multiple MSc and PhD students under her supervision at BCKV. She was selected for a training program in Wuhan, China, on Seed production and crop improvement. She is an active member of several science academies and societies, IAHS, Society for Promotion of Horticulture, Society for Plant Research, Crop and Weed Science Society, and an editorial board member of Journal of Crop and Weed and Medicinal Plants.

Uma Rani Sinniah is a Professor at the Department of Crop Science, Faculty of Agriculture, Universiti Putra Malaysia (UPM). She has more than 40 years of teaching and research experience in the fields of seed technology, cryobiology, germplasm conservation, and plant secondary metabolites, and has successfully completed several projects supported by various national and international funding bodies. She has published >200 peer-reviewed research/review articles, 12 book chapters, and edited 2 books (Springer) to date. Her current research is focused on seed science and technology, cryopreservation, plant cell and tissue culture, and secondary metabolite production.

Contributors

Ghada M. Abbas Department of Pharmacognosy, Horus University—Egypt, New Damietta, Egypt

Mostafa A. Asmaey Department of Chemistry, Faculty of Science, Al-Azhar University, Assiut, Egypt

Bella Belinga Merveilles Aurore Department of Chemistry, Higher Training College, University of Yaounde 1, Yaounde, Cameroon

Gastón Barreto Centro de Investigaciones en Física e Ingeniería del Centro de la Provincia de Buenos Aires CIFICEN (UNCPBA-CICPBA-CONICET), Buenos Aires, Argentina

Facultad de Ingeniería, Universidad Nacional del Centro de la Provincia de Buenos Aires (UNCPBA), Buenos Aires, Argentina

Consejo Nacional de Investigaciones Científicas y Técnicas (CONICET), Buenos Aires, Argentina

Rohit Bharati Plant Virus and Vector Interactions, Czech Agrifood Research Center, Prague, Czech Republic

Faculty of Economics and Management, Department of Economic Theories, Czech University of Life Sciences Prague, Prague 6, Czech Republic

Vinod Chhokar Department of Biotechnology, Guru Jambheshwar University of Science and Technology, Hisar, Haryana, India

Luciano Protti Cosenza Centro de Investigaciones en Física e Ingeniería del Centro de la Provincia de Buenos Aires CIFICEN (UNCPBA-CICPBA-CONICET), Buenos Aires, Argentina

Consejo Nacional de Investigaciones Científicas y Técnicas (CONICET), Buenos Aires, Argentina

Agustina Fernández Centro de Investigaciones en Física e Ingeniería del Centro de la Provincia de Buenos Aires CIFICEN (UNCPBA-CICPBA-CONICET), Buenos Aires, Argentina

Comisión de Investigaciones Científicas de la Provincia de Buenos Aires (CICPBA), Buenos Aires, Argentina

Facultad de Ingeniería, Universidad Nacional del Centro de la Provincia de Buenos Aires (UNCPBA), Buenos Aires, Argentina

Eloy Fernández-Cusimamani The Faculty of Tropical AgriSciences, Department of Crop Sciences and Agroforestry, Czech University of Life Sciences Prague, Suchdol, Czech Republic

Luisa Franchi Centro de Investigaciones en Física e Ingeniería del Centro de la Provincia de Buenos Aires CIFICEN (UNCPBA-CICPBA-CONICET), Buenos Aires, Argentina

Facultad de Ingeniería, Universidad Nacional del Centro de la Provincia de Buenos Aires (UNCPBA), Buenos Aires, Argentina

Goabaone Gaobotse School of Life Sciences, Department of Biological Sciences and Biotechnology, Botswana International University of Science and Technology, Palapye, Botswana

Nedim Emre Gök Faculty of Arts and Science, Department of Chemistry, Bolu Abant Izzet Baysal University, Bolu, Türkiye

Ekrem Gürel Faculty of Arts and Science, Department of Biology, Bolu Abant Izzet Baysal University, Bolu, Türkiye

Faculty of Agriculture, Department of Agricultural Biotechnology, Iğdır University, Iğdır, Türkiye

Songül Gürel Faculty of Arts and Science, Department of Biology, Bolu Abant Izzet Baysal University, Bolu, Türkiye

Faculty of Agriculture, Department of Agricultural Biotechnology, Iğdır University, Iğdır, Türkiye

Abdullah Haikal Department of Pharmacognosy, Mansoura University, Mansoura, Egypt

Shan Huang College of Food Science and Engineering, West Coast Campus of Ocean University, Qingdao City, Shandong Province, China

Santiago-Hernández Iliana Postgraduate College-Campus Veracruz, Veracruz, Mexico

Bello-Bello Jericó Jabín SECIHTI-Postgraduate College-Campus Cordoba, Veracruz, Mexico

Alka Jangra Department of Agricultural Biotechnology, College of Biotechnology, Chaudhary Charan Singh Haryana Agricultural University, Hisar, Haryana, India

David R. Katerere Department of Pharmaceutical Sciences, Tshwane University of Technology, Pretoria, South Africa

Masaki Kita Department of Applied Biosciences, Graduate School of Bioagricultural Sciences, Nagoya University, Nagoya, Japan

Keletso Lashani School of Life Sciences, Department of Biological Sciences and Biotechnology, Botswana International University of Science and Technology, Palapye, Botswana

Department of Biological Sciences, K.K. Birla Institute of Technology and Science, PILANI, Goa, India

Chunyang Li Institute of Agro-Product Processing, Jiangsu Academy of Agricultural Sciences, Nanjing, People's Republic of China

Nguyen Hoang Loc Institute of Bioactive Compounds and Department of Biotechnology, University of Sciences, Hue University, Hue, Vietnam

Abdullah Makhzoum School of Life Sciences, Department of Biological Sciences and Biotechnology, Botswana International University of Science and Technology, Palapye, Botswana

Sonia Malik Department of Biotechnology, Baba Farid College of Engineering and Technology, Bathinda, Punjab, India

Physiology, Ecology and Environment (P2E) Laboratory, University of Orleans, Orleans, France

Kabo Masisi School of Life Sciences, Department of Biological Sciences and Biotechnology, Botswana International University of Science and Technology, Palapye, Botswana

Luciana Mentasti Centro de Investigaciones en Física e Ingeniería del Centro de la Provincia de Buenos Aires CIFICEN (UNCPBA-CICPBA-CONICET), Buenos Aires, Argentina

Facultad de Ingeniería, Universidad Nacional del Centro de la Provincia de Buenos Aires (UNCPBA), Buenos Aires, Argentina

Consejo Nacional de Investigaciones Científicas y Técnicas (CONICET), Buenos Aires, Argentina

Marcos Edel Martínez Montero Universidad de Ciego de Ávila "Máximo Gómez Báez", Ciego de Ávila, Cuba

Iliassou Lah Mouafon Department of Pharmaceutical Sciences, Tshwane University of Technology, Pretoria, South Africa

Gaetan Bayiha Ba Njock Faculty of Science, Department of Chemistry, University of Maroua, Maroua, Cameroon

Mehmet Örgeç Faculty of Arts and Science, Department of Biology, Bolu Abant Izzet Baysal University, Bolu, Türkiye

Faculty of Agriculture, Department of Agricultural Biotechnology, Iğdır University, Iğdır, Türkiye

Harshid Pulparambil KSCSTE–Malabar Botanical Garden and Institute for Plant Sciences, Kozhikode, Kerala, India

P. E. Rajasekharan Division of Flower and Medicinal Crops, ICAR-Indian Institute of Horticultural Research (IIHR), Bengaluru, India

Ebede Guy Roland Department of Chemistry, Higher Training College, University of Yaounde 1, Yaounde, Cameroon

Ignacio Scandura Centro de Investigaciones en Física e Ingeniería del Centro de la Provincia de Buenos Aires CIFICEN (UNCPBA-CICPBA-CONICET), Buenos Aires, Argentina

Comisión de Investigaciones Científicas de la Provincia de Buenos Aires (CICPBA), Buenos Aires, Argentina

Amira E. Shaaban Department of Pharmacognosy, Horus University—Egypt, New Damietta, Egypt

Garima Sharma Department of Biotechnology, Guru Jambheshwar University of Science and Technology, Hisar, Haryana, India

Oksana Sytar Agronomic and Agri-Food Center of the Province of Hainaut (CARAH), Ath, Belgium

Phytopathology, Microbial and Molecular Farming Lab, Haute Ecole Provinciale du Hainaut – Condorcet, Ath, Belgium

Botalys – Research & Development Department, Botalys, Ghislenghien, Belgium

Le Thi Ha Thanh Institute of Bioactive Compounds and Department of Biotechnology, University of Sciences, Hue University, Hue, Vietnam

Ndongo Joseph Thierry Department of Chemistry, Higher Training College, University of Yaounde 1, Yaounde, Cameroon

Nguyen Hoang Tue Institute of Bioactive Compounds and Department of Biotechnology, University of Sciences, Hue University, Hue, Vietnam

Özlem Ünlü Faculty of Arts and Science, Department of Chemistry, Düzce University, Düzce, Türkiye

Faculty of Arts and Science, Department of Chemistry, Bolu Abant İzzet Baysal University, Bolu, Türkiye

Ömer Can Ünüvar Faculty of Arts and Science, Department of Biology, Bolu Abant Izzet Baysal University, Bolu, Türkiye

Faculty of Health Sciences, Department of Nutrition and Dietetics, Iğdır University, Iğdır, Türkiye

Alexander Aguila Wharton Centro de Investigaciones en Física e Ingeniería del Centro de la Provincia de Buenos Aires CIFICEN (UNCPBA-CICPBA-CONICET), Buenos Aires, Argentina

Consejo Nacional de Investigaciones Científicas y Técnicas (CONICET), Buenos Aires, Argentina

Ahmed Zayed Pharmacognosy Department, Tanta University, Tanta, Egypt

Xue Zhao Key Laboratory of Aquatic Product Processing, Ministry of Agriculture and Rural Affairs, National R&D Center for Aquatic Product Processing, South China Sea Fisheries Research Institute, Chinese Academy of Fishery Sciences, Guangzhou City, Guangdong Province, China

Yongqiang Zhao Key Laboratory of Aquatic Product Processing, Ministry of Agriculture and Rural Affairs, National R&D Center for Aquatic Product Processing, South China Sea Fisheries Research Institute, Chinese Academy of Fishery Sciences, Guangzhou City, Guangdong Province, China

The Roots of Youth and the Current Status of Anti-aging Medicinal Plants

1

Ömer Can Ünüvar, Mehmet Örgeç, Nedim Emre Gök, Ekrem Gürel, and Songül Gürel

Abstract

Since humanity's earliest days, the desire to remain youthful has been central to myths and, subsequently, to the scientific world. Plants have been used since ancient times to heal and prevent aging and are now studied by modern science. This section presents the cultural and scientific evolution of anti-aging concepts, highlighting the shift from the symbolic world of narratives to molecular mechanisms. Using the Web of Science database, we analyzed studies published from 2015 to 2025 on global research trends in plant-based anti-aging studies. The results showed that phytochemicals, dermatology, and longevity were the main citation topics at the meso-level. Antioxidant activity, cellular aging, and skin protection were the main attribution topics at the micro-level. Analyses are presented for *Panax ginseng*, *Centella asiatica*, *Ginkgo biloba*, *Curcuma longa*, *Cinnamomum* spp., *Rosmarinus officinalis*, *Hibiscus* spp., *Zingiber officinale*, *Phyllanthus emblica*, and *Aloe vera*. Analyses indicate an interdisciplinary shift

Ö. C. Ünüvar
Faculty of Arts and Science, Department of Biology, Bolu Abant Izzet Baysal University, Bolu, Türkiye

Faculty of Health Sciences, Department of Nutrition and Dietetics, Iğdır University, Iğdır, Türkiye

M. Örgeç · E. Gürel · S. Gürel (✉)
Faculty of Arts and Science, Department of Biology, Bolu Abant Izzet Baysal University, Bolu, Türkiye

Faculty of Agriculture, Department of Agricultural Biotechnology, Iğdır University, Iğdır, Türkiye
e-mail: songul.gurel@ibu.edu.tr

N. E. Gök
Faculty of Arts and Science, Department of Chemistry, Bolu Abant Izzet Baysal University, Bolu, Türkiye

S. Gantait et al. (eds.), *Medicinal Plants with Anti-Ageing Properties*,
https://doi.org/10.1007/978-981-95-8064-4_1

that combines traditional knowledge with modern biotechnology and nanoscience. China and India are leaders among the studies, reflecting a strong biocultural continuity. These findings highlight the central role of plants in modern anti-aging strategies. The roots of youth are not just underground like ancient narratives; they grow inside cells, genes, and molecules.

Keywords

Medicinal plants · Meso-level · Modern medicine · Research areas · Scientific trends · Web of Science database

1.1 Introduction

Humanity's thoughts about aging are one of the oldest important themes. It is deeply interlinked with philosophy and science. The desire to delay or reverse aging is defined as the "prolongevity hypothesis". This hypothesis represents one of the ancient intellectual traditions of civilization. In various cultures, people have shown attitudes toward mortality that are sometimes accepting and at other times challenging; they have viewed long life as a divine blessing, a moral virtue, or a scientific potentiality. Ancient narratives expressed this existential tension through symbolic representations. In the Epic of Gilgamesh, the hero's lost "plant of life" symbolizes humanity's collective disappointment when facing biological limitations (Gruman 1966). In Greek myths, Tithonus's eternal old age or the punishment of Prometheus are cautionary tales that curb humanity's desire for immortality. However, these myths also contain the first steps of scientific inquiry: if death is not a divine curse, it can be a subject for rational examination.

In the subsequent centuries, this line of thought gained a pragmatic dimension. In the Middle Ages and the Renaissance, physicians worked on "elixirs of life" and potions that promised both spiritual and physical renewal. Taoist alchemists, Islamic scholars, and European hygienists all contributed to what Gruman referred to as proto-scientific prolongevitism (Gruman 1966). This view argued that aging could be minimized by lifestyle, dietary habits, and biochemical processes. These intellectual frames gradually grew into systems of medicinal plants, such as *Panax ginseng* and *Centella asiatica*, which became central to this tradition. Before, what ancient physicians described as "restoring balance" is now defined in modern science via antioxidant and regenerative mechanisms. The myths about immortality have gained a moral and psychological dimension. The Western belief of the vampire transformed the desire for immortality from a sacred longing to an aesthetic obsession. The vampire myth reflects an obsession that preserves beauty, on the other hand, at the cost of one's humanity. This transformation acts as a precursor to the modern psychological phenomenon referred to as the Dorian Gray Syndrome. This syndrome can be described as a fear of aging. It results in a behavior of constantly preserving an individual's external appearance. Like the character in Oscar Wilde's 1890 novel, modern mankind attempts to hide, filter, or surgically eliminate aging rather than adopting it (Mehta Ambalal et al. 2023). This phenomenon, particularly in the age of social media, is associated with a distorted view of the body. Constant exposure to unrealistic representations reduces an individual's tolerance for the natural aging process and blurs their perception of beauty and youth (Mehta Ambalal et al. 2023).

Old age, once a symbol of wisdom, is increasingly perceived in today's societies as a "flaw" that must be concealed. The "fountain of youth", as in old narratives, is no longer sought in nature but in dermatology clinics or, in the New Age, through digital filters. Instead of myths, people are seeking immortality (eliminating or reducing the visibility of the effects of aging) through biotechnology. At the biochemical level, it has become measurable. Bjørklund et al. (2022) and Unlu and Koc (2007) identify oxidative stress, mitochondrial dysfunction, and chronic inflammation as the fundamental mechanisms of aging. New perspectives have transformed aging from an unchangeable condition into a controllable biological process. Natural compounds such as polyphenols and flavonoids, known as symbols of longevity in pharmacology, are now recognized as molecules capable of regulating aging processes at the genetic and cellular levels. Al Amin et al.'s (2024) study places this transformation in a holistic context. Their research reveals that today's anti-aging research lies at the intersection of molecular development and traditional wisdom. This research redefines the harmony between nature and humanity.

The phenomenon of "Dorian Gray Syndrome" highlights the psychological limitations of anti-aging research. The desire for youth no longer exists in temples but in laboratories, through aesthetic practices or through digital platforms, serving as the most modern version of humanity's binary response to mortality.

The history of anti-aging can be read as a single narrative, extending from Gilgamesh's lost plant to gene repair with modern techniques like CRISPR: the perpetual conflict between humanity's desire for longevity and the inevitability of change. Myths offered symbols, early humans studied elixirs, and modern science has identified molecular approaches. For thousands of years, one fundamental principle has remained unchanged: humans are searching not only for ways to live longer but also for ways to preserve wholeness in a lifetime.

Since the first days of humanity, the desire to remain young and alive has been at the center of both myths and the medical world. Plants have been used since ancient times to find healing and to stop time by preventing aging and are now being studied by modern science and medicine. In supernatural stories, these plants are depicted as both a source of healing and a curse. Today, science is trying to understand aging on a biological level, examining the anti-aging properties of many plants. There are many studies showing the effects of herbal ingredients used in anti-aging products.

In this chapter, we aim to share with readers the statistical data of recent scientific studies. The Web of Science database was used to perform bibliometric analysis. Data from the last decade (2015–2025) was analyzed utilizing the keywords "plant" and "anti-aging", presenting the focus areas of the studies as "citation topics meso" and "citation topics micro". The analysis was then conducted using the keywords "relevant plant name" and "anti-aging", and the resulting results were analyzed. The analysis results for the 10 most frequently used plants were reported, visualized, and presented to readers.

1.2 Citation Topics Analysis

1.2.1 Citation Topics Meso Analysis

The analysis of the results of publications is presented between 2015 and 2025 using the keywords "plant" and "anti-aging" in the Web of Science "Advanced Search" section at the Citation Topics Meso and Citation Topics Micro levels. From the analysis, a total of 1359 publications were evaluated, and the most common themes were obtained under the titles "Phytochemicals", "Dermatology", and "Longevity" (Table 1.1).

These data show us that "anti-aging plant research" has a multidisciplinary nature. Approximately 62% of the studies are based on biochemistry and dermatology, while 15% intersect with engineering (nanotechnology, biomaterials, and polymer science). From 2015 to 2025, research on the topic of "plants and anti-aging" focused on phytochemistry and dermatology areas. This situation tells us the increasing interest potential of anti-aging research in the cosmetics, pharmaceutical, and biotechnology sectors.

Table 1.1 "Citation topics meso" results of the studies between 2015 and 2025 (from Web of Science database)

Meso subject area	Number of publications	Percentage (%)	Explanation
Phytochemicals	517	38.0	Effect of plant bioactive compounds on anti-aging
Dermatology—general	167	12.3	Skin aging, collagen synthesis, UV damage
Longevity	158	11.6	Life-longing mechanisms and molecular pathways
Crop science	55	4.0	Agricultural production optimization of plant-derived compounds related to anti-aging
Photo productivity	24	1.7	Relations between photosynthetic compounds and their antioxidant capacities
Micro and long noncoding RNA	19	1.4	The role of RNAs in the aging process
Drug delivery chemistry	16	1.1	Usage of plant chemicals with nanotechnological delivery systems
Nanoparticles	15	1.1	New formulations of anti-aging compounds

Table 1.2 "Citation topics micro" results of the studies between 2015 and 2025 (from Web of Science Database)

Micro subject area	Number of publications	Percentage (%)	Explanation
Antioxidant activity	199	14.6	The most dominant topic is reactive removal of the oxygen species and oxidative stress reduction
Skin aging processes	144	10.6	Skin aging, UV protection, collagen destruction
Aging mechanisms	100	7.3	Cellular aging, telomere shortening, DNA damage
Polysaccharide bioactivities	63	4.6	Immunomodulatory and antioxidant effects of plantal polysaccharides
Resveratrol and sirtuins	58	4.2	Activation of sirtuin genes and resveratrol-like compounds
Essential oils	46	3.4	Antioxidant and/or anti-inflammatory potentials of essential oils
Ginsenosides	30	2.2	Anti-aging effect of biochemicals of *Panax ginseng*
Triterpenoids	27	1.9	Relations of triterpenes to cellular protection and cell regeneration
Advanced glycation end products (AGEs)	24	1.7	The role of protein glycation
Telomere dynamics	21	1.5	Telomerase activity and cellular lifetime extension

1.2.2 Citation Topics Micro Analysis

Micro-level analysis of the same dataset shows which micro topics are important in anti-aging plant research (Table 1.2).

When we examine Table 1.2, it can be interpreted that anti-aging-related studies focus on antioxidant systems, cellular aging, and polyphenolic compounds. Citation topics micro-analysis results based on the Web of Science database point out that herbal anti-aging studies are progressing along three fundamental axes between 2015 and 2025: strengthening of the antioxidant defense systems, cellular renewal and genetic balance, and cosmetic-dermatological applications.

When both levels of analysis are considered together, it is seen that publications with the theme of "plants and anti-aging" increased both their biological depth and application diversity from 2015 to 2025:

One-third of scientific production is concentrated in the fields of "phytochemistry" and "dermatology." This indicates that anti-aging compounds of plant origin have gained a central role in cosmetology and pharmaceutical R&D at meso level. "Antioxidant activity" and "skin aging processes" emerged as the two dominant sub-areas. This indicates that biochemical evidence of anti-aging effects has been studied at the micro level. Moreover, data suggest that biotechnology, nanotechnology, and cellular regulatory mechanisms are increasingly being integrated into botanical anti-aging research. The subheadings of "Drug Delivery Chemistry,"

"Nanoparticles," and "Transdermal Delivery" comprise approximately 3.3% of the total. This data indicates that in the future, herbal compounds will be utilized not only in their natural state but also with nanoformulations and/or biocompatible delivery systems.

Overall, the period from 2015 to 2025 can be regarded as a "scientific maturation phase" for herbal anti-aging research: Firstly, the traditional concept of the "natural antioxidant" has evolved to multilayered models based on cellular mechanisms. Secondly, an expanding field of influence has arisen from cosmetics to gene editing and from pharmacology to biotechnology. And finally, when we examine the trends, we observe that "herbal anti-aging approaches" are no longer just an aesthetic desire but have evolved into a holistic strategy for preserving biological youth.

1.3 Bibliometric Analysis of Publications on Anti-aging Plants Between 2015 and 2025

1.3.1 Ginseng

Panax ginseng is a perennial plant that belongs to the Araliaceae family (Su et al. 2023). Ginseng is a medicinal plant grown in 35 countries, particularly in South Korea and China (Potenza et al. 2023). Ginseng is an important natural source because of its anti-inflammatory and anti-aging effects (Attele et al. 1999; Xia et al. 2022). Its important active compound is ginsenoside. It is an important pharmacological compound obtained from several parts of the plants, such as the roots, stems, and leaves. It can enhance immunity and has anti-aging effects (Lai et al. 2018). Moreover, the ginseng plant has various essential secondary metabolites such as phenolic acids, gallic acid, salicylic acid, p-coumaric acid, ferulic acid, cinnamic acid and caffeic acid, flavonoids, saponins, ginseng oils, amino acids, and peptides (Chung et al. 2016; Metwaly et al. 2019). Ginseng has been shown to fight fatigue (Lu et al. 2021), regulate blood sugar levels (Hernández-García et al. 2019), promote heart health (Fan et al. 2020), and possess antitumor activities (Li et al. 2023).

Anti-aging studies related to plants have gradually increased, with plants containing pharmacologically active compounds. Ginseng is commonly studied in the scientific literature with its huge history in traditional usage. From 2015 to 2025, the data from the Web of Science database shows us a fluctuating increase in publication quantities. It began with 9 records in 2015 and increased to 11 in 2017 and 15 in 2018. While a relative decline (5 publications) was observed in 2019, the overall trend indicates continued interest in ginseng in anti-aging research. This distribution indicates that research activities on the topic gained significant growth in the 2017–2018 period and maintained a stable level of interest (Fig. 1.1). This trend can be interpreted as parallel to the increasing mechanistic investigations on ginseng's antioxidant, cell regenerative, and metabolic balancing effects; yet the primary purpose here is solely to observe the publication patterns. Consequently, the period from 2015 to 2018 can be defined as the "development phase" of ginseng anti-aging studies, while the period after 2019 can be defined as the "maturation phase". On

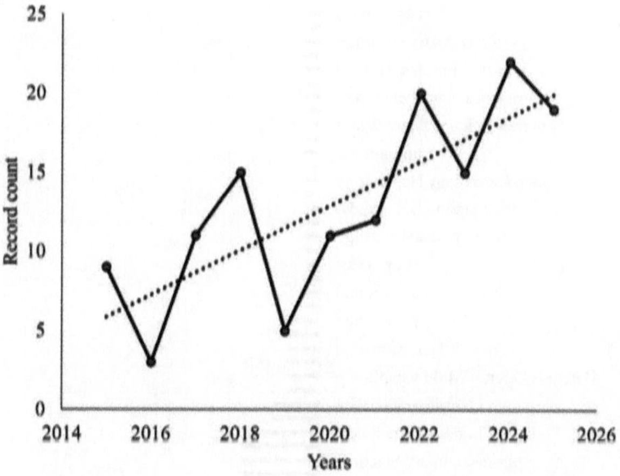

Fig. 1.1 Distribution of publications on Ginseng by year (2015–2025). (Data were visualized using the Web of Science database)

the other hand, it can be strongly speculated that the COVID pandemic has affected the studies and shifted the focus of researchers.

During the analyzed period, the predominant proportion of publications (99 publications) was in the "Research Article" category. This signifies that the subject is primarily supported by original research articles. And it is a research area based on empirical data generation. The low number of publications (1–2 publications) in other types (Early Access, Proceeding Paper, Retracted Publication, Retraction) shows that this area has not yet expanded into reviews or conference proceedings but is primarily driven by laboratory-based studies.

When we examine the distribution of publications by Web of Science categories for Ginseng, the categories with the highest number of publications include "Chemistry Medicinal" (37 publications), "Pharmacology Pharmacy" (35 publications), and "Integrative and Complementary Medicine" (30 publications). It shows us that the anti-aging effects of ginseng are primarily investigated through pharmacological and biochemical aspects. The significant representations of the categories "Plant Sciences" (29 publications) and "Biochemistry Molecular Biology" (25 publications) signify that the plant's basic biological functions are being analyzed at the molecular level (Fig. 1.2). The publication profile over the last decade shows that research has concentrated on mechanistic foundations and applied pharmaceutical effects.

The People's Republic of China (80 publications) is the clear leader in ginseng-related anti-aging research. It is followed by South Korea (40 publications) and the United States (12 publications). Contributions from countries such as India (8 publications) and Italy (4 publications) are limited (Fig. 1.3). This distribution is consistent with the origins and cultural history of ginseng usage; China and Korea provide continuity and depth in research production in this research area. The higher publication numbers of these countries reflect the academic foundation for the widespread use of ginseng in the pharmaceutical and cosmetic industries. The United States' contribution, though less significant but consistent, demonstrates the enduring research interest in this plant within the Western world.

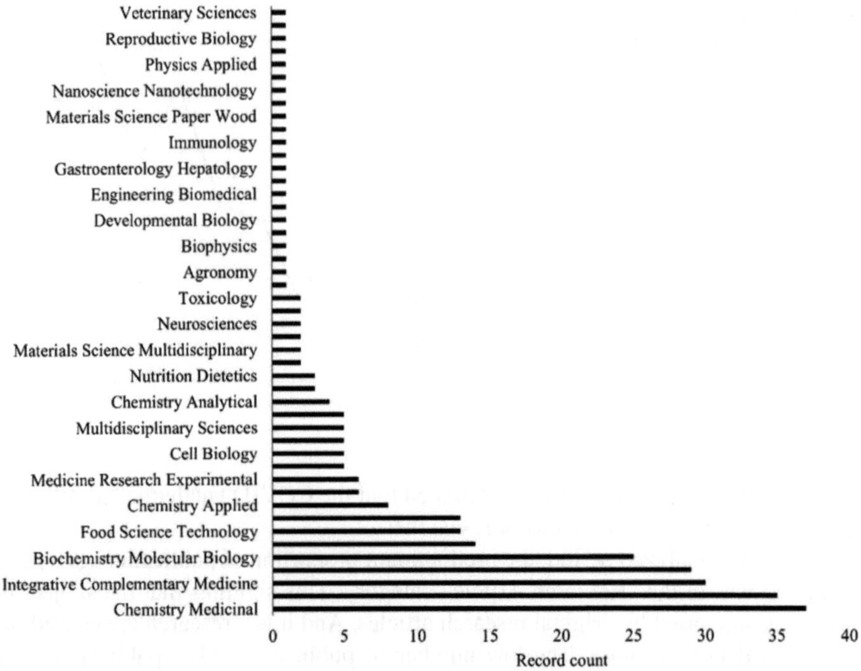

Fig. 1.2 Distribution of publications on Ginseng according to Web of Science categories. (Data were visualized using the Web of Science database)

Bibliometric trends between 2015 and 2025 clearly illustrate the centrality of ginseng in anti-aging research. When publication types, scientific fields, and geographic distribution are considered together, it can be concluded that this field has a particularly Asian-centric emphasis on knowledge production, but global interest is expanding. Despite fluctuations over the years, the overall trend is favorable, and the role of ginseng in anti-aging studies seems to be becoming increasingly systematic.

1.3.2 *Centella asiatica*

Centella asiatica L. Urban is a plant belonging to the Apiaceae family (Witkowska et al. 2024). It is a perennial plant and native to over 20 geographical locations globally (Min et al. 2024). The tropical regions of Southeast Asia, some parts of China, India, South Africa, South America, and Eastern Europe are the main places where it is grown (Das 2011). *C. asiatica* has small green leaves and three-oval fruits, but it has no flavor or aroma (Kandasamy et al. 2023). The height of an adult *C. asiatica* plant is 10–35 cm (Min et al. 2024).

The roots, stems, and leaves of this plant are used medicinally. *C. asiatica* is used medicinally due to its numerous compounds. The most important of these

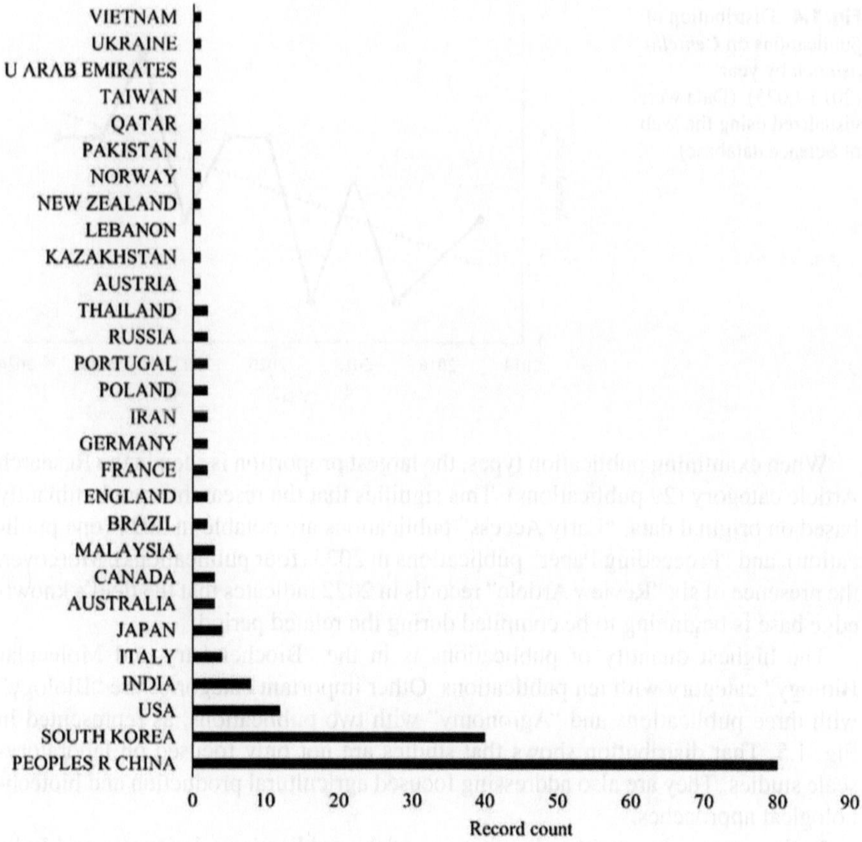

Fig. 1.3 Geographical distribution of publications on Ginseng by country and region. (Data were visualized using the Web of Science database)

compounds are the triterpene acids known as sapogenins, such as asiatic, madecassic, terminolic, synthellic, brahmic, isobramic, betulinic and madasiatic acids, and triterpene glycosides such as asiaticoside A, asiaticoside B, madecassoside, synthelloside, brahmoside, indocentelloside and vitamin C, vitamin B1, and vitamin B2 (Chandrika and Prasad Kumara 2015). Thanks to these important compounds, it is effective in skin and wound healing, antimicrobial, antioxidant, anti-inflammatory activity (Witkowska et al. 2024) and neurodegenerative diseases, digestive disorders (Banerjee et al. 2024). The search of the Web of Science database using the keywords "ginseng" and "anti-aging" between 2015 and 2025 allows us to examine its global research trends in this field.

When the data examined, five publications were published in each of 2025 and 2024, and the number of publications is seven in 2023. This trend, which continued with three publications in 2022 and five in 2021, demonstrates that anti-aging-focused studies on *C. asiatica* have continued steadily over the past 5 years (Fig. 1.4).

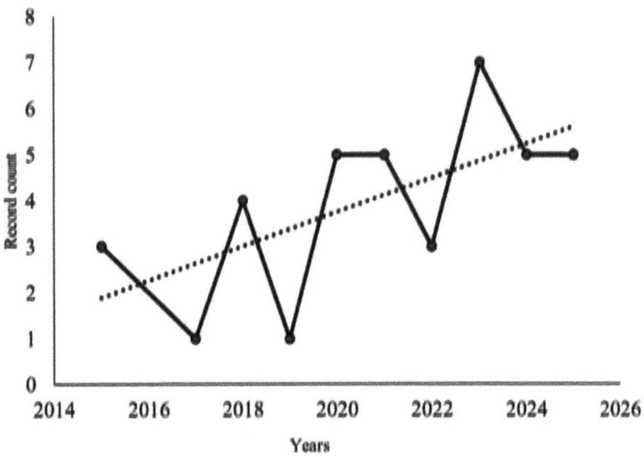

Fig. 1.4 Distribution of publications on *Centella asiatica* by year (2015–2025). (Data were visualized using the Web of Science database)

When examining publication types, the largest proportion is seen in the Research Article category (29 publications). This signifies that the research is predominantly based on original data. "Early Access" publications are notable in 2024 (one publication), and "Proceeding Paper" publications in 2023 (four publications). Moreover, the presence of six "Review Article" records in 2022 indicates that the field's knowledge base is beginning to be compiled during the related period.

The highest quantity of publications is in the "Biochemistry and Molecular Biology" category with ten publications. Other important categories are "Biology" with three publications and "Agronomy" with two publications, as represented in Fig. 1.5. That distribution shows that studies are not only focused on laboratory-scale studies. They are also addressing focused agricultural production and biotechnological approaches.

In the country-by-country distributions of the publications, Indonesia and India share the top two. These countries stand out due to both the plant's natural distribution and its history of use in traditional medicine. In Europe, Italy, Germany, and Greece contributed to the research area. It demonstrates that while research on *C. asiatica* is primarily focused on Asia, it has also increased interest in European countries (Fig. 1.6).

The trends from 2015 to 2025 show us a consistent increasing trend in anti-aging research involving *C. asiatica*. The publication quantities have increased markedly. And biochemical and molecular studies dominate within scientific categories. The country distribution indicates a research profile heavily Asian-centered, whereas limited but varied contributions from Europe suggest potential interest.

1.3.3 *Ginkgo biloba*

Ginkgo biloba, one of the plants with anti-aging effects, belongs to the Ginkgoaceae family (Morvaridzadeh et al. 2020). It is native to China, Japan, and Korea but is also found throughout much of Europe, America, India, and New Zealand (Barbalho et al. 2022). It has been present for 200 million years and is considered a "living

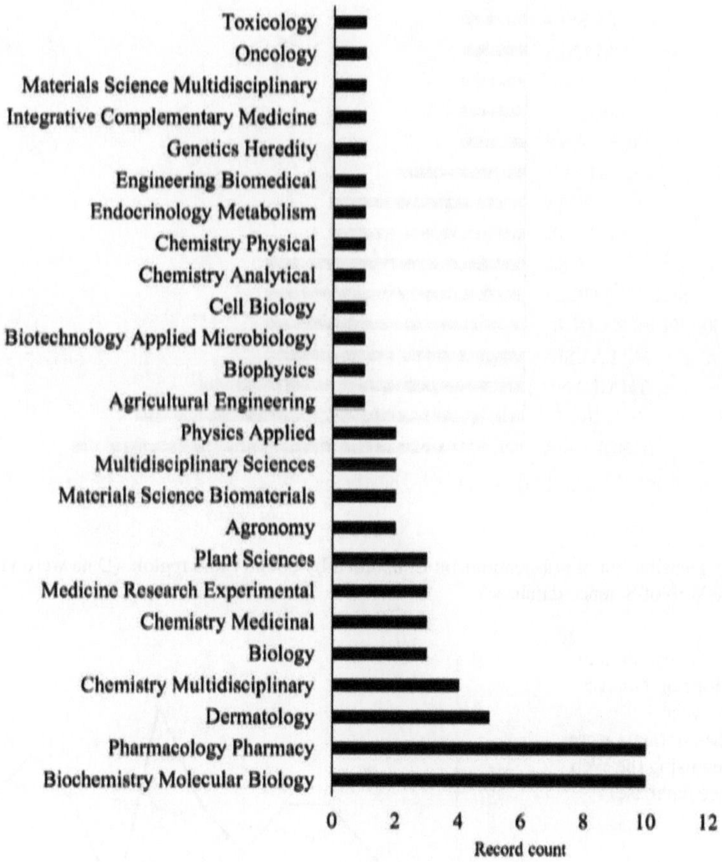

Fig. 1.5 Distribution of publications on *C. asiatica* according to the Web of Science database as categories. (Data were visualized using the Web of Science database)

fossil" (Liu et al. 2021). Ginkgo survived the mass extinctions of the Cretaceous and Paleogene periods and the Pleistocene glaciation to the present day (Chen et al. 2021). The effects of *G. biloba* are primarily because of the bioactive ingredients such as terpenoids, flavonoids, polyphenols, organic acids, alkylphenols, and polysaccharides (Barbalho et al. 2022; Li et al. 2022).

The leaves and seeds of *G. biloba* are among the most preferred phytopharmaceutical supplements in the United States and Europe. *G. biloba* is commercialized as an extract and offers several benefits related to memory, cognition, Alzheimer's disease, Parkinson's disease, and dementia. Extracts from *G. biloba* leaves are also becoming popular as an anti-aging medicinal product to delay aging-related disorders (Zuo et al. 2017; Barbalho et al. 2022). On the other hand, *G. biloba* seed extracts serve a multifunctional effect of anti-aging, anti-atherosclerosis, and anti-fatigue (Yuan et al. 2024). *G. biloba* inhibits ROS and MMP-1 degradation due to its high flavonoid and lactone levels. So, it shows anti-aging effects (Wang et al.

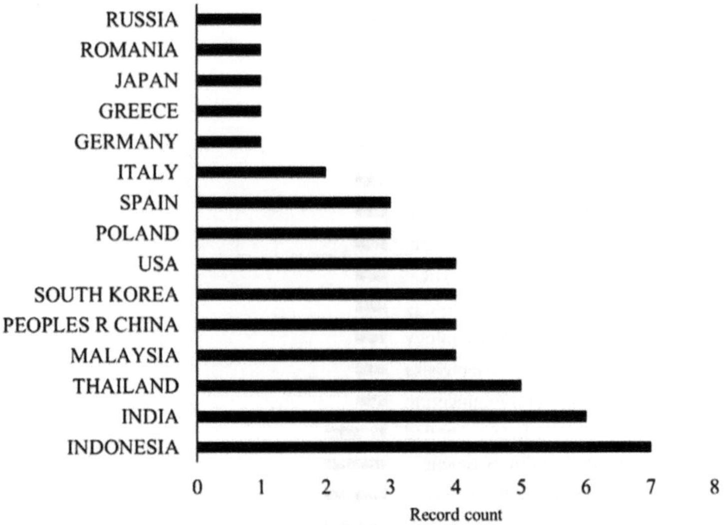

Fig. 1.6 Distribution of publications on *C. asiatica* by country and region. (Data were visualized using the Web of Science database)

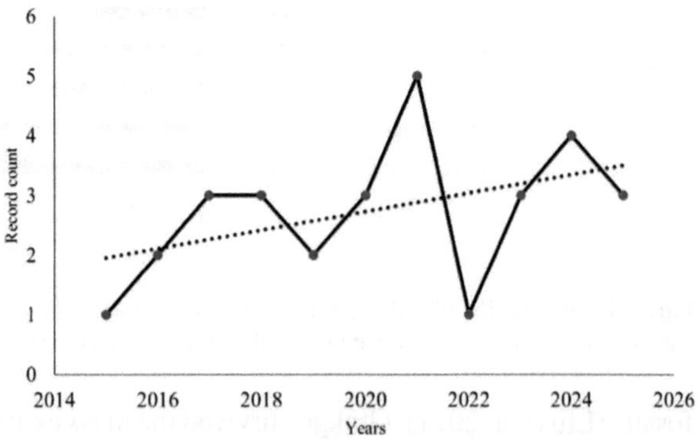

Fig. 1.7 Distribution of publications on *Ginkgo biloba* by year (2015–2025). (Data were visualized using the Web of Science database)

2020). The Web of Science database using the keywords "Ginkgo biloba" and "anti-aging" represents the quantitative and thematic properties of scientific production from 2015 to 2025.

The number of publications shows a relative increase from 2021 to 2025. Three publications were released in 2025, four in 2024, three in 2023, one in 2022, and five in 2021 (Fig. 1.7). This regular distribution indicates that *G. biloba* continues to be a popular subject in anti-aging research and has remained on the scientific agenda during this decade. The observed increase in 2024 and 2025 can be attributed to the proliferation of reviews.

The highest proportion of publications (23 publications) are "Research Articles". The data shows that research activities predominantly rely on original experimental data. The second highest ranking is for "Review Article" publications with seven publications, indicating that the accumulated knowledge in this area has progressed to systematic reviews in recent years. The presence of "Book Chapters" indicates that the topic is also finding a place in academic books. Generally, the publication types suggest that *G. biloba* research is maturing and gaining visibility across various publication formats.

The categories with the highest density of publications are "Food Science and Technology" with eight publications, "Biochemistry and Molecular Biology" with five publications, and "Pharmacology and Pharmacy" with five publications. The sum of these three fields suggests that most of the studies are conducted from biochemical and pharmacological frames. Moreover, the representation of "Chemistry Medicinal" with four publications and "Chemistry Multidisciplinary" with three publications categories demonstrates that chemistry-based studies investigating the phytochemical components of *G. biloba* account for a significant segment. This distribution suggests that the research profile in the anti-aging theme of *G. biloba* is

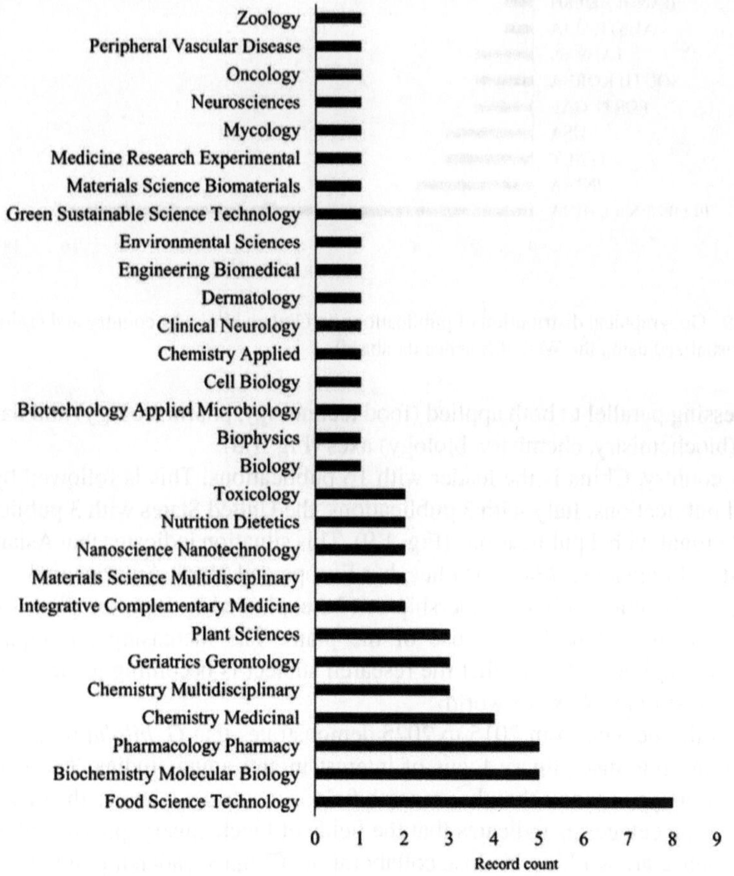

Fig. 1.8 Distribution of publications on *Ginkgo biloba* according to Web of Science categories. (Data were visualized using the Web of Science database)

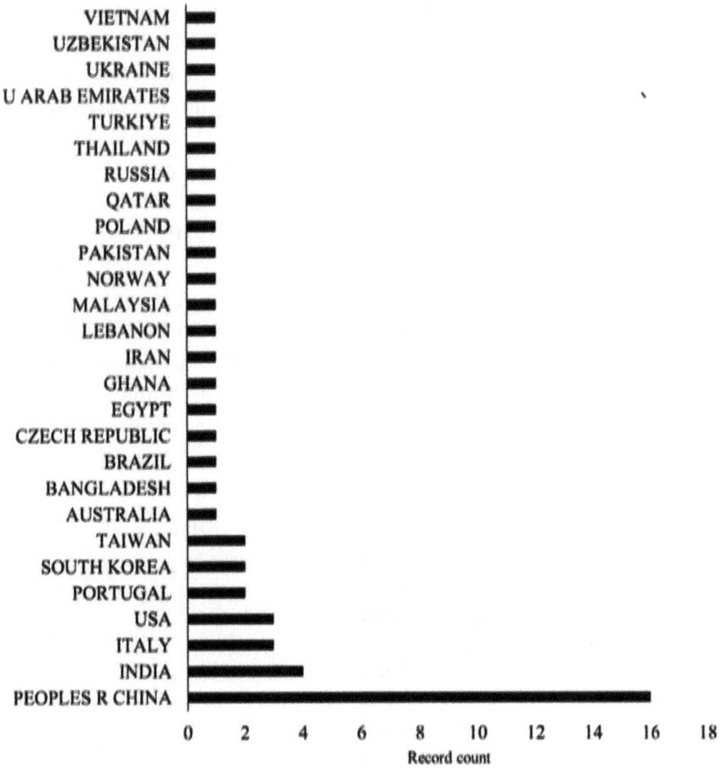

Fig. 1.9 Geographical distribution of publications on *Ginkgo biloba* by country and region. (Data were visualized using the Web of Science database)

progressing parallel to both applied (food technology, pharmacology) and basic science (biochemistry, chemistry, biology) axes (Fig. 1.8).

By country, China is the leader with 16 publications. This is followed by India with 4 publications, Italy with 3 publications, the United States with 3 publications, and Portugal with 2 publications (Fig. 1.9). This situation indicates that Asian countries still dominate *G. biloba* studies, but Europe and North America make significant contributions. China's leadership can be explained by both the plant's origins and its history of traditional use of the plant. The increasing participation of European regions indicates that the research subject is becoming an area of scientific interest in the Western world.

Overall, the data from 2015 to 2025 demonstrates that *G. biloba* remains a persistent and interdisciplinary focus of interest in anti-aging studies. The variety of publication types shows that the research field is maturing, whereas the distribution of scientific categories indicates that the fields of biochemistry, pharmacology, and food science are working in close collaboration. China's leadership in the area suggests that the plant serves as a bridge connecting deep traditional knowledge with modern science.

1.3.4 *Curcuma longa* (Turmeric)

Curcuma longa L. is a rhizomatous, herbaceous, perennial plant belonging to the genus *Curcuma* and the Zingiberaceae family. It can grow up to 1 m tall. The pseudo-stem has a short stem and large leaves. Its flowers are pale yellow and do not bear fruit (Gopinath and Karthikeyan 2018). It is commonly cultivated in India, China, Indonesia, Thailand, and other tropical regions (Trujillo et al. 2013).

The dried root of *C. longa* is known as turmeric (Aggarwal et al. 2013). It is also named "yellow root", "golden spice", and "Indian saffron" and has been used in medicine and religious practices for nearly 6000 years (Gopinath and Karthikeyan 2018).

Curcumin is the most researched polyphenol derived from *C. longa* Linn. Curcumin makes up 2–5% of dried turmeric root (Aggarwal et al. 2013). It holds an important place in the human diet because of its anti-cancer, antioxidant, immune-modulatory, anti-inflammatory, and anti-aging properties (Benameur et al. 2021).

Within the scope of our study, a bibliometric evaluation of research trends between 2015 and 2025 is presented to the readers, with the analysis made using the keywords "turmeric" and "anti-aging" in the Web of Science database. An analysis of the annual publication quantities between 2015 and 2025 indicates a peak of seven publications in 2023. Six publications were recorded in 2024, four in 2022, one in 2025, and one in 2021. This distribution suggests a notable increasing trend in anti-aging studies involving *C. longa* as of 2023. A minor decline was observed in 2024; publication activity is markedly denser compared to the period before 2021 (Fig. 1.10). The topic has been addressed more systematically in scientific communities, particularly over the past 5 years.

The most published type is "Research Article" (17 publications). The 14 publications in the "Review Article" type demonstrate that the body of knowledge in the field is becoming more evident and that new research directions are being identified by evaluating prior studies.

When examining the distribution of publications, the highest number is in the "Pharmacology and Pharmacy" category with eight publications. This is followed by "Geriatrics and Gerontology" with six publications, "Biochemistry and

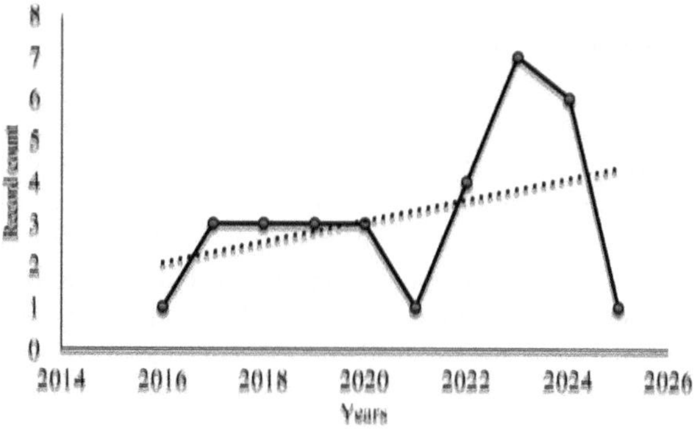

Fig. 1.10 Distribution of publications on *Curcuma longa* (Turmeric) by year (2015–2025). (Data were visualized using the Web of Science database)

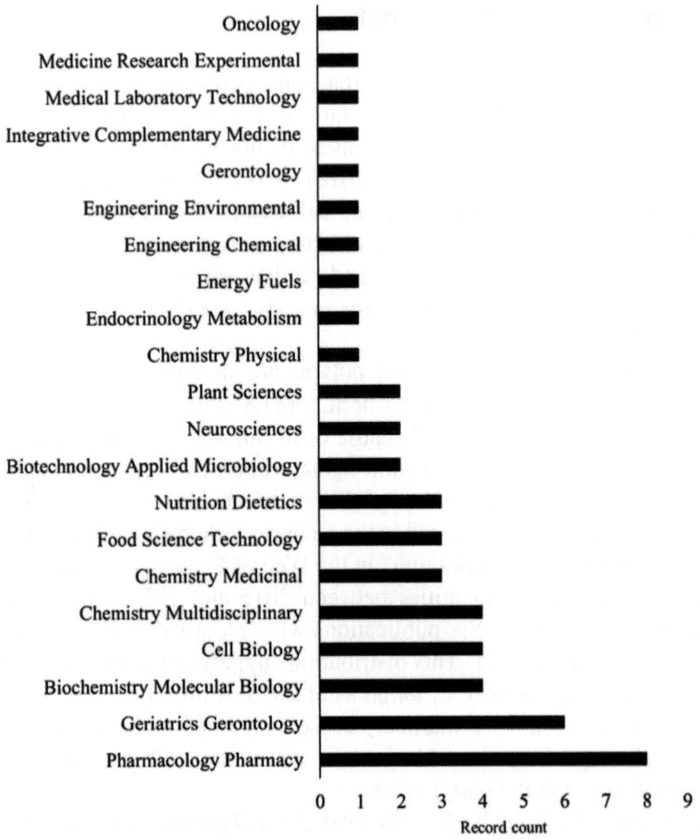

Fig. 1.11 Distribution of publications on *Curcuma longa* (turmeric) according to Web of Science categories. (Data were visualized using the Web of Science database)

Molecular Biology" with four publications, "Cell Biology" with four publications, and "Chemistry Multidisciplinary" with four publications, as shown in Fig. 1.11. This distribution indicates that *C. longa* studies intensify on pharmacology and the biology of aging but also have strong interactions with the disciplines of molecular biology and chemistry. Six publications in the "Geriatrics and Gerontology" category point out that the plant is beginning to be studied at a clinical or applied level, related to aging processes.

In terms of country-based distribution, India (eight publications) and China (eight publications) lead equally. These two countries are followed by the US with six publications, Poland with four publications, and Iran with three publications (Fig. 1.12). The significant superiority of Asian countries in this field can be attributed to their history of *C. longa* usage and their herbal product studies. The contribution of the United States to the studies demonstrates the global scientific acceptance of the subject, and studies conducted in countries such as Poland and

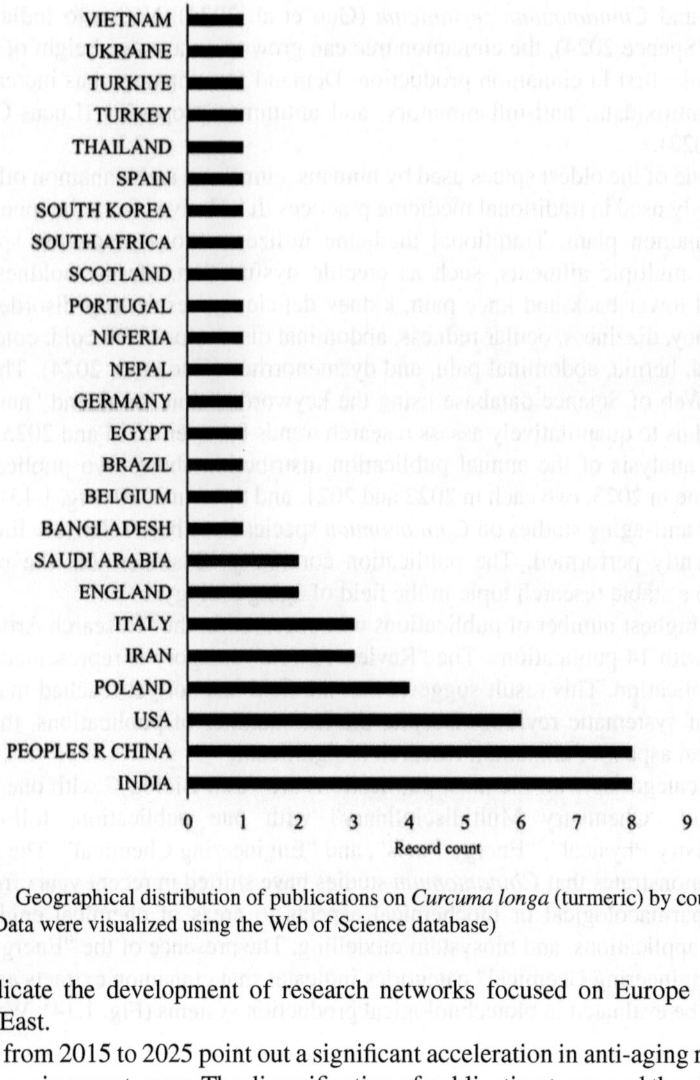

Fig. 1.12 Geographical distribution of publications on *Curcuma longa* (turmeric) by country and region. (Data were visualized using the Web of Science database)

Iran indicate the development of research networks focused on Europe and the Middle East.

Data from 2015 to 2025 point out a significant acceleration in anti-aging research on *C. longa* in recent years. The diversification of publication types and the observed increase in the "Geriatrics and Gerontology" category show that both the experimental and clinical aspects of research are strengthening.

1.3.5 *Cinnamomum* spp. (Cinnamon)

Cinnamomum, also known as cinnamon, belongs to the Lauraceae family, which is an evergreen plant found primarily in tropical and subtropical regions (Mutlu et al. 2023; Guo et al. 2024). Cinnamon originates from the desiccated bark of many species under the *Cinnamomum* genus, such as *Cinnamomum cassia*, *Cinnamomum verum*, and *Cinnamomum zeylanicum* (Guo et al. 2024). Native to India and Sri Lanka (Spence 2024), the cinnamon tree can grow to an average height of 7–10 m. Asia ranks first in cinnamon production. Demand for cinnamon has increased due to its antioxidant, anti-inflammatory, and antitumor properties (Lucas-González et al. 2023).

As one of the oldest spices used by humans, cinnamon and cinnamon oil are also frequently used in traditional medicine practices. It is derived from the inner bark of the cinnamon plant. Traditional medicine utilizes various cinnamon species to address multiple ailments, such as erectile dysfunction, uterine coldness, cold-induced lower back and knee pain, kidney deficiency, respiratory disorders, Yang deficiency, dizziness, ocular redness, abdominal discomfort from cold, cold-related diarrhea, hernia, abdominal pain, and dysmenorrhea (Guo et al. 2024). The search of the Web of Science database using the keywords "cinnamon" and "anti-aging" allowed us to quantitatively assess research trends between 2015 and 2025.

The analysis of the annual publication distribution shows two publications in 2025, one in 2023, two each in 2022 and 2021, and three in 2020 (Fig. 1.13). We can see that anti-aging studies on *Cinnamomum* species have been relatively limited but consistently performed. The publication continuity illustrates that the plant has become a stable research topic in the field of aging biology.

The highest number of publications was observed in the "Research Article" category, with 14 publications. The "Review Article" category is represented by only one publication. This result suggests that the field has not yet reached maturity in terms of systematic reviews. Despite the low number of publications, the strong empirical aspect of cinnamon research is significant.

The categories with the most publications are "Cell Biology" with one publication and "Chemistry Multidisciplinary" with one publication, followed by "Chemistry Physical", "Energy Fuels", and "Engineering Chemical". The distribution demonstrates that *Cinnamomum* studies have shifted in recent years from classical pharmacological or biochemical aspects to areas of chemical engineering, energy applications, and biosystem modelling. The presence of the "Energy Fuels" and "Engineering Chemical" categories indicates that cinnamon extracts are beginning to be evaluated in biotechnological production systems (Fig. 1.14). We can say

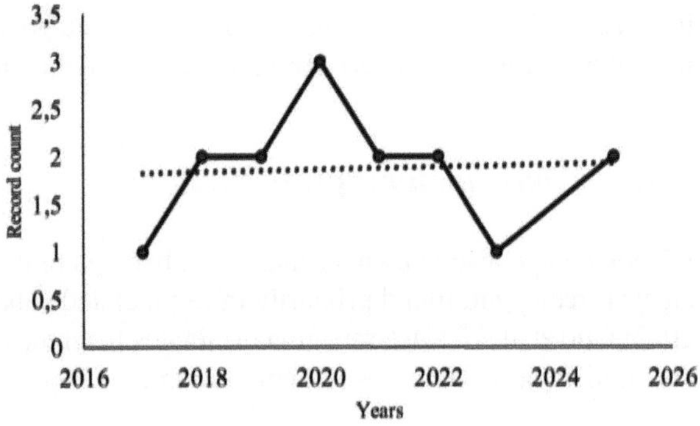

Fig. 1.13 Distribution of publications on cinnamon by year (from 2015 to 2025). (Data were visualized using the Web of Science database)

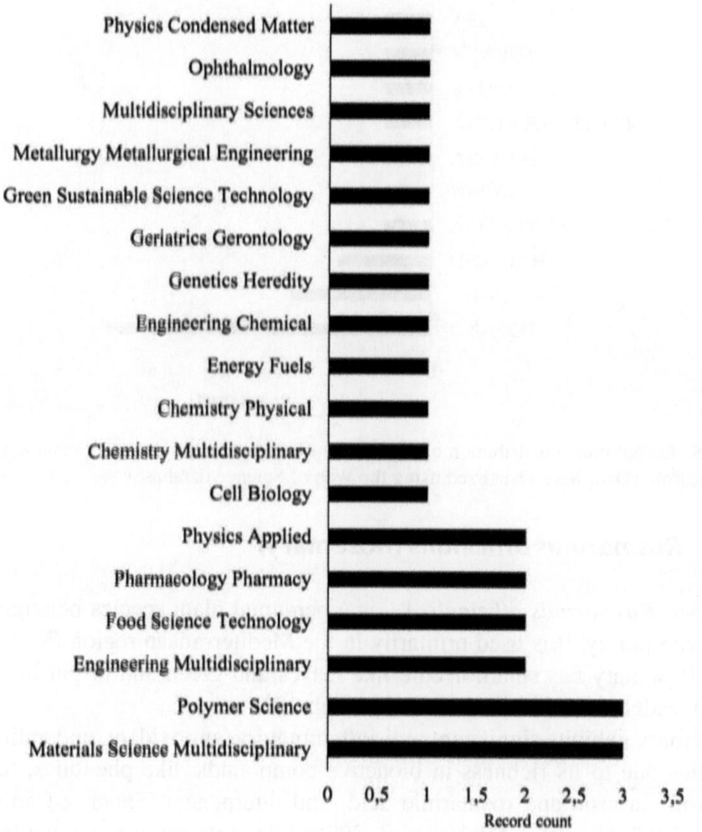

Fig. 1.14 Distribution of publications on *Cinnamomum* spp. (Cinnamon) according to Web of Science categories. (Data were visualized using the Web of Science database)

that anti-aging research encompasses not only biological processes but also the potential for industrial applications of bioactive compounds.

India has the highest number of publications by country with three publications, followed by Australia, Japan, Nigeria, and China (Fig. 1.15). While research is becoming increasingly multicentric on a global scale, Asian countries remain leaders in this field. India's leading position can be explained by both its history of traditional plant usage in its cuisine and its institutional interest in phytochemical studies.

Cinnamomum species hold a relatively limited but consistent place in anti-aging studies from the data. Publication quantities show that the field is still in its early stages, but studies from various scientific fields point out that the topic is gaining an interdisciplinary dimension. The growing focus on chemistry- and engineering-based categories suggests that the bioactive properties of cinnamon compounds are being assessed for both health and commercial applications.

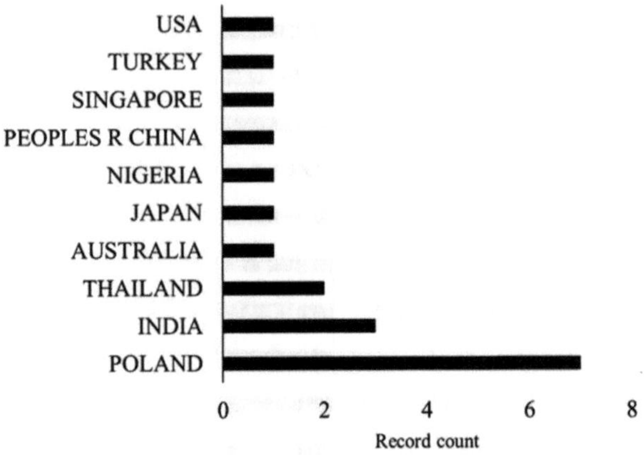

Fig. 1.15 Geographical distribution of publications on *Cinnamomum* spp. (Cinnamon) by country and region. (Data were visualized using the Web of Science database)

1.3.6 *Rosmarinus officinalis* (Rosemary)

Rosemary, *Rosmarinus officinalis* L., is a perennial plant species belonging to the Lamiaceae family. It is used primarily in the Mediterranean region (Moliner et al. 2020). Rosemary has small, needle-like leaves and green and bright blue flowers (Ghasemzadeh Rahbardar and Hosseinzadeh 2025).

Rosemary exhibits significant anti-inflammatory, antioxidant, and antimicrobial properties due to its richness in bioactive compounds, like phenolics, flavonoids (carnosol, carnosic and rosmarinic acid, and diterpenes) (Sánchez-Camargo and Herrero 2017; González-Minero et al. 2020). Since its earliest use, the leaves and essential oil have been valued for their therapeutic effects. Furthermore, edible flowers contain rosmarinic acid, the main phenolic compound. Some studies show that flowers possess both anti-aging and antioxidant activities. It has also been shown to counteract aging processes by reducing oxidative stress and inflammation (Moliner et al. 2020; Di Carlo and Sorrentino 2024; Farsani et al. 2024). The search of the Web of Science database using the keywords "rosemary" and "anti-aging" allowed us to quantitatively assess research trends between 2015 and 2025.

As seen in Fig. 1.16, there were four publications in 2025, two in 2024, two in 2023, two in 2022, and five in 2020. This indicates a significant increase in publications in 2020, followed by a stabilization in the number of publications between 2022 and 2024. The recent increase in 2025 can be considered a sign of a resurgence of interest in rosemary in anti-aging studies. This distribution highlights the varying but steady progress of research on rosemary's antioxidant and anti-aging potential.

When the data is analyzed by publication type, the "Research Article" has the highest number of publications, with 14. Publications in the "Early Access" (one publication) and "Review Article" (five publications) categories show that review studies have begun to increase. The presence of reviews is significant in the area. It

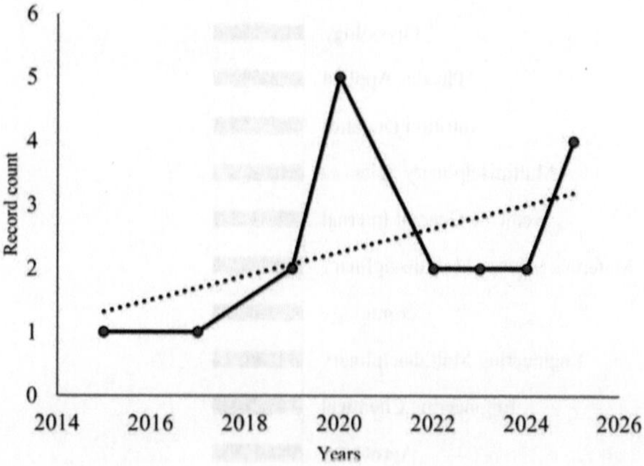

Fig. 1.16 Distribution of publications on rosemary by year (from 2015 to 2025). (Data were visualized using the Web of Science database)

is an important sign that the studies on *R. officinalis* are accumulating and transforming into a synthesis of knowledge.

The categories with the most publications are "Biochemistry and Molecular Biology" with six publications, "Chemistry Multidisciplinary" with five publications, and "Chemistry Medicinal" with three publications. The studies on rosemary focus primarily on biochemical processes and the chemical analysis of phytochemicals. Moreover, publications in "Agricultural Engineering" and "Agronomy" point out that the plant is being evaluated not only in pharmaceutical but also in agricultural production and commercial applications (Fig. 1.17).

When the data is evaluated country by country, India, Spain, and the United States, with three publications for each, stand out with an equal number of publications. These countries are followed by Egypt with two publications and Germany with two publications (Fig. 1.18). A geographically balanced distribution of *R. officinalis* research is seen, with scientific communities based in Asia, Europe, and North America contributing simultaneously. European-based studies exhibit a trend consistent with the plant's Mediterranean origin.

Based on data collected between 2015 and 2025, *R. officinalis* has maintained a steady level of interest in anti-aging studies. Publications primarily consist of experimental articles, suggesting that research remains laboratory-based; however, the increasing number of review studies suggests that the field is maturing. The interdisciplinary category distribution reveals that the anti-aging effects of rosemary are being investigated from both biochemical and chemical perspectives and that it has geographically multicentered research.

1.3.7 Hibiscus

Hibiscus sabdariffa, known as roselle, is a tropical plant that belongs to the Malvaceae family. Although it is native to West Africa, the commercial varieties are those cultivated in China, Thailand, Mexico, and Africa, particularly Sudan,

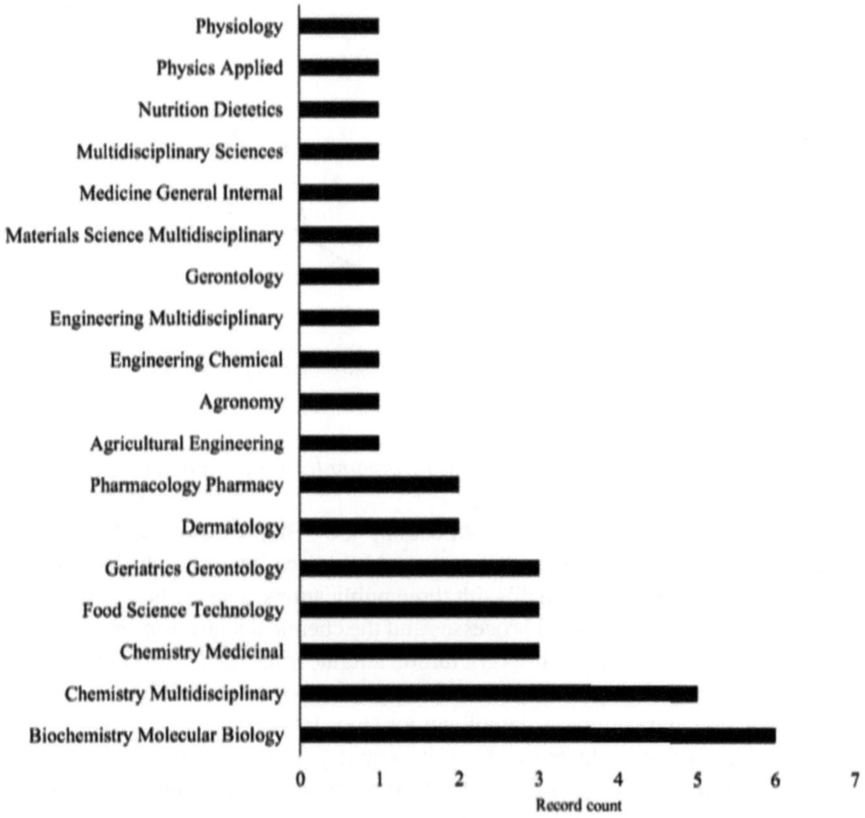

Fig. 1.17 Distribution of publications on *Rosmarinus officinalis* (rosemary) according to Web of Science categories. (Data were visualized using the Web of Science database)

Senegal, and Mali. The plant can grow to four meters and has a cylindrical trunk, usually dark green to red. The leaves are green, and the flowers, which open singly, are yellow with reddish eyes, later turning pink (Sindhu Bharadhi et al. 2023).

Hibiscus is a medicinal plant with therapeutic effects due to its anti-hypercholesteremic, anti-hyperglycemic, wound-healing, anti-aging, and thermogenesis-enhancing properties (Mohamed et al. 2014; Alharbi et al. 2024). It has numerous bioactive chemicals, like delphinidin-3-glucoside, cyanidin-3-glucoside, delphinidin-3-sambubioside, and cyanidin-3-sambubioside (Jabeur et al. 2017). Delphinidins are an important compound with their antioxidant, anti-inflammatory, anti-cancer, and anti-obesity characteristics (Patel et al. 2013). The extract of hibiscus could suppress matrix metalloproteinase activity, protect against oxidative stress, and promote collagen formation. With properties like these, hibiscus becomes a suitable candidate for skincare applications (Li et al. 2020; Prasanth et al. 2025).

As mentioned above, Hibiscus species have received significant attention in recent scientific studies on anti-aging, particularly due to their antioxidant potential.

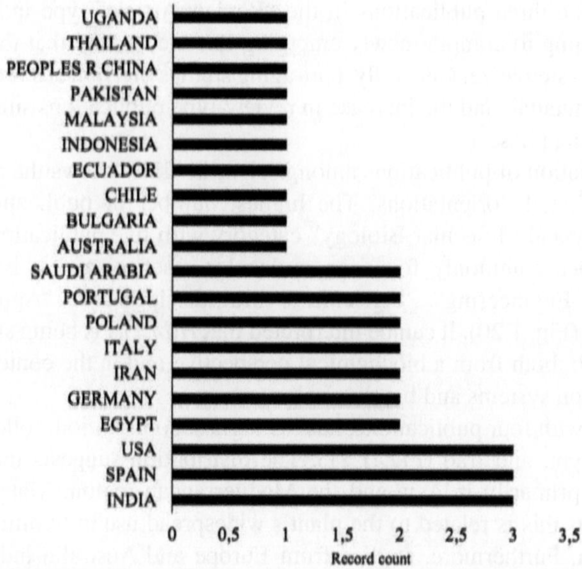

Fig. 1.18 Geographical distribution of publications on *Rosmarinus officinalis* (rosemary) by country and region. (Data were visualized using the Web of Science database)

Fig. 1.19 Distribution of publications on *Hibiscus* spp. by year (2015–2025). (Data were visualized using the Web of Science database)

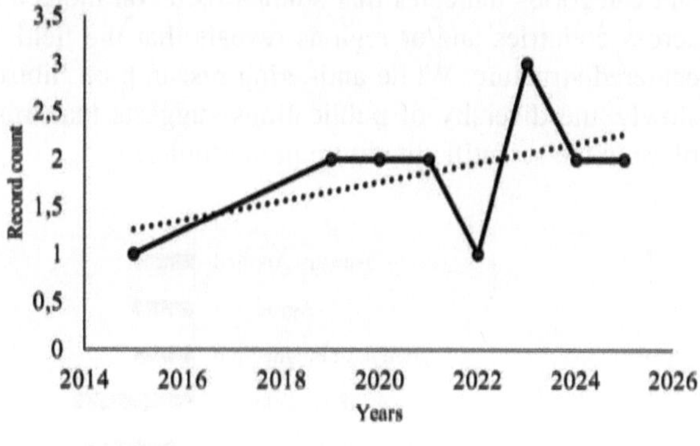

To quantify this interest, we use the search terms "Hibiscus" and "anti-aging" in the Web of Science database to illustrate how research on this plant has progressed between 2015 and 2025.

Publications on the anti-aging effects of *Hibiscus* from 2015 to 2025, however limited, indicate consistency. Two publications were published in 2025 and 2024, three in 2023, one in 2022, and two in 2021, as presented in Fig. 1.19. The graph demonstrates that *Hibiscus* is a developing field of anti-aging studies. The increase observed in 2023 is likely attributable to increased interest in the plant's biochemical profile.

When looking at publication types, the "Research Article" stands out by far with 12 publications. This indicates that most of the research is based on original data production. The three publications in the "Review Article" type indicate that the field is beginning to compile newly emerging knowledge and that the literature is beginning to systematize. Generally, anti-aging studies on *Hibiscus* remain predominantly experimental, and the increase in review-type publications suggests that the field is maturing phase.

The distribution of publications among scientific fields shows the interdisciplinary nature of study orientations. The highest number of publications is in the "Biochemistry and Molecular Biology" category with five publications, indicating that the studies commonly focus on molecular mechanisms. It is followed by "Agricultural Engineering", "Agriculture Multidisciplinary", "Agronomy", and "Biophysics" (Fig. 1.20). It can be interpreted that *Hibiscus* is being studied in anti-aging research, both from a biochemical perspective and in the context of agricultural production systems and biophysical manners.

Malaysia, with four publications, has the highest contribution, followed by Italy, Australia, Egypt, and Iraq (Fig. 1.21). The distribution suggests that research is concentrated primarily in Asia and the Mediterranean region. Malaysia's prominence suggests this is related to the plant's widespread use in traditional medicine in that region. Furthermore, studies from Europe and Australia indicate that the topic is beginning to enter global research activities.

As a final assessment, trends from 2015 to 2025 indicate that hibiscus remains a stable area of research in anti-aging research. The classification of publication types and categories indicates that studies focus on molecular biology. The distribution across countries and/or regions reveals that the field is evolving toward a multi-centered structure. While anti-aging research on hibiscus is currently progressing slowly, the diversity of publications suggests that this plant will gain a stronger place in the scientific literature in the future.

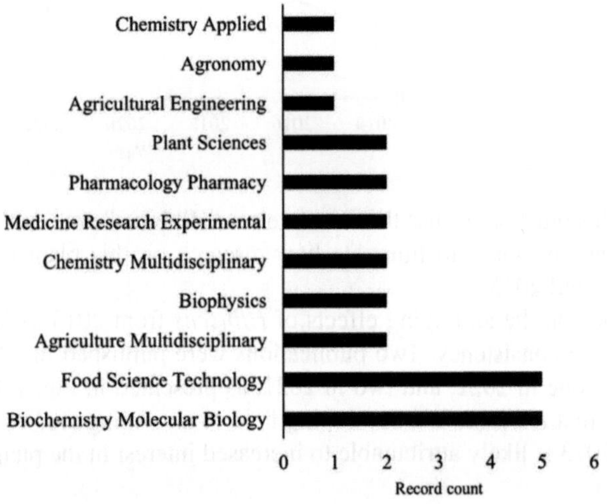

Fig. 1.20 Distribution of publications on *Hibiscus* spp. according to Web of Science categories. (Data were visualized using the Web of Science database)

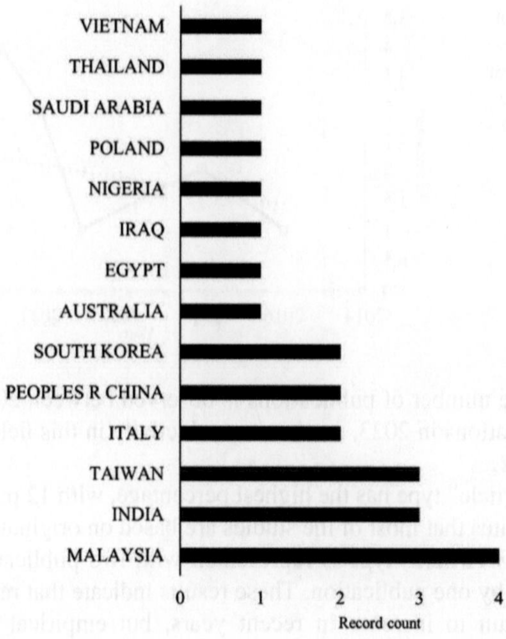

Fig. 1.21 Distribution of publications on *Hibiscus* spp. by country and region. (Data were visualized using the Web of Science database)

1.3.8 *Zingiber officinale* (Ginger)

Zingiber officinale, commonly known as ginger, belongs to the Zingiberaceae family. Ginger is an aromatic herbaceous plant native to Northeast India (Baliga et al. 2011). It is produced in approximately 50 countries, mostly in tropical and subtropical regions. The major producers are India, China, and Nigeria.

The part of ginger used medicinally is the underground stems or rhizomes (Zhang et al. 2021). Ginger contains a wide variety of bioactive compounds, including 6-gingerol, 6-shogaol, 10-gingerol, gingerdiones, gingerdioles, paradols, 6-dehydrogingerols, 5-acetoxy-6-gingerol, 3,5-diacetoxy-6-gingerdioal, and 12-gingerol, which give it medicinal importance (Tanaka et al. 2015; Sahardi and Makpol 2019).

Due to its antioxidant and anti-inflammatory properties, it has the potential to be used as an anti-aging agent (Mohd Sahardi et al. 2020). It is also effective against cancer and exhibits gastroprotective and anti-inflammatory properties. Ginger is among the plants associated with long lifespan and metabolic balance in traditional medicine practices. It has received increasing attention in anti-aging research over the last decade (2015–2025). In this section, using the Web of Science database and the search terms "Ginger" and "anti-aging", we aim to determine how research on this plant has progressed between 2015 and 2025.

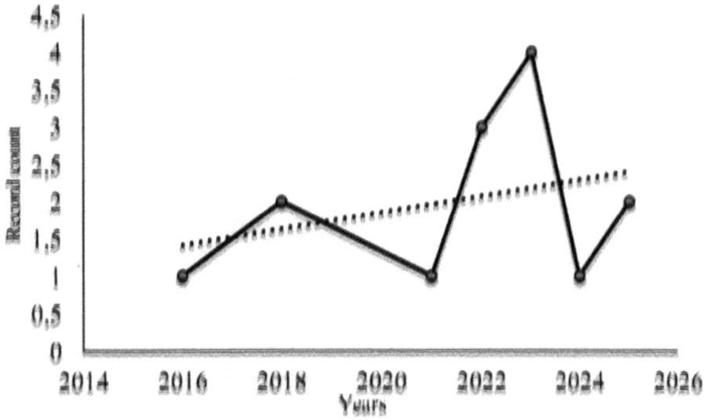

Fig. 1.22 Distribution of publications on *Zingiber officinale* (Ginger) by year (2015–2025). (Data were visualized using the Web of Science database)

An increase in the number of publications is observed between 2015 and 2025, and with four publications in 2023, academic productivity in this field has peaked, as shown in Fig. 1.22.

The "Research Article" type has the highest percentage, with 12 publications, as expected. This indicates that most of the studies are based on original experimental studies. The "Review Article" type is represented with two publications, and the "Early Access" type by one publication. These results indicate that review publications have also begun to increase in recent years, but empirical research still dominates.

The categories with the highest number of publications are "Food Science and Technology" with four publications, "Nutrition and Dietetics" with three publications, "Pharmacology and Pharmacy" with three publications, "Medicinal Chemistry" with two publications, and "Biochemistry and Molecular Biology" with one publication (Fig. 1.23). The observed distribution demonstrates that ginger holds an interdisciplinary position in anti-aging studies, particularly encompassing the fields of nutrition, pharmacology, and chemistry. When examining the categories, the dominance of "Food Science and Technology" and "Nutrition and Dietetics" suggests that ginger's anti-aging effects are linked to food and nutrition studies.

PR of China leads with three publications, followed by India, Egypt, Italy, Japan, and Malaysia (Fig. 1.24). This distribution represents that research is conducted in a balanced manner across both Asia and the Mediterranean.

1.3.9 *Phyllanthus emblica* (Amla)

Phyllanthus emblica L., also known as amla or Indian gooseberry, belongs to the Phyllanthaceae family (Fujii et al. 2008). *P. emblica* is widely distributed in tropical and subtropical regions. Its primary producers are China, India, Sri Lanka, and Thailand (Sripanidkulchai and Junlatat 2014). Amla has a tree-like structure that can reach 8–18 m tall. Its leaves are 7–10 cm long, and its flowers range from yellow to green (Scartezzini et al. 2006). 100 g of amla fruit contains about 445 mg of ascorbic acid, so its taste is sour. The fruit is used as the main ingredient of some traditional medicines (Sripanidkulchai and Junlatat 2014). The fruit has various

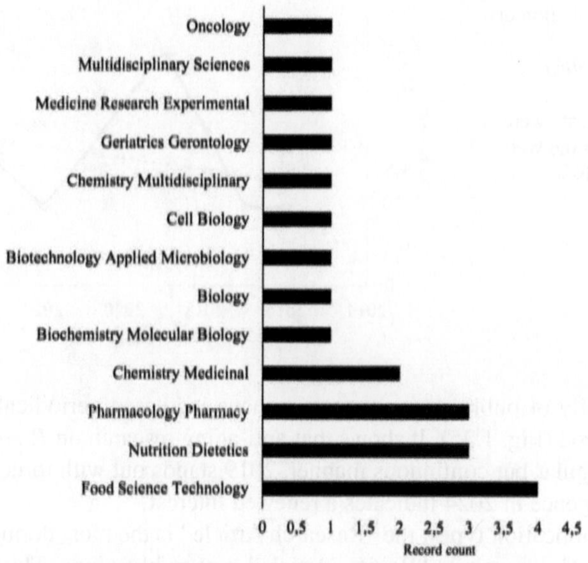

Fig. 1.23 Distribution of publications on Ginger according to Web of Science categories between 2015 and 2025. (Data were visualized using the Web of Science database)

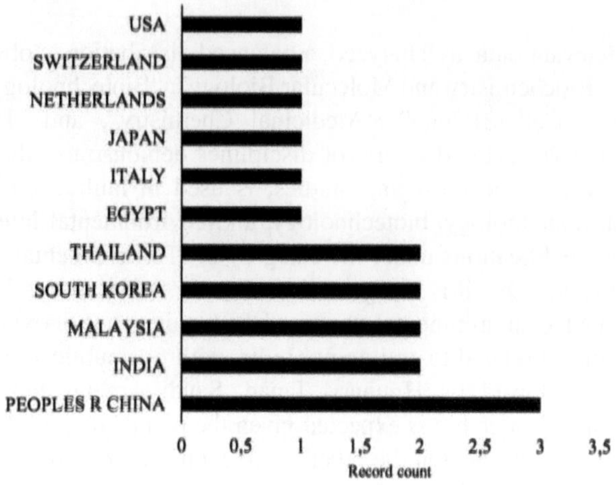

Fig. 1.24 Geographical distribution of publications on *Zingiber officinale* (Ginger) by country and region. (Data were visualized using the Web of Science database)

medicinal properties, including antioxidant, antitumor, hepatoprotective, immunomodulatory, and anti-diabetic agents and protection against photoaging (Rajeshkumar et al. 2003; Achliya et al. 2004; Adil et al. 2010; Saha and Verma 2015). Using the search terms "Amla" and "anti-aging" in the Web of Science database, we can see how research on this plant is progressing between 2015 and 2025.

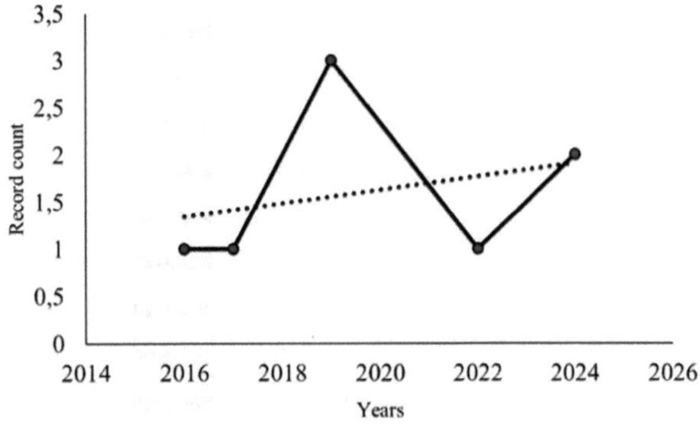

Fig. 1.25 Distribution of publications on *Phyllanthus emblica* (Amla) by year (2015–2025). (Data were visualized using the Web of Science database)

The quantity of publications appears to have increased periodically during the analyzed period (Fig. 1.25). It shows that anti-aging research on *P. emblica* continues in an irregular but continuous manner. 2019 stands out with three publications, while a resurgence in 2024 indicates a renewed interest.

Among publication types, the "Research Article" is the most dominant, with six publications. There are two "Review Article"-type publications. This situation can be interpreted to mean that the literature is being compiled, and systematic analyses are beginning. The trend points out that studies on *P. emblica* are still in the early research phase but are beginning to evolve into compilations as knowledge accumulates.

When the relevant data are analyzed, a balanced distribution is observed among the categories "Biochemistry and Molecular Biology", "Biotechnology and Applied Microbiology", "Cell Biology", "Medicinal Chemistry", and "Environmental Sciences" (Fig. 1.26). This diversity of disciplines demonstrates that amla, while relatively rarely used in anti-aging studies, is used in multidisciplinary studies involving molecular biology, biotechnology, and environmental interactions. The presence of two publications under the category of "Environmental Sciences" can be evaluated to show that it is being studied not only for biomedical aims but also for aspects related to environmental stress and the maintenance of oxidative balance.

For Amla studies related to anti-aging, India, with five publications, is the clear leader. This is followed by Hungary, Japan, Saudi Arabia, and South Korea (Fig. 1.27). India's leadership is expected given the plant's region of origin and its history of traditional usage. On the other hand, contributions from regions such as Europe, Japan, and South Korea demonstrate that *P. emblica* has gained scientific recognition on a global scale.

Overall data from 2015 to 2025 indicate that *P. emblica* has gained a small but consistent part in anti-aging studies. The variety of publication types and the distribution of interdisciplinary categories indicate that research intersects biochemistry, biotechnology, and environmental sciences.

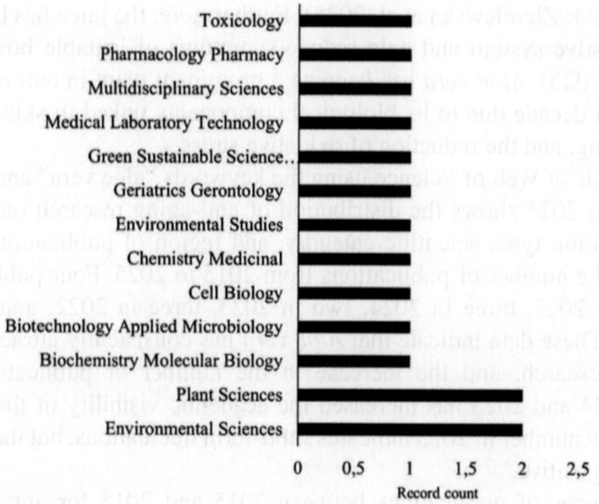

Fig. 1.26 Distribution of publications on *Phyllanthus emblica* (Amla) according to Web of Science categories. (Data were visualized using the Web of Science database)

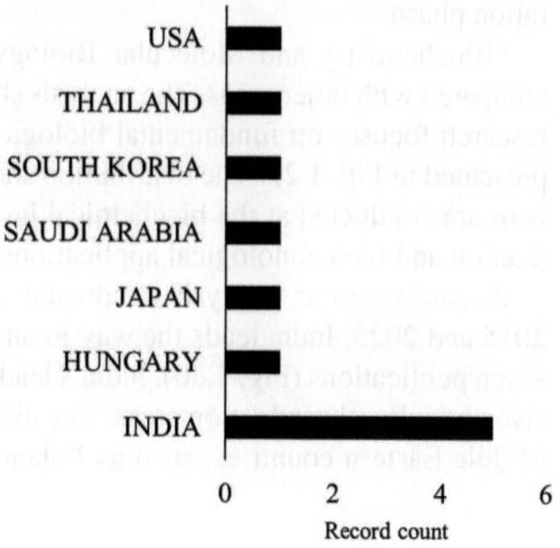

Fig. 1.27 Geographical distribution of publications on *Phyllanthus emblica* (Amla) by country and region. (Data were visualized using the Web of Science database)

1.3.10 *Aloe vera*

Aloe vera has been used in traditional medicine across various cultures and time periods for centuries. The gel obtained from its leaves is used in the food, pharmaceutical, and cosmetic industries and is a plant with a wide range of biological activities (Akaberi et al. 2016). *Aloe vera* is important for its rich content of bioactive compounds, such as vitamins, enzymes, polysaccharides, and amino acids (Hamman 2008). These compounds enhance its moisturizing, anti-inflammatory, antimicrobial, and healing properties, making it an important ingredient for skin care (Surjushe et al. 2008). Acemannan polysaccharide contributes positively to wound healing by accelerating the immune response and fibroblast proliferation

(Pal et al. 2024; Ziemlewska et al. 2025). Furthermore, the juice has been shown to aid the digestive system and help reduce symptoms of irritable bowel syndrome (Abid et al. 2025). *Aloe vera* has become a prominent plant in anti-aging research over the past decade due to its biological components linked to skin regeneration, wound healing, and the reduction of oxidative stress.

An analysis of Web of Science using the keywords "aloe vera" and "anti-aging" from 2015 to 2025 shows the distribution of anti-aging research on this plant by year, publication type, scientific category, and region of publication. There is an increase in the number of publications from 2015 to 2025. Four publications were published in 2025, three in 2024, two in 2023, three in 2022, and two in 2021 (Fig. 1.28). These data indicate that *Aloe vera* has consistently attracted interest in anti-aging research, and the increase in the number of publications observed between 2024 and 2025 has increased the academic visibility of the subject. The relatively low number in 2023 indicates short-term fluctuations, but the overall trend is distinctly positive.

Among types of publications between 2015 and 2015 for anti-aging studies related to *Aloe vera* in the Web of Science database, the highest proportion is in the "Research Article" category with 6 publications, indicating that the research largely relies on original data. 15 "review articles" indicate that the field is entering a maturation phase.

"Biochemistry and Molecular Biology" with seven publications leads the list compared with other areas. The analysis shows that anti-aging and *Aloe vera*-related research focuses on fundamental biological mechanisms. Other key categories are presented in Fig. 1.29. The distribution shows that anti-aging studies related to *Aloe vera* are conducted at the biochemical level, within the context of agricultural production and biotechnological applications.

According to an analysis conducted in the Web of Science database between 2015 and 2025, India leads the way in anti-aging studies related to *Aloe vera*, with seven publications (Fig. 1.30). India's leadership can be directly explained by extensive agricultural production areas. The distribution across European, American, and Middle Eastern countries, such as Poland, the United States, Italy, Iran, and Iraq,

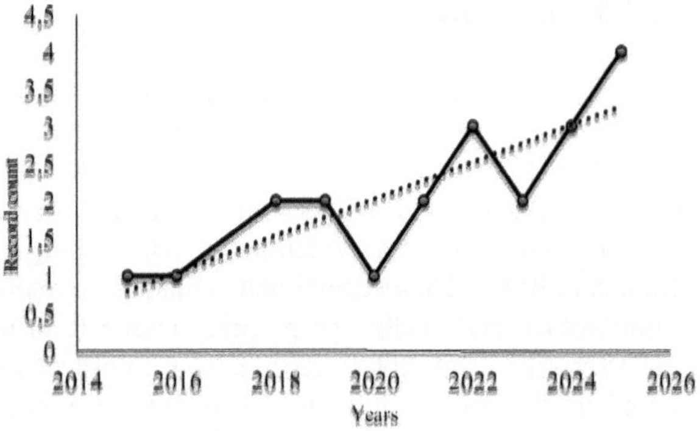

Fig. 1.28 Distribution of publications on *Aloe vera* by year (2015–2025). (Data were visualized using the Web of Science database)

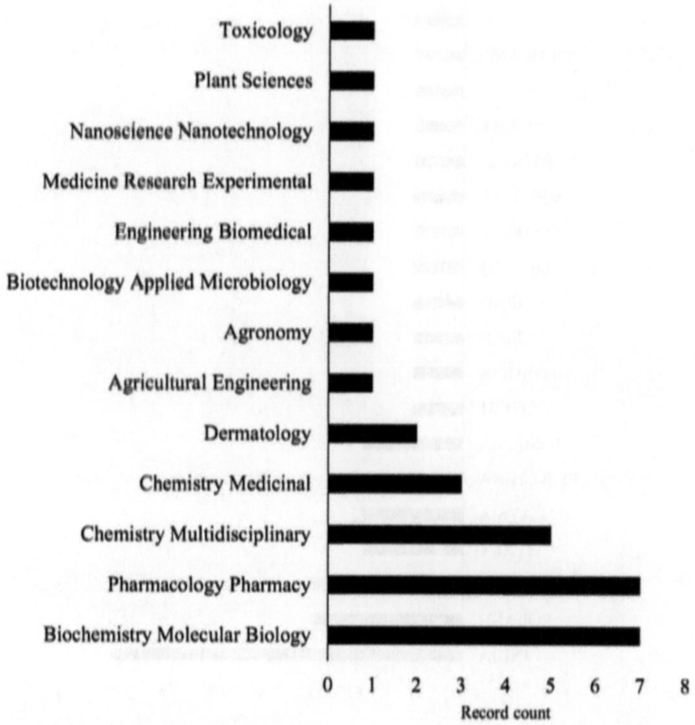

Fig. 1.29 Distribution of publications on *Aloe vera* according to the Web of Science categories. (Data were visualized using the Web of Science database)

highlights the popularity of *Aloe vera* studies. *Aloe vera*-related anti-aging studies are attracting interest worldwide.

Based on the analyses shared above, *Aloe vera*-related anti-aging studies are showing strong growth in both the number of publications and the diversity of disciplines. The variety and number of publications demonstrate the rapid growth of knowledge in the field, and the distribution of categories demonstrates the emergence of a multidisciplinary research field at the intersection of biochemistry, biotechnology, and agricultural sciences. Indian leadership supports the dominance of Asian countries in research on medicinal plants used for anti-aging.

1.4 Conclusion

A glance at myths, folk tales, novels, history, beliefs, and science reveals that the pursuit of eternal youth, or the pursuit of youthful appearance, has existed as both a biological desire and an existential battleground. In ancient myths, the desire to remain eternally youthful or to appear youthful manifests itself through symbols such as the "herb of immortality", the "elixir of life", or the "plant of renewal". In

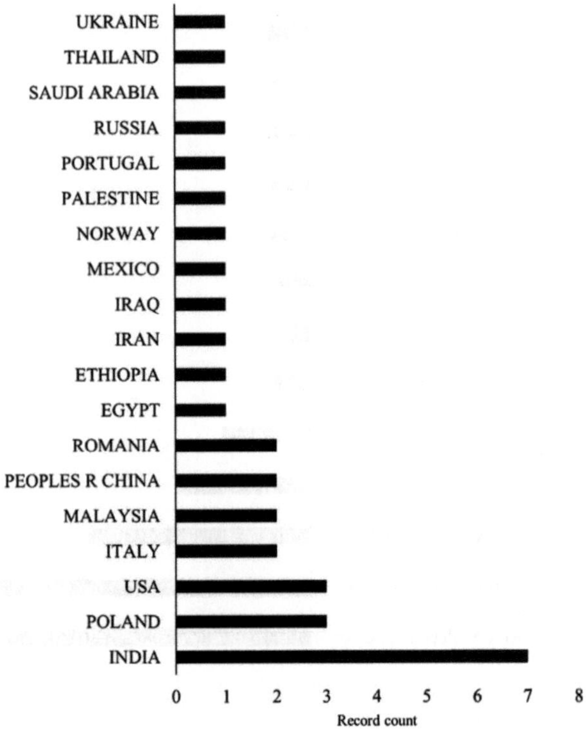

Fig. 1.30 Distribution of publications on *Aloe vera* by country and region. (Data were visualized using the Web of Science database)

modern times, it has been redefined in scientific terms and acquired a new concept. Today, these narratives have evolved into experimentally based hypotheses, measurable biochemical markers, and statistical models.

Our section analyses the distribution of herbal anti-aging research conducted using the Web of Science database between 2015 and 2025. A review of the studies reveals a shift in focus from investigating the antioxidant effects of plants used for anti-aging to analyzing complex biological processes involving cellular aging, genetic regulation, telomere dynamics, and mitochondrial function. This shift can be attributed to our interest in scientific depth and the rapid advancement of technology.

Phytochemicals, dermatology, pharmacology, and biotechnology appear to be at the core of anti-aging research. This trend allows us to assert that herbal ingredients are not merely cosmetic or supplemental elements; they are also a crucial part of the biology of aging. Ginseng, *Centella asiatica*, *Curcuma longa*, *Ginkgo biloba*, *Rosmarinus officinalis*, *Zingiber officinale*, *Hibiscus* species, *Phyllanthus emblica*, and *Aloe vera* are leading species in anti-aging research. These plants demonstrate the integration of both the important knowledge of traditional use and modern molecular biology. The abundance of publications from Asian-based teams points to a new "biocultural continuity" where our cultural heritage is integrated with scientific output. The data demonstrates that plant-based anti-aging research is entering a scientifically mature phase. These studies began at the phytochemical level. With developing technology, they are expanding into an interdisciplinary field, encompassing cellular regeneration, gene activation, telomere stabilization, and nanotechnology.

Ultimately, plant-based anti-aging research is a tool for representing the scientific expression of humanity's historical connection with nature. Ancient myths of the "fountain of youth" are being reconstructed at the molecular level in laboratories. Metabolomics, bioinformatics, and AI-powered analyses are anticipated to further deepen understanding in this area in the future. Science, in a sense, is following ancient narratives and exploring ways to extend quality of life, even if they don't promise immortality.

The living roots of youth are no longer just underground but growing within cells, genes, and molecules.

1.5 Competing Interests

The authors have no relevant financial or non-financial interests to disclose.

References

Abid A, Javed M, Zafar S, Hamdani SAZ, Shah SHBU, Abid J, Ahmad AMR (2025) The green healer: an updated review on the phytochemical profile and therapeutic potential of *Aloe vera*. Front Nutr 12:1–10

Achliya GS, Wadodkar SG, Dorle AK (2004) Evaluation of hepatoprotective effect of Amalkadi Ghrita against carbon tetrachloride-induced hepatic damage in rats. J Ethnopharmacol 90:229–232

Adil MD, Kaiser P, Satti NK, Zargar AM, Vishwakarma RA, Tasduq SA (2010) Effect of *Emblica officinalis* (fruit) against UVB-induced photo-aging in human skin fibroblasts. J Ethnopharmacol 132:109–114

Aggarwal BB, Yuan W, Li S, Gupta SC (2013) Curcumin-free turmeric exhibits anti-inflammatory and anticancer activities: identification of novel components of turmeric. Mol Nutr Food Res 57:1529–1542

Akaberi M, Sobhani Z, Javadi B, Sahebkar A, Emami SA (2016) Therapeutic effects of *Aloe spp.* in traditional and modern medicine: a review. Biomed Pharmacother 84:759–772

Al Amin A, Abyaz R, Siam SA, Alom S, Imran A, Lodh S, Azad AK (2024) Unveiling the fountain of youth: a review of chronological journey from ancient folklore to modern anti-ageing science. Int J Sci Res Eng Dev 7:408–423

Alharbi AE, AlHussaini AM, Alshami I (2024) A comprehensive review of the antimicrobial effects of *Hibiscus* species. Cureus. https://doi.org/10.7759/cureus.73062

Attele AS, Wu JA, Yuan CS (1999) Ginseng pharmacology: multiple constituents and multiple actions. Biochem Pharmacol 58:1685–1693

Baliga MS, Haniadka R, Pereira MM, D'Souza JJ, Pallaty PL, Bhat HP, Popuri S (2011) Update on the chemopreventive effects of ginger and its phytochemicals. Crit Rev Food Sci Nutr 51:499–523

Banerjee O, Singh S, Paul T, Maji BK, Mukherjee S (2024) *Centella asiatica* mitigates the detrimental effects of Bisphenol-A (BPA) on pancreatic islets. Sci Rep 14:1–17

Barbalho SM, Direito R, Laurindo LF, Marton LT, Guiguer EL, Goulart RA, Tofano RJ, Carvalho ACA, Flato UAP, Tofano VAC, Detregiachi CRP, Santos Bueno PC, Girio RSJ, Araújo AC (2022) *Ginkgo biloba* in the aging process: a narrative review. Antioxidants 11:1–25

Benameur T, Soleti R, Panaro MA, La Torre ME, Monda V, Messina G, Porro C (2021) Curcumin as prospective anti-aging natural compound: focus on brain. Molecules 26:1–17

Bjørklund G, Shanaida M, Lysiuk R, Butnariu M, Peana M, Sarac I, Strus O, Smetanina K, Chirumbolo S (2022) Natural compounds and products from an anti-ageing perspective. Molecules 27:7084

Chandrika UG, Prasad Kumara PAA (2015) Gotu Kola (*Centella asiatica*): nutritional properties and plausible health benefits. Adv Food Nutr Res 76:125–157

Chen Y, Fu C, Wu Z, Xu H, Liu H, Schneider H, Lin J (2021) Ginkgo biloba. Trends Genet 37:488–489

Chung IM, Lim JJ, Ahn MS, Jeong HN, An TJ, Kim SH (2016) Comparative phenolic compound profiles and antioxidative activity of the fruit, leaves, and roots of korean ginseng (*Panax ginseng meyer*) according to cultivation years. J Ginseng Res 40:68–75

Das AJ (2011) Review on nutritional, medicinal and pharmacological properties of *Centella asiatica* (Indian pennywort). J Biol Act Prod Nat 1:216–228

Sánchez-Camargo AD, Herrero M (2017) Rosemary (*Rosmarinus officinalis*) as a functional ingredient: recent scientific evidence. Curr Opin Food Sci 14:13–19

Di Carlo E, Sorrentino C (2024) Oxidative stress and age-related tumors. Antioxidants 13:1–19

Fan W, Huang Y, Zheng H, Li S, Li Z, Yuan L, Cheng X, He C, Sun J (2020) Ginsenosides for the treatment of metabolic syndrome and cardiovascular diseases: pharmacology and mechanisms. Biomed Pharmacother 132:110915

Farsani ME, Razavi S, Rasoolijazi H, Esfandiari E, Seyedebrahimi R, Ababzadeh S (2024) Neuroprotective effects of rosemary extract on white matter of prefrontal cortex in old rats. Iran J Basic Med Sci 27:518–523

Fujii T, Wakaizumi M, Ikami T, Saito M (2008) Amla (*Emblica officinalis Gaertn.*) extract promotes procollagen production and inhibits matrix metalloproteinase-1 in human skin fibroblasts. J Ethnopharmacol 119:53–57

Ghasemzadeh Rahbardar M, Hosseinzadeh H (2025) Toxicity and safety of rosemary (*Rosmarinus officinalis*): a comprehensive review. Naunyn Schmiedeberg's Arch Pharmacol 398:9–23

González-Minero FJ, Bravo-Díaz L, Ayala-Gómez A (2020) *Rosmarinus officinalis L.* (rosemary): an ancient plant with uses in personal healthcare and cosmetics. Cosmetics 7:77

Gopinath H, Karthikeyan K (2018) Turmeric: a condiment, cosmetic and cure. Indian J Dermatol Venereol Leprol 84:16–21

Gruman GJ (1966) A history of ideas about the prolongation of life: the evolution of prolongevity hypotheses to 1800. Trans Am Philos Soc 56:1–102

Guo J, Jiang X, Tian Y, Yan S, Liu J, Xie J, Zhang F, Yao C, Hao E (2024) Therapeutic potential of cinnamon oil: chemical composition, pharmacological actions, and applications. Pharmaceuticals 17:1–24

Hamman JH (2008) Composition and applications of *Aloe vera* leaf gel. Molecules 13:1599–1616

Hernández-García D, Granado-Serrano AB, Martín-Gari M, Naudí A, Serrano JC (2019) Efficacy of *Panax ginseng* supplementation on blood lipid profile. A meta-analysis and systematic review of clinical randomized trials. J Ethnopharmacol 243:112090

Jabeur I, Pereira E, Barros L, Calhelha RC, Soković M, Oliveira MBPP, Ferreira ICFR (2017) *Hibiscus sabdariffa* L. as a source of nutrients, bioactive compounds and colouring agents. Food Res Int 100:717–723

Kandasamy A, Aruchamy K, Rangasamy P, Varadhaiyan D, Gowri C, Oh TH, Ramasundaram S, Athinarayanan B (2023) Phytochemical analysis and antioxidant activity of *Centella Asiatica* extracts: an experimental and theoretical investigation of flavonoids. Plants 12:1–19

Lai M, Zhang HJ, Wang F, Shao YL, Yang MW, Hong FF, Yang SL (2018) Anti-aging effects of ginseng and ginsenosides on the nervous system. Int J Pharmacol 14:1188–1197

Li J, Lu YR, Lin IF, Kang W, Chen HB, Lu HF, Wang HM (2020) Reversing UVB-induced photoaging with *Hibiscus sabdariffa calyx* aqueous extract. J Sci Food Agric 100:672–681

Li Y, Sheng Y, Liu J, Xu G, Yu W, Cui Q, Lu X, Du P, An L (2022) Hair-growth promoting effect and anti-inflammatory mechanism of *Ginkgo biloba* polysaccharides. Carbohydr Polym 278:118811

Li J, Li F, Jin D (2023) Ginsenosides are promising medicine for tumor and inflammation: a review. Am J Chin Med 51:883–908

Liu H, Wang X, Wang G, Cui P, Wu S, Ai C, Hu N, Li A, He B, Shao X, Wu Z, Feng H, Chang Y, Mu D, Hou J, Dai X, Yin T, Ruan J, Cao F (2021) The nearly complete genome of *Ginkgo biloba* illuminates gymnosperm evolution. Nat Plants 7:748–756

Lu G, Liu Z, Wang X, Wang C (2021) Recent advances in *panax ginseng C.A. Meyer* as a herb for anti-fatigue: an effects and mechanisms review. Foods 10:1–17

Lucas-González R, Yilmaz B, Mousavi Khaneghah A, Hano C, Shariati MA, Bangar SP, Goksen G, Dhama K, Lorenzo JM (2023) Cinnamon: an antimicrobial ingredient for active packaging. Food Packag Shelf Life 35:101026

Mehta Ambalal S, Monga V, Kashyap R (2023) Exploring the Dorian Gray trait: unveiling the complexities of perceived ageing and self-image. Cureus 15:e47326

Metwaly AM, Lianlian Z, Luqi H, Deqiang D (2019) Black ginseng and its saponins: preparation, phytochemistry and pharmacological effects. Molecules 24:1856

Min DH, Yu YB, Kim TH, Kim H, Lee S (2024) Pharmacological effects of pentacyclic triterpenoids isolated from *Centella asiatica*. Hortic Environ Biotechnol 65:189–197

Mohamed GA, Ibrahim SRM, Elkhayat ES, El Dine RS (2014) Natural anti-obesity agents. Bull Fac Pharm Cairo Univ 52:269–284

Mohd Sahardi NFN, Jaafar F, Mad Nordin MF, Makpol S (2020) *Zingiber officinale* Roscoe prevents cellular senescence of myoblasts in culture and promotes muscle regeneration. Evid Based Complement Alternat Med 2020:1787342

Moliner C, López V, Barros L, Dias MI, Ferreira ICFR, Langa E, Gómez-Rincón C (2020) Rosemary flowers as edible plant foods: phenolic composition and antioxidant properties in caenorhabditis elegans. Antioxidants 9:811

Morvaridzadeh M, Fazelian S, Agah S, Khazdouz M, Rahimlou M, Agh F, Potter E, Heshmati S, Heshmati J (2020) Effect of ginger (*Zingiber officinale*) on inflammatory markers: a systematic review and meta-analysis of randomized controlled trials. Cytokine 135:155224

Mutlu M, Bingol Z, Uc EM, Köksal E, Goren AC, Alwasel SH, Gulcin İ (2023) Comprehensive metabolite profiling of cinnamon (*Cinnamomum zeylanicum*) leaf oil using LC-HR/MS, GC/MS, and GC-FID: determination of antiglaucoma, antioxidant, anticholinergic, and antidiabetic profiles. Life 13:1

Pal S, Raj M, Singh M, Saurav K, Paliwal C, Saha S, Sharma AK, Singh M (2024) Thee of *Aloe vera* on skin and its commensals: contribution of acemannan in curing acne caused by propionibacterium acnes. Microorganisms 12:1–10

Patel K, Jain A, Patel DK (2013) Medicinal significance, pharmacological activities, and analytical aspects of anthocyanidins 'delphinidin': a concise report. J Acute Dis 2:169–178

Potenza MA, Montagnani M, Santacroce L, Charitos IA, Bottalico L (2023) Ancient herbal therapy: a brief history of *Panax ginseng*. J Ginseng Res 47:359–365

Prasanth MI, Malar DS, Verma K, Prasansuklab A, Tencomnao T (2025) *Hibiscus sabdariffa calyx* extract protects human keratinocyte cells from fluoranthene-induced ferroptosis via the repression of aryl hydrocarbon receptor. Ecotoxicol Environ Saf 291:117871

Rajeshkumar NV, Radahakrishna Pillai M, Kuttan R (2003) Induction of apoptosis in mouse and human carcinoma cell lines by *Emblica officinalis* polyphenols and its effect on chemical carcinogenesis. J Exp Clin Cancer Res 22:201–212

Saha S, Verma RJ (2015) Antioxidant activity of polyphenolic extract of *Phyllanthus emblica* against lead acetate induced oxidative stress. Toxicol Environ Heal Sci 7:82–90

Sahardi NFNM, Makpol S (2019) Ginger (*Zingiber officinale Roscoe*) in the prevention of ageing and degenerative diseases: review of current evidence. Evid Based Complement Alternat Med 2019:5054395

Scartezzini P, Antognoni F, Raggi MA, Poli F, Sabbioni C (2006) Vitamin C content and antioxidant activity of the fruit and of the Ayurvedic preparation of *Emblica officinalis Gaertn*. J Ethnopharmacol 104:113–118

Sindhu Bharadhi SA, Karthikeyan N, Thangaiah A, Balasubramaniam R, Thiyagarajan A, Jacob R (2023) Active compound analysis of ethanolic extract of roselle calyces (*Hibiscus sabdariffa L.*). Qual Assur Saf Crops Foods 15:117–128

Spence C (2024) Cinnamon: the historic spice, medicinal uses, and flavour chemistry. Int J Gastron Food Sci 35:100858

Sripanidkulchai B, Junlatat J (2014) Bioactivities of alcohol based extracts of *Phyllanthus emblica* branches: antioxidation, antimelanogenesis and anti-inflammation. J Nat Med 68:615–622

Su J, Su Q, Hu S, Ruan X, Ouyang S (2023) Research progress on the anti-aging potential of the active components of ginseng. Nutrients 15:1–19

Surjushe A, Vasani R, Saple D (2008) *Aloe vera*: a short review. Indian J Dermatol 53:163–166

Tanaka K, Arita M, Sakurai H, Ono N, Tezuka Y (2015) Analysis of chemical properties of edible and medicinal ginger by metabolomics approach. Biomed Res Int 2015:671058

Trujillo J, Chirino YI, Molina-Jijón E, Andérica-Romero AC, Tapia E, Pedraza-Chaverrí J (2013) Renoprotective effect of the antioxidant curcumin: recent findings. Redox Biol 1:448–456

Unlu ES, Koc A (2007) Effects of deleting mitochondrial antioxidant genes on life span. Ann N Y Acad Sci 1100:505–509

Wang X, Gong X, Zhang H, Zhu W, Jiang Z, Shi Y, Li L (2020) In vitro anti-aging activities of *ginkgo biloba* leaf extract and its chemical constituents. Food Sci Technol 40:476–482

Witkowska K, Paczkowska-Walendowska M, Garbiec E, Cielecka-Piontek J (2024) Topical application of *Centella asiatica* in wound healing: recent insights into mechanisms and clinical efficacy. Pharmaceutics 16:1–19

Xia P, Mao Y, Liang Z (2022) Two important ginseng plants in Southeast Asia: a systematic review of their traditional uses, botany, phytochemistry, and pharmacology. Acta Physiol Plant 44:1–14

Yuan Y, Xiang X, Jiang X, Liu Y, Zhang M, Lu L, Zhang X, Liu X, Tan Q, Zhang J (2024) *Ginkgo biloba* bioactive phytochemicals against age-related diseases: evidence from a stepwise, high-throughput research platform. Antioxidants 13:1–19

Zhang M, Zhao R, Wang D, Wang L, Zhang Q, Wei S, Lu F, Peng W, Wu C (2021) Ginger (*Zingiber officinale Rosc.*) and its bioactive components are potential resources for health beneficial agents. Phytother Res 35:711–742

Ziemlewska A, Zagórska-Dziok M, Nowak A, Muzykiewicz-Szymańska A, Wójciak M, Sowa I, Szczepanek D, Nizioł-Łukaszewska Z (2025) Enhancing the cosmetic potential of *Aloe Vera* gel by kombucha-mediated fermentation: phytochemical analysis and evaluation of antioxidant, anti-aging and moisturizing properties. Molecules 30:1–24

Zuo W, Yan F, Zhang B, Li J, Mei D (2017) Advances in the studies of *Ginkgo biloba* leaves extract on aging-related diseases. Aging Dis 8:812–826

Clinical Evidence and Significance of Plant Secondary Metabolites in Anti-aging Therapies

2

Oksana Sytar

Abstract

Aging is a multifactorial biological process marked by the progressive decline in physiological function and increased susceptibility to chronic diseases, including neurodegenerative, cardiovascular, and metabolic disorders. Plant-derived secondary metabolites—such as flavonoids, alkaloids, terpenoids, and glycosides—have emerged as promising therapeutic agents due to their antioxidant, anti-inflammatory, neuroprotective, and cytoprotective properties. This review explores the clinical and experimental evidence supporting the anti-aging potential of key phytochemicals, focusing on their mechanistic roles in modulating oxidative stress, cholinesterase activity, mitochondrial health, and inflammatory signaling. Particular emphasis is placed on bioactive compounds from diverse plant families, including *Apiaceae, Orobanchaceae, Solanaceae, Orchidaceae, Malvaceae, Alliaceae, Asparagaceae, Fabaceae, Cucurbitaceae, Asteraceae, Elaeagnaceae, Ranunculaceae, Apocynaceae, Juglandaceae, Moringaceae, Araliaceae, Lythraceae, Amaranthaceae, Menispermaceae, Lamiaceae,* and *Santalaceae*. By choosing multiple biological pathways simultaneously, specific secondary metabolites propose a multi-pronged strategy to enhance the duration of healthy living and reduce the advancement of age-related diseases. The available preclinical and clinical data highlight the importance of plant secondary metabolites as promising candidates for integrative and preventive approaches in anti-aging therapy.

O. Sytar (✉)
Agronomic and Agri-Food Center of the Province of Hainaut (CARAH), Ath, Belgium

Phytopathology, Microbial and Molecular Farming Lab, Haute Ecole Provinciale du Hainaut – Condorcet, Ath, Belgium

Botalys – Research & Development Department, Botalys, Ghislenghien, Belgium
e-mail: o.sytar@carah.be

S. Gantait et al. (eds.), *Medicinal Plants with Anti-Ageing Properties*,
https://doi.org/10.1007/978-981-95-8064-4_2

Keywords

Alkaloids · Flavonoids · Inflammation · Mitochondrial health · Oxidative stress · Terpenes

2.1 Introduction

Aging is a complex, multifactorial biological process defined by the reduction of some physiological functions and enhanced sensitivity to chronic diseases such as neurodegenerative diseases, metabolic syndromes, and cardiovascular disorders (Rodríguez-Rodero et al. 2011; Guo et al. 2022). Mitochondria represent cells' principal endogenous source of reactive oxygen species (ROS). During aging, mitochondrial efficiency declines, leading to reduced adenosine triphosphate (ATP) production and an increase in ROS generation. Concurrently, the endogenous antioxidant defense mechanisms become less effective. Elevated ROS levels induce oxidative stress, significantly damaging cellular structures, including organelle membranes, DNA, lipids, and proteins (Schieber and Chandel 2014). This cumulative oxidative damage is a key contributor to the onset and progression of the aging phenotype (Maldonado et al. 2023).

Telomere shortening, cellular senescence, inflammation, and oxidative stress are cellular-level changes that collectively contribute to tissue dysfunction, reduced regenerative capacity, and the acceleration of the aging process (Kirkwood 2005). Given the growing aging population worldwide, there is increasing scientific and clinical interest in interventions that may prolong or alleviate the effects of aging to refine both lifespan and healthspan—the period of life spent in good health.

Anti-aging therapies encompass a broad spectrum of strategies, including lifestyle interventions, pharmacological agents, nutraceuticals, and botanical extracts. Among the most promising approaches that could be tested in humans for their potential to extend healthspan are: (i) dietary interventions that mimic the effects of chronic dietary restriction, such as periodic fasting-mimicking diets and protein restriction; (ii) pharmacological agents that inhibit the growth hormone/IGF-I signaling axis; (iii) compounds targeting the mTOR–S6K pathway to suppress age-related anabolic signaling; and (iv) agents that activate energy-sensing pathways such as AMP-activated protein kinase (AMPK) or specific sirtuins, particularly SIRT1 (Longo et al. 2015). These measures aim to regulate and decrease oxidative stress and refine cellular resistance with major aging mechanisms, thereby delaying the onset of age-related diseases and supporting a longer, improved longevity (Kirkland 2016).

Plant secondary metabolites have gained considerable attention as promising candidates for anti-aging interventions due to their diverse bioactivities, including antioxidant, anti-inflammatory, and cytoprotective effects, which can modulate multiple aging-related pathways simultaneously (Sytar and Hajihashemi 2024; Khanam et al. 2025). In addition to well-studied phytochemicals such as resveratrol, epicatechin, quercetin, and curcumin, recent research has uncovered novel compounds with potential anti-aging properties. Notably, the amino acid L-theanine from green tea and the lignans arctigenin and matairesinol isolated from *Arctium lappa* seeds have emerged as bioactive agents (Corrêa et al. 2016). Furthermore, the vast biodiversity of plants continues to be a rich source for discovering new secondary metabolites with promising therapeutic potential.

Currently, many anti-aging compounds are under development that target calorie-restriction mimetics, autophagy induction, enhancement of cell regeneration, and epigenetic modulation of gene activity, including inhibition of histone deacetylases and DNA methyltransferases (Liu 2022). To develop effective and significant evidence for clinical research is important for safe anti-aging interventions.

2.2 Classification and Types of Major Plant Secondary Metabolites

2.2.1 Polyphenols

Polyphenols are a big and variable class of natural metabolites presented in plants, known for their potent antioxidant and anti-inflammatory properties. They are essential for protecting cells against oxidative stress, modulating signaling pathways, and supporting mitochondrial function—mechanisms that are crucial in the context of aging and age-related diseases (Zahra et al. 2024). Among polyphenols, flavonoids—including flavonols, flavones, flavanones, and anthocyanins—are particularly interesting in anti-aging research. Owing to their capacity to penetrate the blood-brain barrier, modulate critical enzymes, and affect gene regulation, flavonoids have shown promising neuroprotective, cardioprotective, and anti-inflammatory effects. This makes them especially relevant for developing multi-targeted strategies to delay aging processes and extend healthspan.

2.2.2 Flavonoids

Flavonoids are a subclass of polyphenols classified into several subclasses based on their chemical structure, including flavonols, flavones, flavan-3-ols (flavanols), isoflavones, anthocyanidins, and chalcones (Busch et al. 2015). These subclasses differ in their degree of oxidation, pattern of hydroxylation, and types of sugar attachments, which influence their biological activity and distribution in plants (Dias et al. 2021).

Flavonoids, a major class of polyphenolic compounds, are widely distributed among medicinal and edible plants and play significant roles in plant defense, pigmentation, and human health due to their antioxidant, anti-inflammatory, and therapeutic properties (Ullah et al. 2020; Zahra et al. 2024). Numerous plant species from various botanical families exhibit diverse flavonoid profiles.

Among the members of *Malvaceae, Althaea officinalis* contains hypolaetin-8-glucoside, isoquercitrin, and kaempferol (Al-Snafi 2013). *Firmiana simplex* also accumulates kaempferol, luteolin, epicatechin, apigenin, and isorhamnetin and quercetin (Ghareeb et al. 2018), while *Malva sylvestris* is rich in rutin, delphinidin, malvidin, and other anthocyanins (Paul et al. 2024).

The *Alliaceae* members such as *Allium sativum* and *Allium cepa* are notable for catechins and quercetin glycosides (Bouhenni et al. 2021). Similarly, *Asparagus officinalis* from the *Asparagaceae* family contains rutin, quercetin, and isoquercetin, highlighting the prevalence of flavonol derivatives in monocots (Guo et al. 2020).

Alpina galanga (*Zingiberaceae*) and *Cinnamomum cassia* (*Lauraceae*) both show the presence of quercetin and kaempferol (Suzery et al. 2019; Hussain et al. 2021). These compounds are frequent across species in aromatic and spice-producing families.

Several species from the *Asteraceae* family, including A*rtemisia dracunculus, Anacyclus pyrethrum, Cichorium intybus,* and *Echinacea purpurea,* contain quercetin, kaempferol, and related derivatives (Janda et al. 2021; Sissi et al. 2022; Truzzi et al. 2025). These flavonoids contribute to the strong antioxidant potential characteristic of this family.

Flavonoids are highly diverse in *Fabaceae*, where *Astragalus membranaceus* and *Glycyrrhiza glabra* accumulate isoflavonoids like calycosin, formononetin, and glabridin, in addition to kaempferol and quercetin (Wahab et al. 2021; Imtiaz et al. 2025). These species illustrate the common occurrence of both flavonols and isoflavones in legumes. Fruits of plants such as *Citrullus vulgaris* and *Citrullus colocynthis (Cucurbitaceae)* contain kaempferol, isoorientin, and quercetin derivatives, indicating the richness of cucurbits in bioactive flavonoids (Borecka and Karaś 2025). *Cucurbita maxima* seeds add luteolin and vanillin to this profile (Faturoti and Ogidi 2025).

Plants like *Elaeagnus angustifolia* and *Hippophae rhamnoides (Elaeagnaceae)* are distinguished by high levels of flavan-3-ols such as catechin, epicatechin, and gallocatechin, along with flavonols like kaempferol and quercetin (Amiri Tehranizadeh et al. 2016; Raclariu-Manolică and Socaciu 2023).

Tropical and subtropical species such as *Mangifera indica* (*Anacardiaceae*), *Eugenia jambolana (Myrtaceae)*, and *Anacardium occidentale* are rich in both flavonols and anthocyanins including cyanidin, delphinidin, and myricetin (Ayyanar and Subash-Babu 2012; Salehi et al. 2019; Kim et al. 2021).

Notably, many plants possess multiple subclasses of flavonoids. For example, *Gymnema sylvestre* contains flavonols (kaempferol), flavanols (quercitol), and additional phenolics such as curcumin (Parveen et al. 2019). *Hypericum perforatum* exhibits unique flavonoid glycosides (rutin, quercetin) and biflavonoids like biapigenin and amentoflavone (Baljak et al. 2024).

The presence of flavones such as apigenin and luteolin is also widespread, as seen in *Daucus carota, Lavandula angustifolia,* and *Helianthus annuus* (Sytar and Hajihashemi 2024). These compounds contribute to anti-inflammatory and neuroprotective effects.

Overall, the distribution of flavonoids across different plant taxa reveals a pattern of both commonality and specialization. While compounds such as quercetin and kaempferol are nearly ubiquitous, others like isoflavonoids (in legumes) or anthocyanins (in fruits and flowers) exhibit more selective occurrence, reflecting the ecological and physiological adaptations of each plant species. Table 2.1 summarizes the identified flavonoids with potential anti-aging properties.

2.2.3 Alkaloids

Another variable group of natural organic compounds is alkaloids. This group mostly contains basic nitrogen atoms (Ferreira 2022). They are known for a wide range of pharmacological effects and are classified based on their chemical structure and biosynthetic origin. Among the 27,000 known alkaloids, over 17,000 have demonstrated diverse pharmacological properties, including anticancer activity (Mohammed et al. 2024).

Alkaloids are generally classified into three major categories: true alkaloids, protoalkaloids, and pseudoalkaloids. Both true alkaloids and protoalkaloids are biosynthesized from amino acids, while pseudoalkaloids originate from non–amino acid precursors (Dey et al. 2020). In addition to this core classification, alkaloids can be classified according to their biosynthetic origin and chemical structure. The scheme of classification of alkaloids is presented in Fig. 2.1.

In nowadays screening, alkaloids were detected in most of the plants, but their specific types were not identified in many cases, indicating a need for further phytochemical investigation (Sytar and Hajihashemi 2024). The presence and identification of alkaloids in the screened plants with potential anti-aging properties are presented in Table 2.2.

2.2.4 Terpenes: Classification, Bioactivity, and Distribution in Medicinal Plants

Terpenes represent the largest and most functionally diverse class of secondary metabolites, comprising over 30,000 known compounds with significant biological and industrial relevance (Kennedy and Wightman 2011). They are biosynthetically derived from isoprene (C_5H_8) units and are classified based on the number of these units: monoterpenes ($C_{10}H_{16}$), sesquiterpenes ($C_{15}H_{24}$), diterpenes ($C_{20}H_{32}$), sesterterpenes ($C_{25}H_{40}$), triterpenes ($C_{30}H_{48}$), tetraterpenes ($C_{40}H_{64}$), and polyterpenes [$(C_5H_8)_n$] (Mishra et al. 2024).

Natural terpenes have garnered interest as potential therapeutic agents due to their dual attributes of high efficacy and low toxicity, particularly in the treatment of microbial infections and cancers (Khanam et al. 2025). Compounds such as β-caryophyllene have demonstrated cytoprotective, antioxidant, and anti-inflammatory effects. Nevertheless, some terpenes—especially

Table 2.1 Presence and identification of flavonoids in screened plants with possible anti-aging potential

Flavonoid subclass	Common compounds	Representative plant families and species	Notes/key points	Anti-aging and therapeutic properties
Flavonols	Quercetin, Kaempferol, Myricetin, Rutin, Isoquercitrin	*Malvaceae*: *Althaea officinalis* (hypolaetin-8-glucoside, isoquercitrin, kaempferol), *Firmiana simplex* (kaempferol, quercetin), *Malva sylvestris* (rutin, delphinidin, malvidin) *Alliaceae*: *Allium sativum, Allium cepa* (catechins, quercetin glycosides) *Asparagaceae*: *Asparagus officinalis* (rutin, quercetin, isoquercetin) *Zingiberaceae*: *Alpina galanga* (quercetin, kaempferol) *Lauraceae*: *Cinnamomum cassia* (quercetin, kaempferol) *Asteraceae*: *Artemisia dracunculus, Anacyclus pyrethrum, Cichorium intybus, Echinacea purpurea* (quercetin, kaempferol) *Fabaceae*: *Astragalus membranaceus, Glycyrrhiza glabra* (kaempferol, quercetin) *Cucurbitaceae*: *Citrullus vulgaris, Citrullus colocynthis* (kaempferol, isoorientin, quercetin derivatives), *Cucurbita maxima* (luteolin, vanillin) *Elaeagnaceae*: *Elaeagnus angustifolia, Hippophae rhamnoides* (catechin, epicatechin, gallocatechin, kaempferol, quercetin) *Anacardiaceae, Myrtaceae: Mangifera indica, Eugenia jambolana, Anacardium occidentale* (flavonols, anthocyanins like cyanidin, delphinidin, myricetin), *Gymnema sylvestre*: Flavonols (kaempferol), flavanols (quercitol), phenolics (curcumin) *Hypericaceae*: *Hypericum perforatum* (rutin, quercetin, biflavonoids: Biapigenin, amentoflavone)	Widely distributed; key antioxidants and anti-inflammatory agents	Potent antioxidants reduce oxidative stress linked to skin aging and cellular senescence. Quercetin and kaempferol improve collagen synthesis, UV protection, and reduce wrinkles. Also neuroprotective and cardiovascular benefits.
Flavones	Apigenin, Luteolin, Chrysoeriol, Diosmetin	*Apiaceae: Daucus carota* *Lamiaceae*: *Lavandula angustifolia* *Asteraceae: Helianthus annuus* Also present in other families	Anti-inflammatory, neuroprotective effects	Anti-inflammatory and anti-glycation effects help prevent skin aging and cognitive decline. Luteolin supports mitochondrial function and protects against photoaging.

Flavonoid subclass	Common compounds	Representative plant families and species	Notes/key points	Anti-aging and therapeutic properties
Flavan-3-ols (Flavanols)	Catechin, Epicatechin, Gallocatechin	*Alliaceae*: *Allium sativum* Elaeagnaceae: *Elaeagnus angustifolia, Hippophae rhamnoides-Gymnema sylvestre* (quercitol)	Strong antioxidants; cardiovascular benefits	Powerful free radical scavengers, improve skin elasticity and hydration, protect against UV-induced damage, and reduce inflammation.
Isoflavones	Formononetin, Calycosin, Genistein, Daidzein	*Fabaceae: Astragalus membranaceus, Glycyrrhiza glabra* —often found in legumes	Mostly in legumes; phytoestrogens	Phytoestrogenic activity mimics estrogen, promoting skin firmness, reducing wrinkles, and supporting bone health during aging.

Flavonoid subclass	Common compounds	Representative plant families and species	Notes/key points	Anti-aging and therapeutic properties
Anthocyanidins	Cyanidin, Delphinidin, Malvidin, Peonidin	*Malvaceae*: *Malva sylvestris* (delphinidin, malvidin) *Anacardiaceae: Mangifera indica, Anacardium occidentale* *Myrtaceae: Eugenia jambolana*	Responsible for pigmentation in fruits and flowers	Strong antioxidants that protect skin from oxidative damage, enhance microcirculation, and help prevent age-related macular degeneration.
Chalcones and other flavonoids	Licochalcone A, Glabridin, Biflavonoids (biapigenin, amentoflavone)	*Fabaceae: Glycyrrhiza glabra* *Hypericaceae: Hypericum perforatum* (biflavonoids such as biapigenin and amentoflavone)	Specialized flavonoids with unique bioactivities	Anti-inflammatory, anti-collagenase, and antimicrobial effects support skin barrier function and slow skin aging. Some biflavonoids show neuroprotective properties.

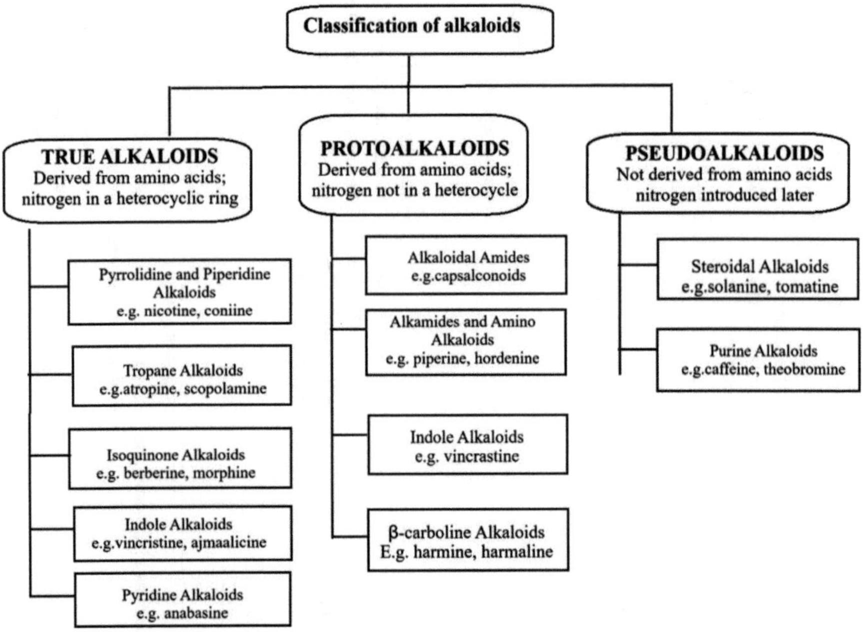

Fig. 2.1 The scheme of classification of alkaloids

Table 2.2 Presence and identification of alkaloids in screened plants with possible anti-aging potential

Plant name	Family	Plant part used	Alkaloids	References
Althaea officinalis	*Malvaceae*	Whole plant	Present, not specified	Al-Snafi (2013)
Allium sativum	*Alliaceae*	Bulbs	β-Carboline alkaloids	Baek et al. (2019)
Asparagus officinalis	*Asparagaceae*	Whole plant	Asparagamine	Negi et al. (2010)
Astragalus membranaceus	*Fabaceae*	Roots	Indolizine alkaloids	Liu et al. (2018)
Capsicum annuum	*Solanaceae*	Fruits	Capsaicinoids	Estrada et al. (2002)
Citrullus colocynthis	*Cucurbitaceae*	Fruit	Choline	Bhasin et al. (2020)
Echinacea purpurea	*Asteraceae*	Whole plant	Pyrrolizidine alkaloids	Lim (2013)
Elaeagnus angustifolia	*Elaeagnaceae*	Aerial part, fruit	β-Carboline derivatives (e.g., harmane)	Tolkachev et al. (2009)

(continued)

Table 2.2 (continued)

Plant name	Family	Plant part used	Alkaloids	References
Hydrastis canadensis	*Ranunculaceae*	Leaves	Isoquinoline alkaloids (e.g., berberine)	Mandal et al. (2020)
Gymnema sylvestre	*Apocynaceae*	Whole plant	Various alkaloids (e.g., ajmalicine, quinine)	Tiwari et al. (2014)
Hippophae rhamnoides	*Elaeagnaceae*	Fruits	Pyridoindole alkaloid (hippophamide)	OuYang et al. (2015)
Juglans mandshurica	*Juglandaceae*	Fruit	Quinoline alkaloids	Luan et al. (2021)
Malva silvestris	*Malvaceae*	Flower	Sanguinarine, berberine	Mohajer et al. (2016)
Moringa oleifera	*Moringaceae*	Whole plant	Marumoside A/B, Niaziminin	Pareek et al. (2023)
Nigella sativa	*Ranunculaceae*	Whole plant	Pyrazole, nigellicimine	Ferizi et al. (2023)
Panax ginseng	*Araliaceae*	Whole plant	β-Carbolines (e.g., perlolyrine)	Ru et al. (2015)
Punica granatum	*Lythraceae*	Fruits	Pelletierine, hygrine, pseudopelletierine	Bravo and Vila (2022)
Rhazya stricta	*Apocynaceae*	Whole plant	Indole alkaloids (e.g., ajmalicine, razyamide, rhazizine)	Abdul-Hameed et al. (2022)
Spinacia oleracea	*Amaranthaceae*	Whole plant	Tyramine group, deoxypeganine	Perez Gutierrez et al. (2019)
Stephania tetrandra	*Menispermaceae*	Roots	Various isoquinoline alkaloids (Tetrandrine, fangchinoline)	Chan et al. (2021)
Stachys schtschegleevii	*Lamiaceae*	Leaves	Stachydrine	Gören (2014)
Trigonella foenum-graecum	*Fabaceae*	Seeds	Trigonelline, choline	Rahman and Husen (2023)
Viscum album	*Santalaceae*	Whole plant	Novel aminoalkaloids	Amer et al. (2013)

monoterpenes—exhibit notable cytotoxicity in various biological systems (Khanam et al. 2025). A wide spectrum of terpenes and terpenoids is distributed across medicinal and aromatic plants, encompassing several botanical families. These compounds, localized in diverse plant parts (bulbs, rhizomes, leaves, flowers, fruits, or entire plants), often contribute to stress responses and possess pharmacological potential.

Within the *Alliaceae* family, *Allium sativum* (garlic) bulbs contain a diverse volatile profile, including nerolidol, phytol, squalene, α-pinene, terpinolene, limonene, 1,8-cineole, and γ-terpinene. Similarly, *Allium cepa* (onion) also contains terpenoids, although specific constituents remain unidentified (Pontin et al. 2015; Sytar and Hajihashemi 2024).

Members of the *Zingiberaceae* family, such as *Alpinia galanga*, are particularly rich in terpenoids. Their rhizomes contain a complex mixture of monoterpenes (camphene, α–/β-pinene, phellandrene, terpinene, cymene, cineole, camphor, borneol, and their acetate esters) and sesquiterpenes (zerumbone, bisabolene, farnesene, sesquiphellandrene) (Arambewela et al. 2007). This underscores the biochemical diversity typical of secondary plant metabolism.

A prominent and extensively studied member of this family is *Curcuma caesia*. Essential oil analysis of its rhizomes has identified 30 compounds, accounting for 97% of total oil content. Major constituents include monoterpenes such as camphor (28%), 1,8-cineole (5%), borneol (4%), and bornyl acetate (3%), and sesquiterpenes like ar-turmerone (12%), (Z)-ocimene (8%), ar-curcumene (7%), elemene (5%), and curcumene (3%) (Ibrahim et al. 2023). These compounds collectively contribute to its demonstrated antioxidant, antimicrobial, and anti-inflammatory properties.

Among these, ar-turmerone and β-turmerone, both sesquiterpenes, have shown potential in dermatological applications. They exhibit strong antioxidant activity, inhibit NF-κB signaling, suppress pro-inflammatory cytokines, and modulate melanin synthesis—contributing to reduced hyperpigmentation, improved skin tone, and enhanced wound healing through stimulated collagen production (Park et al. 2012; Kebede et al. 2021; Goenka et al. 2021).

Other plant families also showcase rich terpene diversity. For example, *Aloysia citrodora (Verbenaceae)* contains monoterpenes (limonene, ocimene, camphene, myrcene, cineole) and sesquiterpenes (e.g., humulene), contributing to its distinctive aroma and bioactivity (Rezig et al. 2019). In the *Lamiaceae* family, *Lavandula angustifolia* (lavender) flowers are rich in α-pinene, camphene, thujene, cymene, limonene, eucalyptol, ocimene, caryophyllene, terpinolene isomers, and linalool derivatives, which are responsible for its therapeutic effects (Pokajewicz et al. 2021; Kozuharova et al. 2023). *Rosmarinus officinalis* (rosemary) is another notable example, with essential oils containing 1,8-cineole, camphor, α-pinene, as well as ursolic and betulinic acids—compounds that exhibit both aromatic and health-promoting properties (Andrade et al. 2018; Aziz et al. 2022).

Sytar and Hajihashemi (2024) identified several plant species reported to contain terpenes or terpenoids, though individual compounds were not specified. These species span diverse botanical families and merit further phytochemical investigation.

From the *Asteraceae* family, *Artemisia* sp., *Centaurea cyanus, Tussilago farfara,* and *Taraxacum officinale* are traditionally used in herbal medicine and are likely to contain bioactive terpenes, mainly sesquiterpene lactones and other terpenoids (Nadaf et al. 2023; Sytar and Hajihashemi 2024). *Echinacea purpurea* (Asterid clade) is similarly noted for its immunomodulatory potential (Lim 2013).

Primula veris (*Primulaceae*) and *Viola odorata* (*Violaceae*) are appreciated for their aromatic and therapeutic properties, with terpenoids contributing significantly to their pharmacological potential (Motti et al. 2022). *Agrimonia eupatoria* (*Rosaceae*) is also traditionally employed in herbal medicine, primarily due to its diverse array of secondary metabolites. Notably, specific triterpenoid constituents such as ursolic acid, euscapic acid, and 28-β-D-glucopyranosyl euscapic acid have been identified in this species (Paluch et al. 2020). *Betula pendula* (*Betulaceae*) is rich in essential oils and may serve as sources of pharmacologically active terpenoids (Rastogi et al. 2015).

The plants may synthesize a vast spectrum of terpenes and terpenoids that have significant roles in the development of plant stress reaction and ecological adaptation. The different structure of these metabolites which may affect various range of bioactivities highlight their potential in advancing nutraceuticals and anti-aging treatments. Continued exploration of plant-derived terpenes may facilitate the discovery of novel bioactive agents for human health and disease prevention. The presence and identification of terpenes in the screened plants with potential anti-aging properties are presented in Table 2.3.

2.2.5 Other Relevant Compounds

Other specific relevant secondary metabolites that are actively studied nowadays are phenylethanoid glycoside and steroidal lactones. Steroidal lactones are often found in *Solanaceae* family of plants and are known for anti-inflammatory, anticancer, and adaptogenic properties. As an example, *Withania somnifera* (L.) Dunal, commonly known as ashwagandha or "Indian ginseng," is rich in steroidal lactones, particularly withanolides such as withaferin A, withanolide D, and withanolide V. These compounds show a huge range of pharmacological activities. Withaferin A is well-documented for its potent antioxidant and anti-inflammatory activities, contributing to its protective role against oxidative stress and inflammation—key factors in skin aging and degenerative diseases (Paul et al. 2021). Withanolide D enhances the

Table 2.3 Presence and identification of terpenes in selected plants with possible anti-aging potential

Family	Plant species	Key terpenes	Plant parts
Alliaceae	*Allium sativum* (garlic), *Allium cepa* (onion)	α-Pinene, terpinolene, limonene, cineole, γ-terpinene	Bulbs
Apiaceae	*Eryngium maritimum*	Germacrene D, spathulenol, oleanolic acid	Whole plant
Zingiberaceae	*Alpinia galanga*, *Curcuma caesia*	Camphor, cineole, borneol, ar-turmerone, β-turmerone	Rhizomes
Verbenaceae	*Aloysia citrodora*	Limonene, camphene, myrcene, humulene	Leaves
Lamiaceae	*Lavandula angustifolia*, *Rosmarinus officinalis*	Linalool, camphor, α-pinene, betulinic acid	Flowers, leaves
Asteraceae	*Artemisia* sp., *Centaurea cyanus*, *Taraxacum officinale*	Sesquiterpene lactones (varied)	Various parts
Primulaceae	*Primula veris*	Various terpenoids	Flowers/leaves
Violaceae	*Viola odorata*	Terpenoids (not specified)	Flowers
Rosaceae	*Agrimonia eupatoria*	Ursolic acid, euscapic acid	Aerial parts
Betulaceae	*Betula pendula*	Essential oils, triterpenoids	Leaves/bark

Table 2.4 Representative plants containing phenylethanoid glycosides and steroidal lactones

Compound class	Representative compounds	Common plant sources
Phenylethanoid glycosides	Acteoside, Echinacoside, Verbascoside	*Cistanche, Plantago, Rehmannia, Olea*
Steroidal lactones	Withanolides, Physalins	*Withania, Physalis, Solanum, Datura*

radiosensitivity of human cancer cells by disrupting non-homologous end-joining (NHEJ) DNA repair pathways, thereby promoting cancer cell death. Notably, withanolide V has been identified as a potential inhibitor of the SARS-CoV-2 main protease (Mpro), offering therapeutic promise against COVID-19 (Bashir et al. 2023). Representative plants containing phenylethanoid glycosides and steroidal lactones are listed in Table 2.4.

These are phenylpropanoid-derived compounds typically associated with antioxidant, neuroprotective, and anti-inflammatory activities. Phenylethanoid glycosides (PhGs), including acteoside, echinacoside, tubuloside A, and tubuloside B, are among the principal bioactive constituents of *Cistanche salsa*. These compounds have attracted considerable interest for their neuroprotective, antioxidant, and anti-apoptotic properties.

Echinacoside, a key PhG, has been associated with anti-aging effects by promoting controlled apoptosis of damaged or precancerous cells, thereby reducing chronic inflammation and tumorigenic risk (Wang et al. 2016; Xiao et al. 2022a, b). It also influences the MAPK signaling cascade, enhancing its neuroprotective properties.

2.3 Mechanisms of Action in Anti-aging

The aging process is defined by continual physiological decline focused by cumulative cellular and molecular damage. Key hallmarks include increased oxidative stress, mitochondrial dysfunction, chronic low-grade inflammation, cellular senescence, and impaired metabolic regulation (Baechle et al. 2023). Natural products rich in secondary metabolites have emerged as promising anti-aging agents due to their ability to modulate these underlying mechanisms (Dixit et al. 2025). This section outlines the principal biological pathways through which such compounds exert anti-aging effects.

2.3.1 Antioxidant Properties and Oxidative Stress Reduction

Oxidative damage, caused by excessive generation of reactive oxygen species (ROS) and a reduction in the body's natural antioxidant defenses, plays a central role in aging and age-associated diseases (Dixit et al. 2025). ROS can damage

DNA, lipids, and proteins, leading to cellular dysfunction and death (Zorov et al. 2014; Juan et al. 2021). Plant-derived secondary metabolites, particularly phenolic compounds, phenylethanoid glycosides (PhGs), and withanolides, exert strong antioxidant effects by scavenging free radicals, upregulating antioxidant enzymes (such as SOD, CAT, and GPx), and reducing lipid peroxidation markers like malondialdehyde (MDA) (Bhattacharya et al. 2001; Bolat et al. 2024; Liu et al. 2024).

For example, acteoside, echinacoside, and gastrodin have been shown to neutralize ROS in neuronal and fibroblast models, thereby protecting cellular integrity (Xia et al. 2018; Hu et al. 2025). The antioxidant capacity of polysaccharides from *Gastrodia elata*, though primary metabolites, also contributes to redox balance, supporting overall anti-aging potential (Bai et al. 2022).

2.3.2 Modulation of Cellular Senescence and Apoptosis

Cellular senescence is a condition of permanent cessation of cell proliferation accompanied by metabolic changes and a pro-inflammatory secretory phenotype (SASP) (Kumari and Jat 2021). While senescence prevents malignant transformation, its accumulation contributes to tissue dysfunction and systemic aging. Apoptosis, also known as programmed cell elimination, when improperly regulated, can hasten aging by reducing the number of viable cells (Elmore 2007.

Several natural compounds modulate these processes. Tubuloside A and withaferin A have demonstrated the ability to delay senescence in human cell models by reducing oxidative burden and modulating p53/p21 and p16INK4a pathways (Fitsiou et al. 2021; Varesi et al. 2022). Such secondary metabolites further exhibit anti-apoptotic effects by stabilizing mitochondrial membranes and inhibiting the activation of caspases, particularly caspase-3 and caspase-9 (Brentnall et al. 2013; Anchimowicz et al. 2025). These actions preserve cellular homeostasis and prolong functional cell lifespan (Anchimowicz et al. 2025).

2.3.3 Anti-inflammatory Effects

Aging is frequently associated with an ongoing low-level inflammation known as "inflammaging," characterized by elevated levels of pro-inflammatory cytokines (e.g., TNF-α, IL-6, IL-1β) (Ferrucci and Fabbri 2018; Tylutka et al. 2024). Persistent inflammation contributes to age-related pathologies, including neurodegeneration, metabolic syndrome, and sarcopenia (Antuña et al. 2022; Li et al. 2023).

Secondary metabolites from medicinal plants can effectively mitigate inflammatory signaling (Gonfa et al. 2023). Withanolides as major compounds of extracts of *Whitania somnifera* have been shown to inhibit NF-κB and MAPK pathways, thereby reducing cytokine expression and immune cell infiltration (Tancreda et al. 2025). Similarly, gastrodin and 4-hydroxybenzyl alcohol downregulate microglial activation and cytokine release in neuroinflammatory models (Dai et al. 2024; Wang et al. 2025). By restoring immune balance and reducing tissue damage, these compounds contribute to systemic resilience and healthier aging.

2.3.4 Enhancement of Mitochondrial Function and Energy Metabolism

Mitochondrial impairment is a key feature of aging, resulting in reduced ATP synthesis, increased ROS generation, and impaired cellular metabolism (Srivastava 2017; Amorim et al. 2022). Enhancing mitochondrial health is thus a key strategy in anti-aging interventions (Phua et al. 2024).

Certain plant-derived metabolites improve mitochondrial biogenesis and function (Anchimowicz et al. 2025). Withaferin A and β-Caryophyllene have been reported to increase the mitochondrial complexes I and IV activities, stabilize mitochondrial membranes, and increase ATP synthesis (Zhao et al. 2022; Ullah et al. 2021). Numerous studies have demonstrated that flavonoids improve glycolipid metabolism by supporting the glucose transporter 4 (GLUT4) translocation and activating the AMP-activated protein kinase (AMPK) signaling pathway, thereby enhancing glycemic regulation in diabetic models. In the context of obesity, flavonoids have shown notable inhibitory effects on lipogenesis, suppressed adipocyte differentiation, and reduced the secretion of pro-inflammatory cytokines, primarily through AMPK-mediated mechanisms (Saikia et al. 2025). Improved mitochondrial efficiency not only supports cellular vitality but also extends tissue health and organismal lifespan.

2.4 Clinical Evidence of Efficacy

2.4.1 Effects on Skin Health and Collagen Preservation

Armeria maritima (Plumbaginaceae), or sea thrift, is a coastal perennial rich in protective metabolites adapted to saline and windy habitats. Its main compounds include phenolic acids (gallic, ferulic, caffeic, p-coumaric), flavonoids (quercetin, kaempferol, luteolin glycosides), triterpenoids (ursolic and oleanolic acids), and anthocyanins (Gourguillon et al. 2016). It also accumulates osmoprotectants like proline and glycine betaine that defend against oxidative and osmotic stress (Hanson et al. 1994).

These metabolites give *A. maritima* strong antioxidant, anti-inflammatory, and photoprotective properties. Phenolics and flavonoids neutralize free radicals and protect from UV damage, while triterpenoids boost collagen synthesis, inhibit MMPs, and improve skin elasticity (Nowak-Perlak et al. 2025). Osmoprotectants further help maintain hydration and barrier function under stress. Recent research shows that *A. maritima* extracts enhance fibroblast activity and protect keratinocytes from UV-induced apoptosis, confirming its anti-aging and skin-rejuvenating potential.

Saussurea involucrata (*Asteraceae*), commonly known as snow lotus, is an alpine medicinal plant native to the high-altitude regions of the Himalayas and Tianshan Mountains. Thriving in extreme cold, low oxygen, and intense UV radiation, it produces a rich array of secondary metabolites such as flavonoids (apigenin, acacetin), lignans, and sesquiterpene lactones (Chik et al. 2015). These compounds exhibit strong antioxidant and anti-inflammatory properties that protect cellular components from oxidative stress and support tissue regeneration (Zhao et al. 2017).

Phytochemical studies have demonstrated that extracts of *S. involucrata* can attenuate UV-induced damage, enhance collagen synthesis, and improve skin elasticity, thereby counteracting visible signs of aging (Lin et al. 2025). Its potent anti-photoaging activity, coupled with its symbolic resilience to environmental stress, positions *S. involucrata* as a valuable natural source for anti-aging and photoprotective skincare formulations.

Silene acaulis (Caryophyllaceae), commonly known as moss campion, is a perennial cushion-forming plant native to the Arctic and alpine zones of Europe and North America. Its exceptional adaptation to harsh, cold, and UV-intense environments has resulted in the accumulation of diverse bioactive metabolites, notably triterpenoid saponins, flavonoids, and phenolic acids (Mamadalieva et al. 2014). These compounds demonstrate potent antioxidant, anti-inflammatory, and photoprotective activities that enhance cellular defense mechanisms and promote skin resilience under environmental stress.

Recent investigations indicate that extracts of *S. acaulis* can mitigate oxidative damage by scavenging free radicals and stimulating endogenous antioxidant enzyme systems (Donà et al. 2013). The plant's metabolites also exhibit anti-photoaging effects by preventing UV-induced lipid peroxidation, supporting collagen stabilization, and maintaining skin hydration and barrier integrity. Owing to its ecological resilience, unique secondary metabolism, and suitability for tissue culture, *S. acaulis* represents a promising and sustainable plant for research and development of natural anti-aging and skin-protective formulations.

Annona purpurea Moc. and Sessé ex Dunal (*Annonaceae*) is native to tropical Central and South America and is known for its diverse secondary metabolites with potential anti-aging activity. The plant contains alkaloids such as aporphine, anonamine, and liriodenine, which exhibit antioxidant and physiological effects (Anaya-Esparza et al. 2020). Its seed essential oil, rich in β-eudesmol (68.9%), shows strong antioxidant activity (Muñoz-Acevedo et al. 2016). In addition, methanolic leaf extracts contain flavonoids like quercetin-3-O-rhamnoside and kaempferol-3-O-rhamnoside with UV-protective and anti-inflammatory effects, while β-sitosterol contributes to barrier repair and skin hydration (Chang et al. 1998). These properties make *A. purpurea* a promising source of multifunctional natural compounds for anti-aging and skin-protective applications.

Ferula communis subsp. *cardonae (Apiaceae)*, commonly known as the Sicilian giant fennel, is a rare subspecies distributed mainly in the Mediterranean basin, particularly in Sicily and southern Italy (Dettori et al. 2016). This perennial herb is recognized for its rich content of bioactive secondary metabolites, including sesquiterpene coumarins, ferulic acid derivatives, and phenolic compounds. These constituents contribute to the plant's antioxidant, anti-inflammatory, and photoprotective properties.Among the key metabolites, ferulenol, ferprenin, and umbelliferone derivatives have demonstrated significant free radical scavenging activity (Wang et al. 2023, Kornicka et al. 2023), while ferulic acid is widely known for its anti-aging and UV-protective potential (Purushothaman 2024). Due to these properties, *F. communis* subsp. *cardonae* extracts are considered promising natural ingredients for cosmetic applications aimed at skin protection, regeneration, and soothing effects.

Xerophyta retinervis (*Velloziaceae*) is a medicinal plant known for its diverse array of bioactive compounds. Significant levels of phytosterols have been detected, including γ-sitosterol (34.46 μg/g), stigmasterol (2.36 μg/g), and stigmast-4-en-3-one (16.9 μg/g) (Nyalo et al. 2023). γ-sitosterol has been shown to stimulate collagen synthesis, supporting skin firmness and elasticity (Aguayo-Morales et al. 2023). Triterpenoids like olean-12-en-3-one present in *Xerophyta retinervis* promote wound healing, support the skin barrier, and offer anti-inflammatory and remyelinating effects (Senol et al. 2023). Fatty acids identified in *Xerophyta retinervis* including palmitic acid (24.41 μg/g) play a critical role in skin hydration, texture improvement, and anti-aging. Palmitic acid enhances barrier function and supports the delivery of active ingredients (Kim et al. 2010). Hexacosane (4.65 μg/g), a long-chain hydrocarbon, acts as an effective occlusive and emollient, protecting the skin from dehydration and external stressors while remaining non-comedogenic (Lee et al. 2011; Diao et al. 2021). Additionally, a novel compound, methyl xerophytolate A, a 3-geranyl-4-hydroxybenzoic acid derivative, has been isolated from a related species, *Xerophyta plicata* (Riehl et al. 2006), suggesting further pharmacological potential including antioxidant and antimicrobial properties.

Curcuma caesia (*Zingiberaceae*) contains key bioactives like curcumin, curcumenol, and β-elemene, known for their anti-aging and skin regenerative effects. Curcumenol induces ferroptosis in cancer cells (Zhang et al. 2021), and curcumin reduces oxidative stress and inflammation, enhancing collagen synthesis and wound healing (Thangapazham et al. 2007). β-elemene supports skin repair, collagen production, and is used in treating skin cancers (Balavandi et al. 2020). These compounds show promise in clinical settings for skin rejuvenation, anti-inflammatory, and anti-cancer therapies.

Scutellaria baicalensis is rich in flavonoids such as baicalein, baicalin, wogonin, and scutellarin, which collectively target major aging hallmarks. These compounds reduce chronic inflammation via NF-κB inhibition, scavenge ROS, and boost mitochondrial function, protecting cells from oxidative damage (Song et al. 2020; Lee et al. 2015). Wogonin activates autophagy (AMPK/mTOR) and baicalein enhances ATP production and PGC-1α, supporting cellular energy (Hu et al. 2021; Fang et al. 2018). Flavonoids and polysaccharides also modulate the gut microbiome, reducing systemic inflammation. Additionally, sesquiterpenes like β-caryophyllene and germacrene D contribute to skin protection, UV defense, and barrier repair—highlighting the plant's strong anti-aging and dermo-cosmetic potential.

Dendrobium nobile (Orchidaceae), a rare epiphytic orchid used in Traditional Chinese Medicine, shows remarkable anti-aging, immunomodulatory, and antioxidant activities. Its main bioactive constituents—dendrobine, gigantol, and polysaccharides—exhibit strong neuroprotective and cytoprotective effects by reducing oxidative damage and enhancing mitochondrial function. The biological activity of *Dendrobium nobile* is indicating similar rejuvenating and restorative potential (Ng et al. 2012; Chen et al. 2022).

Cistanche salsa (*Orobanchaceae*), a herb used in Traditional Chinese Medicine, shows strong potential for anti-aging applications due to its immune-boosting, neuroprotective, and antioxidant effects (Zhou et al., 2019). Its activity is comparable to PSK (a known immunomodulator), suggesting similar cytoprotective and rejuvenating properties (Maruyama et al. 2009; Fu et al. 2017).

Key compounds such as echinacoside, acteoside, and tubulosides A/B—classified as phenylethanoid glycosides (PhGs)—support mitochondrial integrity, inhibit apoptosis, and reduce oxidative stress, making them valuable for neurodegenerative disease prevention and cellular longevity (Liu et al. 2025; Yao et al. 2024). Echinacoside also aids in the removal of precancerous cells and modulates MAPK pathways, while acteoside offers cardiovascular protection and may lower vascular dementia risk (Zhou et al. 2020). Additionally, polysaccharides from *C. salsa* enhance immune responses and may positively modulate the gut microbiota, contributing to systemic anti-aging effects via SCFA production (Zhou et al. 2023). Together, these compounds make *Cistanche salsa* (*Orobanchaceae*) a promising multi-target anti-aging herb, with applications in brain health, immune balance, and skin vitality.

Clinical evidence supporting the efficacy of certain rare medicinal plants with presence of specific secondary metabolite is presented in Table 2.5.

2.4.2 Neuroprotective Benefits of Plant-Derived Compounds and Their Role in Cognitive Health

Neurodegenerative diseases and age-related cognitive decline represent a growing global health burden. Central to the pathophysiology of conditions such as Alzheimer's disease (AD) and Parkinson's disease (PD) are oxidative stress, mitochondrial dysfunction, chronic inflammation, and impaired synaptic signaling. Increasing evidence supports the potential of plant-derived secondary metabolites to counteract these pathogenic mechanisms and promote neuroprotection through diverse molecular pathways.

Alzheimer's disease, a progressive neurodegenerative disorder, is characterized by memory deficits and cognitive deterioration, often associated with cholinergic dysfunction and oxidative stress. Within the *Apiaceae* family—known for its rich phytochemical diversity—numerous genera have emerged as promising candidates for therapeutic intervention in AD due to their anticholinesterase, antioxidant, and neuroprotective properties.

Table 2.5 Bioactive compounds and clinical anti-aging potential of selected xerophytic and medicinal plants for cosmeceutical applications

Plant	Key compounds	Classification	Anti-aging action	Cosmeceutical use/ mechanism	References and clinical study results
Xerophyta retinervis (*Velloziaceae*)	γ-Sitosterol, stigmasterol, stigmast-4-en-3-one, palmitic acid, hexacosane, methyl xerophytolate A	Phytosterols, triterpenoids, fatty acids, hydrocarbons	Collagen stimulation, hydration, anti-inflammatory	Skin repair, moisturization, barrier enhancement; supports collagen synthesis and wound healing	Nyalo et al. (2023) Aguayo-Morales et al. (2023) Senol et al. (2023) Kim et al. (2010) Diao et al. (2021)
Curcuma caesia (*Zingiberaceae*)	Curcumin, curcumenol, β-elemene	Curcuminoids, sesquiterpenes	Antioxidant, regenerative, anti-inflammatory	Reduces UV damage, enhances collagen synthesis and wound healing	Thangapazham et al. (2007) (clinical) Zhang et al. (2021) Balavandi et al. (2020)
Scutellaria baicalensis (*Lamiaceae*)	Baicalein, baicalin, wogonin, scutellarin, β-caryophyllene, germacrene D	Flavonoids, sesquiterpenes	Antioxidant, mitochondrial support, anti-inflammatory	UV protection, barrier repair, ROS scavenging, NF-κB inhibition, autophagy activation	Song et al. (2020) Lee et al. (2015) (clinical) Hu et al. (2021) Fang et al. (2018)
Dendrobium nobile (*Orchidaceae*)	Dendrobine, gigantol, polysaccharides	Alkaloids, phenolics	Cytoprotective, antioxidant, mitochondrial support	Promotes skin vitality, longevity; modulates gut–skin axis	Long et al. (2023) Sarsaiya et al. (2025) Jain et al. (2025)
Armeria maritima (*Plumbaginaceae*)	Gallic, ferulic, caffeic, p-coumaric acids; quercetin, kaempferol, luteolin glycosides; ursolic and oleanolic acids; proline, glycine betaine	Phenolic acids, flavonoids, triterpenoids, osmoprotectants	Antioxidant, anti-inflammatory, photoprotective	Enhances collagen synthesis, inhibits MMPs, improves elasticity, maintains hydration and barrier function	Gourguillon et al. (2016) Hanson et al. (1994) Nowak-Perlak et al. (2025)

Saussurea involucrata (*Asteraceae*)	Apigenin, acacetin, lignans, sesquiterpene lactones	Flavonoids, phenolic compounds, terpenoids	Antioxidant, regenerative, anti-photoaging	Enhances collagen synthesis, protects from UV damage, improves skin elasticity	Chik et al. (2015) Zhao et al. (2017) Lin et al. (2025)
Silene acaulis (*Caryophyllaceae*)	Triterpenoid saponins, flavonoids, phenolic acids	Saponins, flavonoids, phenolics	Antioxidant, photoprotective, anti-inflammatory	Prevents lipid peroxidation, stabilizes collagen, supports hydration	Mamadalieva et al. (2014) Donà et al. (2013)
Annona purpurea (*Annonaceae*)	Aporphine, anonamine, liriodenine, β-eudesmol, quercetin-3-O-rhamnoside, kaempferol-3-O-rhamnoside, β-sitosterol	Alkaloids, sesquiterpenes, flavonoids, phytosterols	Antioxidant, anti-inflammatory, UV-protective	Enhances barrier repair, moisturization, and skin regeneration	Anaya-Esparza et al. (2020) Muñoz-Acevedo et al. (2016) Chang et al. (1998)
Ferula communis subsp. *cardonae* (*Apiaceae*)	Ferulenol, ferprenin, umbelliferone derivatives, ferulic acid	Sesquiterpene coumarins, phenolics	Antioxidant, photoprotective, anti-inflammatory	Skin protection, regeneration, soothing, UV defense	Dettori et al. (2016) Wang et al. (2023) Kornicka et al. (2023) Rizwanullah (2024)
Cistanche salsa (*Orobanchaceae*)	Echinacoside, acteoside, tubulosides A/B, polysaccharides	Phenylethanoid glycosides, polysaccharides	Antioxidant, neuroprotective, immunomodulatory	Supports mitochondrial integrity, reduces apoptosis, enhances gut–skin axis via SCFAs	Maruyama et al. (2009) Fu et al. (2018) Liu et al. (2025) Yao et al. (2024) Zhou et al. (2020) Zhou et al. (2023)

***Apiaceae* Family and Cholinesterase Inhibition:** Several members of the *Apiaceae* family, including *Astrodaucus*, *Ferula*, *Leiotulus*, *Angelica*, and *Anthriscus*, have demonstrated significant bioactivities relevant to neurodegenerative disease management:

Astrodaucus orientalis exhibits multifaceted biological effects, including inhibition of acetylcholinesterase (AChE) and butyrylcholinesterase (BChE), suppression of human carbonic anhydrase isoenzymes I and II (hCA I and II), and strong antioxidant activity. Essential oil from the fruits showed the highest AChE inhibition (44.13%), while root dichloromethane extract displayed potent BChE inhibition (86.13%). Major active constituents include octyl acetate, α-pinene, and sabinene, supported by LC-MS/MS and GC-MS/MS analyses that revealed a complex array of phenolics and monoterpenes (Yuca et al. 2024).

Ferula longipedunculata demonstrated potent anticholinesterase and antioxidant effects, particularly in root-derived methanolic and hexane extracts. The hexane fraction inhibited AChE by 37.99% and BChE by 81.66%, correlating with high phenolic content and notable free-radical scavenging activity (Karakaya et al. 2021).

Leiotulus dasyanthus roots were identified as rich sources of coumarins such as bergapten, pimpinellin, and umbelliferone, with the dichloromethane root extract showing remarkable AChE (66.55%) and BChE (92.21%) inhibitory activity (Karakaya et al. 2019b).

Angelica purpurascens fruits and flowers contain high levels of β-phellandrene, limonene, octyl acetate, stigmasterol, and bergapten. These constituents contributed to significant inhibition of AChE (up to 39.86%) and BChE (up to 84.02%), alongside effective suppression of hCA I and II (Karakaya et al. 2020).

Anthriscus nemorosa root essential oil exhibited potent BChE inhibition (88.51%), mainly attributed to α-pinene (72.09%), myristicin, and related monoterpenes. Phenolic-rich ethyl acetate fractions showed the highest antioxidant activity in DPPH assays. GC–MS and GC–FID analyses confirmed a strong correlation between chemical composition and biological efficacy (Karakaya et al. 2021).

Collectively, these *Apiaceae* species demonstrate a dual mechanism of action—cholinesterase inhibition and antioxidative defense—positioning them as promising phytopharmaceutical candidates for AD treatment. Bioactive compounds such as coumarins, monoterpenes, sesquiterpenes, and phenolics from roots and fruits are particularly implicated in their neuroprotective effects.

Phenylethanoid Glycosides (PhGs): Broad-Spectrum Neuroprotection Among non-*Apiaceae*-derived phytochemicals, phenylethanoid glycosides (PhGs) such as acteoside, echinacoside, and tubuloside B, primarily sourced from *Cistanche species* and related Orobanchaceae species, have shown potent neuroprotective effects. These compounds are effective in both in vitro and in vivo models, including 6-hydroxydopamine (6-OHDA) and MPTP-induced neurotoxicity. Their neuroprotective mechanisms include:

- Inhibition of caspase-dependent apoptosis.
- Reduction of oxidative stress.
- Preservation of mitochondrial integrity.
- Enhancement of neuronal survival and delay of neurodegeneration (Li et al. 2015).

Recent research further corroborates these findings. Liu et al. (2025) demonstrated robust antioxidant activity of PhGs in neuronal systems. Tureyen et al. (2023) reported that tubuloside A delays cellular senescence via oxidative stress mitigation. Yao et al. (2024) showed that tubuloside B protects neurons through mitochondrial stabilization. Furthermore, acteoside has been shown to enhance brain blood flow and increase resistance to low oxygen conditions, indicating possible advantages for vascular cognitive impairment (Zhou et al. 2020).

Rehmannia glutinosa (*Plantaginaceae*), commonly known as Chinese foxglove, represents another rich source of phenylethanoid glycosides, notably acteoside, isoacteoside, and echinacoside, together with iridoid glycosides and polysaccharides (Zhang et al. 2008). Its PhGs exert neuroprotective, antioxidant, and anti-apoptotic effects, helping maintain mitochondrial stability and supporting neuronal and dermal cell survival under oxidative stress (Deng et al. 2004; Yang et al. 2017). In dermatological applications, *R. glutinosa* extracts enhance skin repair, promote collagen synthesis, and prevent UV-induced aging by modulating ROS pathways and activating antioxidant enzymes (Guo et al. 2024).

Other plant sources rich in phenylethanoid glycosides (PhGs) such as acteoside, isoacteoside, and echinacoside include several genera across diverse botanical families. *Plantago major* and *P. lanceolata* (*Plantaginaceae*) accumulate acteoside and isoacteoside, contributing to their traditional use in wound healing and skin-soothing applications (Budzianowska et al. 2023). In the *Oleaceae* family, *Olea europaea* (olive leaves) and Syringa vulgaris (lilac) contain verbascoside and isoacteoside, compounds with strong antioxidant, anti-inflammatory, and UV-protective effects (Marčetić et al. 2025). Members of the *Scrophulariaceae* family, such as *Buddleja officinalis, B. davidii,* and *Verbascum thapsus*, are also significant sources of acteoside and echinacoside, showing notable photoprotective and dermo-cosmetic potential (Chen et al. 2025; Marčetić et al. 2025). Additionally, *Pedicularis* spp. (*Orobanchaceae*) (Zhang et al. 2011) and several *Lamiaceae* plants, including *Stachys officinalis* and *Leonurus cardiaca,* produce high levels of PhGs that support collagen stability (Di Stasi et al. 2023; Fierascu et al. 2019), enhance stress tolerance, and protect against oxidative damage.

2.4.3 Impact of Specific Secondary Metabolites on Metabolic and Cardiovascular Health

Cardiovascular and metabolic disorders, including atherosclerosis, hypertension, dyslipidemia, type 2 diabetes mellitus (T2DM), and metabolic syndrome, represent major global health burdens (Vaduganathan et al. 2022; Chew et al. 2023). These complex disorders are frequently linked by common underlying mechanisms, including persistent inflammation, oxidative damage, impaired endothelial function, and insulin resistance. In this context, plant-derived secondary metabolites have emerged as promising candidates for the prevention and management of such disorders due to their multi-targeted bioactivity and favorable safety profiles.

Flavonoids like quercetin and resveratrol improve vascular function, lower blood pressure, and increased sensitivity to insulin by stimulating signaling pathways (e.g., AMPK, SIRT1). Quercetin, a widely studied flavonol found in *Allium cepa, Camellia sinensis*, and *Apocynaceae* family species, reduces systolic blood pressure by enhancing nitric oxide (NO) bioavailability and suppressing NADPH oxidase activity (Salehi et al. 2020; Aghababaei and Hadidi 2023). It additionally enhances insulin responsiveness by regulating the PI3K/Akt signaling cascades in skeletal muscle and adipose tissue.

Resveratrol, a stilbene primarily found in *Vitis vinifera* (*Vitaceae*), activates SIRT1 and AMPK pathways (Öztürk et al. 2017), promoting mitochondrial biogenesis, glucose uptake, and lipid metabolism. These effects collectively promote upgraded glycemic control, attenuation of fatty liver, and reduced cardiovascular risk in both animal and clinical studies (Heebøll et al. 2014; Chen et al. 2015).

The phenylethanoid glycosides such as acteoside support glucose and lipid regulation and protect pancreatic cells (Xiao et al. 2022a, b). Shi et al. (2023) found that Phenylethanoid glycosides from traditional Mongolian medicine *Cymbaria daurica* alleviate alloxan-induced INS-1 cells oxidative stress and apoptosis (Shi et al. 2023). Lignans such as secoisolariciresinol diglucoside (SDG), commonly found in flaxseed (*Linum usitatissimum, Linaceae*) (Kezimana et al. 2018), have demonstrated lipid-lowering, antihypertensive, and anti-inflammatory properties. SDG modulates hepatic lipid metabolism and inhibits aortic plaque formation via downregulation of VCAM-1 and MCP-1 expression, thereby reducing vascular inflammation (Wei et al. 2020).

Among phenylethanoid glycosides (PhGs), acteoside and echinacoside, abundant in *Cistanche* spp. (*Orobanchaceae*), have shown remarkable effects on glucose and lipid homeostasis (Morikawa et al. 2014; Zhou et al. 2023). These compounds exert antioxidant and anti-inflammatory activity while enhancing insulin receptor sensitivity and GLUT4 translocation in peripheral tissues. Acteoside also protects pancreatic β-cells from oxidative injury, suggesting a potential role in T2DM prevention and management (Galli et al. 2020).

Alkaloids like berberine act on AMPK to reduce blood glucose and lipids, with efficacy comparable to some pharmaceuticals (Utami et al. 2023). Anthocyanins from berries (*Vaccinium, Rubus, Sambucus*) enhance endothelial function and reduce oxidative damage, improving insulin sensitivity (Belwal et al. 2017). Higenamine, a benzylisoquinoline alkaloid from *Aconitum* and *Nandina* species (*Ranunculaceae* and *Berberidaceae*), acts as a β-adrenergic receptor agonist, inducing vasodilation, increasing cardiac output, and facilitating lipid mobilization (Zhang et al. 2017). However, its stimulatory effects necessitate cautious use, especially in hypertensive or arrhythmic patients.

Saponins, particularly ginsenosides from *Panax* species (*Araliaceae*), and triterpenoids such as oleanolic acid and ursolic acid, demonstrate significant anti-hyperglycemic, anti-obesity, and cardioprotective properties (Attele et al. 2002; Arulnangai et al. 2025). Ginsenoside Rg1 improves endothelial function, suppresses vascular inflammation, and reduces myocardial oxidative damage through Nrf2 activation and NF-κB inhibition (Zhang et al. 2025a, b).

Other bioactives like withanolides from *Withania somnifera* and cardiac glycosides like digoxin (in low doses) also show promise in metabolic regulation and heart function support (Wiciński et al. 2024; Fender et al. 2024). Cardenolides from *Digitalis* species (*Plantaginaceae*) have a long history of use in managing heart failure (Bal Krishna et al. 2015). Although their clinical use is now limited to specific cardiac conditions, recent studies suggest low-dose digoxin may modulate systemic metabolism by altering Na^+/K^+-ATPase-mediated signaling and influencing adipocyte metabolism (Gurel et al. 2017).

A growing body of evidence underscores the therapeutic potential of plant-derived secondary metabolites in managing metabolic and cardiovascular disorders. Through pleiotropic mechanisms—including antioxidant defense, anti-inflammatory signaling, modulation of lipid and glucose metabolism, and endothelial function enhancement—compounds such as flavonoids, phenylethanoids, alkaloids, and triterpenoids offer a multi-dimensional approach to disease prevention and management. Continued investigation into their pharmacokinetics, safety profiles, and synergistic combinations is essential for translating these bioactives into evidence-based phytopharmaceuticals for cardiometabolic health.

2.5 Safety, Bioavailability, and Pharmacokinetics

2.5.1 Challenges in Clinical Application

Despite the promising therapeutic potential of numerous phytochemicals, their clinical translation remains limited by key challenges, particularly related to safety, bioavailability, and pharmacokinetics (Lagoa et al. 2020). The Fig. 2.2 has been shown major key challenges which we are going to discuss. A major barrier is poor bioavailability, often resulting from low aqueous solubility, rapid metabolism, and limited gastrointestinal absorption (Hu et al. 2023). For example, well-studied compounds such as curcumin, quercetin, and resveratrol exhibit low plasma concentrations following oral administration, which restricts their therapeutic efficacy (Islam et al. 2024; Warias et al. 2025).

Nevertheless, these phytochemicals have demonstrated beneficial effects on bone metabolism and show potential as adjunctive treatments for conditions such as osteoporosis, bone tumors, and periodontitis. Furthermore, their application to titanium surfaces in orthopedic implants has been proposed to enhance osteogenesis and improve osseointegration (Inchingolo et al. 2022).

Herbal medicines, as a broader category of phytochemical-based therapies, remain the primary form of healthcare for many in developing countries and are also widely used as complementary or alternative treatments in developed nations. These treatments are crucial in addressing numerous health conditions, with more than

50% of drugs in Western pharmacopoeias originating from botanical sources (Sivasankari et al. 2013). Nevertheless, herbal preparations usually comprise intricate blends of biologically active molecules, making it challenging to standardize them and to pinpoint the individual components responsible for their therapeutic efficacy. Although numerous studies report pharmacological activity, few isolate or characterize the active phytochemicals. Understanding the pharmacokinetic behavior of these constituents is essential, yet many remain unexplored in this context (Mehta et al. 2015).

In addition, considerable pharmacokinetic variability exists among individuals, influenced by genetic polymorphisms, gut microbiota composition, and co-administration with other substances (Rombolà et al. 2020; Favari et al. 2024). For instance, genetic polymorphisms affecting enzymes involved in phenolic metabolism can significantly alter the bioavailability and systemic activity of flavanones and flavan-3-ols (Favari et al. 2024). Likewise, gut microbiota plays a crucial role in modulating the absorption, metabolism, and excretion of phenolic compounds, contributing further to inter-individual variability. This intricacy obstructs the formulation of uniform dosage protocols and poses challenges to gaining regulatory authorization.

Safety continues to be a major issue. Although numerous secondary metabolites are regarded as safe when consumed in dietary quantities, their administration at medicinal levels can cause undesirable effects or interfere with standard pharmaceutical treatments (Fasinu et al. 2012; Ramírez-Rendon et al. 2022). Alkaloids, in particular, raise safety concerns. For example, berberine has been shown to inhibit cytochrome P450 enzymes (Etheridge et al. 2007), potentially altering drug metabolism and leading to drug–herb interactions. Guo et al. (2011) demonstrated a dose-dependent effect of berberine on hepatic cytochrome P450 mRNA expression and activity in mice. In contrast, Jiang et al. (2017) reported induction of CYP1A2 in rats via intraperitoneal administration. However, oral administration of high-dose berberine failed to induce significant changes in CYP1A2 activity, likely due to its extremely low bioavailability. Similar safety considerations apply to other phytochemicals, including glycosides, saponins, and essential oils, which require thorough pharmacokinetic and toxicological evaluation.

The integration of specific secondary metabolites into evidence-based clinical practice is further limited by the lack of comprehensive clinical trials assessing their long-term safety and therapeutic efficacy. Although some compounds have progressed to clinical investigation, available data remain fragmented. For example, Xie et al. (2016) reviewed clinical progress with nanoformulations of paclitaxel and vincristine, both of which are in current use. Similarly, Lagoa et al. (2020) summarized trials of catechin- and anthocyanin-rich oral formulations aimed at chemoprevention of premalignant lesions. Despite these advances, there is an urgent need for well-designed, large-scale, and controlled clinical studies to confirm preclinical findings, elucidate pharmacokinetic and pharmacodynamic profiles, determine optimal dosing strategies, and ensure long-term safety. Scheme with challenges and considerations in the clinical application of phytochemicals has been presented in Fig. 2.2.

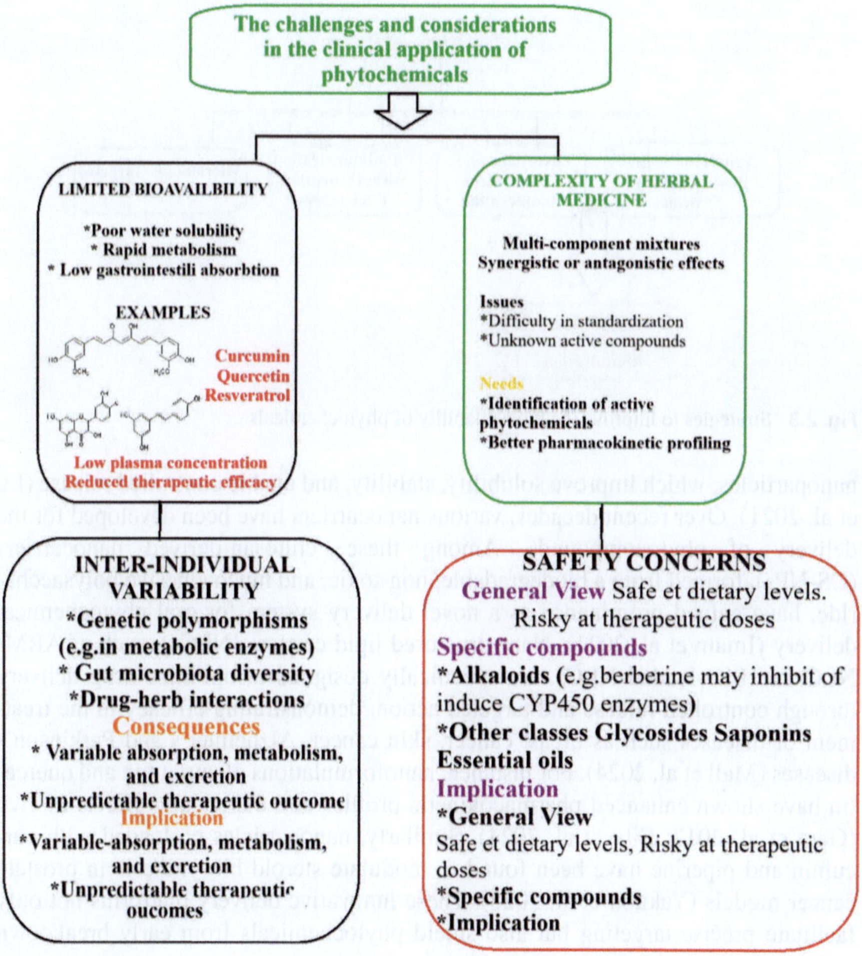

Fig. 2.2 Scheme with challenges and considerations in the clinical application of phytochemicals

2.5.2 Strategies to Improve Bioavailability

Improving the bioavailability of phytochemicals remains a major challenge in translating their preclinical efficacy into clinical relevance. Many promising secondary metabolites suffer from poor aqueous solubility, low membrane permeability, rapid metabolism, and poor gastrointestinal absorption. To address these limitations, several innovative strategies have been explored in Fig. 2.3.

Nanotechnology-Based Delivery Systems One of the most promising strategies for enhancing the therapeutic potential of phytochemicals is the use of nanocarrier systems, including nanoparticles, liposomes, nanoemulsions, and solid lipid

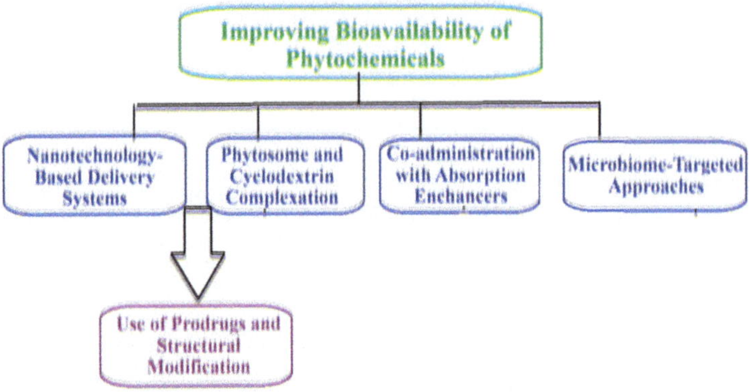

Fig. 2.3 Strategies to improve the bioavailability of phytochemicals

nanoparticles, which improve solubility, stability, and enable controlled release (Lu et al. 2021). Over recent decades, various nanocarriers have been developed for the delivery of phytocompounds. Among these, chitosan-derived nanocarriers (CS-NPs), formed from a biodegradable, non-toxic, and mucoadhesive polysaccharide, have gained prominence as a novel delivery system for oral phytochemical delivery (Imam et al. 2021). Nanostructured lipid carriers (NLCs), such as ARM-NLC and PIO-loaded NLCs, are specifically designed to optimize drug delivery through controlled release and targeted action, demonstrating efficacy in the treatment of diseases such as breast cancer, skin cancer, Alzheimer's and Parkinson's diseases (Mall et al. 2024). For instance, nanoformulations of curcumin and quercetin have shown enhanced pharmacokinetic profiles and therapeutic effects in vivo (Gera et al. 2017; Eity et al. 2024). Similarly, nanoparticles co-loaded with curcumin and piperine have been found to modulate steroid biosynthesis in prostate cancer models (Yakubu et al. 2025). These innovative delivery platforms not only facilitate precise targeting but also shield phytochemicals from early breakdown and metabolic conversion.

Phytosome and Cyclodextrin Complexation Phytosomes are lipid-compatible molecular complexes formed by combining plant-derived bioactive compounds with phospholipids. This structure enhances the compounds' membrane permeability and bioavailability by promoting stronger interactions with lipid-rich biological membranes. The formation of phytosomes relies on hydrogen bonding between active phytoconstituents and phospholipids, resulting in stable complexes that integrate into lipid environments (Lu et al. 2019).

Similarly, cyclodextrin inclusion complexes can encapsulate hydrophobic molecules, significantly improving their water solubility and chemical stability (Musuc 2024; Zhang et al. 2025a, b). Cyclodextrins are cyclic oligosaccharides composed of D-glucose units linked by α-(1 → 4) glycosidic bonds (Freudenberg and

Meyer-Delius 1938). Among these, α-, β-, and γ-cyclodextrins are the most commonly studied and utilized in pharmaceutical and industrial applications due to their ability to form host–guest complexes that enhance the solubility, stability, and bioavailability of encapsulated substances (Musuc 2024).

Co-administration with Absorption Enhancers The bioavailability of certain phytochemicals can be markedly enhanced through co-administration with bioenhancers (Kesarwani et al. 2013; Peterson et al. 2019). Piperine, a major active component of black pepper, is one such bioenhancer that inhibits hepatic and intestinal glucuronidation, thereby improving systemic exposure to co-administered compounds. Efflux transporters also play a critical role in the pharmacokinetics of flavolignans, such as silybin, by restricting their intestinal absorption and overall bioavailability. Piperine has been shown to modulate the activity of these transporters, resulting in increased in vivo absorption and improved therapeutic potential of silybin (Bi et al. 2019). This strategy has also proven effective in enhancing the oral bioavailability of other phytochemicals, including curcumin and resveratrol (Ghobadi and Asoodeh 2023).

Use of Prodrugs and Structural Modification Chemical modification of phytochemicals into prodrugs or semi-synthetic derivatives offers a promising strategy to enhance their pharmacokinetic profiles (Markovic et al. 2020; Rassu et al. 2023). Such structural modifications can improve aqueous solubility, reduce susceptibility to metabolic degradation, and facilitate targeted drug delivery, thereby enhancing overall therapeutic efficacy. Prodrugs, defined as bioreversible and pharmacologically inactive derivatives of active compounds, are designed to undergo enzymatic or chemical conversion in the body to release the active drug. While conventional prodrug strategies primarily aim to optimize physicochemical and biopharmaceutical attributes—such as solubility, permeability, and stability—contemporary approaches increasingly focus on incorporating cellular and molecular targeting to achieve more efficient and site-specific therapeutic outcomes (Markovic et al. 2020).

Microbiome-Targeted Approaches Recent research suggests that the human gut microbiome plays a crucial role in metabolizing phytochemicals into bioactive forms (Sharma et al. 2022). Sharma et al. (2022) explored how bioactive compounds influence gut microbiota composition, highlighting their potential role in the prevention and control of type 2 diabetes. Modulating the microbiota through prebiotics, probiotics, or co-formulated compounds may enhance the biotransformation and systemic availability of these molecules (Bedu-Ferrari et al. 2022).

In conclusion, combining formulation innovations with biotechnological and microbiome-targeted strategies offers a comprehensive approach to overcoming bioavailability barriers and advancing phytochemicals toward effective and clinically viable therapeutics.

2.6 Synergistic Effects and Combination Therapies

2.6.1 Combination with Other Phytochemicals

The healing effects of plant secondary metabolites in anti-aging interventions are greatly enhanced when these compounds are used in combination with other phytochemicals, reflecting the natural synergy present in whole plant extracts and traditional medicinal formulations (Corrêa et al. 2016). This polypharmacological approach can target multiple aging-related pathways simultaneously, offering broader biological activity and improved efficacy compared to isolated compounds.

Phytochemicals such as polyphenols, terpenoids, alkaloids, and saponins exert diverse biological effects including antioxidant, anti-inflammatory, immunomodulatory, and neuroprotective actions (de Lima et al. 2025). A few clinical trials have studied the effects of plant secondary metabolites in humans. Following a rigorous screening process conducted in accordance with the Preferred Reporting Items for Systematic Reviews and Meta-Analyses (PRISMA) guidelines, 13 clinical trials were ultimately included in this review. The phytocompounds studied encompass polyphenols—including flavonoids such as luteolin and quercetin, phenolic acids such as caffeic acid, rosmarinic acid, and ferulic acid as well as other polyphenols like resveratrol—alkaloids such as berberine, huperzine A, and caffeine, and terpenoids such as ginkgolides and limonene (de Lima et al. 2025).

When combined, these molecules often demonstrate additive or synergistic effects in modulating oxidative stress, mitochondrial function, cellular senescence, and age-related inflammatory responses (Wang et al. 2013; Ebrahimirad et al. 2025). For instance, the combination of resveratrol and quercetin has been shown to enhance SIRT1 activation and AMPK signaling more effectively than either compound alone, leading to improved mitochondrial biogenesis and metabolic balance (Peredo-Escárcega et al. 2015; Iside et al. 2020).

Similarly, formulations combining curcumin with piperine—an alkaloid from black pepper—have demonstrated significantly increased bioavailability and amplified antioxidant activity in both preclinical and clinical studies (Tripathi et al. 2022). These combinations not only improve the absorption and bioavailability of individual metabolites but may also allow for lower effective doses, thereby diminishing the risk of adverse reactions while optimizing outcomes of medicaments.

A recent study by Lelong et al. (2026) evaluated the adaptogenic effects of *Rhodiola rosea* root powder obtained through a novel biotechnological process, which differs from typical plant extract mixtures. The root powder was enriched with 3% salidroside, a key bioactive component. In a mouse model, this novel formulation showed notable anxiety-reducing and stress-adaptive effects, indicating that such *Rhodiola rosea* novel extract could function as a natural and sustainable approach to stress alleviation via decreasing level of cortisole. Such results highlight the importance of advanced extraction techniques and optimized formulations in boosting the therapeutic potential of plant-based compounds. Future research should explore the long-term effects of such standardized formulations on stress-related pathways and their roles in promoting mental resilience and healthy aging.

Complex botanical preparations used in traditional medicine systems such as Ayurveda, Traditional Chinese Medicine (TCM), and Kampo often incorporate multiple herbs rich in diverse secondary metabolites. These combinations are formulated based on empirical evidence to achieve balanced therapeutic outcomes and are now being validated by modern pharmacological and clinical research. For example, the *Ginkgo biloba* and *Panax ginseng* combination has been reported to improve cognitive function and delay neurodegeneration through complementary mechanisms involving neurotransmitter regulation and oxidative stress reduction (Wesnes et al. 2000).

Moreover, recent advances in systems biology and metabolomics have enabled the identification of network-based interactions between phytochemicals and molecular targets involved in aging. This has paved the way for designing rational combinations of plant-derived compounds that can modulate key aging-related pathways such as NF-κB, Nrf2, mTOR, and FOXO more precisely (Mavrogonatou and Kletsas 2024).

In conclusion, the combination of plant secondary metabolites with other phytochemicals represents a promising strategy to enhance their clinical relevance in anti-aging therapies. These synergistic effects can enhance therapeutic potency, absorption, and safety parameters, facilitating the advancement of multi-targeted plant-based pharmaceuticals designed to support healthy aging and reduce the risk of age-related disorders.

2.6.2 Integration with Lifestyle and Dietary Interventions

The clinical significance of plant secondary metabolites in anti-aging therapies is increasingly appreciated within the broader framework of integrative medicine, particularly when combined with evidence-based lifestyle and dietary strategies. Lifestyle factors, including diet, physical activity, stress management, and sleep hygiene, are all recognized as key modulators of biological aging (Lu 2024; Núñez-Cortés et al. 2025). Nutritional interventions, in particular, are powerful tools in mitigating age-related cellular decline, and when enriched with plant-derived bioactive compounds, they can provide synergistic benefits.

Caloric restriction (CR), one of the most extensively studied non-genetic interventions for lifespan extension, has been shown to attenuate oxidative damage, improve mitochondrial function, and modulate key metabolic pathways such as the insulin/IGF-1 and mTOR axes. However, long-term adherence to strict CR regimens may not be feasible or desirable for most individuals. This has led to interest in CR mimetics—compounds that can replicate the molecular and physiological benefits of caloric restriction without reducing caloric intake (Chiba et al. 2010).

Plant secondary metabolites such as resveratrol, quercetin, curcumin, and epigallocatechin gallate (EGCG) are considered natural CR mimetics (Yessenkyzy et al. 2020). These compounds activate longevity-related pathways, such as SIRT1, AMPK, and FOXO transcription factors, which regulate autophagy, mitochondrial biogenesis, and stress resistance (Ren et al. 2025; Brimson et al. 2021). When included as part of a balanced diet—particularly within the context of Mediterranean, DASH, or plant-based dietary patterns—these phytochemicals may significantly enhance healthspan outcomes.

Furthermore, polyphenol-rich diets have been associated with lower markers of systemic inflammation and improved cognitive performance in older adults, suggesting that their regular intake through dietary means, rather than supplementation alone, may offer sustained benefits (Naomi et al. 2023). Clinical trials have also demonstrated that flavonoid-containing interventions, such as those using extracts from blueberries, grapes, or green tea, improve vascular function, glucose metabolism, and memory performance in aging populations (Cheatham et al. 2022).

Importantly, the bioavailability of some specific secondary metabolites may differ significantly based on food matrix, host metabolism, and gut microbiota composition. Therefore, incorporating plant secondary metabolites into whole-food-centered nutritional treatment strategies may propose an efficient and sustainable method compared to isolated supplementation. Progress in nutrigenomics and personalized nutrition is projected to further optimize these treatments, enabling customized anti-aging plans tailored to an individual's genetic profile, microbiota, and metabolic condition.

In summary, the clinical utility of plant secondary metabolites in anti-aging therapies is best realized when aligned with comprehensive lifestyle modifications. This integrative approach holds promise not only for delaying the biological aging process but also for improving quality of life and reducing the burden of age-associated diseases.

2.7 Conclusion

Plant secondary metabolites represent a vast and largely untapped pharmacological resource for addressing the multifactorial nature of aging and age-related diseases. Through diverse mechanisms—including the modulation of oxidative stress, mitochondrial function, inflammation, synaptic integrity, and enzymatic activity—these compounds offer a promising avenue for promoting neuroprotection, cardiovascular resilience, and metabolic balance. Evidence from both traditional medicine and modern pharmacology supports the therapeutic relevance of phytochemicals such as acteoside, germacrene, tubuloside B, withaferin A, gastrodin, flavonoids baicalin, wogonin, and various coumarins. Botanicals from a broad range of families—including *Apiaceae, Orobanchaceae, Solanaceae, Orchidaceae, Malvaceae, Alliaceae, Asparagaceae, Fabaceae, Cucurbitaceae, Asteraceae, Elaeagnaceae, Ranunculaceae, Apocynaceae, Juglandaceae, Moringaceae, Araliaceae, Lythraceae, Amaranthaceae, Menispermaceae, Lamiaceae,* and *Santalaceae*—display significant potential to combat neurodegenerative disorders, enhance cognitive health, and improve systemic function. Ongoing investigations into the clinical uses, safety evaluations, and cooperative effects of these natural substances are crucial for converting their biological potential into effective anti-aging treatments.

Incorporating phytochemicals into therapeutic frameworks may ultimately support not only lifespan extension but also the promotion of healthier aging.

Acknowledgement This work was supported by the BEWARE grant.

Competing Interests The authors have no relevant financial or non-financial interests to disclose.

References

Abdul-Hameed ZH, Bawakid NO, Alorfi HS, Sobahi TR, Alburae NA, Abdel-Lateff A, Elbehairi SEI, Alfaifi MY, Alhakamy NA, Alarif WM (2022) Monoterpene indole alkaloids from the aerial parts of *Rhazya stricta* induce cytotoxicity and apoptosis in human adenocarcinoma cells. Molecules 27(4):1422

Aghababaei F, Hadidi M (2023) Recent advances in potential health benefits of quercetin. Pharmaceuticals (Basel) 16(7):1020

Aguayo-Morales H, Sierra-Rivera CA, Claudio-Rizo JA, Cobos-Puc LE (2023) Horsetail (*Equisetum hyemale*) extract accelerates wound healing in diabetic rats by modulating IL-10 and MCP-1 release and collagen synthesis. Pharmaceuticals 16:514

Al-Snafi AE (2013) The pharmaceutical importance of *Althaea officinalis* and *Althaea rosea*: a review. Int J PharmTech Res 5(3):1378–1385

Amer B, Juvik OJ, Francis GW, Fossen T (2013) Novel GHB-derived natural products from European mistletoe (*Viscum album*). Pharm Biol 51(8):981–986

Amiri Tehranizadeh Z, Baratian A, Hosseinzadeh H (2016) Russian olive (*Elaeagnus angustifolia*) as a herbal healer. Bioimpacts 6(3):155–167

Amorim JA, Coppotelli G, Rolo AP, Palmeira CM, Ross JM, Sinclair DA (2022) Mitochondrial and metabolic dysfunction in ageing and age-related diseases. Nat Rev Endocrinol 18(4):243–258

Anaya-Esparza LM, García-Magaña ML, Domínguez-Ávila JA, Yahia EM, Salazar-López NJ, González-Aguilar GA, Montalvo-González E (2020) Annonas: underutilized species as a potential source of bioactive compounds. Food Res Int 138(Part A):109775

Anchimowicz J, Zielonka P, Jakiela S (2025) Plant secondary metabolites as modulators of mitochondrial health: an overview of their anti-oxidant, anti-apoptotic, and mitophagic mechanisms. Int J Mol Sci 26(1):380

Andrade JM, Faustino C, Garcia C, Ladeiras D, Reis CP, Rijo P (2018) *Rosmarinus officinalis* L.: an update review of its phytochemistry and biological activity. Future Sci OA 4(4):FSO283

Antuña E, Cachán-Vega C, Bermejo-Millo JC, Potes Y, Caballero B, Vega-Naredo I, Coto-Montes A, Garcia-Gonzalez C (2022) Inflammaging: implications in sarcopenia. Int J Mol Sci 23(23):15039

Arambewela LS, Arawwawala M, Owen NL, Jarvis B (2007) Volatile oil of *Alpinia galanga* Willd. Of Sri Lanka. J Essential Oil Res 19(5):455–456

Arulnangai R, Thabassoom A, Banu VH, Thirugnanasambandham K, Ganesamoorthy R (2025) Recent developments on ursolic acid and its potential biological applications. Toxicol Rep 14:101900

Attele AS, Zhou YP, Xie JT, Wu JA, Zhang L, Dey L, Pugh W, Rue PA, Polonsky KS, Yuan CS (2002) Antidiabetic effects of *Panax ginseng* berry extract and the identification of an effective component. Diabetes 51(6):1851–1858

Ayyanar M, Subash-Babu P (2012) *Syzygium cumini* (L.) Skeels: a review of its phytochemical constituents and traditional uses. Asian Pacific. J Trop Biomed 2(3):240–246

Aziz E, Batool R, Akhtar W, Shahzad T, Malik A, Shah MA, Iqbal S, Rauf A, Zengin G, Bouyahya A, Rebezov M, Dutta N, Khan MU, Khayrullin M, Babaeva M, Goncharov A, Shariati MA, Thiruvengadam M (2022) Rosemary species: a review of phytochemicals, bioactivities and industrial applications. S Afr J Bot 151(B):3–18

Baechle JJ, Chen N, Makhijani P, Winer S, Furman D, Winer DA (2023) Chronic inflammation and the hallmarks of aging. Mol Metab 74:101755

Baek SC, Nam KH, Yi SA, Jo MS, Lee KH, Lee YH, Lee J, Kim KH (2019) Anti-adipogenic effect of β-carboline alkaloids from garlic (*Allium sativum*). Foods 8(12):673

Bai L, Xu D, Zhou YM, Zhang YB, Zhang H, Chen YB, Cui YL (2022) Antioxidant activities of natural polysaccharides and their derivatives for biomedical and medicinal applications. Antioxidants 11(12):2491

Bal Krishna A, Manikyam HK, Sharma VK, Sharma N (2015) Plant cardenolides in therapeutics. Int J Indigenous Med Plant 48(2):1871–1896

Balavandi Z, Neshasteh-Riz A, Koosha F, Eynali S, Hoormand M, Shahidi M (2020) The use of β-Elemene to enhance radiosensitization of A375 human melanoma cells. Cell J 21(4):419–425

Baljak J, Bogavac M, Karaman M, Srđenović Čonić B, Vučković B, Anačkov G, Kladar N (2024) Chemical composition and biological activity of *Hypericum* species—*H. Hirsutum, H. Barbatum, H. Rochelii*. Plants 13(20):2905

Bashir A, Nabi M, Tabassum N, Afzal S, Ayoub M (2023) An updated review on phytochemistry and molecular targets of *Withania somnifera* (L.) Dunal (Ashwagandha). Front Pharmacol 14:1049334

Bedu-Ferrari C, Biscarrat P, Langella P, Cherbuy C (2022) Prebiotics and the human gut microbiota: from breakdown mechanisms to the impact on metabolic health. Nutrients 14(10):2096

Belwal T, Nabavi SF, Nabavi SM, Habtemariam S (2017) Dietary anthocyanins and insulin resistance: when food becomes a medicine. Nutrients 9(10):1111

Bhasin A, Singh S, Garg R (2020) Nutritional and medical importance of Citrullus colocynthis — a review. Plant Arch 20(Suppl 2):3400–3406

Bhattacharya A, Ghosal S, Bhattacharya SK (2001) Anti-oxidant effect of *Withania somnifera* glycowithanolides in chronic footshock stress-induced perturbations of oxidative free radical scavenging enzymes and lipid peroxidation in rat frontal cortex and striatum. J Ethnopharmacol 74(1):1–6

Bi X, Yuan Z, Qu B, Zhou H, Liu Z, Xie Y (2019) Piperine enhances the bioavailability of silybin via inhibition of efflux transporters BCRP and MRP2. Phytomedicine 54:98–108

Bolat E, Sarıtaş S, Duman H, Eker F, Akdaşçi E, Karav S, Witkowska AM (2024) Polyphenols: secondary metabolites with a biological impression. Nutrients 16(15):2550

Borecka M, Karaś M (2025) A comprehensive review of the nutritional and health-promoting properties of edible parts of selected *Cucurbitaceae* plants. Foods 14(7):1200

Bouhenni H, Doukani K, Hanganu D, Olah NK, Sekeroglu N, Gezici S (2021) Analysis of bioactive compounds and antioxidant activities of cultivated garlic (*Allium sativum* L.) and red onion (Allium cepa L.) in Algeria. Int J Agric Environ Food Sci 5(4):550–560

Bravo JA, Vila JL (2022) A collateral approach on the ^{13}C NMR analysis of pseudopelletierine, an alkaloid from pomegranate. Bolivian J Chem 39(5):175–182

Brentnall M, Rodriguez-Menocal L, De Guevara RL et al (2013) Caspase-9, caspase-3 and caspase-7 have distinct roles during intrinsic apoptosis. BMC Cell Biol 14:32

Brimson JM, Prasanth MI, Malar DS, Thitilertdecha P, Kabra A, Tencomnao T, Prasansuklab A (2021) Plant polyphenols for aging health: implication from their autophagy modulating properties in age-associated diseases. Pharmaceuticals 14(10):982

Budzianowska A, Derda M, Budzianowski J, Szopa A, Kikowska M (2023) Comparative study of *Plantago media* extracts in the treatment of *Acanthamoeba* sp. trophozoites. Appl Sci 13:7075

Busch C, Burkard M, Leischner C, Lauer UM, Frank J, Venturelli S (2015) Epigenetic activities of flavonoids in the prevention and treatment of cancer. Clin Epigenet 7(1):64

Chan EWC, Wong SK, Chan HT (2021) An overview on the chemistry, pharmacology and anticancer properties of tetrandrine and fangchinoline (alkaloids) from *Stephania tetrandra* roots. J Integr Med 19(4):311–316

Chang F-R, Wei J-L, Teng C-M, Wu Y-C (1998) Antiplatelet aggregation constituents from *Annona purpurea*. J Nat Prod 61(12):1457–1461

Cheatham CL, Nieman DC, Neilson AP, Lila MA (2022) Enhancing the cognitive effects of flavonoids with physical activity: is there a case for the gut microbiome? Front Neurosci 16:833202

Chen S, Zhao X, Ran L, Wan J, Wang X, Qin Y, Shu F, Gao Y, Yuan L, Zhang Q, Mi M (2015) Resveratrol improves insulin resistance, glucose and lipid metabolism in patients with non-alcoholic fatty liver disease: a randomized controlled trial. Digest Liver Dis 47(3):226–232

Chen CY, Wang GH, Chang YC, Liu S, Lin YT, Lai YL, Chung YC (2025) Cosmeceutical application of extracts from the flowers, stems, and leaves of *Buddleja davidii* grown at different altitudes. Front Pharmacol 16:1551134

Chew NWS, Ng CH, Tan DJH, Kong G, Lin C, Chin YH, Lim WH, Huang DQ, Quek J, Fu CE, Xiao J, Syn N, Foo R, Khoo CM, Wang JW, Dimitriadis GK, Young DY, Siddiqui MS, Lam CSP, Wang Y, Figtree GA, Chan MY, Cummings DE, Noureddin M, Wong VW, Ma RCW, Mantzoros CS, Sanyal A, Muthiah MD (2023) The global burden of metabolic disease: data from 2000 to 2019. Cell Metab 35(3):414–428.e3

Chiba T, Tsuchiya T, Komatsu T, Mori R, Hayashi H, Shimokawa I (2010) Development of calorie restriction mimetics as therapeutics for obesity, diabetes, inflammatory and neurodegenerative diseases. Curr Genomics 11(8):562–567

Chik WI, Zhu L, Fan LL, Yi T, Zhu GY, Gou XJ, Tang YN, Xu J, Yeung WP, Zhao ZZ, Yu ZL, Chen HB (2015) *Saussurea involucrata*: a review of the botany, phytochemistry and ethnopharmacology of a rare traditional herbal medicine. J Ethnopharmacol 172:44–60

Corrêa RCG, Peralta RM, Haminiuk CWI, Ferreira ICFR et al (2016) New phytochemicals as potential human anti-aging compounds: reality, promise, and challenges. Crit Rev Food Sci Nutr 58(6):942–957

Chen H, Shi X, Zhang L, Yao L, Cen L, Li L, Lv Y, Wei C (2022) Ultrasonic extraction process of polysaccharides from *Dendrobium nobile* Lindl.: optimization, physicochemical properties and anti-inflammatory activity. Foods 11(19):2957. https://doi.org/10.3390/foods11192957

Dai Y, Ban W, Yang Z (2024) Gastrodin, a promising natural small molecule for the treatment of central nervous system disorders, and its recent progress in synthesis, pharmacology and pharmacokinetics. Int J Mol Sci 25(17):9540

de Lima EP, Laurindo LF, Catharin VCS, Direito R, Tanaka M, Jasmin Santos German I, Lamas CB, Guiguer EL, Araújo AC, Fiorini AMR, Barbalho SM (2025) Polyphenols, alkaloids, and terpenoids against neurodegeneration: evaluating the neuroprotective effects of phytocompounds through a comprehensive review of the current evidence. Metabolites 15(2):124

Deng M, Zhao JY, Tu PF, Jiang Y, Li ZB, Wang YH (2004) Echinacoside rescues the SH-SY5Y neuronal cells from TNFα-induced apoptosis. Euro J Pharmacol 505(1–3):11–18

Dettori CA, Loi MC, Brullo S, Fraga i Arguimbau P, Tamburini E, Bacchetta G (2016) The genetic diversity and structure of the *Ferula communis* L. complex (*Apiaceae*) in the Tyrrhenian area. Flora 223:138–146

Dey P, Kundu A, Kumar A, Gupta M, Lee BM, Bhakta T, Dash S, Kim HS (2020) Analysis of alkaloids (indole alkaloids, isoquinoline alkaloids, tropane alkaloids). In: Recent advances in natural products analysis, pp 505–567

Di Stasi M, Dolci D, Peruzzi L, Braca A, De Leo M (2023) Phytochemical study of *Stachys sylvatica* (Lamiaceae) aerial parts. Plant Biosyst 157(3):569–583

Diao P, He H, Tang J, Xiong L, Li L (2021) Natural compounds protect the skin from airborne particulate matter by attenuating oxidative stress. Biomed Pharmacother 138:111534

Dias MC, Pinto DCGA, Silva AMS (2021) Plant flavonoids: chemical characteristics and biological activity. Molecules 26(17):5377

Dixit V, Chaubey KK, Dayal D, Chole PB, Manjunath BT, Bachheti RK, Gupta PC, Bachheti A, Worku LA (2025) Natural products in aging: cellular mechanisms and emerging therapeutics for age-related disorders. J Aging Res 2025:6868732

Donà M, Balestrazzi A, Mondoni A, Rossi G, Ventura L, Buttafava A, Macovei A, Sabatini ME, Valassi A, Carbonera D (2013) DNA profiling, telomere analysis and antioxidant properties as tools for monitoring ex situ seed longevity. Ann Bot 111(5):987–998

Ebrahimirad F, Mirmahdizade SE, Mahmoodieh B, Najafi S, Banihashemian SM, Nikakhtar S, Mobaraki H, Sadeghi A, Kossari N, SadatRafiei SK, Ghodsi Moghadam M, Mashkani A, Asadi Anar M, Khosravi F (2025) Antioxidant strategies against cellular senescence: unveiling the power of synthetic versus natural antioxidants in a systematic review. Front Aging 6:1543360

Eity TA, Bhuia MS, Chowdhury R, Ahmmed S, Sheikh SS, Akter R, Islam MT (2024) Therapeutic efficacy of quercetin and its nanoformulation both the mono- or combination therapies in the management of cancer: An update with molecular mechanisms. J Trop Med 2024:5594462

Elmore S (2007) Apoptosis: a review of programmed cell death. Toxicol Pathol 35(4):495–516

Estrada B, Bernal MA, Díaz J, Pomar F, Merino F (2002) Capsaicinoids in vegetative organs of *Capsicum annuum* L. in relation to fruiting. J Agric Food Chem 50(5):1188–1191

Etheridge AS, Black SR, Patel PR, So J, Mathews JM (2007) An in vitro evaluation of cytochrome P450 inhibition and P-glycoprotein interaction with goldenseal, *Ginkgo biloba*, grape seed, milk thistle, and ginseng extracts and their constituents. Planta Med 73(7):731–741

Fang P, Yu M, Min W, Wan D, Han S, Shan Y, Wang R, Shi M, Zhang Z, Bo P (2018) Effect of baicalin on GLUT4 expression and glucose uptake in myotubes of rats. Life Sci 196:156–161

Fasinu PS, Bouic PJ, Rosenkranz B (2012) An overview of the evidence and mechanisms of herb–drug interactions. Front Pharmacol 3:69

Faturoti OA, Ogidi CO (2025) Medicinal utilization and suitability of pumpkin (*Cucurbita maxima*) seeds for the development of value-added products. IntechOpen. https://doi.org/10.5772/intechopen.1008975

Favari C, Rinaldi de Alvarenga JF, Sánchez-Martínez L, Tosi N, Mignogna C, Cremonini E, Manach C, Bresciani L, Del Rio D, Mena P (2024) Factors driving the inter-individual variability in the metabolism and bioavailability of (poly)phenolic metabolites: a systematic review of human studies. Redox Biol 71:103095

Fender J, Klöcker J, Boivin-Jahns V, Ravens U, Jahns R, Lorenz K (2024) "Cardiac glycosides"–quo vaditis?–Past, present, and future? Naunyn Schmiedeberg's Arch Pharmacol 397(12):9521–9531

Ferizi R, Ramadan MF, Maxhuni Q (2023) Black seeds (*Nigella sativa*): medical application and pharmaceutical perspectives. J Pharm Bioallied Sci 15(2):63–67

Ferreira MU (2022) Alkaloids in future drug discovery. Molecules 27(4):1347

Ferrucci L, Fabbri E (2018) Inflammageing: chronic inflammation in ageing, cardiovascular disease, and frailty. Nature Rev Cardiol 15(9):505–522

Fierascu RC, Fierascu I, Ortan A, Fierascu IC, Anuta V, Velescu BS, Pituru SM, Dinu-Pirvu CE (2019) *Leonurus cardiaca* L. as a source of bioactive compounds: An update of the European medicines agency assessment report (2010). BioMed Res Int 2019:4303215

Fitsiou E, Pulido T, Campisi J, Alimirah F, Demaria M (2021) Cellular senescence and the senescence-associated secretory phenotype as drivers of skin photoaging. J Invest Dermatol 141(4S):1119–1126

Freudenberg K, Meyer-Delius M (1938) Über die Schardinger-Dextrine aus Stärke. Ber Dtsch Chem Ges 71:1596–1600

Fu Z, Fan X, Wang X, Gao X (2018) Cistanches Herba: an overview of its chemistry, pharmacology, and pharmacokinetic properties. J Ethnopharmacol 219:233–247. https://doi.org/10.1016/j.jep.2017.10.015

Galli A, Marciani P, Marku A, Ghislanzoni S, Bertuzzi F, Rossi R, Di Giancamillo A, Castagna M, Perego C (2020) Verbascoside protects pancreatic β-cells against ER-stress. Biomedicine 8(12):582

Gera M, Sharma N, Ghosh M, Huynh DL, Lee SJ, Min T, Kwon T, Jeong DK (2017) Nanoformulations of curcumin: an emerging paradigm for improved remedial application. Oncotarget 8(39):66680–66698

Ghareeb MA, Mohamed T, Saad AM, Refahy LA, Sobeh M, Wink M (2018) HPLC-DAD-ESI-MS/MS analysis of fruits from *Firmiana simplex* (L.) and evaluation of their antioxidant and antigenotoxic properties. J Pharm Pharmacol 70(1):133–142

Ghobadi N, Asoodeh A (2023) Co-administration of curcumin with other phytochemicals improves anticancer activity by regulating multiple molecular targets. Phytother Res 37(4):1688–1702

Goenka S, Johnson F, Simon SR (2021) Novel chemically modified curcumin (CMC) derivatives inhibit tyrosinase activity and melanin synthesis in B16F10 mouse melanoma cells. Biomolecules 11(5):674

Gonfa YH, Tessema FB, Bachheti A, Rai N, Tadesse MG, Singab AN, Chaubey KK, Bachheti RK (2023) Anti-inflammatory activity of phytochemicals from medicinal plants and their nanoparticles: a review. Curr Res Biotechnol 6:100152

Gören AC (2014) Use of *Stachys* species (mountain tea) as herbal tea and food. Rec Nat Prod 8(2):71–82

Gourguillon L, Cattuzzato L, Lavaud C, Lobstein A (2016) Biological properties and phytochemical analysis of the halophyte *Armeria maritima* Willd. (*Plumbaginaceae*). Planta Med 82(S01):S1–S381

Guo Y, Pope C, Cheng X, Zhou H, Klaassen CD (2011) Dose-response of berberine on hepatic cytochromes P450 mRNA expression and activities in mice. J Ethnopharmacol 138(1):111–118

Guo Q, Wang N, Liu H, Li Z, Lu L, Wang C (2020) The bioactive compounds and biological functions of *Asparagus officinalis* L.—A review. J Funct Foods 65:103727

Guo J, Huang X, Dou L et al (2022) Aging and aging-related diseases: from molecular mechanisms to interventions and treatments. Signal Transduct Target Ther 7:391

Guo X, He L, Sun J, Ye H, Yin C, Zhang W, Han H, Jin W (2024) Exploring the potential of anthocyanins for repairing photoaged skin: a comprehensive review. Foods 13:3506

Gurel E, Karvar S, Yucesan B, Eker I, Sameeullah M (2017) An overview of cardenolides in digitalis—more than a cardiotonic compound. Curr Pharm Des 23(34):5104–5114

Hanson AD, Rathinasabapathi B, Rivoal J, Burnet M, Dillon MO, Gage DA (1994) Osmoprotective compounds in the *Plumbaginaceae*: a natural experiment in metabolic engineering of stress tolerance. Proc Natl Acad Sci USA 91(1):306–310

Heebøll S, Thomsen KL, Pedersen SB, Vilstrup H, George J, Grønbæk H (2014) Effects of resveratrol in experimental and clinical non-alcoholic fatty liver disease. World J Hepatol 6(4):188–198

Hu P, Wang J, Qing Y et al (2021) FV-429 induces autophagy blockage and lysosome-dependent cell death of T-cell malignancies via lysosomal dysregulation. Cell Death Dis 12:80

Hu Y, Lin Q, Zhao H, Li X, Sang S, McClements DJ, Long J, Jin Z, Wang J, Qiu C (2023) Bioaccessibility and bioavailability of phytochemicals: influencing factors, improvements, and evaluations. Food Hydrocoll 135:108165. https://doi.org/10.1016/j.foodhyd.2022.108165

Hu L, Lin C, Li R, Xu S, Xu Q, An Z, Tang C (2025) Gastrodin protects neuronal cells against oxidative stress through miRNA-125b-5p/Mamdc2 axis. NeuroMolecular Med 27(1):29

Hussain Z, Khan JA, Arshad MI, Muhammad F, Abbas RZ (2021) Comparative characterization of cinnamon, cinnamaldehyde and kaempferol for phytochemical, antioxidant and pharmacological properties using acetaminophen-induced oxidative stress mouse model. Bol Latinoam Caribe Plant Med Aromát 20(4):339–350

Ibrahim NNA, Wan Mustapha WA, Sofian-Seng NS, Lim SJ, Mohd Razali NS, Teh AH, Rahman HA, Mediani A (2023) A comprehensive review with future prospects on the medicinal properties and biological activities of *Curcuma caesia* Roxb. Evid Based Complement Alternat Med 2023:7006565

Imam SS, Alshehri S, Ghoneim MM, Zafar A, Alsaidan OA, Alruwaili NK, Gilani SJ, Rizwanullah M (2021) Recent advancement in chitosan-based nanoparticles for improved oral bioavailability and bioactivity of phytochemicals: challenges and perspectives. Polymers 13(22):4036

Imtiaz I, Schloss J, Bugarcic A (2025) Interplay between traditional and scientific knowledge: phytoconstituents and their roles in lung and colorectal cancer signaling pathways. Biomolecules 15(3):380

Inchingolo AD, Inchingolo AM, Malcangi G, Avantario P, Azzollini D, Buongiorno S, Viapiano F, Campanelli M, Ciocia AM, De Leonardis N, de Ruvo E, Ferrara I, Garofoli G, Montenegro V, Netti A, Palmieri G, Mancini A, Patano A, Piras F, Marinelli G, Di Pede C, Laudadio C, Rapone B, Hazballa D, Corriero A, Fatone MC, Palermo A, Lorusso F, Scarano A, Bordea IR, Di Venere D, Inchingolo F, Dipalma G (2022) Effects of resveratrol, curcumin and quercetin supplementation on bone metabolism: a systematic review. Nutrients 14(17):3519

Iside C, Scafuro M, Nebbioso A, Altucci L (2020) SIRT1 activation by natural phytochemicals: an overview. Front Pharmacol 11:1225

Islam MR, Rauf A, Akash S, Trisha SI, Nasim AH, Akter M, Dhar PS, Ogaly HA, Hemeg HA, Wilairatana P, Thiruvengadam M (2024) Targeted therapies of curcumin focus on its therapeutic benefits in cancers and human health: molecular signaling pathway-based approaches and future perspectives. Biomed Pharmacother 170:116034

Jain A, Sarsaiya S, Gong Q, Wu Q, Shi J (2025) Chemical constituents and pharmacological activities of *Dendrobium nobile*. In: CABI Books. CABI, Wallingford, pp 29–44. https://doi.org/10.1079/9781800627987.0003

Janda K, Gutowska I, Geszke-Moritz M, Jakubczyk K (2021) The common chicory (*Cichorium intybus* L.) as a source of extracts with health-promoting properties—a review. Molecules 26(6):1814

Jiang B, Meng L, Zhang F, Jin X, Zhang G (2017) Enzyme-inducing effects of berberine on cytochrome P450 1A2 in vitro and in vivo. Life Sci 189:1–7

Juan CA, Pérez de la Lastra JM, Plou FJ, Pérez-Lebeña E (2021) The chemistry of reactive oxygen species (ROS) revisited: outlining their role in biological macromolecules (DNA, lipids and proteins) and induced pathologies. Int J Mol Sci 22(9):4642

Karakaya S, Yılmaz SV, Koca M, Demirci B, Sytar O (2019) Screening of non-alkaloid acetylcholinesterase inhibitors from extracts and essential oil's of *Anthriscus nemorosa* (M.Bieb.) Spreng. (Apiaceae). S Afr J Bot 125:261–269

Karakaya S, Koca M, Sytar O, Duman H (2021) Determination of natural phenolic compounds of Ferula longipedunculata Peşmen and assessment of their antioxidant and anticholinesterase potentials. Nat Prod Res 35(10):1654–1656. https://doi.org/10.1080/14786419.2019.1619728

Karakaya S, Koca M, Sytar O, Duman H (2019b) The natural phenolic compounds and their antioxidant and anticholinesterase potential of herb *Leiotulus dasyanthus* (K. Koch) Pimenov & Ostr. Nat Prod Res 34(9):1303–1305

Karakaya S, Bingol Z, Koca M, Dagoglu S, Pınar NM, Demirci B, Gulcin İ, Brestic M, Sytar O (2020) Identification of non-alkaloid natural compounds of *Angelica purpurascens* (Avé-Lall.) Gilli. (Apiaceae) with cholinesterase and carbonic anhydrase inhibition potential. Saudi Pharm J 28(1):1–14

Kebede BH, Forsido SF, Tola YB, Astatkie T (2021) Free radical scavenging capacity, antibacterial activity and essential oil composition of turmeric (*Curcuma domestica*) varieties grown in Ethiopia. Heliyon 7(2):e06239

Kennedy DO, Wightman EL (2011) Herbal extracts and phytochemicals: plant secondary metabolites and the enhancement of human brain function. Adv Nutr 2:32–50

Kesarwani K, Gupta R, Mukerjee A (2013) Bioavailability enhancers of herbal origin: an overview. Asian Pac J Trop Biomed 3(4):253–266

Kezimana P, Dmitriev AA, Kudryavtseva AV, Romanova EV, Melnikova NV (2018) Secoisolariciresinol diglucoside of flaxseed and its metabolites: biosynthesis and potential for nutraceuticals. Front Genet 9:641

Khanam S, Mishra P, Faruqui T, Alam P, Albalawi T, Siddiqui F, Rafi Z, Khan S (2025) Plant-based secondary metabolites as natural remedies: a comprehensive review on terpenes and their therapeutic applications. Front Pharm 16:1587215

Kim EJ, Kim MK, Jin XJ, Oh JH, Kim JE, Chung JH (2010) Skin aging and photoaging alter fatty acids composition, including 11,14,17-eicosatrienoic acid, in the epidermis of human skin. J Korean Med Sci 25(6):980–983

Kim H, Castellon-Chicas MJ, Arbizu S, Talcott ST, Drury NL, Smith S, Mertens-Talcott SU (2021) Mango (*Mangifera indica* L.) polyphenols: anti-inflammatory intestinal microbial health benefits, and associated mechanisms of actions. Molecules 26(9):2732

Kirkland JL (2016) Translating the science of aging into therapeutic interventions. Cold Spring Harb Perspect Med 6(3):a025908

Kirkwood TB (2005) Time of our lives. What controls the length of life? EMBO Rep 6:S4–S8

Kozuharova E, Simeonov V, Batovska D, Stoycheva C, Valchev H, Benbassat N (2023) Chemical composition and comparative analysis of lavender essential oil samples from Bulgaria in relation to the pharmacological effects. Pharmacia 70(2):395–403

Kumari R, Jat P (2021) Mechanisms of cellular senescence: cell cycle arrest and senescence associated secretory phenotype. Front Cell Dev Biol 9:645593

Kornicka A, Balewski Ł, Lahutta M, Kokoszka J (2023) Umbelliferone and its synthetic derivatives as suitable molecules for the development of agents with biological activities: a review of their pharmacological and therapeutic potential. Pharmaceuticals 16(12):1732. https://doi.org/10.3390/ph16121732

Lagoa R, Silva J, Rodrigues JR, Bishayee A (2020) Advances in phytochemical delivery systems for improved anticancer activity. Biotechnol Adv 38:107382

Lee E, An S, Cho SA, Yun Y, Han J, Hwang YK, Kim HK, Lee TR (2011) The influence of alkane chain length on the skin irritation potential of 1,2-alkanediols. Int J Cosmetic Sci 33(5):421–425

Lee W, Ku SK, Bae JS (2015) Anti-inflammatory effects of Baicalin, Baicalein, and Wogonin in vitro and in vivo. Inflammation 38(1):110–125

Lelong C, Ris L, Sytar O, Defrere S, Villers A (2026) Rhodiola rosea L. roots powder strongly reduces anxiety and corticosterone levels induced by chronic stress in a murine model. J Pharm Health Care Sci 12(1):21. https://doi.org/10.1186/s40780-025-00532-4

Li N, Wang J, Ma J, Gu Z, Jiang C, Yu L, Fu X (2015) Neuroprotective effects of Cistanches Herba therapy on patients with moderate Alzheimer's disease. Evid-Based Complement Altern Med 2015:103985. https://doi.org/10.1155/2015/103985

Li X, Li C, Zhang W et al (2023) Inflammation and aging: signaling pathways and intervention therapies. Signal Transduct Target Ther 8:239

Lim TK (2013) Echinacea purpurea. In: Edible medicinal and non-medicinal plants, Volume 7: Flowers. Springer, Dordrecht, pp 340–371. https://doi.org/10.1007/978-94-007-7395-0_23

Lin YK, Liu YJ, Shen JW, Chiang HM, Lin YH, Chan ST, Chiang CF (2025) Anti-aging and skin-improving effects of *Saussurea involucrata* stem cell extract: cellular mechanisms and clinical validation. Acta Sci Nutr Health 9(2):55–62

Liu JK (2022) Antiaging agents: safe interventions to slow aging and healthy life span extension. Nat Prod Bioprospec 12:18

Liu R, Li H, Yang J et al (2018) Indole diketopiperazines from endophytic fungus of *Astragalus membranaceus* and biological evaluation. Chem Nat Compd 54(6):1196–1198

Liu M, Tan H, Xie H (2024) Phenylethanoid glycosides from *Michelia champaca* leaves. Phytochemistry 226:114118

Liu X, Yang Z, Han M, Zhang Y, Muhammad H, Zhong H, Guan R (2025) Bioactive components, pharmacological properties, and applications of *Cistanche deserticola* Y. C Ma: a comprehensive review. Nutrients 17(9):1501

Long Y, Wang W, Zhang Y, Du F, Zhang S, Li Z, Deng J, Li J (2023) Photoprotective effects of *Dendrobium nobile* Lindl. polysaccharides against UVB-induced oxidative stress and apoptosis in HaCaT cells. Int J Mol Sci 24:6120

Longo VD, Antebi A, Bartke A et al (2015) Interventions to slow aging in humans: are we ready? Aging Cell 14(4):497–510

Lu WH (2024) Effect of modifiable lifestyle factors on biological aging. JAR Life 13:88–92

Lu M, Qiu Q, Luo X, Liu X, Sun J, Wang C, Lin X, Deng Y, Song Y (2019) Phyto-phospholipid complexes (phytosomes): a novel strategy to improve the bioavailability of active constituents. Asian J Pharm Sci 14(3):265–274

Lu H, Zhang S, Wang J, Chen Q (2021) A review on polymer and lipid-based nanocarriers and its application to nano-pharmaceutical and food-based systems. Front Nutr 8:783831

Luan F, Wang Z, Yang Y, Ji Y, Lv H, Han K, Liu D, Shang X, He X, Zeng N (2021) *Juglans mandshurica* Maxim.: a review of its traditional usages, phytochemical constituents, and pharmacological properties. Front Pharmacol 11:569800

Maldonado E, Morales-Pison S, Urbina F, Solari A (2023) Aging hallmarks and the role of oxidative stress. Antioxidants (Basel) 12(3):651

Mall J, Naseem N, Haider MF, Rahman MA, Khan S, Siddiqui SN (2024) Nanostructured lipid carriers as a drug delivery system: a comprehensive review with therapeutic applications. Intelligent Pharm 3:243–255

Mamadalieva NZ, Lafont R, Wink M (2014) Diversity of secondary metabolites in the genus *Silene* L. (*Caryophyllaceae*)—structures, distribution, and biological properties. Diversity 6:415–499

Mandal SK, Maji AK, Mishra SK, Ishfaq PM, Devkota HP, Silva AS, Das N (2020) Goldenseal (*Hydrastis canadensis* L.) and its active constituents: a critical review of their efficacy and toxicological issues. Pharmacol Res 160:105085

Marčetić M, Bufan B, Drobac M, Antić Stanković J, Arsenović Ranin N, Milenković MT, Božić DD (2025) Multifaceted biological properties of verbascoside/acteoside: antimicrobial, cytotoxic, anti-inflammatory, and immunomodulatory effects. Antibiotics 14:697

Markovic M, Ben-Shabat S, Dahan A (2020) Prodrugs for improved drug delivery: lessons learned from recently developed and marketed products. Pharmaceutics 12(11):1031

Maruyama S, Akasaka T, Yamada K, Tachibana H (2009) Bioactivity of high-molecular-weight constituents of Cistanche tubulosa. Pharmacologyonline 1:913–921

Mavrogonatou E, Kletsas D (2024) Plant-derived senotherapeutics for the prevention and treatment of intervertebral disc degeneration and aging. Meta 14(3):146

Mehta P, Shah R, Lohidasan S, Mahadik KR (2015) Pharmacokinetic profile of phytoconstituent(s) isolated from medicinal plants—a comprehensive review. J Trad Complement Med 5(4):207–227

Mishra P, Khan S, Rafi Z, Faruqui T, Mansoor S, Ahmad I et al (2024) Exploring the significance and cutting-edge applications of terpenes and terpenoid-derived inorganic nanoparticles. Sci Adv Mater 16:665–681

Mohajer S, Mat Taha R, Ramli R, Mohajer M (2016) Phytochemical constituents and radical scavenging properties of *Borago officinalis* and *Malva sylvestris*. Ind Crop Prod 94:673–681

Mohammed NJ, Ali W, Al-Fahham A (2024) The chemical structure, classification and clinical significance of alkaloids. Int J Health Med Res 3(10):10

Morikawa T, Ninomiya K, Imamura M, Akaki J, Fujikura S, Pan Y, Yuan D, Yoshikawa M, Jia X, Li Z, Muraoka O (2014) Acylated phenylethanoid glycosides, echinacoside and acteoside from *Cistanche tubulosa*, improve glucose tolerance in mice. J Nat Med 68(3):561–566

Motti R, Paura B, Cozzolino A, Falco B (2022) Edible flowers used in some countries of the Mediterranean Basin: an ethnobotanical overview. Plants (Basel) 11(23):3272

Muñoz-Acevedo A, Aristizábal-Córdoba S, Rodríguez JD, Torres EA, Molina AM, Gutiérrez RG, Kouznetsov VV (2016) Citotoxicidad/capacidad antiradicalaria in vitro y caracterización estructural por GC-MS/^{1}H-^{13}C-RMN de los aceites esenciales de hojas de árboles joven/adulto de *Annona purpurea* Moc. & Sessé ex Dunal de Repelón (Atlántico, Colombia). Boletín Latinoamericano y del Caribe de Plantas Medicinales y Aromáticas 15:99–111

Musuc AM (2024) Cyclodextrins: advances in chemistry, toxicology, and multifaceted applications. Molecules 29(22):5319

Ng TB, Liu J, Wong JH, Ye X, Sze SCW, Tong Y, Zhang KY (2012) Review of research on Dendrobium, a prized folk medicine. Appl Microbiol Biotechnol 93:1795–1803. https://doi.org/10.1007/s00253-011-3829-7

Nadaf M, Khalil Abad MH, Omidipour R, Soorgi H, Riahi-Madvar A, Saleh Ghamari E (2023) Ethnobotanical knowledge, chemistry, and pharmacology of the *Asteraceae* family in Iran: a review. Ethnobot Res Appl 30:1–27

Naomi R, Yazid MD, Teoh SH, Balan SS, Shariff H, Kumar J, Bahari H, Embong H (2023) Dietary polyphenols as a protection against cognitive decline: evidence from animal experiments; mechanisms and limitations. Antioxidants (Basel) 12(5):1054

Negi JS, Singh P, Joshi GP, Rawat MS, Bisht VK (2010) Chemical constituents of *Asparagus*. Pharmacogn Rev 4(8):215–220

Nowak-Perlak M, Olszowy M, Woźniak M (2025) The natural defense: anti-aging potential of plant-derived substances and technological solutions against photoaging. Int J Mol Sci 26(16):8061

Núñez-Cortés R, Salazar-Méndez J, Nijs J (2025) Physical activity as a central pillar of lifestyle modification in the management of chronic musculoskeletal pain: a narrative review. J Funct Morphol Kinesiol 10(2):183

Nyalo PO, Omwenga GI, Ngugi MP (2023) Antibacterial properties and GC-MS analysis of ethyl acetate extracts of *Xerophyta spekei* (Baker) and *Grewia tembensis* (Fresen). Heliyon 9(3):e14461

OuYang J, Zhou WN, Li G, Wang XY, Ding CX, Suo YR, Wang HL (2015) Three new alkaloids from *Hippophae rhamnoides* Linn. Subsp. sinensis Rousi. Heterocycles 98:1287–1291

Öztürk E, Arslan AKK, Yerer MB, Bishayee A (2017) Resveratrol and diabetes: a critical review of clinical studies. Biomed Pharmacother 95:230–234

Paluch Z, Biriczová L, Pallag G, Carvalheiro Marques E, Vargová N, Kmoníčková E (2020) The therapeutic effects of *Agrimonia eupatoria* L. Physiol Res 69(Suppl 4):S555–S571

Pareek A, Pant M, Gupta MM, Kashania P, Ratan Y, Jain V, Pareek A, Chuturgoon AA (2023) *Moringa oleifera*: an updated comprehensive review of its pharmacological activities, ethnomedicinal, phytopharmaceutical formulation, clinical, phytochemical, and toxicological aspects. Int J Mol Sci 24(3):2098

Park SY, Jin ML, Kim YH, Kim Y, Lee SJ (2012) Anti-inflammatory effects of aromatic-turmerone through blocking of NF-κB, JNK, and p38 MAPK signaling pathways in amyloid β-stimulated microglia. Int Immunopharmacol 14(1):13–20

Parveen S, Ansari M, Parveen R, Khan W, Ahmad S, Husain SA (2019) Chromatography based metabolomics and in silico screening of *Gymnema sylvestre* leaf extract for its antidiabetic potential. Evid Based Complement Alternat Med 2019:7523159

Paul S, Chakraborty S, Anand U, Dey S, Nandy S, Ghorai M et al (2021) *Withania somnifera* (L.) Dunal (ashwagandha): a comprehensive review on ethnopharmacology, pharmacotherapeutics, biomedicinal and toxicological aspects. Biomed Pharmacother 143:112175

Paul ZA, Malla AT, Dar MA, Masoodi MH (2024) Phytochemistry and pharmacological activity of *Malva sylvestris* L: a detailed insight. Comb Chem High Throughput Screen 27(16):2309–2322

Peredo-Escárcega AE, Guarner-Lans V, Pérez-Torres I, Ortega-Ocampo S, Carreón-Torres E, Castrejón-Tellez V et al (2015) The combination of resveratrol and quercetin attenuates metabolic syndrome in rats by modifying the serum fatty acid composition and by upregulating SIRT 1 and SIRT 2 expression in white adipose tissue. Evid Based Complement Alternat Med 2015:474032

Perez Gutierrez RM, Gutierrez Velazquez E, Paredes-Carrera SP (2019) *Spinacia oleracea* Linn considered as one of the most perfect foods: a pharmacological and phytochemical review. Mini Rev Med Chem 19(20):1666–1680

Peterson B, Weyers M, Steenekamp JH, Steyn JD, Gouws C, Hamman JH (2019) Drug bioavailability enhancing agents of natural origin (bioenhancers) that modulate drug membrane permeation and pre-systemic metabolism. Pharmaceutics 11(1):33

Phua QH, Ng SY, Soh BS (2024) Mitochondria: a potential rejuvenation tool against aging. Aging Dis 15(2):503–516

Pokajewicz K, Białoń M, Svydenko L, Fedin R, Hudz N (2021) Chemical composition of the essential oil of the new cultivars of *Lavandula angustifolia* Mill. bred in Ukraine. Molecules 26(18):5681

Pontin M, Bottini R, Burba JL, Piccoli P (2015) *Allium sativum* produces terpenes with fungistatic properties in response to infection with *Sclerotium cepivorum*. Phytochemistry 115:152–160

Purushothaman JR, Rizwanullah M (2024) Ferulic acid: a comprehensive review. Cureus 16(8):e68063

Raclariu-Manolică AC, Socaciu C (2023) In search of authenticity biomarkers in food supplements containing sea buckthorn: a metabolomics approach. Foods 12(24):4493

Rahman S, Husen A (2023) *Trigonella foenum-graecum* L.: a useful medicinal plant for diabetes and other diseases. In: Antidiabetic medicinal plants and herbal treatments, 1st edn. CRC Press, pp 1–13

Ramírez-Rendon D, Passari AK, Ruiz-Villafán B, Rodríguez-Sanoja R, Sánchez S, Demain AL (2022) Impact of novel microbial secondary metabolites on the pharma industry. Appl Microbiol Biotechnol 106(5–6):1855–1878

Rassu G, Sorrenti M, Catenacci L, Pavan B, Ferraro L, Gavini E, Bonferoni MC, Giunchedi P, Dalpiaz A (2023) Conjugation, prodrug, and co-administration strategies in support of nano-technologies to improve the therapeutic efficacy of phytochemicals in the central nervous system. Pharmaceutics 15(6):1578

Rastogi S, Pandey MM, Kumar Singh Rawat A (2015) Medicinal plants of the genus Betula—traditional uses and a phytochemical-pharmacological review. J Ethnopharmacol 159:62–83

Ren ZQ, Zheng SY, Sun Z, Luo Y, Wang YT, Yi P, Li YS, Huang C, Xiao WF (2025) Resveratrol: molecular mechanisms, health benefits, and potential adverse effects. MedComm 6(6):e70252

Rezig L, Saada M, Trabelsi N, Tammar S, Limam H, Bettaieb Rebey I, Smaoui A, Sghaier G, Del Re G, Ksouri R, Msaada K (2019) Chemical composition, antioxidant and antimicrobial activities of *Aloysia triphylla* L. essential oils and methanolic extract. Ital J Food Sci 31:556–572

Riehl CA, Pinto AC, Figueroa-Villar JD (2006) Isolation and structural determination of xerophytolic acid a, a 3-geranyl-4-hydroxybenzoate derivative from *Xerophyta plicata*. Nat Prod Res 20(13):1225–1230

Rodríguez-Rodero S, Fernández-Morera JL, Menéndez-Torre E, Calvanese V, Fernández AF, Fraga MF (2011) Aging genetics and aging. Aging Dis 2(3):186–195

Rombolà L, Scuteri D, Marilisa S, Watanabe C, Morrone LA, Bagetta G, Corasaniti MT (2020) Pharmacokinetic interactions between herbal medicines and drugs: their mechanisms and clinical relevance. Life (Basel) 10(7):106

Ru W, Wang D, Xu Y, He X, Sun YE, Qian L, Zhou X, Qin Y (2015) Chemical constituents and bioactivities of *Panax ginseng* (C.A. Mey). Drug Discov Ther 9(1):23–32

Saikia L, Talukdar NC, Dutta PP (2025) Exploring the therapeutic role of flavonoids through AMPK activation in metabolic syndrome: a narrative review. Phytother Res 39(3):1403–1421

Salehi B, Gültekin-Özgüven M, Kırkın C, Özçelik B, Morais-Braga MFB, Carneiro JNP, Bezerra CF, Silva TGD, Coutinho HDM, Amina B, Armstrong L, Selamoglu Z, Sevindik M, Yousaf Z, Sharifi-Rad J, Muddathir AM, Devkota HP, Martorell M, Jugran AK, Martins N, Cho WC (2019) *Anacardium* plants: chemical, nutritional composition and biotechnological applications. Biomolecules 9(9):465

Salehi B, Azzini E, Zucca P, Maria Varoni E, Anil Kumar NV, Dini L, Panzarini E, Rajkovic J, Valere Tsouh Fokou P, Peluso I, Prakash Mishra A, Nigam M, El Rayess Y, El Beyrouthy M, Setzer WN, Polito L, Iriti M, Sureda A, Quetglas-Llabrés M, Sharifi-Rad J (2020) Plant-derived bioactives and oxidative stress-related disorders: a key trend towards healthy aging and longevity promotion. Appl Sci 10(3):947

Sarsaiya S, Jain A, Gong Q, Wu Q, Shi J (2025) *Dendrobium nobile* and skin health: from traditional use to modern applications. In: CABI books. CABI, pp 218–236. https://doi.org/10.1079/9781800627987.0013

Schieber M, Chandel NS (2014) ROS function in redox signaling and oxidative stress. Curr Biol 24(10):R453–R462

Semwal P, Painuli S, Abu-Izneid T, Rauf A, Sharma A, Daştan SD, Kumar M, Alshehri MM, Taheri Y, Das R, Mitra S, Emran TB, Sharifi-Rad J, Calina D, Cho WC (2022) Diosgenin: an updated pharmacological review and therapeutic perspectives. Oxidative Med Cell Longev 2022:1035441

Senol H, Ozgun-Acar O, Dağ A, Eken A, Guner H, Aykut ZG, Topcu G, Sen A (2023) Synthesis and comprehensive in vivo activity profiling of olean-12-en-28-ol, 3β-pentacosanoate in experimental autoimmune encephalomyelitis: a natural remyelinating and anti-inflammatory agent. J Nat Prod 86(1):103–118

Sharma BR, Jaiswal S, Ravindra PV (2022) Modulation of gut microbiota by bioactive compounds for prevention and management of type 2 diabetes. Biomed Pharmacother 152:113148

Shi R, Li X, Gao B, Zhang C, Li M (2023) Phenylethanoid glycosides from traditional Mongolian medicine *Cymbaria daurica* alleviate alloxan-induced INS-1 cells oxidative stress and apoptosis. Food Sci Human Wellness 12:1580–1589

Sissi S, Di Giacomo S, Ferrante C, Angelini P, Macone A, Giusti AM, Toniolo C, Vitalone A, Abdellah A, Larhsini M, Menghini L, Markouk M, Mazzanti G, Di Sotto A (2022) Characterization of the phytochemical composition and bioactivities of *Anacyclus maroccanus* ball. And *Anacyclus radiatus* Loisel aerial parts: preliminary evidence for the possible development of Moroccan plants. Molecules 27(3):692

Sivasankari B, Pitchaimani S, Anandharaj M (2013) A study on traditional medicinal plants of Uthapuram, Madurai District, Tamilnadu, South India. Asian Pac J Trop Biomed 3(12):975–979

Song J, Zhou YZ, Pang YY, Gao L, Du GH, Qin XM (2020) The anti-aging effect of *Scutellaria baicalensis* Georgi flowers extract by regulating the glutamine-glutamate metabolic pathway in d-galactose induced aging rats. Exp Gerontol 134:110843

Srivastava S (2017) The mitochondrial basis of aging and age-related disorders. Genes (Basel) 8(12):398

Suzery M, Ningrum AN, Nudin B, Mulyani NS, Cahyono B (2019) Determination of quercetin and rutin in red galangal rhizomes (*Alpinia purpurata*) and white galangal (*Alpinia galanga*) using high performance liquid chromatography method. IOP Conf Ser: Earth Environ Sci 292:012064

Sytar O, Hajihashemi S (2024) Specific secondary metabolites of medicinal plants and their role in stress adaptation. In: Plant secondary metabolites and abiotic stress. Wiley, pp 425–479

Tancreda G, Ravera S, Panfoli I (2025) Preclinical evidence of *Withania somnifera* and Cordyceps spp.: neuroprotective properties for the management of Alzheimer's disease. Int J Mol Sci 26(11):5403

Thangapazham RL, Sharma A, Maheshwari RK (2007) Beneficial role of curcumin in skin diseases. Adv Exp Med Biol 595:343–357

Tiwari P, Mishra BN, Sangwan NS (2014) Phytochemical and pharmacological properties of *Gymnema sylvestre*: an important medicinal plant. Biomed Res Int 2014:830285

Tolkachev ON, Abizov EA, Abizova EV, Mal'tsev SD (2009) Phytochemical study of the bark of some plants of the Elaeagnaceae family as a natural source of β-carboline indole alkaloids. Pharm Chem J 42(11):630–632

Tripathi AK, Ray AK, Mishra SK (2022) Molecular and pharmacological aspects of piperine as a potential molecule for disease prevention and management: evidence from clinical trials. Beni Suef Univ J Basic Appl Sci 11(1):16

Truzzi E, Bertelli D, Catellani B, Jazi DD, Benvenuti S (2025) Recovery of bioactive compounds from the biomass of aromatic plants after distillation using NADES: a sustainable alternative extraction method. Molecules 30(5):1120

Tureyen A, Demirel HH, Demirkapi EN, Eryavuz AM, Ince S (2023) Tubuloside a, a phenylethanoid glycoside, alleviates diclofenac-induced hepato-nephro oxidative injury via Nrf2/HO-1. J Cell Mol Med 27:3404–3413

Tylutka A, Walas Ł, Zembron-Lacny A (2024) Level of IL-6, TNF, and IL-1β and age-related diseases: a systematic review and meta-analysis. Front Immunol 15:1330386

Ullah H, Di Minno A, Santarcangelo C, Khan H, Daglia M (2021) Improvement of oxidative stress and mitochondrial dysfunction by β-caryophyllene: a focus on the nervous system. Antioxidants 10(4):546

Ullah A, Munir S, Badshah SL, Khan N, Ghani L, Poulson BG, Emwas AH, Jaremko M (2020) Important flavonoids and their role as a therapeutic agent. Molecules 25:5243. https://doi.org/10.3390/molecules25225243

Utami AR, Maksum IP, Deawati Y (2023) Berberine and its study as an antidiabetic compound. Biology (Basel) 12(7):973

Vaduganathan M, Mensah GA, Turco JV, Fuster V, Roth GA (2022) The global burden of cardiovascular diseases and risk: a compass for future health. J Am Coll Cardiol 80(25):2361–2371

Varesi A, Chirumbolo S, Campagnoli LIM, Pierella E, Piccini GB, Carrara A, Ricevuti G, Scassellati C, Bonvicini C, Pascale A (2022) The role of antioxidants in the interplay between oxidative stress and senescence. Antioxidants (Basel) 11(7):1224

Wahab S, Annadurai S, Abullais SS, Das G, Ahmad W, Ahmad MF, Kandasamy G, Vasudevan R, Ali MS, Amir M (2021) *Glycyrrhiza glabra* (licorice): a comprehensive review on its phytochemistry, biological activities, clinical evidence and toxicology. Plants 10(12):2751

Wang CH, Wu SB, Wu YT, Wei YH (2013) Oxidative stress response elicited by mitochondrial dysfunction: implication in the pathophysiology of aging. Exp Biol Med 238(5):450–460

Wang W, Luo J, Liang Y, Li X (2016) Echinacoside suppresses pancreatic adenocarcinoma cell growth by inducing apoptosis via the mitogen-activated protein kinase pathway. Mol Med Rep 13(3):2613–2618

Wang J, Zheng Q, Wang H, Shi L, Wang G, Zhao Y, Fan C, Si J (2023) Sesquiterpenes and sesquiterpene derivatives from ferula: their chemical structures, biosynthetic pathways, and biological properties. Antioxidants (Basel) 13(1):7

Wang D, Liu W, Lu M, Xu Q (2025) Neuropharmacological effects of *Gastrodia elata* Blume and its active ingredients. Front Neurol 16:1574277

Warias P, Plewa P, Poniewierska-Baran A (2025) Resveratrol, piceatannol, curcumin, and quercetin as therapeutic targets in gastric cancer—mechanisms and clinical implications for natural products. Molecules 30(1):3

Wei L, Zhao C, Dong S, Yao S, Ji B, Zhao B, Liu Z, Liu X, Wang Y (2020) Secoisolariciresinol diglucoside alleviates hepatic lipid metabolic misalignment involving the endoplasmic reticulum-mitochondrial axis. Food Funct 11(5):3952–3963

Wesnes KA, Ward T, McGinty A, Petrini O (2000) The memory enhancing effects of a Ginkgo biloba/ Panax ginseng combination in healthy middle-aged volunteers. Psychopharmacology 152(4):353–361

Wiciński M, Fajkiel-Madajczyk A, Kurant Z, Liss S, Szyperski P, Szambelan M, Gromadzki B, Rupniak I, Słupski M, Sadowska-Krawczenko I (2024) Ashwagandha's multifaceted effects on human health: impact on vascular endothelium, inflammation, lipid metabolism, and cardiovascular outcomes—a review. Nutrients 16(15):2481

Xia D, Zhang Z, Zhao Y (2018) Acteoside attenuates oxidative stress and neuronal apoptosis in rats with focal cerebral ischemia-reperfusion injury. Biol Pharm Bull 41(11):1645–1651

Xiao Y, Ren Q, Wu L (2022a) The pharmacokinetic property and pharmacological activity of acteoside: a review. Biomed Pharmacother 153:113296

Xiao L, Yao J, Miao Y, Ou B, Wang J, Huang Y, Zhou B, Ge L, Tian J, Zeng X (2022b) Tubuloside B, isolated from *Cistanche tubulosa*, a promising agent against M1 macrophage activation via synergistically targeting Mob1 and ERK1/2. Biomed Pharmacother 153:113414

Xie J, Yang Z, Zhou C, Zhu J, Lee R, Teng L (2016) Nanotechnology for the delivery of phytochemicals in cancer. Biotechnol Adv 34:343–353

Yakubu J, Natsaridis E, du Toit T et al (2025) Nanoparticles with curcumin and piperine modulate steroid biosynthesis in prostate cancer. Sci Rep 15:13613

Yang J, Ju B, Yan Y, Xu H, Wu S, Zhu D, Cao D, Hu J (2017) Neuroprotective effects of phenylethanoid glycosides in an *in vitro* model of Alzheimer's disease. Exp Ther Med 13(5):2423–2428

Yao J, Wan H, Zhang J, Shen W, Wei X, Shi C, Ou B, Liu D, Ge L, Fei J et al (2024) Tubuloside B, a major constituent of *Cistanche deserticola*, inhibits migration of hepatocellular carcinoma by inhibiting Hippo-YAP pathway. Phytomedicine 129:155552

Yessenkyzy A, Saliev T, Zhanaliyeva M, Masoud AR, Umbayev B, Sergazy S, Krivykh E, Gulyayev A, Nurgozhin T (2020) Polyphenols as caloric-restriction mimetics and autophagy inducers in aging research. Nutrients 12(5):1344

Yuca H, Aydin B, Karakaya S, Goger G, Bingöl Z, Civas A, Koca M, Demirci B, Sytar O, Gulcin I, Guvenalp Z (2024) Exploring *Astrodaucus orientalis* (L.) Drude: phytochemical analysis and its biological potential against Alzheimer's and diabetes. Chem Biodivers 21(2):e202301753

Zahra M, Abrahamse H, George BP (2024) Flavonoids: antioxidant powerhouses and their role in nanomedicine. Antioxidants 13(8):922

Zhang RX, Li MX, Jia ZP (2008) *Rehmannia glutinosa*: review of botany, chemistry and pharmacology. J Ethnopharmacol 117(2):199–214

Zhang BB, Shi K, Liao ZX, Dai Y, Zou ZH (2011) Phenylpropanoid glycosides and triterpenoid of *Pedicularis kansuensis* maxim. Fitoterapia 82(6):854–860

Zhang N, Lian Z, Peng X, Li Z, Zhu H (2017) Applications of higenamine in pharmacology and medicine. J Ethnopharmacol 196:242–252

Zhang R, Pan T, Xiang Y, Zhang M, Xie H, Liang Z, Chen B, Xu C, Wang J, Huang X, Zhu Q, Zhao Z, Gao Q, Wen C, Liu W, Ma W, Feng J, Sun X, Duan T, Leung EH, Xie T, Wu Q, Sui X (2021) Curcumenol triggered ferroptosis in lung cancer cells via lncRNA H19/miR-19b-3p/FTH1 axis. Bioact Mater 13:23–36

Zhang L, Chen P, Xu L, Guo M (2025a) Roles and mechanisms of ginsenoside Rg1 in coronary artery disease: progress and perspectives. J Ginseng Res 49(3):225–236

Zhang C, Yi L, Bai Y, Yang X, He J, Bai Y, Zhang Y, Li B, Zhang J (2025b) Cyclodextrin inclusion complex enhances solubility and antimicrobial activity of chlortetracycline hydrochloride. Talanta 292:127977

Zhao T, Li SJ, Zhang ZX, Zhang ML, Shi QW, Gu YC, Dong M, Kiyota H (2017) Chemical constituents from the genus *Saussurea* and their biological activities. Heterocycl Commun 23(5):331–358

Zhao R, Xu Y, Wang X, Zhou X, Liu Y, Jiang S, Zhang L, Yu Z (2022) Withaferin a enhances mitochondrial biogenesis and BNIP3-mediated mitophagy to promote rapid adaptation to extreme hypoxia. Cells 12(1):85

Zhou XL, Xu MB, Jin TY, Rong PQ, Zheng GQ, Lin Y (2019) Preclinical evidence and possible mechanisms of extracts or compounds from cistanches for Alzheimer's disease. Aging Dis 10(5):1075–1093

Zhou F, Huang W, Xu T, Wu L, Chen Q, Peng J, Liu X, Lu B (2020) Natural P-gp inhibitor EGCG improves the acteoside absorption in Caco-2 cell monolayers and increases the oral bioavailability of acteoside in rats. Food Chem Toxicol 146:111827

Zhou S, Feng D, Zhou Y, Duan H, Jiang Y, Yan W (2023) Analysis of the active ingredients and health applications of cistanche. Front Nutr 10:1101182

Zorov DB, Juhaszova M, Sollott SJ (2014) Mitochondrial reactive oxygen species (ROS) and ROS-induced ROS release. Physiol Rev 94:909–950

Chemical Diversity and Mechanistic Insights into Anti-aging Phytocompounds from Indigenous Southern African Species

3

Keletso Lashani, Sonia Malik, Chunyang Li, Goabaone Gaobotse, Abdullah Makhzoum, and Kabo Masisi

Abstract

The rich biodiversity of plants that are typically native to Southern Africa has long been an integral part of conventional medicine due to their numerous health benefits, encompassing over 3000 plant species. Remarkably, about 20–25% of these species have been documented to possess significant dermatological effects, such as anti-aging, skin rejuvenation qualities, and wound healing, hence their usage in traditional medicine. With an emphasis on their bioactive compounds and underlying mechanisms of action, this chapter explores the potential of these plants as natural resources for combating the aging process. Through a

K. Lashani
Department of Biological Sciences and Biotechnology, Faculty of Sciences, Botswana International University of Science and Technology, Palapye, Botswana

Department of Biological Sciences, K.K. Birla Institute of Technology and Science, PILANI, Goa, India

S. Malik
Department of Biotechnology, Baba Farid College of Engineering and Technology, Bathinda, Punjab, India

Physiology, Ecology and Environment (P2E) Laboratory, University of Orleans, Orleans, France

C. Li
Institute of Agro-Product Processing, Jiangsu Academy of Agricultural Sciences, Nanjing, People's Republic of China

G. Gaobotse · A. Makhzoum (✉) · K. Masisi (✉)
Department of Biological Sciences and Biotechnology, Faculty of Sciences, Botswana International University of Science and Technology, Palapye, Botswana
e-mail: makhzouma@biust.ac.bw; masisik@biust.ac.bw

S. Gantait et al. (eds.), *Medicinal Plants with Anti-Ageing Properties*,
https://doi.org/10.1007/978-981-95-8064-4_3

comprehensive investigation of plant species such as *Clerodendrum glabrum, Ximenia caffra, Sutherlandia frutescens, Schotia brachypetala, Psychotria capensis,* and *Peltophorum africanum*, we outline the antioxidants, which include but are not limited to tannins, saponins, phenols, terpenoids, quinones, glycosides, steroids, alkaloids, anti-inflammatory, and collagen-stimulating properties of these plant species, all of which are imperative for maintaining skin firmness and delaying skin aging. We also highlight the molecular pathways that these specific plant species impact, together with the regulation of cellular senescence, extracellular matrix disintegration, and oxidative stress. This book chapter highlights the potential of nature's pharmacy, which may lead to the production of new anti-aging medications based on traditional knowledge of Southern African flora.

Keywords

Cellular senescence · Collagen-stimulating · Phytochemicals · Skin firmness · Traditional knowledge

3.1 Introduction

Age is a normal part of life for all living things. The part of the body most significantly impacted in humans is the skin. There are two types of skin aging: premature aging, often referred to as photoaging, and age-dependent, or chronological aging (Mukherjee et al. 2011). External causes trigger the latter, manifesting as deep furrows, dark/light pigmentation, and a leathery appearance (Fisher et al. 2002; Maity et al. 2011). The skin's natural aging process is seen through wrinkles. Aging is accompanied by a gradual, complex biological process that slowly decreases physiological stability, reducing most body functions and increasing the risk of death (López-Otín et al. 2013; Lenart and Krejci 2016). Reactive oxygen species (ROS) production in the skin is one of the many biological processes that contribute to aging. Reactive oxygen species are continually produced at low levels in the epidermis and serve as a protective mechanism against microbial invaders. They are also naturally produced as part of the body's reaction to inflammation (Birsan et al. 2022). The body employs antioxidant mechanisms to prevent excessive oxidative damage, as a buildup of ROS has been linked to skin aging and other skin conditions (Tsuchida and Kobayashi 2020; Birsan et al. 2022). Significant human illnesses, including cancer, diabetes, cardiovascular disease, and neurodegenerative diseases, are mostly caused by this gradual aging stage (Corella and Ordovas 2014).

There is a great demand for the use of healthier skin products, especially those derived from natural sources, thus prompting to uncover novel, user-friendly, and efficient anti-aging products (Goyal and Jerold 2023). Numerous scientific studies have been conducted on plants indigenous to Southern Africa, such as *Clerodendrum glabrum, Schotia brachypetala, Psychotria capensis*, and *Peltophorum africanum, Sutherlandia frutescens,* and *Ximenia caffra,* thus validating their anti-aging properties, such as inhibition of enzymes linked to aging, e.g., elastase, collagenase, and hyaluronidase. The efficacy of these plants in slowing down the aging process is demonstrated by their phytochemicals like tannins, saponins, phenols, terpenoids, quinones, glycosides, steroids, and alkaloids, which have been shown to inhibit 80% of these enzymes (Ndlovu et al. 2013; Tundis et al. 2015). Though satisfactory anti-aging results have been recorded from these secondary metabolites, certain hiccups have emerged due to certain toxicity when used at high levels (Michalak 2022; Pillay et al. 2022). For these bioactive compounds to be used in the cosmetic industry, it's imperative that we fully understand and appreciate their molecular mechanisms, thus necessitating the need to further extract and study them. This can be facilitated by integrating conventional knowledge with advanced scientific research, thereby enabling the manufacture of marketable and effective anti-aging skin products (Pillay et al. 2022; Lashani et al. 2024). Not only are these medicinal plants vital in the cosmetic industries in Sub-Saharan Africa, but their usage also extends into medicine and nutritional properties. In countries like Mali and Cameroon, remedies for infections, injuries, and inflammations commonly coexist with plant-based anti-aging applications due to similar biochemical mechanisms (Kazeem et al. 2024).

In brief, the vast richness of anti-aging phytochemicals found in Southern Africa's rich ethnopharmacological legacy holds a key to both beauty products and therapeutic applications. Their importance in enhancing cellular longevity and promoting healthy skin is attributed to their anti-inflammatory, anti-collagenase, and antioxidant qualities.

3.2 Key Indigenous Plant Species with Anti-aging Properties

3.2.1 *Clerodeundrum glabrum* E. Mey. var. *glabrum*

Within the Verbenaceae family, the genus *Clerodendrum* has over 580 species and is extensively dispersed in Asia, Australia, America, and Africa (South Africa, Eswatini, Botswana, Namibia) (Fig. 3.1) (Shrivastava and Patel 2007). *Clerodendrum glabrum* E. Mey. var. glabrum, often called tinder wood or white cat's whiskers and referred to as Munukhatshilongwe in Venda and Zimbabwe, is a small to medium-sized deciduous tree that is extensively dispersed in the Bushveld and at the edge of the forest (Van Wyk 2013). These are small trees, lianas, and shrubs that generally develop fully to a height ranging from 1 to 12 m, possessing leaves that are either opposite or spirally arranged (Fig. 3.2) (Adefegha and Oboh 2011). These leaves, when crushed, tend to produce a foul smell. Apart from their odor, they are green in color with a smooth margin. They have captivating flowers, which are characterized by a mauve to pink color that makes a thick terminal head comprising 4 pink stamens with yellow anthers (Fig. 3.2). The blossoming of the flowers usually occurs between November and April. The genus has been noted for the production of important secondary metabolites (Table 3.1). Ethnobotanical studies have indicated that the species are well known for their range of traditional importance in

Fig. 3.1 Geographical distribution of *Clerodeundrum glabrum* in Southern Africa (Source: Google Earth maps 2025) Δ = Species present

Fig. 3.2 *Clerodendrum* species characterized by large masses of pink starred flowers with long stamens protruding from axillary buds. (https://www.google.com/search?sca_esv=bf302389b4c6e1ea&sxsrf=AE3TifPm0kqCQ7SY3zBCKnZ0Yeie6ZGe6Q)

Table 3.1 Ethnobotany of medicinal plants from Southern Africa with reported anti-aging properties

Plant species	Family	Major phytochemicals	Pharmacological activities	Plant part used	References
Peltophorum africanum	Fabaceae	Flavonoids, betulinic acid, bergenin, tannins, quercetin, rutin, saponins, steroids, and benzoic acid.	Treating diarrhea, wounds, medicine for livestock, infertility, eye infections, back pains, impotency, STIs, asthma	Roots, stem bark, leaves	Amusan et al. (2002), Steenkamp (2003), Chinsembu et al. (2011)
Clerodendrum glabrum	Verbenaceae	Flavonoids, phenylethanoid glycosides, steroids, triterpenoids	Antihypertensive, antioxidant, sedative, analgesic, and anti-inflammatory, sedative, analgesic, and anti-asthma characteristics	Roots, leaves, bark, stem	Cheng et al. (2001), Kang et al. (2003), Choi et al. (2004), Chae et al. (2005), Shrivastava and Patel (2007), Abdel Wahab et al. (2008), Adebayo et al. (2022)
Schotia brachypetala	Fabaceae	Flavonoids, tannins, pro-anthocyanidins, methyl-5, 11, 14, 17-17-eicosatetraenoate, flavanols acyl glucosides, anthocyanins, dihydrochalcones, gallic acid, Myricetin-3-O-α-L-1C4-rhamnoside, Quercetin-3-O-L-1C4-rhamnoside	Treat flu symptoms, diarrhea, Alzheimer's disease, heartburn, dysentery, heart and nervous system	Roots, bark	Wyk and Gericke (2000), McGaw et al. (2002), Paiva et al. (2010), Adewusi et al. (2011), Du et al. (2014), Hassaan et al. (2014)
Psychotria capensis	Rubiaceae	Tannins, toluene, furan-2-methoxy, benzyl alcohol, phenolic, flavonoids, indole and monoterpene indole alkaloids, quinolone, isoquinoline alkaloids, coumarins, terpenoids, and cyclic peptides	Stomach pains, tuberculosis, respiratory infections, worm infections, emetic use	Roots, stem, bark, leaves	Nepokroeff et al. (1999), Benevides et al. (2005), Porto et al. (2009), Aro et al. (2019), Aro et al. (2022)
Ximenia caffra	Olacaceae	Polyphenols, flavonoids, catechin, and rutin	Anti-inflammatory, antibacterial, and antiproliferative benefits	Stems, roots, pulp, leaves	Zhen et al. (2015), Makhubele et al. (2024), Nyila (2025)
Sutherlandia frutescens	Fabaceae	Tannins, saponins, phenols, terpenoids, quinones, glycosides, steroids, alkaloids,	Treats wounds, a cold, eye infections, digestive disorders, diabetes, cancer, diabetes, autoimmune disorders, and depression	Leaves, stems, flowers, pods, roots	Wyk et al. (1997), Van Wyk and Albrecht. (2008), Aboyade et al. (2014), Tobwala et al. (2014), Mathe et al. (2024)

traditional medicine (Table 3.1). It has been stated that the chemical components found in these species are related to the biological activity of these species, as documented in old literature (Wang et al. 2015).

This plant also serves as an important source of food to a butterfly, *Spindasis natalensis*, commonly known as Natal Bar. The appearance of its beautiful flowers also acts as an attractive site for many insects, especially butterflies. One of the traditional uses of *C. glabrum* is as an insect repellent in rural areas where commercial insect repellents are scarce to find or expensive to buy. The Vhavenda people believed that using the tree helps to scare away bad spirits and witchcraft. Due to the popular belief that witches use polecats and hyenas for witchcraft, it was believed that the strong smell from the leaves would scare them away.

Several scientific studies have shown that *Clerodendrum glabrum* is known for its ability to function as an anti-aging agent through a series of enzymatic reactions. The plant species demonstrates strong anti-elastase, anti-hyaluronidase, and anti-collagenase properties, which are essential for maintaining skin suppleness and delaying the appearance of wrinkles. Various studies revealed that the stem ethyl acetate extracts remarkably inhibited elastase and collagenase by 89.38% and 88.64%, respectively, whereas a low inhibitory activity was recorded for hyaluronidase at 8.11% (Ndlovu et al. 2013). Furthermore, the ability of *Clerodendrum glabrum* to scavenge free radicals due to its antioxidant properties is what makes it possible to improve skin health (Table 3.2) (Ndlovu et al. 2013; Ogundajo and Ashafa 2019).

3.2.2 *Schotia brachypetala* Sond

Schotia brachypetala Sond, commonly known as the Weeping Boer-bean due to its dripping nectar from its flowers, is a flowering leguminous tree classified within the Fabaceae family, commonly found in Southern Africa (Fig. 3.3) (Du et al. 2014). Reaching up to 25 feet in height, this medium-sized, semi-deciduous tree is recognized by its broadly spreading crown made up of glossy green, alternating pinnately compound leaves with leaflets that are 4–6 inches long. The plant species comprises of visible, appealing flowers that are cup-shaped (Fig. 3.4), which are responsible for attracting a variety of bird species due to their nectar and, at the same time, aid in efficient pollination. During spring, that's when the plant starts flowering, lasting a period of 2 weeks (Tucker 2001; Kiepiel et al. 2022).

A decoction of the bark was administered by traditional healers to strengthen the body and cure a wide variety of ailments (Table 3.1). The plant's traditional usage for treating diarrhea and its activity against Alzheimer's disease can be explained by the evidence that extracts from different portions of *S. brachypetala* were effective against bacteria that cause gastrointestinal illnesses. The seeds may be roasted and consumed (Wyk and Gericke 2000). A variety of secondary metabolites were extracted and characterized from various parts of the *Schotia* plant (Table 3.1). These compounds were initially isolated from the plant's leaves. The potent antioxidant activity of these phenolic compounds was evidenced by in-vivo and in-vitro

Table 3.2 Antioxidant activity of major phytochemicals in relation to anti-aging

Compound	Representative compounds	Antioxidant activity	Role in anti-aging	References
Flavonoids	Quercetin, kaempferol, luteolin, apigenin, rutin, morin, myricetin, genistin, arbutin	Disruption of the production of free radicals, chelates free metals, and inhibits lipid peroxidation	Inflammatory agent, protects the skin from ultraviolet rays, helps to moisturize and make the skin soft, cofactors of enzymes	Jain et al. (2010), Malinowska (2013), Panche et al. (2016)
Phenolics	Gallic acid, ferulic acid, caffeic acid, sinapic acid, benzoic acid, vanillic acid, syringic acid, veratric acid, salicylic acid	Scavenging of free radicals, donating hydrogen to free radicals, quenching of singlet oxygen,	Regulates tyrosinase activity, thereby aiding in depigmentation, promotes the production of collagen and elastin fibers, keeps the skin moisturized and soft, and prevents the formation of erythema induced by UV.	Robbins (2003), Tsao (2010), Zillich et al. (2015)
Tannins	Ellagic acid, tannic acid, punicalagin	Scavenging free radicals, prevention of lipid peroxidation	Induces the synthesis of tropoelastin, inhibits elastase activity, exhibits antimicrobial and anti-inflammatory properties, and promotes wound healing.	Sieniawska (2015), Pizzi (2019)
Terpenoids	Geraniol, ursolic acid, paclitaxel, perillyl alcohol, artemisinin, cucurbitacin, costunolide, triptonide, oleanolic acid	Radical scavenging activity protects plants from oxidative stress, quenching of singlet oxygen, and inhibition of acetyl- and butyrylcholinesterase, pro-inflammatory activity, and antitumor activity	Possess antifungal, antibacterial, and anti-inflammatory properties which help to maintain the integrity of the skin, protect skin from UV damage, inhibit melanogenesis, and inhibit metalloproteinase induced by UV, regulate tyrosinase activity, thus helping in depigmenting, improves autophagy	Tetali (2019), Yang et al. (2020), Zhang et al. (2021), Trepa et al. (2024)
Carotenoids	Lycopene, lutein, β-carotene, astaxanthin, neoxanthin, capsanthin, synechoxanthin, crocin, zeaxanthin	Quenching of singlet oxygen, enhancement of antioxidant enzymes (CAT, SOD, GPx), anti-cancer activity, lowers chances of drug withdrawal syndrome, anti-inflammatory activity, boosts the immune system	Protection of the skin from the UV rays, protects against skin damage brought about by free radicals	Alavizadeh and Hosseinzadeh (2014), Johnson (2014), Kaczor et al. (2016)

(continued)

Table 3.2 (continued)

Compound	Representative compounds	Antioxidant activity	Role in anti-aging	References
Alkaloids	Caffeine, nicotine, berberine, indole, agelasine, isoquinoline, polyamine, pyridine, piperazine, thiazole.	Antimalarial activity, antibacterial activity, radical scavenging activity, anticancer activity, antiviral activity, antifungal activity	Enhances glucose metabolism, improves autophagy, improves mitochondrial biosynthesis, antiglycation activity	Kim et al. (2012), Arai et al. (2014), Cushnie et al. (2014), Thawabteh et al. (2019)
Saponins	Glycyrrhetinic acid, oleanolic acid, quillaic acid, soyasapgenol A,B,C, gypsogenin	Enhancement of antioxidant enzymes (CAT, SOD, GPx), prevents the production of free radicals, inhibits lipid peroxidation, metal chelating activity, and antimicrobial activity	Reduction of malondialdehyde, antiglycation activity, inhibition of tyrosinase	Güçlü-Üstündağ and Mazza (2007), Desai et al. (2009)

Fig. 3.3 Geographical distribution of *Schotia brachypetala* in Southern Africa. (Source: Google Earth maps 2025) Δ = Species present

Fig. 3.4 *Schotia brachypetala* characterized by shiny compound leaves and a thick cluster of reddish flowers. (https://www.google.com/search?q=Schotia+brachypetala+Sond&sca_esv=bf302389b4c6e1ea&udm=2&biw=1522&bih=722&sxsr)

studies via the activation of DAF-16/FOXA transcription factor in *C. elegans*, which revealed their ability to scavenge free radicals and, at the same time, increase longevity. A satisfactory IC_{50} value as low as 9 µg/mL was recorded during the DPPH assay, along with notable FRAP and ABTS assays (Sobeh et al. 2016).

3.2.3 *Psychotria capensis* (Eckl) Vatke

Psychotria species are typically the major genus found in the flowering plant in Rubiaceae. *Psychotria* and other members of the tribe *Psychotrieae* are usually characterized by small, whitish flowers that may be pollinated by bees, flies, and moths, although showy inflorescences have been derived in various neotropical lineages (Nepokroeff et al. 1999; Porto et al. 2009). Despite being extensively dispersed in tropical and pantropical regions, these species are primarily shrubs; however, they are also known to be vines, herbaceous, and epiphytes, which are commonly found in South Africa, Zimbabwe, Mozambique, and Eswatini (Fig. 3.5) (Taylor 1996). These plants usually grow well at sea elevation of approximately 1500 m in altitude. It is commonly found in swampy thick woodlands, the edges of

Fig. 3.5 Geographical distribution of *Psychotria capensis* in Southern Africa. (Source: Google Earth maps 2025) Δ = Species present

rivers, and in bushes and grasslands that experience a substantial amount of rainfall. *Psychotria capensis* has a thick trunk with perpendicular branches and light brown bark. When mature, it can reach a total height of 3–8 m. The leaves, which are big, oval to obovate, usually measure between 70 and 150 mm in length and 15–60 mm in width. The leaves feel leathery to the touch with a deep green, shiny upper side, while the underside is usually pale. Sometimes the lower side of the leaves may be smooth or less hairy. The inflorescences are usually terminal with breaching heads measuring up to 80 mm across. The flowers are conspicuous due to their golden to yellow color and are tiny (Fig. 3.6). The tubular corollas are short, recording an approximate measurement of 4 mm with 5 lobes and a hairy throat (Nepokroeff et al. 1999). The leaves, roots, barks, and rhizomes are frequently employed in traditional medicine to improve bronchial and gastrointestinal conditions, including cough, bronchitis, ulcer, and stomachache. Additionally, they are important in treating female reproductive system infections (Lajis et al. 1993; Formagio et al. 2014). Over the past decades, there has been a recognizable surge in phytochemical research on *Psychotria* plants, which has significantly advanced the ethnobotanical, pharmacological, and chemotaxonomic knowledge in the scientific field, as well as molecular phylogenetic analysis. The *Psychotria* genus is well-known for its richness in various bioactive compounds (Table 3.1). These identified compounds have been linked to the wide array of positive effects found in these plants, which include cytoprotective, pain-relieving, antimicrobial, and their positive effect on the central nervous system (Nepokroeff et al. 1999; Benevides et al. 2005). A study conducted by Ndlovu et al. (2013) demonstrated over 90% inhibition of elastase activity, a vital enzyme responsible for skin aging. Moreover, another research involving a related species, *P. capollacea,* showed neuroprotective activity, thus improving memory functions and at the same time ensuring acetylcholinesterase activity, which is

Fig. 3.6 Flowering inflorescence of *Psychotria capensis*, displaying compact clusters of yellow, star-shaped flowers and glossy opposite leaves. (https://www.gardensonline.com.au/global/gallery/Default.aspx?ItemId=4325&ItemType=Plant&srsltid=AfmBOooYAAx9_EZBQR4OP-ICFsVCWHmtTBXf-2ZWzoVICt7_SLsXcP4o)

associated with cognitive decline (Formagio et al. 2024). Alkaloids extracted from *Psychotria* species possess the ability to inhibit sirtuin enzymes, which are responsible for aging and neurological diseases, as verified by Sacconnay et al. (2015). Even though these discoveries highlight the anti-aging potential of *Psychotria capensis*, additional in vivo experiments are required to thoroughly study its effectiveness and underlying mechanisms of action.

3.2.4 *Peltophorum africanum* Sond

The genus *Peltrophorum* is widely distributed across the tropics, with *P. africanum* being the only species from its family to be found in Southern Africa (Fig. 3.7). It is commonly known as the Weeping Wattle (Mongalo 2013; Mazimba 2014). The tree is deciduous/semi-deciduous with an estimated height of 9–15 m when fully mature, with primarily greyish to brown branches extending from the base. Each pinna is comprised of approximately 8–22 pairs of leaflets, which are bipinnate and hairy, dull green in color. From November to February, the bisexual flowers become brightly yellow with creased petals at the tips (Fig. 3.8). The brown to black seeds

Fig. 3.7 Geographical distribution of *Peltophorum africanum* in Southern Africa. (Source: Google Earth maps 2025) Δ = Species present

of the plant are dispersed by birds and animals and are contained in a flat dark brown fruit. *Peltophorum africanum* is easily cultivated from the seeds around the summer period, which is the best time for its growth (Mazimba 2014).

The Fabaceae family mostly comprises different species of trees, shrubs, and herbs. These plants play a vital role in nitrogen fixation, all thanks to their symbiotic relationship with bacteria. The medicinal qualities that *Peltophorum africanum* possesses for both humans and animals are what make it stand out among other plants in this family (Bizimenyera et al. 2005; Mongalo and Raletsena 2025). The tree is categorized by its widespread presence in the Southern African landscape (Fig. 3.7), where it is easily detectable due to its distinguishing features (Theo et al. 2009). Due to its adaptability to diverse environmental conditions, it is prevalent across various habitats, from savannahs and woods to riverbanks and rocky riverbanks. Due to its versatility, the tree tends to have a significant role in the biological balance of these ecosystems, providing shelter, sunlight, and food resources (Mongalo and Raletsena 2025).

The satisfactory antioxidant activity of *Peltophorum africanum* can be beneficial in protecting against oxidative stress, which is a key contributor to premature aging (Bizimenyera et al. 2007; Ndlovu et al. 2013; Okeleye et al. 2015). Oxidative stress has garnered attention for its significant influence on cell damage and premature aging (Sharma 2014; Engwa et al. 2022). The ability of *Peltophorum africanum* to successfully scavenge free radicals has been tested through in-vitro assays such as DPPH and ABTS (Bizimenyera 2008; Okeleye et al. 2015; Tundis et al. 2015). The anti-inflammatory properties of *Peltophorum africanum* are responsible for conditions linked to inflammation, which has been found to cause aging (Bizimenyera et al. 2007; Adebayo et al. 2015). Additionally, chronic conditions such as heart diseases, arthritis, and neurological diseases are also caused by chronic inflammation (Rea et al. 2018; Chung et al. 2019). By reducing inflammation, extracts from *Peltophorum africanum* may assist in alleviating these age-related problems at the same time promoting general health during aging (Bizimenyera 2008). It has been

Fig. 3.8 Morphological characteristics of *Peltophorum africanum* characterized by brightly yellow flowers with creased petals at the tips. (https://www.plantbook.co.za/peltophorum-africanum/)

demonstrated that the anti-inflammatory activity of *Peltophorum africanum* involves the suppression of pro-inflammatory enzymes such as lipoxygenase and inducible nitric oxide synthase (iNOS) (Adebayo et al. 2015, 2017).

3.2.5 *Sutherlandia frutescens* (L.)

Sutherlandia frutescens is a leguminous plant that belongs to the Fabaceae family and is commonly known as cancer bush, phetol (meaning "change"), and insiswa (meaning "drive away darkness"). Its medicinal properties have been documented (Table 3.1). Different ethnic groups, inclusive of Zulu, Xhosa, Sotho, and Khoi-San, have used the plant to treat a wide range of ailments (Table 3.1). The plant is widely distributed across Africa, and in Southern Africa is mostly found in the southern parts of Namibia, the southeast part of Botswana, the center of South Africa, and many parts of Lesotho (Fig. 3.9). This small shrub is characterized by reddish to orange flowers (Fig. 3.10), distinctive seed pods shaped like a balloon, and is drought-tolerant, which makes it easier for it to grow and survive under harsh conditions. Its antioxidant, anti-inflammatory, and anti-mutagenic properties have been

Fig. 3.9 Geographical distribution of *Sutherlandia frutescens* (L.) in Southern Africa. (Source: Google Earth maps 2025) Δ = Species present

Fig. 3.10 *Sutherlandia frutescens* (L.) with well-displayed reddish to orange flowers with hairy leaves that make it look silver-like. (https://www.google.com/search?sa=X&sca_esv=8b213484c977a2a9&hl=en&sxsrf=AE3TifNwVT8Vsa35236bBSXwegNigKFMXA:1761285809557&udm=2&fbs)

reported to be linked to slowing or delaying the process of mature skin-aging (Fernandes et al. 2004; Kundu et al. 2005). *Sutherlandia frutescens* is a prostrate to erect bush plant with an estimated height ranging from 0.2 to 2.5 m when fully grown. It is perennial but has a relatively short life span.

The tree is characterized by conspicuous leaves that have short stalks, are pinnate, stipulate, and comprise about 8 pairs of opposite terminal leaflets. The shape of the leaflets depends on their origin, with each being narrowly oblong or elliptic in shape. The texture of the leaflets is usually smooth to silky-hairy. Captivating red blossoms, occasionally white, are found in the axillary racemes. The striking beauty of each flower can be noticed by the tubular and a pair of small wing petals, and a large petal that is recurved, streaked with white lines. The plant bears unique, large fruits, which are usually papery and bladder-shaped, varying from broadly oval to oblong in shape. The smooth seeds of the fruit are shaped like kidneys and have a light to dark brown color with surfaces that are rough to rough (Van Wyk and Albrecht 2008) (Fig. 3.10). In-vitro analysis of crude extracts isolated using different solvents has been shown to exhibit a significant antioxidant potency in cell lines due to their ability to scavenge free radicals such as superoxide, hydrogen peroxide, and superoxide anions, which have been found to have a substantial contribution to skin tissue degradation and premature skin aging. An experiment led by Tobwala et al. (2014) using cell lines extracted from Chinese hamster ovary (CHO), human hepatoma (HepaRG), and human lung carcinoma (A549), the plant extracts exhibited their cytoprotectivity by protecting the cell lines from oxidative stress through the reduction of intracellular ROS, simultaneously maintaining the glutathione redox balance (GSH/GSSG), which is imperative for detoxifying cells and improving longevity. In a similar case, the proliferation of many human tumor cell lines, including breast cancer cells such as MCF7 and MDA-MB-468, along with Jurkat and HL60 leukemia cells, was successfully inhibited by ethanol extracts of *S. frutescens*, thus highlighting its anti-cancer and anti-aging properties through the regulation of cell growth (Tai et al. 2004).

3.2.6 *Ximenia caffra* Sond. *var. caffra.*

Ximenia caffra Sond *var. caffra* is a small, roundish, deciduous tree found in most parts of Africa, belonging to the family Olacaceae. The fruit is commonly known as sour plum in English or Moretlologa-wa-kgomo in Setswana (Majinda and Motswaledi 1998). The plant makes its mark in the community due to its vital applications in food and traditional medicines. The plant habitat stretches from Tanzania in East Africa via South Africa, Namibia, and Botswana (Fig. 3.11). Additional tropical places, such as India, Australia, New Zealand, and the West Indies, also provide shelter to the plant (Mabberley 2008). *Ximenia caffra* Sond is a deciduous tree with a height of about 6 m when fully mature. The leaves are clustered and silky green flowers that bloom between August and October (Fig. 3.12). Appearing singly

Fig. 3.11 Geographical distribution of *Ximenia caffra* Sond. var. caffra in Southern Africa. (google.com/maps) Δ = Species available

on a stem, these flowers later develop into round, deep red fruits that become a source of food for both animals and humans (Nair et al. 2013).

This plant has been praised for its usage as a herb in traditional medicine for many years. Several health benefits have been documented from various parts of the plant, including antibacterial, anti-inflammatory, antioxidant, and antidiabetic properties. In relation to anti-aging, *X. caffra* has been shown to exhibit anti-aging properties due to a significant number of fatty acids, several bioactive compounds (Table 3.1), which have been shown to possess the ability to neutralize free radicals (Table 3.2), which are responsible for oxidative stress that is linked to premature aging (Zhen et al. 2015). The activity of enzymes involved in premature aging, such as collagenase, tyrosinase, hyaluronidase, and elastase, is usually inhibited by *X. caffra* extracts (Bakrim et al. 2022).

As evidenced by in vivo and animal model studies using the nematode Caenorhabditis, it has been highlighted that the leaf extracts of the fruit possess the ability to shield against oxidative stress through reducing ROS accumulation, promoting genetic factors like DAF-16 that have been associated with stress resistance, and longevity (Bakrim et al. 2022). Additionally, the leaf extracts have been documented to have health benefits (Table 3.1), which are responsible for maintaining healthy skin and delaying premature aging (Zhen et al. 2015; Makhubele et al. 2024; Nyila 2025). In a nutshell, it is noteworthy to mention that the anti-aging properties of *X. caffraa* possess the characteristics to suppress oxidative stress, prevent enzymes responsible for degrading skin tissues, and promote cellular pathways that promote healthy skin and longevity. The antioxidant activity of *X. caffra* to suppress enzymes responsible for premature aging has been documented (Maroyi 2016; Masuku et al. 2020; Chipuwa et al. 2024).

Fig. 3.12 Morphological characteristics of *Ximenia caffra* Sond. var. *caffra* with reddish to orange ripe fruits, along with clustered and silky green flowers and leaves. (https://www.waterberg-bioquest.co.za/Tree%20spp%20pgs/xim_caff.html)

3.3 Mechanisms of Action

3.3.1 Antioxidant Mechanism of Action

The body has established numerous protective mechanisms to reduce the production of free radicals and transform them into less dangerous derivatives, thus avoiding cellular damage brought by oxidative stress. These mechanisms include endogenous and exogenous, which together form a strong antioxidant system (Fig. 3.13). Antioxidants are compounds that donate electrons to free radicals, neutralizing or preventing them from causing harm, thereby helping to prevent premature aging and skin damage. They are naturally present in fruits, vegetables, wines, and tea, and they are considered good for our well-being (Dhalaria et al. 2020; Bjørklund et al. 2022; Aversa et al. 2016). Oxidative stress and skin aging are prevented by the scavenging activity of *Clerodendrum glabrum*, *Schotia brachypetala*, *Psychotria capensis*, *Ximenia caffra*, *Sutherlandia frutescens*, and *Peltophorum africanum* through radical scavenging activity and enzyme inhibition (Table 3.3) (Ndlovu et al. 2013). The strong free radical scavenging activity exhibited by these plants, notably the ethyl acetate extract of *Peltophorum africanum* bark, neutralizes the $ABTS^+$ free radicals. Its antioxidant activity recorded an IC_{50} value of 1.99 ± 0.23 g/mL (Ndlovu et al. 2013). The IC_{50} signifies the concentration of the extract essential to inhibit 50% of the free radical's activity, reflecting its potency as an antioxidant agent. Their activity is attributed to the substantial phenolic content, which can donate electrons or neutralize reactive oxygen species (Mazimba 2014).

A study by Kondratov et al. (2009) was conducted to show that there is a correlation between premature aging and ROS accumulation using Bmal1−/− mice's

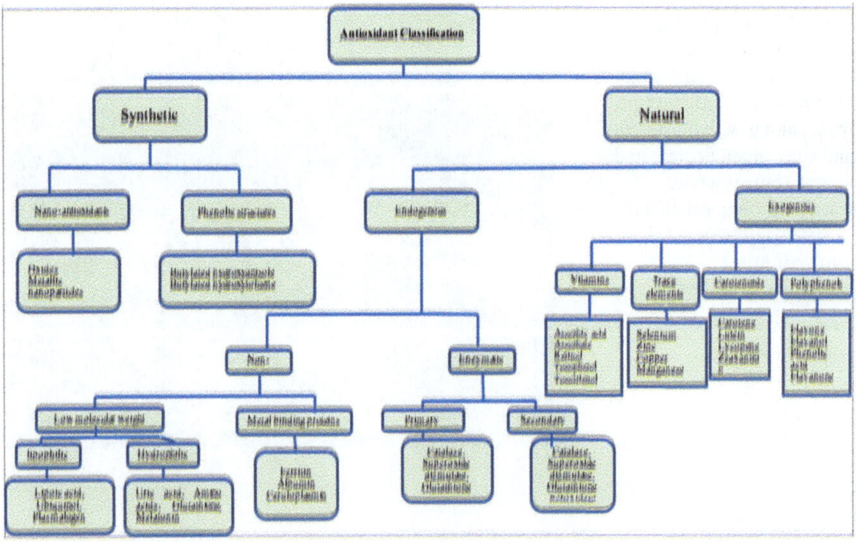

Fig. 3.13 Classification of natural and synthetic antioxidants

Table 3.3 Enzyme inhibition by medicinal plants of Southern Africa and their relevance to anti-aging

Plant species	Phytochemicals involved	Identification method of major phytochemicals	Antioxidant index IC_{50} (μg/mL)	Elastase inhibition (%)	Collagenase inhibition (%)	Hyaluronidase inhibition (%)	Mechanism of action of bioactive compounds	Cell-based analysis	References
Schotia brachypetala	Catechins, tannins, phenolic acid	UV-vis, HPLC, quantified at >110 mg GAE/g	ABTS IC_{50} 1.99 ± 0.2, DPPH-3.4 ± 0.28	90	82	75	Induces biosynthesis of collagen 1 transcription, upregulates the SIRT1-AMPK axis	91% H_2O_2 protection in HaCaT cells	Ndlovu et al. (2013)
Psychotria capensis	Flavonoids, iridoids, chlorogenic acid, kaempferol	FTIR, LC-HRMS	ABTS-35 ± 1.8	92	78	67	Improves mitochondrial SOD-CAT ratio	Fibroblast senescence was reduced by 54% using the β-gal assay	Ndlovu et al. (2013), Desmiaty et al. (2022)
Clerodendrum glabrum	Apigenin, scutellarein	GC-MS fingerprinting	48.2 ± 3.6	60	70	50	Inhibition of AGE-RAGE axis, small activation of MAPK	There was 75% cell viability	Desmiaty et al. (2022)
Peltophorum africanum	Rutin, epicatechin, ferulic acid	TLC, LC-MS	DPPH-2.13 ± 0.20	88	85	68	Activates Nrf2-HO-1 pathway, inhibits synthesis of NF-κB	Improved collagen synthesis by 48% using the Sirius red assay	Ndlovu et al. (2013)

(continued)

Table 3.3 (continued)

Plant species	Phytochemicals involved	Identification method of major phytochemicals	Antioxidant index IC_{50} (μg/mL)	Elastase inhibition (%)	Collagenase inhibition (%)	Hyaluronidase inhibition (%)	Mechanism of action of bioactive compounds	Cell-based analysis	References
Ximenia caffra	Flavanol glycosides, tannins, phenolic acids	HPTLC, GC-MS quantification	DPPH-5 ± 0.4 FRAP-18.3 ABTS-7.6 ± 0.3	83	85	72	Activates insulin /IGF pathway modulation	*Caenorhabditis elegans* lifespan was increased by 27%, 80% reduction of intracellular ROS	Bakrim et al. (2022)
Sutherlandia frutescens	L-canavanine, D-pinitol, quercetin glycosides	NMR, UPLC-QqQ	DPPH-6.2 FRAP-210 μmol Fe^{2+}/g	75	69	61	Inhibits lipid peroxidation, reduction of intracellular ROS	Upregulates the SIRT1-AMPK axis, FoxO3a dependent restoration	Tobwala et al. (2014)

tissue. Suppose the early aging characteristics are brought about by excessive oxidative stress and high levels of ROS accumulation, then the ability of antioxidants to reduce the severity of oxidative stress may delay or slow down the process of premature aging. Likewise, this tactic has been found to delay ROS-induced age-related conditions in nematodes, flies, and mice (Bokov et al. 2004; Kondratov et al. 2009). To drive the point home, a potent low molecular weight (LMW) antioxidant, N-acetyl-L-cysteine (NAC), was found to be effective when administered to mice; in fact, it delayed tumorigenesis in p53−/− mice (Sablina et al. 2005).

Polyphenols play a vital role in the antioxidant activity of medicinal plants. One with satisfactory total phenolic compounds is *Schotia brachypetala,* making it a vital source of natural antioxidants (Sobeh et al. 2016). *Peltophorum africanum*, on the other hand, comprises a variety of phytochemicals, dominated by flavonoids (Mongalo and Raletsena 2025). Flavonoids generated by plants have shown promise as natural antioxidants that can prevent lipid peroxidation and cellular damage (Fig. 3.7) (Hernández-Rodríguez et al. 2019). Their ability to efficiently scavenge free radicals is attributed to distinctive chemical structures (Fig. 3.14), which include hydroxyl groups and conjugated double bonds, enabling them to effectively scavenge free radicals (Fig. 3.15). They can donate to lipid radicals, thus stabilizing

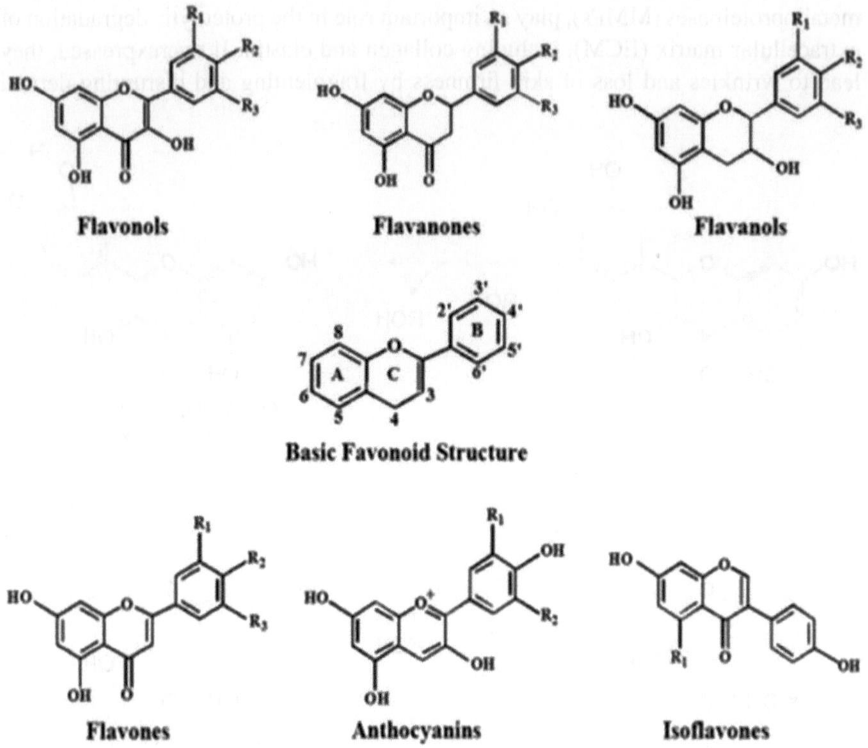

Fig. 3.14 Chemical structure of flavonoids and their sub-groups. (Adapted from Nebo et al. 2014)

lipid radicals and interrupting them from reacting with oxygen, hence effectively stopping the oxidative chain of reaction and maintaining the cellular membrane integrity (Treml and Šmejkal 2016). Furthermore, some flavonoids possess metal-chelating abilities, thus enabling them to successfully bind transition metals such as iron and copper, which are responsible for catalyzing lipid peroxidation through redox cycling (Kejík et al. 2021). Surprisingly, the radical scavenging activity through DPPH assay of *Clerodendrum glabrum* surpassed that of ascorbic acid (Kar et al. 2019). Bioactive compounds isolated from plants can help regulate the body's endogenous antioxidant defense systems. They usually achieve this through upregulation of expression of key enzymes like superoxide dismutase, catalase, and glutathione peroxidase. The presence of these endogenous enzymes ensures that ROS are converted into less harmful molecules (water and oxygen), achieving the goal of alleviating oxidative stress (Na et al. 2011; Jalouli et al. 2025).

3.3.2 Enzyme Inhibition Activity

Not only do these plants possess the ability to scavenge free radicals, but they also inhibit important enzymes involved in degrading vital skin structural components (Fig. 3.16). A group of zinc-containing enzymes with different domains, matrix metalloproteinases (MMPs), play an important role in the proteolytic degradation of extracellular matrix (ECM), including collagen and elastin. If overexpressed, they lead to wrinkles and loss of skin firmness by fragmenting and disrupting dermal

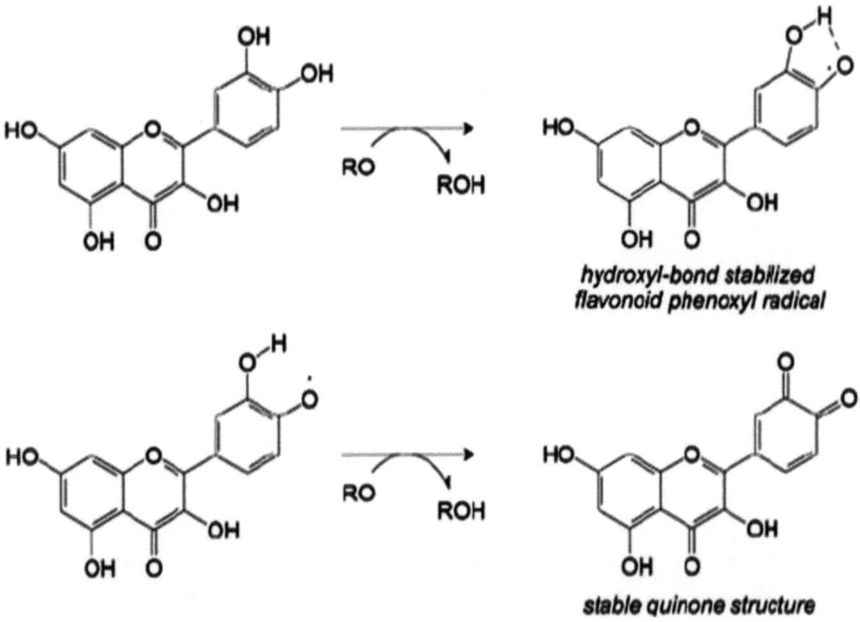

Fig. 3.15 Radical scavenging activity of flavonoids. (Adapted from Katz et al. 2011)

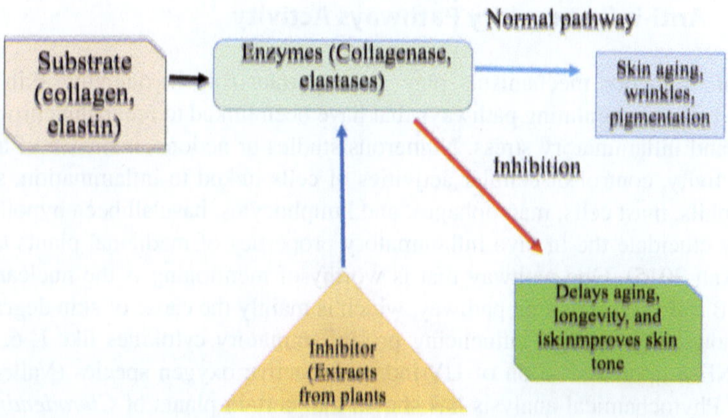

Fig. 3.16 Mechanism of enzyme inhibition activity by medicinal plants

structures, which is usually caused by UV-induced reactive oxygen species (Wittenauer et al. 2015). Shifting focus on the enzymatic activity of extracellular matrix-degrading proteases and elastases, this plant gives a promising tactic for delaying skin damage caused by ultraviolet (UV). This can be achieved by neutralization of ROS, since free radicals activate proteolytic enzymes, which aid in premature aging (Kammerer et al. 2004). The enzyme collagenase breaks down collagen, making up most of the skin. When collagenase activity is inhibited, precollagen fiber formation and wrinkle development are delayed (Fig. 3.16) (Mukherjee et al. 2011). A noteworthy inhibition of collagenase was demonstrated in a study including six plants (Table 3.3): *Clerodendrum glabrum*, *Schotia brachypetala*, *Psychotria capensis*, *Peltophorum africanum*, *Ximenia caffra*, and *Sutherlandia frutescens*. The study revealed that the ethyl acetate extracts of *C. glabrum* stems, *P. africanum* bark and stems, and *S. brachypetala* bark and leaves exhibited substantial activity against collagenase, surpassing 80% inhibition when compared to the positive control EDTA (83.75 ± 2.89%). The methanolic extracts of *S. brachypetala* bark showed 77.51% inhibition, while the ethyl acetate bark extract showed an even higher inhibition of 87.61% thus showing a strong protective effect against the degradation of collagen (Ndlovu et al. 2013). This suggests that these plants may be used in anti-aging skincare products.

In another study led by Lall et al. (2019), it was shown that *Schotia brachypetala* possessed substantial antioxidant and anti-tyrosinase properties. When the inhibitory effects of the tree's bark and leaves were evaluated, it was noted that over 70% of tyrosinase was inhibited. Moreover, its polyphenolic compounds, such as flavonoids and gallic acid, also support the anti-collagenase activity, thus halting the breakdown of collagenase that ultimately leads to wrinkles and loss of firmness.

3.3.3 Anti-inflammatory Pathways Activity

Anti-inflammatory mechanisms play an important role in delaying skin aging, mainly by downregulating pathways that have been linked to recurring chronic oxidative and inflammatory stress. Numerous studies or actions, inclusive of antioxidant activity, control of cellular activities of cells linked to inflammation, such as neutrophils, mast cells, macrophages, and lymphocytes, have all been hypothesized to fully elucidate the in vivo inflammatory properties of medicinal plants (Nworu and Akah 2015). One pathway that is worthy of mentioning is the nuclear factor kappa B (NF-κB) signaling pathway, which is mainly the cause of skin degradation and photoaging through influencing pro-inflammatory cytokines like L-6, IL-1β, and TNF-α upon activation of UV-induced reactive oxygen species (Vallee et al. 2021). Phytochemical analysis has shown that certain plants of *Clerodendrum* sp. possess remarkable levels of bioactive compounds, which include terpenoids, flavonoids, and phenolic glycosides. These bioactive compounds have been shown to influence plants' anti-inflammatory qualities by altering vital cellular pathways involved in the inflammatory response. The principal mechanisms that contribute to the anti-inflammatory activity are predominantly associated with the suppression of tumor necrosis factor (TNF), myeloperoxidase activity, prostaglandin formation, particularly prostaglandin E2 (PGE2), and maintaining the stability of cell membranes (Choi et al. 2004). Studies on similar species (*C. serratum*) have demonstrated a noticeable anti-inflammatory efficacy, as demonstrated by reduction in the growth of granuloma tissue and paw edema caused by carrageenan in experimental animal models. The effects are significant, but not as potent as conventional anti-inflammatory medicines, and they are regarded as moderate compared to other plant extracts that have been tested (Solapure et al. 2016).

The acetone extracts of *Psychotria capensis* and its fractions have been shown to suppress the production of nitric oxide (NO) in lipopolysaccharide-induced macrophages, which is a key player in the inflammatory responses. Interestingly, the extracts can upregulate cytokines like interleukin-12p70, at the same time downregulate cytokine IL-10. Likewise, isolated compounds and partially purified fractions from *Peltophorum africanum* have shown a satisfactory anti-inflammatory activity, particularly through suppression of pro-inflammatory cytokines (Adebayo et al. 2017). Various fractions collected from acetone leaf extracts exhibited strong anti-inflammatory activity with an IC_{50} of 12.42 μg/mL against 15-lipoxygenase while inhibiting close to 100% of nitric oxide. Furthermore, pro-inflammatory cytokines like ILβ, IL-2, and NFA-α were inhibited by acetone leaf extracts, thus making the plant desirable for the development of anti-inflammatory drugs (Adebayo et al. 2017). Likewise, the methanolic extracts exhibited a similar pattern, demonstrating dose-dependent anti-inflammatory activity. Using the rat paw model induced with carrageenan, it was observed that edema was inhibited by up to 85% at 200 mg/kg, which is fairly comparable to the standard medication indomethacin. Pro-inflammatory mediators like PGE2 and TNF-α were noticeably lowered by higher doses of methanolic extracts and simultaneously restored the tissue to a near normal histological state (El-Hawary et al. 2015). In a nutshell, the highest anti-inflammatory potency is displayed by *Schotia brachypetala* and *Peltrophorum africanum,* with *S. brachypetala* showing consistency and effective potency in in vivo analysis. In contrast, *C. glabrum* showed the least anti-inflammatory activity compared to the other plants, but was moderately potent. Although *P. capensis* shows satisfactory in vitro analysis, it is limited by its cytotoxicity at concentrations considered effective (Aro et al. 2016). It's safe to say that these results support the notion that the use of these traditional medicinal plants in anti-inflammatory drug development is desirable, and further scientific research, particularly focusing on *S. brachypetala* and *P. africum* as promising candidates (El-Hawary et al. 2015).

3.4 Impact Assessment of Phytochemicals Used in Anti-aging Formulas

Human aging is more noticeable in the skin, where several changes, such as wrinkling, lack of moisture, loss of skin texture, and an increase in pore size, can be seen (Wang et al. 2015). Green tea catechin (GTC) metabolites were studied to see how they affect humans when supplemented. The authors showed that these bioactive compounds were successfully absorbed by the skin, and their ability to protect against UV-induced skin inflammation was noticeable in conjunction with reduced proinflammatory eicosanoid 12-hydroxyeicosatetraenoic acid (Rhodes et al. 2013). Not only are many anti-aging substances administered orally, but they can also be applied physically to the skin, thus improving the quality of the skin (Singh and Agarwal 2009). Active compounds extracted from medicinal plants can be incorporated into herbal cosmetics. With that said, a great deal of scientific research is currently in the pipeline to design and characterize phytochemicals used in cosmetic formulations to combat premature aging (Jeon et al. 2009).

To successfully protect our skin from oxidative damage and photoaging, the use of premium moisturizers containing silymarin/silibinin, which are organic flavonolignans present in milk and artichokes, is highly recommended (Singh and Agarwal 2009). This was demonstrated by the use of a mouse skin model, which showed the silymarin/silibinin's significant anti-aging abilities by acting as a protective shield against UV-induced damage, alongside environmental pollutants (Singh and Agarwal 2009).

3.5 Clinical Studies

All in all, no in-depth clinical trials have been conducted to assess the safety and side effects of these medicinal plants when used in cosmetic products, except for *Sutherlandia frutescens* (L.). To assess *Sutherlandia frutescens* (L.) safety in humans, a well-thought-out randomized double blind, placebo-controlled design in healthy adults was carried out. It was noted that oral intake of 400 mg of *Sutherlandia* twice a day for a period of 90 days had no adverse effects over the placebo. Fatigue, sleeplessness, and headache were among the insignificant generalized symptoms that were recorded. As per the results published by Memorial Sloan-Kettering Cancer Centre, there were some side effects recorded, which were experienced by sensitive people occasionally, but not life-threatening, which included stomach pains and diarrhea. In conclusion, it's safe to state that when the plant is prescribed at the recommended dosages, it is clinically safe (Johnson et al. 2007).

3.6 Conclusion

The major benefits of medicinal plants are their antioxidant activity, which helps to protect our skin against damage from oxidative stress, thus helping to prevent premature aging and maintain longevity. Since the golden age, plant-based products have been used to treat skin conditions. Even today, natural compounds extracted from plants are still highly preferred in various friendly market formulations for skin care and cleansing. Proper purification and clinical trial of extracted compounds from medicinal plants should be prioritized to produce more potent cosmetic products, since many of them have not been properly studied. There is still a vast array of medicinal plants in Africa that are yet to be discovered, which can be used in cosmetic production.

Competing Interests The authors have no relevant financial or non-financial interests to disclose.

References

Abdel Wahab SI, Mohamed AWH, Mohamed OY, Taha MME, Abdul AB, Al-Zubairi AS (2008) Serotonergic properties of the roots of *Clerodendron capitatum*. Am J Biochem Biotechnol 4(4):425–430

Aboyade OM, Styger G, Gibson D, Hughes G (2014) *Sutherlandia frutescens*: the meeting of science and traditional knowledge. J Altern Complement Med 20(2):71–76

Adebayo SA, Dzoyem JP, Shai LJ, Eloff JN (2015) The anti-inflammatory and antioxidant activity of 25 plant species used traditionally to treat pain in southern African. BMC Complement Altern Med 15:1–10

Adebayo SA, Steel HC, Shai LJ, Eloff JN (2017) Investigation of the mechanism of anti-inflammatory action and cytotoxicity of a semipurified fraction and isolated compounds from the leaf of *Peltophorum africanum* (Fabaceae). J Evid Based Complement Altern Med 22(4):840–845

Adebayo JO, Balogun EA, Zailani HA (2022) In vitro and in vivo antiplasmodial activities and cytotoxicities of *Clerodendrum violaceum* leaf extracts. Afr J Biomed Res 25(2):249–255

Adefegha SA, Oboh G (2011) Cooking enhances the antioxidant properties of some tropical green leafy vegetables. Afr J Biotechnol 10(4):632–639

Adewusi EA, Moodley N, Steenkamp V (2011) Antioxidant and acetylcholinesterase inhibitory activity of selected southern African medicinal plants. S Afr J Bot 77(3):638–644

Alavizadeh SH, Hosseinzadeh H (2014) Bioactivity assessment and toxicity of Crocin: a comprehensive review. Food Chem Toxicol 64:65–80

Amusan OO, Dlamini PS, Msonthi JD, Makhubu LP (2002) Some herbal remedies from Manzini region of Swaziland. J Ethnopharmacol 79(1):109–112

Arai M, Yamano Y, Setiawan A, Kobayashi M (2014) Identification of the target protein of agelasine D, a marine sponge diterpene alkaloid, as an anti-dormant mycobacterial substance. Chem Bio Chem 15(1):117–123

Aro AO, Dzoyem JP, Eloff JN, McGaw LJ (2016) Extracts of six Rubiaceae species combined with rifampicin have good in vitro synergistic antimycobacterial activity and good anti-inflammatory and antioxidant activities. BMC Complement Altern Med 16:1–8

Aro AO, Dzoyem JP, Goddard A, Fonteh P, Kayoka-Kabongo PN, McGaw LJ (2019) In vitro antimycobacterial, apoptosis-inducing potential, and immunomodulatory activity of some Rubiaceae species. Front Pharmacol 10:185

Aro AO, Famuyide IM, Oyagbemi AA, Kabongo-Kayoka PN, McGaw LJ (2022) In vitro potential of the acetone leaf extract and fractions of *Psychotria capensis* (Eckl.) Vatke (Rubiaceae) to combat co-infection of tuberculosis and helminthiasis. Front Pharmacol 12:744137

Aversa R, Petrescu RV, Apicella A, Petrescu FI (2016) One can slow down the aging through antioxidants. Am J Eng Appl Sci 9(4)

Bakrim WB, Nurcahyanti ADR, Dmirieh M, Mahdi I, Elgamal AM, El Raey MA, Wink M, Sobeh M (2022) Phytochemical profiling of the leaf extract of *ximenia americana var. caffra* and its antioxidant, antibacterial, and antiaging activities in vitro and in *Caenorhabditis elegans*: a cosmeceutical and dermatological approach. Oxidative Med Cell Longev 2022(1):3486257

Benevides PJC, Young MCM, Bolzani VDS (2005) Biological activities of constituents from Psychotria spectabilis. Pharm Biol 42(8):565–569

Birsan M, Gore E, Stan CD, Cristofor AC, Palimariciuc M, Scripcariu ȘI, Antonoaea P, Vlad RA, Todoran N, Ciurba A (2022) Development and evaluation of hydrophilic cream bases suitable for custom cosmetic skincare or compounding pharmaceutical dermal products. Farmacia 70(5)

Bizimenyera ES (2008) The potential role of antibacterial, antioxidant and antiparasitic activity of Peltophorum africanum Sond.(Fabaceae) extracts in ethnoveterinary medicine. Doctoral dissertation, University of Pretoria

Bizimenyera SE, Swan GE, Chikoto H, Eloff JN (2005) Rationale for using *Peltophorum africanum* (Fabaceae) extracts in veterinary medicine. J S Afr Vet Assoc 76(2):54–58

Bizimenyera ES, Aderogba MA, Eloff JN, Swan GE (2007) Potential of neuroprotective antioxidant-based therapeutics from *Peltophorum africanum Sond.* (Fabaceae). Afr J Tradit Complement Altern Med 4(1):99–106

Bjørklund G, Shanaida M, Lysiuk R, Butnariu M, Peana M, Sarac I, Strus O, Smetanina K, Chirumbolo S (2022) Natural compounds and products from an anti-aging perspective. Molecules 27(20):7084

Bokov A, Chaudhuri A, Richardson A (2004) The role of oxidative damage and stress in aging. Mech Ageing Dev 125(10–11):811–826

Chae S, Kim JS, Kang KA, Bu HD, Lee Y, Seo YR, Hyun JW, Kang SS (2005) Antioxidant activity of isoacteoside from *Clerodendron trichotomum*. J Toxicol Environ Health A 68(5):389–400

Cheng HH, Wang HK, Ito J, Bastow KF, Tachibana Y, Nakanishi Y, Xu Z, Luo TY, Lee KH (2001) Cytotoxic pheophorbide-related compounds from *Clerodendrum c alamitosum* and C. cyrtophyllum. J Nat Prod 64(7):915–919

Chinsembu KC, Hedimbi M, Mukaru WC (2011) Putative medicinal properties of plants from the Kavango region, Namibia. J Med Plants Res 5(31):6787–6797

Chipuwa M, Muzandu K, Lavina P, Muwowo S (2024) Antioxidant effects of methanol leaf extract of *Ximenia Caffra* (Olacaceae) in cisplatin-induced nephrotoxic rats. Me J Zamb 51(2):110–126

Choi JH, Whang WK, Kim HJ (2004) Studies on the anti-inflammatory effects of *Clerodendron trichotomum* Thunberg leaves. Arch Pharm Res 27:189–193

Chung HY, Kim DH, Lee EK, Chung KW, Chung S, Lee B, Seo AY, Chung JH, Jung YS, Im E, Lee J (2019) Redefining chronic inflammation in aging and age-related diseases: proposal of the senoinflammation concept. Aging Dis 10(2):367

Corella D, Ordovas JM (2014) Aging and cardiovascular diseases: the role of gene–diet interactions. Ageing Res Rev 18:53–73

Cushnie TT, Cushnie B, Lamb AJ (2014) Alkaloids: an overview of their antibacterial, antibiotic-enhancing and antivirulence activities. J Antimicrob Agents 44(5):377–386

Desai SD, Desai DG, Kaur H (2009) Saponins and their biological activities. Pharma Times 41(3):13–16

Desmiaty Y, Faizatun F, Noviani Y, Ratih H, Ambarwati NSS (2022) Potential of natural products in inhibiting premature skin aging. Int J Appl Pharm 14:1–5

Dhalaria R, Verma R, Kumar D, Puri S, Tapwal A, Kumar V, Nepovimova E, Kuca K (2020) Bioactive compounds of edible fruits with their anti-aging properties: a comprehensive review to prolong human life. Antioxidants 9(11):1123

Du K, Marston A, van Vuuren SF, van Zyl RL, Coleman C, Zietsman PC, Bonnet SL, Ferreira D, van der Westhuizen JH (2014) Flavonolacyl glucosides from the aril of Schotia brachypetala Sond. and their antioxidant, antibacterial and antimalarial activities. Phytochem Lett 10:122–128

El-Hawary EA, Abdel-Aziz E, Labib RM, Esmat A, Abdel-Naim AB, Khalifa AE, Singab ANB, Ayoub NA (2015) Anti-inflammatory activity of the phenolic-rich extract of Schotia brachypetalea Sond. Fam. Fabaceae, cultivated in Egypt

Engwa GA, Nweke FN, Nkeh-Chungag BN (2022) Free radicals, oxidative stress-related diseases, and antioxidant supplementation. Altern Ther Health Med 28(1)

Fernandes AC, Cromarty AD, Albrecht C, van Rensburg CEJ (2004) The antioxidant potential of *Sutherlandia frutescens*. J Ethnopharmacol 95(1):1–5

Fisher GJ, Kang S, Varani J, Bata-Csorgo Z, Wan Y, Datta S, Voorhees JJ (2002) Mechanisms of photoaging and chronological skin aging. Arch Dermatol 138(11):1462–1470

Formagio ASN, Volobuff CRF, Santiago M, Cardoso CAL, Vieira MDC, Pereira ZV (2014) Evaluation of antioxidant activity, total flavonoids, tannins and phenolic compounds in *Psychotria* leaf extracts. Antioxidants 3(4):745–757

Formagio ASN, Vilegas W, Kassuya CAL, De Almeida VP, Manfron J, Konkiewitz EC, Ziff EB, Faoro JAM, Dos Santos JM, Cecatto AJ, Sarragiotto MH (2024) A comprehensive description of the anatomy and histochemistry of *Psychotria capillacea* (Müll. Arg.) Standl. and an investigation into its anti-inflammatory effects in mice and role in scopolamine-induced memory impairment. Pharmaceuticals 17(5):564

Goyal N, Jerold F (2023) Biocosmetics: technological advances and future outlook. Environ Sci Pollut Res Int 30(10):25148–25169

Güçlü-Üstündağ Ö, Mazza G (2007) Saponins: properties, applications and processing. Crit Rev Food Sci Nutr 47(3):231–258

Hassaan Y, Handoussa H, El-Khatib AH, Linscheid MW, El Sayed N, Ayoub N (2014) Evaluation of plant phenolic metabolites as a source of Alzheimer's drug leads. Biomed Res Int 2014(1):843263

Hernández-Rodríguez P, Baquero LP, Larrota HR (2019) Flavonoids: potential therapeutic agents by their antioxidant capacity. In: Bioactive compounds, pp 265–288

Jain PK, Kharya MD, Gajbhiye A, Sara UVS, Sharma VK (2010) Flavonoids as nutraceuticals. A review. Herba Polonica 56(2)

Jalouli M, Rahman MA, Biswas P, Rahman H, Harrath AH, Lee IS, Kang S, Choi J, Park MN, Kim B (2025) Targeting natural antioxidant polyphenols to protect neuroinflammation and neurodegenerative diseases: a comprehensive review. Front Phamacol 16:1492517

Johnson EJ (2014) Role of lutein and zeaxanthin in visual and cognitive function throughout the lifespan. Nutr Rev 72(9):605–612

Johnson Q, Syce J, Nell H, Rudeen K, Folk WR (2007) A randomized, double-blind, placebo-controlled trial of Lessertia frutescens in healthy adults. PLoS Clin Trials 2(4):e16

Kaczor A, Baranska M, Czamara K (2016) Carotenoids: overview of nomenclature, structures, occurrence, and functions. In: Carotenoids: nutrition, analysis and technology, pp 1–13

Kammerer D, Claus A, Carle R, Schieber A (2004) Polyphenol screening of pomace from red and white grape varieties (*Vitis vinifera L.*) by HPLC-DAD-MS/MS. J Agric Food Chem 52(14):4360–4367

Kang DG, Lee YS, Kim HJ, Lee YM, Lee HS (2003) Angiotensin converting enzyme inhibitory phenylpropanoid glycosides from *Clerodendron trichotomum*. J Ethnopharmacol 89(1):151–154

Kar P, Dutta S, Chakraborty AK, Roy A, Sen S, Kumar A, Lee J, Chaudhuri TK, Sen A (2019) The antioxidant rich active principles of *Clerodendrum* sp. controls haloalkane xenobiotic induced hepatic damage in murine model. Saudi. J Biol Sci 26(7):1539–1547

Katz DL, Doughty K, Ali A (2011) Cocoa and chocolate in human health and disease. Antioxid Redox Signal 15(10):2779–2811

Kazeem MI, Mellem JJ, Sabiu S (2024) Medicinal foods and plants with antiaging properties: a review of in vitro and in vivo studies. Food Front 5(1):24–45

Kejík Z, Kaplánek R, Masařík M, Babula P, Matkowski A, Filipenský P, Veselá K, Gburek J, Sýkora D, Martásek P et al. (2021) Iron complexes of flavonoids-antioxidant capacity and beyond. Int J Mol Sci 22:646. https://doi.org/10.3390/ijms22020646

Kiepiel I, Brown M, Johnson SD (2022) A generalized bird pollination system in Schotia brachypetala (Fabaceae). Plant Biol 24(5):806–814

Kim H, Lantvit D, Hwang CH, Kroll DJ, Swanson SM, Franzblau SG, Orjala J (2012) Indole alkaloids from two cultured cyanobacteria, Westiellopsis sp. and Fischerella muscicola. Bioorg Med Chem 20(17):5290–5295

Kondratov RV, Vykhovanets O, Kondratova AA, Antoch MP (2009) Antioxidant N-acetyl-L-cysteine ameliorates symptoms of premature aging associated with the deficiency of the circadian protein BMAL1. Aging 1(12):979

Kundu JK, Mossanda KS, Na HK, Surh YJ (2005) Inhibitory effects of the extracts of *Sutherlandia frutescens* (L.) R. Br. and *Harpagophytum procumbens* DC. on phorbol ester-induced COX-2 expression in mouse skin: AP-1 and CREB as potential upstream targets. Cancer Lett 218(1):21–31

Lajis NH, Mahmud Z, Toia RF (1993) The alkaloids of Psychotria rostrata. Planta Med 59(04):383–384

Lall N, Van Staden AB, Rademan S, Lambrechts I, De Canha MN, Mahore J, Winterboer S, Twilley D (2019) Antityrosinase and anti-acne potential of plants traditionally used in the Jongilanga community in Mpumalanga. S Afr J Bot 126:241–249

Lashani K, Malik S, Gaobotse G, Makhzoum A, Masisi K (2024) Exploring nature's pharmacy: indigenous plants of southern Africa with Antiallergic properties and their mechanism of action. In: Biotechnology of medicinal plants with antiallergy properties: research trends and prospects. Springer, Singapore, pp 127–154

Lenart P, Krejci L (2016) Reprint of "DNA, the central molecule of aging". Mutat Res 788:25–31

López-Otín C, Blasco MA, Partridge L, Serrano M, Kroemer G (2013) The hallmarks of aging. Cell 153(6):1194–1217

Mabberley DJ (2008) The plant book: a portable dictionary of plants, their classification and uses. Cambridge University Press, Cambridge

Maity N, Nema NK, Abedy S, Sarkar BK, Mukherjee PK (2011) Exploring Tagetes erecta Linn flower for the elastase, hyaluronidase and MMP-1 inhibitory activity. J Ethnopharmacol 137:1300–1305

Majinda Runner RT, Motswaledi MS (1998) Antibiotic activity of selected Botswana medicinal plants. Botswana Notes Rec 30(1):157–162

Makhubele ME, Odeyemi SW, Masebe TM, Lebelo SL (2024) Investigation of the phytochemical content, antioxidant activities, and antimicrobial properties of the selected South African medicinal plants. Herbmed Pharmacol 13(3):381–389

Malinowska P (2013) Effect of flavonoids content on antioxidant activity of commercial cosmetic plant extracts. Herba Polonica 59(3)

Maroyi A (2016) *Ximenia caffra Sond.* (Ximeniaceae) in sub-Saharan Africa: a synthesis and review of its medicinal potential. J Ethnopharmacol 184:81–100

Masuku NP, Unuofin JO, Lebelo SL (2020) Phytochemical content, antioxidant activities, and androgenic properties of four south African medicinal plants. J Herbmed Pharmacol 9(3):245–256

Mathe E, Sethoga L, Mapfumari S, Adeniran O, Mokgotho P, Shai J, Gololo S (2024) Phytochemical screening and characterization of volatile compounds from three medicinal plants with reported anticancer properties using GC-MS. Life 14(11):1375

Mazimba O (2014) Pharmacology and phytochemistry studies in Peltophorum africanum. Bull Fac Pharm Cairo Univ 52(1):145–153

McGaw LJ, Jäger AK, Van Staden J (2002) Isolation of antibacterial fatty acids from *Schotia brachypetala*. Fitoterapia 73(5):431–433

Michalak M (2022) Plant-derived antioxidants: significance in skin health and the ageing process. Int J Mol Sci 23(2):585

Mongalo NI (2013) *Peltophorum africanum* Sond [Mosetlha]: a review of its ethnomedicinal uses, toxicology, phytochemistry and pharmacological activities. J Med Plants Res 7(48):3484–3491

Mongalo NI, Raletsena MV (2025) Bioactive molecules, ethnomedicinal uses, toxicology, and pharmacology of *Peltophorum africanum Sond* (Fabaceae): systematic review. Plants 14(2):239

Mukherjee PK, Maity N, Nema NK, Sarkar BK (2011) Bioactive compounds from natural resources against skin aging. Phytomedicine 19(1):64–73

Na MK, Thuong PT, Bae KH (2011) Natural compounds with antioxidant activity: recent findings from studies on medicinal plants. Nat Prod Sci 17(2):65–79

Nair JJ, Mulaudzi RB, Chukwujekwu JC, Van Heerden FR, Van Staden J (2013) Antigonococcal activity of *Ximenia caffra* Sond. (Olacaceae) and identification of the active principle. S Afr J Bot 86:111–115

Ndlovu G, Fouche G, Tselanyane M, Cordier W, Steenkamp V (2013) In vitro determination of the anti-aging potential of four southern African medicinal plants. BMC Complement Altern Med 13:1–7

Nebo L, Varela RM, Molinillo JM, Sampaio OM, Severino VG, Cazal CM, das Graças Fernandes MF, Fernandes JB, Macías FA (2014) Phytotoxicity of alkaloids, coumarins and flavonoids isolated from 11 species belonging to the Rutaceae and Meliaceae families. Phytochem Lett 8:226–232

Nepokroeff M, Bremer B, Sytsma KJ (1999) Reorganization of the genus Psychotria and tribe Psychotrieae (Rubiaceae) inferred from ITS and rbcL sequence data. Syst Bot:5–27

Nyila MA (2025) *Ximenia caffra Sond.* The magic wild indigenous plant that offers immense contribution as food and medicine. Nat Prod Res:1–13

Nworu CS, Akah PA (2015) Anti-inflammatory medicinal plants and the molecular mechanisms underlying their activities. Afr J Tradit, Complement Altern Med 12(5):52–61. https://doi.org/10.4314/ajtcam.v12i5.3S

Ogundajo AL, Ashafa AOT (2019) Medicinal properties of *Clerodendrum glaburum E* may leaf extracts: phytochemical constituents, antioxidant, cytotoxicity, and carbohydrate-metabolizing enzyme inhibitory potentials. Comp Clin Path 28:927–936

Okeleye BI, Nongogo V, Mkwetshana NT, Ndip RN (2015) Polyphenolic content and in vitro antioxidant evaluation of the stem bark extract of *Peltophorum africanum Sond* (Fabaceae). Afr J Tradit Complement Altern Med 12(1):1–8

Paiva PM, Gomes FS, Napoleão TH, Sá RA, Correia MTS, Coelho LCBB (2010) Antimicrobial activity of secondary metabolites and lectins from plants. In: Current research, technology and education topics in applied microbiology and microbial biotechnology, vol. 1, issue 2, pp 396–406

Panche AN, Diwan AD, Chandra SR (2016) Flavonoids: an overview. J Nutr Sci 5:e47

Pizzi A (2019) Tannins: prospectives and actual industrial applications. Biomolecules 9:344. https://doi.org/10.3390/biom9080344

Pillay M, Takaidza S, Madike LN, Terblanche U (2022) Conservation of medicinal plant genetic resources in southern Africa. In: Plant genetic resources, inventory, collection and conservation. Springer, Singapore, pp 319–338

Porto DD, Henriques AT, Fett-Neto AG (2009) Bioactive alkaloids from south American *Psychotria* and related species. Open Bioactive Compounds J 2(1):29–36

Rea IM, Gibson DS, McGilligan V, McNerlan SE, Alexander HD, Ross OA (2018) Age and age-related diseases: role of inflammation triggers and cytokines. Front Immunol 9:586

Rhodes LE, Darby G, Massey KA, Clarke KA, Dew TP, Farrar MD, Bennett S, Watson RE, Williamson G, Nicolaou A (2013) Oral green tea catechin metabolites are incorporated into human skin and protect against UV radiation-induced cutaneous inflammation in association with reduced production of pro-inflammatory eicosanoid 12-hydroxyeicosatetraenoic acid. Br J Nutr 110(5):891–900

Robbins RJ (2003) Phenolic acids in foods: an overview of analytical methodology. J Agric Food Chem 51(10):2866–2887

Sablina AA, Budanov AV, Ilyinskaya GV, Agapova LS, Kravchenko JE, Chumakov PM (2005) The antioxidant function of the p53 tumor suppressor. Nat Med 11(12):1306–1313

Sacconnay L, Ryckewaert L, dos Santos PC, Guerra MC, Kato L, de Oliveira CMA, Henriques A, Carrupt PA, Simões-Pires C, Nurisso A (2015) Alkaloids from *Psychotria* target sirtuins: in silico and in vitro interaction studies. Planta Med 81(06):517–524

Sharma N (2014) Free radicals, antioxidants and disease. Biol Med 6(3):1

Shrivastava N, Patel T (2007) *Clerodendrum* and healthcare: an overview. Med Aromat Plant Sci Biotechnol 1(1):142–150

Sieniawska E (2015) Activities of tannins–from in vitro studies to clinical trials. Nat Prod Commun 10(11):1934578

Singh RP, Agarwal R (2009) Cosmeceuticals and silibinin. Clin Dermatol 27(5):479–484

Sobeh M, ElHawary E, Peixoto H, Labib RM, Handoussa H, Swilam N, El-Khatib AH, Sharapov F, Mohamed T, Krstin S, Linscheid MW (2016) Identification of phenolic secondary metabolites from *Schotia brachypetala Sond.*(Fabaceae) and demonstration of their antioxidant activities in Caenorhabditis elegans. PeerJ 4:2404

Solapure P, Pradeep RM, Hegde PL (2016) Comparative anti-inflammatory activity of *Clerodendrum serratum* (Linn) moon and *Solanum xanthocarpum Schrad* and *Wendl* in wistar ablino rats. J Phytopharmacol 5(2):38–44

Steenkamp V (2003) Traditional herbal remedies used by south African women for gynaecological complaints. J Ethnopharmacol 86(1):97–108

Taylor CM (1996) Overview of the *Psychotrieae* (Rubiaceae) in the Neotropics. Opera Bot Belg:261–270

Tetali SD (2019) Terpenes and isoprenoids: a wealth of compounds for global use. Planta 249(1):1–8

Thawabteh A, Juma S, Bader M, Karaman D, Scrano L, Bufo SA, Karaman R (2019) The biological activity of natural alkaloids against herbivores, cancerous cells and pathogens. Toxins 11(11):656

Theo A, Masebe T, Suzuki Y, Kikuchi H, Wada S, Obi CL, Bessong PO, Usuzawa M, Oshima Y, Hattori T (2009) *Peltophorum africanum*, a traditional south African medicinal plant, contains an anti-HIV-1 constituent, betulinic acid. Tohoku J Exp Med 217(2):93–99

Tobwala S, Fan W, Hines CJ, Folk WR, Ercal N (2014) Antioxidant potential of Sutherlandia frutescens and its protective effects against oxidative stress in various cell cultures. BMC Complement Altern Med 14(1):271

Treml J, Šmejkal K (2016) Flavonoids as potent scavengers of hydroxyl radicals. Compr Rev Food Sci Food Saf 15(4):720–738

Trepa M, Sułkowska-Ziaja K, Kała K, Muszyńska B (2024) Therapeutic potential of fungal terpenes and terpenoids: application in skin diseases. Molecules 29(5):1183

Tsuchida K, Kobayashi M (2020) Oxidative stress in human facial skin observed by ultraweak photon emission imaging and its correlation with biophysical properties of skin. Sci Rep 10(1):9626

Tucker SC (2001) Floral development in Schotia and Cynometra (Leguminosae: Caesalpinioideae: Detarieae). Am J Bot 88(7):1164–1180

Tundis R, Loizzo MR, Bonesi M, Menichini F (2015) Potential role of natural compounds against skin aging. Curr Med Chem 22(12):1515–1538

Tai J, Cheung S, Chan E, Hasman D (2004) In vitro culture studies of Sutherlandia frutescens on human tumor cell lines. J Ethnopharmacol. 93(1):9-19. https://doi.org/10.1016/j.jep.2004.02.028.

Tsao R (2010) Chemistry and biochemistry of dietary polyphenols. Nutrients 2;1231–1246. https://doi.org/10.3390/nu2121231

Uriostegui-Pena AG, Torres-Copado A, Ochoa-Sanchez A, Luna-Bárcenas G, Sahare P, Paul S (2025) Nanoformulated phytochemicals in skin anti-aging research: an updated mini review. 3 Biotech 15(1):31. https://doi.org/10.1007/s13205-024-04197-y

Vallee A, Lecarpentier Y, Vallee JN (2021) Interplay of opposing effects of the WNT/β-catenin pathway and PPARγ and implications for SARS-CoV2 treatment. Front Immunol 12:666693

Van Wyk B (2013) Field guide to trees of southern Africa. Penguin Random House South Africa

Van Wyk BE, Albrecht C (2008) A review of the taxonomy, ethnobotany, chemistry and pharmacology of *Sutherlandia frutescens* (Fabaceae). J Ethnopharmacol 119(3):620–629

Wang HMD, Chen CC, Huynh P, Chang JS (2015) Exploring the potential of using algae in cosmetics. Bioresour Technol 184:355–362

Wittenauer J, Mäckle S, Sußmann D, Schweiggert-Weisz U, Carle R (2015) Inhibitory effects of polyphenols from grape pomace extract on collagenase and elastase activity. Fitoterapia 101:179–187

Wyk BV, Gericke N (2000) People's plants: a guide to useful plants of Southern Africa, p 351

Wyk BV, Oudtshoorn BV, Gericke N (1997) Medicinal plants of South Africa, p 304

Yang W, Chen X, Li Y, Guo S, Wang Z, Yu X (2020) Advances in pharmacological activities of terpenoids. Nat Prod Commun 15(3):1934578X20903555

Zhang, Yin M, Feng X, Ibrahim SA, Liu Y, Huang W (2021) Anti-inflammatory activity of four triterpenoids isolated from Poriae cutis. Foods 10(12):3155

Zhen J, Guo Y, Villani T, Carr S, Brendler T, Mumbengegwi DR, Kong ANT, Simon JE, Wu Q (2015) Phytochemical analysis and anti-inflammatory activity of the extracts of the African medicinal plant *Ximenia caffra*. J Anal Chem 2015(1):948262

Zillich OV, Schweiggert-Weisz U, Eisner P, Kerscher M (2015) Polyphenols as active ingredients for cosmetic products. Int J Cosmet Sci 37(5):455–464

Aquatic Plant-Derived Secondary Metabolites with Anti-aging Properties

4

Xue Zhao, Shan Huang, and Yongqiang Zhao

Abstract

As the global population ages at an accelerated pace, research into anti-aging mechanisms has emerged as a pivotal area within the life sciences. Aquatic plants have gained recognition as a notable source of bioactive compounds with anti-aging properties, attributable to their unique ecological adaptability and the variety of metabolic products they produce. This chapter delineates the various constituents abundant in polyphenols, polysaccharides, cellulose, and carotenoids present in aquatic plants, which demonstrate significant anti-aging effects through multiple pathways, including the scavenging of free radicals, the inhibition of oxidative stress responses, the regulation of cellular apoptosis, and the enhancement of immune function. Furthermore, the chapter examines recent advancements in the utilization of bioactive compounds derived from aquatic plants across the food, cosmetic, and pharmaceutical industries for anti-aging applications. Current research emphasizes the sustainable development and high-value utilization of aquatic plant resources, which is anticipated to become a significant trend in the formulation of anti-aging products, thereby offering diverse solutions for promoting healthy aging in humans.

X. Zhao · Y. Zhao (✉)
Key Laboratory of Aquatic Product Processing, Ministry of Agriculture and Rural Affairs, National R&D Center for Aquatic Product Processing, South China Sea Fisheries Research Institute, Chinese Academy of Fishery Sciences,
Guangzhou City, Guangdong Province, China
e-mail: zhaoyq@scsfri.ac.cn

S. Huang
College of Food Science and Engineering, West Coast Campus of Ocean University, Qingdao City, Shandong Province, China

S. Gantait et al. (eds.), *Medicinal Plants with Anti-Ageing Properties*,
https://doi.org/10.1007/978-981-95-8064-4_4

Keywords

Algae · Active ingredient · Oxidative stress · Longevity gene · Metabolic regulators · Functional product development

4.1 Introduction

Aquatic plants are defined as the flora that thrive in aquatic environments (Firmin et al. 2022). These plants exhibit unique structural and physiological adaptations that enable their survival in these habitats (Zhao et al. 2024a, b, c). Notably, the root systems of aquatic plants are often less developed or feature specialized aeration tissues that facilitate the acquisition of oxygen from surrounding water. Additionally, stems and leaves of aquatic plants often display specific adaptations that allow them to thrive in aquatic environments (Nguyen et al. 2021). For example, their leaves are typically characterized by a thin and pliable structure that facilitates gas exchange and photosynthesis under submerged conditions (Koga et al. 2024). Aquatic plants are integral to the functioning of aquatic ecosystems because they contribute to water purification, provide habitats for various organisms, and support biodiversity, among other essential ecological functions (Al-Thani and Yasseen 2021). These plants are critical components of the aquatic habitats. Under controlled growth conditions, aquatic plants contribute significantly to the structure, services, and functions of aquatic ecosystems (Ceschin et al. 2021).

4.2 Classification and Diversity of Aquatic Plants

Aquatic plants can be categorized into three main groups: emergent, floating, and submerged plants (Selvaraj and Velvizhi 2021). In addition, algae are present in lacustrine environments and serve as a fundamental component of the aquatic food chain (González Fernández et al. 2023) (Fig. 4.1). Emergent plants are

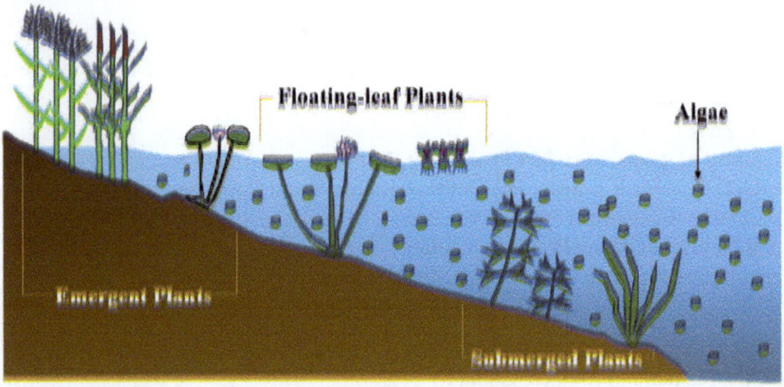

Fig. 4.1 Schematic diagram of the morphological classification of aquatic plants and their growth positions in water bodies

characterized by their root systems, which are situated in the sediment at the bottom of aquatic environments, whereas their stems and foliage rise above the water surface. Notable examples of emergent plants include the lotus (*Nelumbo nucifera*), cattail (*Typha latifolia*), and common reed (*Phragmites australis*) (Wang et al. 2025). Aquatic plants are integral to ecosystem function. These plants have the ability to take up nutrients and contaminants from aquatic environments, thereby contributing to water quality purification. They are essential habitats and breeding sites for various aquatic faunal species. Aquatic plants contribute to the enhancement of landscape diversity in aquatic ecosystems (Mansour et al. 2022).

Floating aquatic plants are defined as species that exist on the surface of water bodies, with their roots typically not extending into the underlying sediment but remain suspended in the water column. Notable examples of floating aquatic plants include water lily (*Nymphaea tetragona*), water hyacinth (*Eichhornia crassipes*), duckweed (*Lemna minor*), and water lettuce (*Pistia stratiotes*) (Jewell and Bell 2023). These plants typically possess broad leaves, which enhance their buoyancy and ability to remain afloat. Floating aquatic plants perform multiple functions in aquatic ecosystems. They can obstruct sunlight, thereby inhibiting the proliferation of excessive algal bloom (Zhao et al. 2022). In addition, their root systems are effective for nutrient absorption from water, which aids in water purification. Nonetheless, certain floating plant species, such as the water hyacinth, exhibit rapid reproductive rates, which may result in ecological imbalances in aquatic environments (Galgali et al. 2023).

Submerged plants are substantial aquatic flora that exist entirely beneath water surfaces. Their root systems may be minimal or may have undergone degeneration, as these organisms are capable of absorbing essential water and nutrients through various anatomical structures (Peng et al. 2022). Submerged plants possess well-developed aerenchyma, which enhances gas exchange efficiency in hypoxic environments. The morphology of their leaves is frequently characterized by ribbon-like or filamentous forms, as exemplified by species such as *Vallisneria natans* (Lour.) Hara, *Myriophyllum spicatum*, *Ceratophyllum demersum*, and *Hydrilla verticillata* (Li et al. 2023a, b, c, d). The stems and leaves of these aquatic plants possess a soft texture that facilitates their ability to float or drift with the current, thereby allowing them to adapt to dynamic fluctuations in the underwater ecosystem. Furthermore, the flowers of submerged plants tend to be diminutive and exhibit a short flowering duration, with some species exhibiting no flowering (Aguzzi et al. 2024). The significance of submerged vegetation in aquatic ecosystems is evident in various dimensions. These plants serve as critical habitats and nutritional resources for aquatic fauna, and contribute to the enhancement of water quality by assimilating nutrients and contaminants present in the water column. Additionally, through photosynthesis, submerged plants generate oxygen, which elevates the levels of dissolved oxygen in water, thereby fostering a more favorable aquatic environment (Yu et al. 2020).

Algae can generally be categorized as aquatic plants because of their predominant growth in aquatic environments and dependence on these ecosystems for survival and reproduction (Ceschin et al. 2021). Nevertheless, when examined through a more precise botanical classification lens, algae are distinct from conventional aquatic plants, such as lotuses and water lilies, which are classified as higher plants characterized by the presence of differentiated roots, stems, and leaves (Singh et al. 2021). Algae possess a comparatively uncomplicated structure and lack true roots, stems, or leaves, categorizing them primarily as lower plants. Nevertheless, owing to their significant role as biological components within aquatic ecosystems, algae are occasionally classified as aquatic plants for research and ecological descriptions (Zhang et al. 2020).

Algae are classified into two main categories: microalgae, which include microscopic photosynthetic eukaryotes and cyanobacteria, and macroalgae, which are commonly referred to as seaweeds (Pereira 2021) (Table 4.1). Seaweeds are further categorized into three groups according to their pigment composition: green, red, and brown algae (Manzoor et al. 2024). Phytoplankton, microscopic algae consisting of one or more diminutive cells, are suspended within the water column of lakes. A significant proliferation of phytoplankton can result in green coloration of the water, suggesting that even lakes that are perceived as visually "cleanest" are not immune to this phenomenon (Forsch et al. 2021). In contrast, filamentous algae exhibit growth patterns characterized by the formation of cell chains that can develop into extensive chain-like structures or mat-like aggregates. Within this category of filamentous algae, certain species possess the ability to float freely and become dispersed throughout the water column in accordance with flow dynamics, whereas others may adhere to aquatic vegetation or establish growth directly on the substrate of the lake bed. However, they are classified within the algal group rather than the plant kingdom (Zhi et al. 2023).

Table 4.1 Classification and description of aquatic plants

Category	Description	Examples
Emergent plants	Plants that grow in the shallowest parts of water bodies with leaves extending above the water surface.	Broadleaf cattail (*Typha latifolia*), common reed (*Phragmites australis*)
Floating-leaf plants or floating aquatic plants	Plants that grow at intermediate water depths, either rooted or free-floating with leaves on the water surface.	White water lily (*Nymphaea lotus*), water lettuce (*Pistia stratiotes*), water hyacinth (*Eichhornia crassipes*)
Submerged plants	Plants that grow at the deepest parts of the littoral zone, with all parts remaining submerged.	Hydralla or waterthyme (*Hydrilla verticillata*), curly-leaf pondweed (*Potamogeton crispus*), vallisneria (*Vallisneria spinulosa*)
Microalgae	Microscopic photosynthetic organisms found in both marine and freshwater habitats.	*Chaetoceros*, *Chlorella*, *Arthrospira* (Spirulina)
Macroalgae (seaweeds)	Macroscopic marine algae that can reach considerable lengths and exhibit rich diversity.	*Ulva intestinalis*, *Porphyra yezoensis*, *Gracilaria gracilis*, *Laminaria japonica*, *Undaria pinnatifida*

4.3 Adaptation Strategies for Aquatic Plants

Aquatic plants adapt to their environment through various strategies, such as synthesizing antioxidants to resist oxidative damage, regulating water and nutrient absorption, optimizing photosynthesis efficiency, and continuing their populations through the regeneration and reproduction of growth points. These mechanisms help them survive and reproduce in aquatic environments (Mapelli-Brahm et al. 2023).

4.3.1 Anti-aging Mechanisms

A comprehensive exploration of the mechanisms underlying the aging process has contributed to the development of effective strategies for its prevention and treatment. These strategies have been instrumental in addressing various age-related diseases, including cardiovascular diseases, diabetes, cancer, neurodegenerative diseases, and osteoporosis (Han et al. 2024). By addressing the multifaceted nature of aging, it is possible to mitigate the adverse effects of aging on human health, thereby enhancing overall quality of life. For instance, increased insulin resistance and attenuation of pancreatic islet function are prevalent in patients with diabetes, and the high prevalence of cancer in older age groups is closely related to the accumulation of genetic mutations and reduced immune system capacity during cellular senescence (Bao et al. 2022). Neurodegenerative disorders, such as Alzheimer's and Parkinson's diseases, which predominantly appear later in life, are directly related to neuronal senescence and degeneration. Despite the recognition of these associations, our understanding of their nature and potential for decelerating the aging process remains inadequate (Zhao et al. 2024a, b, c).

Aging is a multilevel, multidimensional biological phenomenon that involves a wide range of molecular, cellular, and systemic physiological changes. Therefore, its mechanisms are extensive and cannot be reduced to a single factor (Jiang et al. 2025). Over the past decades, there has been a surge in research focusing on nutritional, sensory, and genetic pathways to understand the processes and patterns underlying the aging phenotype (Kassis et al. 2023) (Fig. 4.2). The following pathways have been the focus of these studies: 1. Telomere shortening pathway: telomeres, which are structures composed of repetitive DNA sequences and their associated proteins located at the ends of chromosomes, undergo progressive shortening during cell division. Telomere attrition triggers cellular senescence, which leads to irreversible growth arrest (Li et al. 2025). 2. The DNA damage repair pathway is susceptible to endogenous and exogenous factors, including ultraviolet rays, free radicals, and chemicals, which can inflict DNA damage within cells. Failure to rectify these lesions can lead to their accumulation, thereby compromising cellular function and contributing to hallmarks of cellular senescence, including death. DNA damage repair involves a series of complex mechanisms such as base excision repair, mismatch repair, and nucleotide excision repair. With advancing age, the efficiency of these repair processes gradually declines (Zhao et al. 2023). 3.

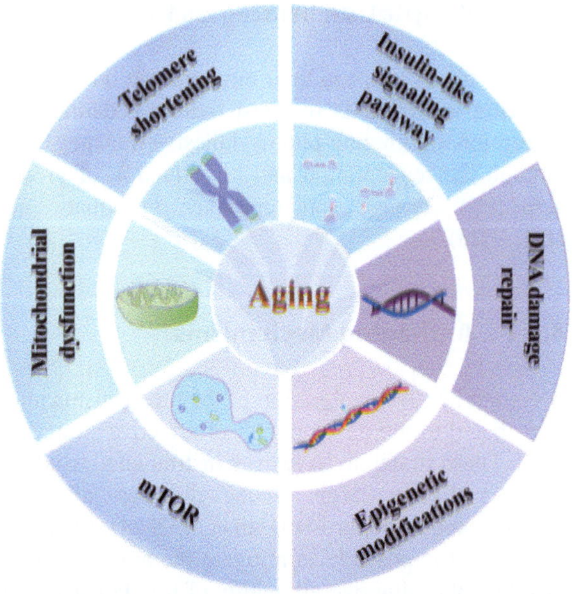

Fig. 4.2 Diagram of the core elements and pathways of action of the anti-aging mechanism

Epigenetic modifications, which do not involve alterations in DNA sequence, play a significant role in regulating gene expression through mechanisms such as DNA methylation, regulation by non-coding RNAs, and histone modifications. These epigenetic alterations can affect the expression of various aging-associated genes (Wang et al. 2022). 4. The mitochondrial dysfunction pathway is characterized by the mitochondrial function as a cellular energy factory. Dysfunction of this pathway is evidenced by increased generation of reactive oxygen species (ROS), accumulation of mtDNA mutations, and further alterations in cell physiology and aging (Rehman et al. 2023). 5. The insulin/insulin-like signaling pathway (IIS) is a vital signaling pathway that modulates various physiological processes in living organisms, including cellular growth, metabolism, reproduction, and aging. Inhibition of IIS signaling enhances antioxidant capacity and mitigates ROS formation, potentially promoting cellular homeostasis and decelerating aging (Sultanova et al. 2025). 6. Mechanistic target of rapamycin (mTOR), a central regulator of cellular processes, plays a pivotal role in various physiological functions, including cell growth, metabolism, survival, and aging. Inhibition of the mTOR signaling pathway has been observed to extend the lifespan of diverse organisms, a phenomenon that is attributed to the enhancement of cellular autophagy, a process whereby damaged organelles and protein aggregates are removed, thereby maintaining cellular health (Mannick and Lamming 2023). Consequently, a comprehensive examination of mTOR offers significant potential for elucidating cell physiology and the mechanisms underlying aging processes as well as for the development of therapeutic interventions for age-related diseases (Marafie et al. 2024).

4.3.2 Anti-aging Mechanisms in Aquatic Plants

The anti-aging mechanisms associated with aquatic plants are complex processes that are influenced by a variety of factors (Seferli et al. 2024). The following are possible anti-aging mechanisms (Fig. 4.3): 1. Antioxidant defense system: during the long evolutionary process, aquatic plants have developed a complex antioxidant defense system. They accumulate large amounts of antioxidant substances, such as vitamin C, vitamin E, flavonoids, and polyphenols, which effectively neutralize free radicals and reduce cellular damage caused by oxidative stress (Song et al. 2023). Water dropwort (*Oenanthe javanica*), for instance, is abundant in vitamin C and flavonoids, which aid in the scavenging of free radicals, and consequently, in the protection of cells from oxidative damage (Liu et al. 2021). 2. Repair of damaged cell structures: aquatic plants possess the capacity to repair damaged cell structures in response to external environmental stresses or damage. This capacity involves the maintenance of the integrity and function of damaged cell membranes (Li et al.

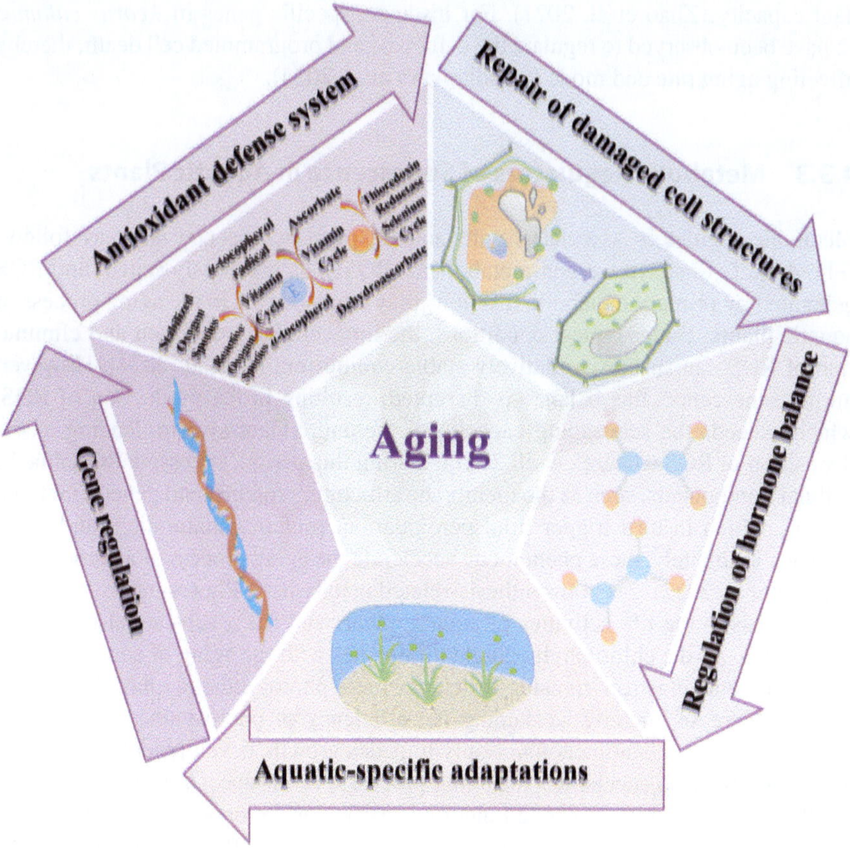

Fig. 4.3 Diagram of anti-aging strategies of aquatic plants

2023a, b, c, d). For example, water lily (*Nymphaea tetragona*) utilizes metabolic pathways to repair damaged proteins and lipids, thereby maintaining the normal physiological functions of cells (Li et al. 2023a, b, c, d). 3. The regulation of hormone balance is of particular interest. A variety of aquatic plants have been shown to possess the capacity to decelerate the aging process by regulating the balance of phytohormones. Ethylene and abscisic acid (ABA) have been shown to play pivotal roles in plant senescence (Peng et al. 2025). The lotus (*Nelumbo nucifera*) plant has been shown to effectively delay petal senescence and apoptosis by regulating ethylene production and response mechanisms (Wang et al. 2023a, b, c, d). 4. Special mechanisms for adaptation to aquatic environments: specific environmental conditions in water bodies have prompted aquatic plants to develop unique adaptive mechanisms that also contribute to anti-aging processes (Hu et al. 2025). For instance, *Scirpus validus* maintains cellular stability and mitigates the effects of senescence by regulating ion channels and transporter proteins under various salinities (Pérez et al. 2023). 5. Gene regulation: the regulation of gene expression in aquatic plants plays a significant role in the anti-aging process, with particular genes capable of controlling the synthesis of antioxidant enzymes and enhancing antioxidant capacity (Zhao et al. 2021). For instance, specific genes in *Acorus calamus* L. have been observed to regulate the cell cycle and programmed cell death, thereby affecting aging rate and mode (Chodkowska et al. 2024).

4.3.3 Metabolic Regulators of Senescence in Aquatic Plants

Metabolic regulators associated with aquatic plant senescence were as follows (Fig. 4.4): 1. Reactive oxygen metabolism (ROM), as previously stated, and ROS generated by primary metabolic reactions play a pivotal role in the aging process of aquatic plants. Under typical conditions, the intracellular production and elimination of ROS can sustain a relatively stable equilibrium (Ji et al. 2024). However, during senescence, this balance is disrupted, resulting in the production of ROS, which exceeds the scavenging capacity of the antioxidant system, leading to the deposition of ROS (Staszek et al. 2021). During this period, excessive ROS attacks cellular components, such as the membrane structure, proteins, and genetic material (DNA), which in turn trigger lipid peroxidation, protein denaturation, and DNA damage. Ultimately, these phenomena accelerate the aging process of aquatic plants (Anetor et al. 2022). 2. Photosynthesis-related metabolism: Photosynthesis is a critical process in the life activities of aquatic plants and has a substantial impact on senescence. As the chlorophyll content diminishes with the aging of aquatic plants, photosynthetic electron transfer efficiency and photosynthetic phosphorylation activity decrease, thereby weakening the efficiency of photosynthesis (Pramanic et al. 2023). This phenomenon not only impedes growth, development, and energy absorption but also accelerates the aging process. For instance, a decline in chlorophyll content and photosynthetic function is particularly pronounced in certain submerged plants during the aging stage (Peng et al. 2022). 3. Protein metabolism: the process of senescence in aquatic plants is attributed to an imbalance between

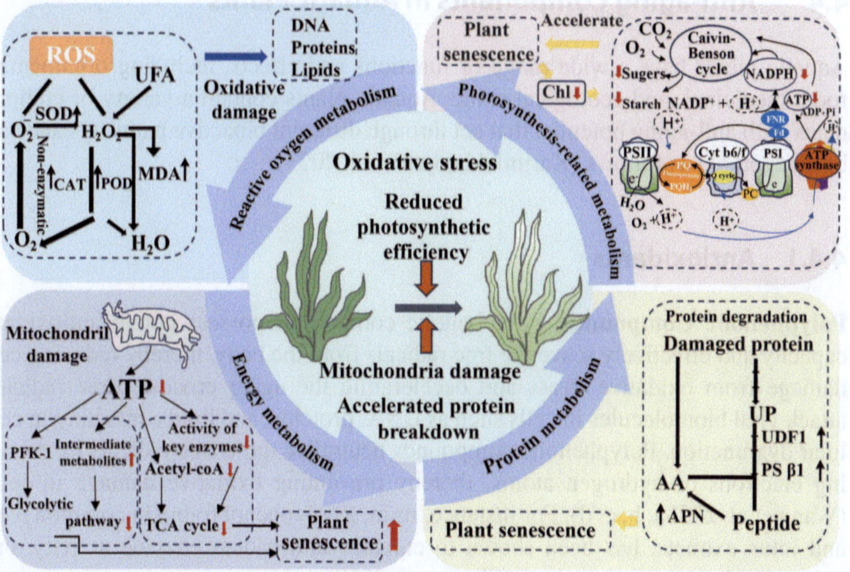

Fig. 4.4 Analysis of key factors of metabolic regulation in the senescence process of aquatic plants

protein synthesis and degradation. As senescence progresses, there is a gradual decrease in protein synthesis capacity, concurrent with an increase in the activity of specific proteases (e.g., cysteine proteases and serine proteases). This acceleration of protein decomposition leads to the destruction of cell structure and function. Abnormalities in protein modification and folding contribute to the onset of senescence in aquatic plants (Gao et al. 2020). 4. Energy metabolism: the energy metabolism of aquatic plants changes during aging. For instance, the efficiency of respiration involving NAD+ (nicotinamide adenine dinucleotide) may be reduced, leading to a decrease in energy production (Kristanti and Hadibarata 2023). Conversely, it affects normal physiological functions, growth, and development of aquatic plants. Impairment of mitochondrial function has been shown to result in a decrease in ATP production, as well as a reduction in the efficiency of intracellular glycolysis, the tricarboxylic acid cycle, and other pathways (Shi et al. 2023). This, in turn, affects the metabolism and physiological functions of normal cells and accelerates the aging process. Moreover, senescence in aquatic plants is affected by various environmental factors, such as light, temperature, and nutrient levels. These factors may indirectly influence the senescence process by modulating the activity of metabolic regulators. A comprehensive investigation of the anti-aging mechanisms and associated metabolic regulators of aquatic plants promises to offer insights into their growth and development. Such research may significantly contribute to the preservation and management of aquatic ecosystems (Li et al. 2021).

4.4 Anti-aging Components in Aquatic Plants

Aquatic plants have a wide range of functions and effects, including ornamental, food, medicinal, and ecological value. Aquatic plants contain a variety of components with anti-aging potential that act through different bioactive pathways and can be categorized as follows (Karimi Alavijeh et al. 2023).

4.4.1 Antioxidants

Polyphenolic Compounds Polyphenolic compounds possess potent antioxidant capacity and effectively scavenge free radicals from the body, thereby reducing cell damage from oxidative stress and decelerating the aging process. Free radicals attack vital biomolecules in cells such as DNA, proteins, and lipids, resulting in cellular dysfunction. Polyphenolic compounds neutralize these free radicals by donating electrons or hydrogen atoms, thereby preventing oxidative damage to cells (Wang et al. 2023a, b, c, d). For instance, rutin, a polyphenol found in common reed and lotus extracts, has been shown to enhance antioxidant enzyme activity and reduce oxidative stress markers (Chen et al. 2019).

Flavonoids Isoflavones and flavonols found in aquatic plants have demonstrated remarkable antioxidant activities. For instance, the substantial isoflavones present in gorgonian extracts have been shown to effectively activate the Nrf2 signaling pathway, thereby enhancing the antioxidant properties of cells and potentially decelerating the aging process (Gravandi et al. 2023). Flavonols, present in *Isoetes sinensis*, have been shown to scavenge free radicals within the body, thereby reducing cell damage caused by oxidative stress. This property contributes to the anti-aging potential of flavanols and, to a certain extent, the prevention of chronic diseases. The flavonoids investigated in this study have been shown to inhibit the production of free radicals as well as enhance the activity of antioxidant enzymes and repair oxidation-induced damage (Wang et al. 2020). They have been observed to promote the expression and activity of antioxidant enzymes by regulating intracellular signaling pathways, such as activating the Nrf2 signaling pathway. In addition, these compounds have been observed to form chelates with metal ions and reduce the generation of free radicals catalyzed by metal ions. A comprehensive examination of these antioxidants offers insights into their role in promoting health and longevity through diverse biochemical pathways, thereby providing a scientific foundation for the development of natural plant-based anti-aging strategies (Ferdous and Balia Yusof 2021).

4.4.2 Polysaccharides

Polysaccharides present in aquatic plants safeguard cellular integrity by neutralizing free radicals within the body and attenuating oxidative stress. These polysaccharides have been shown to stimulate immune cell activity, thereby amplifying the body's immune response and fortifying its defense against diseases (Sun et al. 2024). Furthermore, these polysaccharides can influence a variety of metabolic activities within the cell, thereby regulating metabolic pathways, including glucose, lipid, and protein metabolism. This is achieved through the activation or inhibition of specific key enzymes, ensuring the maintenance of normal physiological functions in the cell and, consequently, promoting a healthy lifespan (Li et al. 2023a, b, c, d). For instance, the administration of kelp polysaccharides to mice substantially enhances their antioxidant capacity and immune function, thereby extending their lifespan (Michalak et al. 2022). Fucoidan has antioxidant properties and a strong ability to neutralize superoxide free radicals. Its potential to combat diseases caused by free radicals is significant, as its antioxidant strength correlates with the molecular weight and sulfate levels of fucoidan. Additionally, it inhibits matrix metalloproteinase-1 (MMP-1), an enzyme that breaks down collagen, which helps to effectively slow the aging process. (Achmad et al. 2024). These findings not only expand our knowledge of the application of aquatic plant polysaccharides in the biomedical field but also provide a scientific foundation for the development of novel natural anti-aging products. For instance, the polysaccharide endemic to *Ulva lactuca* (Chlorophyta) not only provides a distinctive gustatory experience but also contributes to anti-aging by hydrating and enhancing skin elasticity (Liu et al. 2019).

4.4.3 Fibers

Aquatic plants are endowed with structural integrity that enables them to maintain an upright position in the water, a feat made possible by their robust cell walls, which are composed of various polysaccharides, including cellulose, hemicellulose, and lignin (Koley et al. 2023). Cellulose, in particular, confers upon aquatic plants remarkable qualities of mechanical strength and stability. Hemicellulose is a heteroglycan composed of a variety of monosaccharides. When combined with cellulose, it constitutes the cell wall of aquatic plants. This polysaccharide possesses a complex and diverse chemical structure and functions.

Lignin, a complex aromatic polymer, is prevalent in both woody and aquatic plants, providing structural integrity and protection (Tanpichai et al. 2023).

Fibers, which undergo degradation during the growth and reproductive cycles of these plants, are subsequently utilized to satisfy the energy demands of aquatic ecosystems. These fibers possess the capacity to adsorb heavy metal ions and organic pollutants from water, thereby facilitating water purification and safeguarding aquatic ecosystems (Arivendan et al. 2022). Consumption of dietary fiber by the human body has been shown to diminish the absorption of fats and cholesterol within the intestines, support the equilibrium of gut microbiota, facilitate intestinal motility, and mitigate the occurrence of constipation. Effective management of body weight is associated with a reduced risk of various diseases and may contribute to the aging process (Seferli et al. 2024). For instance, aquatic plants that are high in fiber, such as lotus root (*Nelumbo nucifera*) (Yang et al. 2024a, b) and water bamboo (*Phyllostachys heteroclada Oliver*), serve dual purposes as both food sources and animal feed, thereby offering significant nutritional and health advantages (Liang et al. 2024).

4.4.4 Carotenoids

Aquatic plants are typically abundant in carotenoids, which possess antioxidant properties and have been shown to mitigate oxidative stress, thereby contributing to the prevention of aging and the development of numerous chronic diseases. Representative carotenoids include β-carotene, lutein, zeaxanthin, and astaxanthin (Chen et al. 2022). β-carotene, in particular, is commonly found in algae and, upon ingestion, can be converted into vitamin A, a nutrient that plays a pivotal role in visual function, skin health, and the immune system (Ritu et al. 2023). Lutein and zeaxanthin are primarily found in edible aquatic plants such as lotus roots (*Nelumbo nucifera*). As the predominant pigment in the macula of the retina, they possess the capacity to absorb harmful blue light and prevent retinopathy (Yang et al. 2024a, b). Astaxanthin is identified in aquatic microalgae. For instance, it accumulates in substantial quantities in the *Haematococcus Pluvialis* under conditions of stress. In general, carotenoids have been shown to enhance body resistance by promoting macrophage activity and enhancing the function of natural killer cells. They have also been demonstrated to reduce blood cholesterol and triglyceride levels, which can contribute to the prevention of atherosclerosis and consequently reduce the risk of cardiovascular disease (Wilawan et al. 2024). Furthermore, carotenoids have been observed to influence the secretion of cytokines, such as interleukin (IL) and tumor necrosis factor (TNF), thereby balancing the immune response and inflammation levels, and reducing chronic inflammation (Bayram and Samur 2025). Consequently, it can significantly slow the aging process. It promotes the proliferation and differentiation of stem cells, accelerates tissue regeneration and cellular repair mechanisms, and facilitates the restoration of both structural integrity and functionality of bodily tissues. By stimulating collagen synthesis, carotenoids enhance skin elasticity and firmness while reducing the visible signs of aging (Kumar et al. 2024).

4.4.5 Other Ingredients

Aquatic plants contain a diverse array of bioactive compounds, including triterpenoid saponins, phycocyanin, vitamins, and essential trace elements, in addition to the aforementioned anti-aging constituents (Hosseini et al. 2022). Triterpenoid saponins play a crucial role in mitigating the cellular damage caused by free radicals because of their ability to bind to these reactive species, thereby neutralizing their harmful effects. Furthermore, these compounds have been shown to facilitate nerve regeneration, regulate blood glucose levels, prevent the onset of metabolic syndrome, modulate hormone levels, enhance immune cell functionality, regulate cytokine secretion, and maintain a balance between immune responses and inflammatory processes (Asahi and Sawada 2024). This multifaceted action contributes to the reduction of chronic inflammation, which is associated with aging. For instance, the triterpenoid saponins found in water chestnut (*Eleocharis dulcis*) have demonstrated efficacy in alleviating inflammatory responses in the body, providing significant relief from cough and other respiratory conditions (Dávid et al. 2021). Additionally, saponins offer protective benefits for cardiovascular health, potentially delaying the progression of age-related cardiovascular diseases. There is evidence of trace amounts of triterpenoid saponins in certain algal species; however, the existing body of research is limited (Giuliani et al. 2024).

Phycocyanin, a pigment found in certain algae, has been shown to stimulate the proliferation and differentiation of stem cells, thereby contributing to the maintenance of normal tissue structure and function and potentially slowing the aging process. Although the anti-aging effects of phycocyanin have been primarily explored in cellular and animal models, further clinical studies are necessary to ascertain its long-term safety and efficacy in humans (Nova et al. 2024). Notably, phycocyanin confers a distinctive blue color to Spirulina, a property attributable to its unique chromophores that reflect blue light. This characteristic makes the product highly sought-after in the food and cosmetic industries, where it is valued for its antioxidant and anti-aging properties (Braga-Souto et al. 2025).

Phycocyanin plays a pivotal role in photosynthesis by facilitating the transfer of light energy to chlorophyll and enhancing the efficiency of light energy conversion. This process contributes to the maintenance of cell viability, thereby counteracting the effects of aging (Buecker et al. 2025). Aquatic plants are rich sources of vitamins. This vitamin exhibits potent antioxidant activity, protecting cells from damage caused by free radicals and slowing the aging process. Furthermore, the presence of vitamin C in green algae acts as an antioxidant, enhancing the skin's defense mechanisms and protecting it from damage caused by ultraviolet radiation and adverse environmental conditions, thereby delaying the aging process (Li et al. 2024). Aquatic plants also contain essential trace elements, such as iodine, iron, zinc, phosphorus, and magnesium, which are vital for maintaining human health and play a significant role in cellular metabolism and repair processes, thereby contributing to anti-aging effects (Polechońska and Klink 2021).

4.5 Application of Aquatic Plants in Anti-aging

The application of aquatic plants in anti-aging has gradually attracted widespread attention, covering various fields such as food, cosmetics, and pharmaceuticals (Seferli et al. 2024). In terms of food, aquatic plants are rich in antioxidants, vitamins, and minerals, which can effectively neutralize free radicals, enhance the body's immune system, and promote metabolism. These plants are often used as health foods and nutritional supplements, providing natural nutritional support to help delay the aging process (Plaskova and Mlcek 2023). In the cosmetics field, extracts from aquatic plants are widely used in skincare products due to their significant moisturizing, anti-wrinkle, and skin-repairing effects. These ingredients can improve skin elasticity, reduce fine lines and wrinkles, thereby enhancing the skin's glow and health, making them popular among consumers (Nutho and Tungmunnithum 2024). In pharmaceuticals, research has found that certain active components in aquatic plants possess anti-inflammatory properties and promote cell regeneration, and are being developed as anti-aging drugs. The extracts from these plants can be used not only in traditional medicine but may also become a new option for anti-aging treatments in modern medicine, providing a diversified solution for delaying aging (Wang et al. 2021).

4.5.1 Applications in the Food Industry

The increasing popularity of healthy eating has led to growing interest in edible aquatic plants, which are becoming more prevalent among consumers. This interest can be attributed to the plants' high nutritional value, distinctive flavor, and diverse culinary applications. These aquatic plants are notable for their high nutrient content, including proteins, vitamins, and minerals (Table 4.2). The consumption of these plants has been associated with numerous health benefits, including anti-aging effects (Suraiya et al. 2024).

Aquatic Plants in Daily Diets Aquatic plants have long been a staple of traditional diets and have unique applications in the modern food industry. Some aquatic plants are directly consumed as food. For example, lotus root (*Nelumbo nucifera*) can be heated and consumed or dried and ground into a powder, which is convenient and quick to prepare (Zhao et al. 2024a, b, c). Lotus root (*Nelumbo nucifera*) juice is processed with pear juice and other juices to produce mixed drinks, such as fruit and vegetable juices (Melini et al. 2019). *Brasenia schreberi*, notable for its smooth texture, has applications in soups and salads (Hong and Gruda 2020). Water chestnut (*Eleocharis dulcis*) is suitable for consumption in raw and cooked foods (Lu et al. 2022). Water caltrop (*Trapa natans*) kernels can be consumed either raw or cooked and are commonly used in making pastries, added to soups, or eaten as a snack (Naseem et al. 2024). Seaweeds, such as kelp and nori, can be utilized in soups (Figueroa and Aguilera 2023).

Aquatic Plant-Based Products Aquatic plants with anti-aging effects can be further processed into edible products. For instance, lotus root or lotus seed can be processed into snacks or desserts through frying or baking. Lotus root (*Nelumbo nucifera*) (Seferli et al. 2024) and water caltrops (*Trapa natans*) (Lü et al. 2022) can be canned through the candying process to preserve their taste and nutrition, extend

Table 4.2 Common aquatic plants and their scientific name, edible parts, main uses, key anti-aging components, and mechanisms of action

Aquatic plant	Scientific name	Edible part	Main uses	Key anti-aging components	Mechanisms of action
Water chestnut	*Eleocharis dulcis*	Corm	Eaten raw, stir-fried, salads, desserts, canned	Crisp and sweet, rich in starch and vitamin C	Antioxidant (neutralizes free radicals), hydrates skin, reduces oxidative stress (Lu et al. 2022).
Water caltrop	*Trapa natans*	Kernel (seed)	Eaten raw, cooked, pastries, soups, snacks	Angular fruit, rich in starch and protein	Anti-inflammatory, regulates blood sugar, inhibits collagen degradation (Lü et al. 2022).
Lotus root	*Nelumbo nucifera*	Rhizome	Stir-fried, soups, cold dishes, processed into lotus root powder	Crisp and tender, rich in dietary fiber and vitamins	Antioxidant (scavenges ROS), inhibits melanogenesis, protects against UV-induced aging (Zhao et al. 2024a, b, c).
Lotus seed	*Nelumbo nucifera*	Seed	Porridge, soups, desserts	Nutrient-rich, known for its nourishing properties	Enhances skin elasticity, reduces wrinkles, improves sleep quality (Wang et al. 2023a, b, c, d).
Water shield	*Brasenia schreberi*	Tender leaves and stems	Cold dishes, soups, side dishes	Smooth and tender, rich in mucilage and polysaccharides	Anti-inflammatory, repairs mucosal tissues, delays cellular senescence (Wan et al. 2023).
Water bamboo	*Zizania latifolia*	Tender stem	Stir-fried, soups, cold dishes	Crisp and flavorful, rich in dietary fiber and minerals	Lowers lipid oxidation, supports gut health, reduces metabolic aging (Nguyen et al. 2024).
Water celery	*Oenanthe javanica*	Stems and leaves	Stir-fried, cold dishes, soups	Refreshing aroma, rich in vitamins and minerals	Antioxidant, reduces inflammation, calms sensitive skin (Kumar et al. 2022).

(continued)

Table 4.2 (continued)

Aquatic plant	Scientific name	Edible part	Main uses	Key anti-aging components	Mechanisms of action
Kelp	*Saccharina japonica*	Thallus	Cold dishes, soups, side dishes	Rich in iodine, dietary fiber, and polysaccharides	Stimulates collagen synthesis, regulates thyroid function, inhibits MMP-1 activity (Nie et al. 2023).
Nori	*Pyropia* spp.	Thallus	Sushi, soup ingredients, snacks	Rich in protein, vitamins, and minerals	Strengthens skin barrier, boosts immunity, mitigates fatigue-related aging (Yang et al. 2024a, b).
Water spinach	*Ipomoea aquatica*	Tender stems and leaves	Stir-fried, soups, cold dishes	Crisp and tender, rich in vitamins and iron	Antioxidant, improves skin texture, supports iron-dependent metabolic processes (John et al. 2025).

the preservation time, and facilitate convenient consumption at any time. The addition of dietary fiber, which is abundant in lotus root, bread, cookies, and other foods, has been shown to enhance both the sensory appeal and digestibility of these foods, while concurrently conferring a variety of health benefits (Sopawong et al. 2022). *Ulva lactuca* polysaccharides can be developed into food or health products with specific anti-aging and other health functions (Liu et al. 2022).

4.5.2 Applications in the Cosmetics Industry

Aquatic plants have been shown to possess significant potential in the realm of beauty and skincare, particularly with regard to their anti-aging properties (Table 4.3). Water lily (*Nymphaea tetragona*) extract is notable for its abundance of natural moisturizing factors that are capable of forming a protective film on the surface of the skin. This film impedes water loss and maintains skin hydration. Moreover, the ability of the extract to penetrate deep into the skin has been demonstrated to replenish moisture, thereby contributing to its hydrated appearance (Li et al. 2024). Polysaccharides present in the water shield (*Brasenia schreberi*) exhibit effective moisturizing characteristics, enabling them to absorb moisture from the atmosphere. They facilitate skin metabolism, improve the skin's capacity for water retention, and contribute to softer and smoother skin texture (Feng et al. 2019). Lotus flowers contain abundant flavonoids and polyphenolic compounds, which exhibit significant antioxidant properties. These compounds effectively neutralize detrimental free radicals and mitigate oxidative damage, consequently contributing to the delay in skin aging and reduction of wrinkle formation (Wang et al. 2023a, b,

Table 4.3 Aquatic plants and their applications in cosmetics and pharmaceuticals, anti-aging active components, and their effects

Aquatic plant	Scientific name	Anti-aging active components	Cosmetic applications	Cosmetic benefits	Pharmaceutical applications	Pharmaceutical benefits	References
Seaweed	*Sargassum* spp.	Polysaccharides, alginic acid, iodine, vitamins, minerals	Moisturizer, face masks, serums	Hydration, antioxidant, anti-inflammatory, promotes skin repair	Anti-tumor, antiviral, lipid-lowering, immune modulation	Antioxidant, anti-inflammatory, anti-tumor, immune regulation	Michalak et al. (2022), Figueroa and Aguilera (2023), Wang et al. (2021)
Nori	*Pyropia* spp.	Proteins, polysaccharides, vitamin B12	Face masks, serums, lotions	Hydration, soothing, strengthens skin barrier	Nutritional supplements, anti-fatigue, anemia treatment	Boosts immunity, reduces fatigue, enhances hematopoiesis	Yanagido et al. (2018), Pereira (2022)
Water shield	*Brasenia schreberi*	Mucilage, polysaccharides	Gels, face masks, moisturizers	Hydration, anti-aging, skin repair	Anti-inflammatory, anti-ulcer, gastric mucosa protection	Anti-inflammatory, mucosal repair, promotes digestive health	Wan et al. (2023), Feng et al. (2019)
Lotus root	*Nelumbo nucifera*	Polyphenols, flavonoids	Antioxidant serums, anti-aging creams	Antioxidant, anti-aging, brightening	Antioxidant, anti-aging, blood sugar regulation	Antioxidant, hypoglycemic, protects cardiovascular health	Wang et al. (2023a, b, c, d), Moon et al. (2023)
Lotus seed	*Nelumbo nucifera*	Polyphenols, polysaccharides	Anti-aging creams, eye serums	Antioxidant, anti-wrinkle, soothing	Sedative, anti-aging, sleep improvement	Calming, anti-aging, improves sleep quality	Sopawong et al. (2022), Moon et al. (2023)

(continued)

Table 4.3 (continued)

Aquatic plant	Scientific name	Anti-aging active components	Cosmetic applications	Cosmetic benefits	Pharmaceutical applications	Pharmaceutical benefits	References
Water caltrop	*Trapa natans*	Starch, polyphenols	Face masks, scrubs	Cleansing, exfoliation, hydration	Anti-inflammatory, antioxidant, blood sugar regulation	Anti-inflammatory, antioxidant, regulates blood glucose	Rehman et al. (2024), Lü et al. (2022), Matovic et al. (2025), Yasuda et al. (2021)
Water chestnut	*Eleocharis dulcis*	Starch, vitamin C	Moisturizers, face masks	Hydration, soothing, brightening	Heat-clearing, diuretic, relieves throat inflammation	Detoxifies, promotes diuresis, alleviates throat swelling	Li et al. (2023a, b, c, d), Li et al. (2024)
Kelp	*Saccharina japonica*	Fucoidan, iodine, minerals	Anti-aging serums, moisturizers	Antioxidant, hydration, promotes collagen synthesis	Anti-tumor, lipid-lowering, thyroid function regulation	Anti-tumor, lipid-lowering, regulates thyroid hormone levels	Liu et al. (2023), Zhang et al. (2025), Park et al. (2025)
Water celery	*Oenanthe javanica*	Vitamin C, flavonoids	Antioxidant serums, soothing masks	Antioxidant, anti-inflammatory, calms sensitive skin	Antioxidant, anti-inflammatory, blood pressure regulation	Antioxidant, anti-inflammatory, regulates blood pressure	Bae et al. (2022), Michalak (2023), Gębalski et al. (2022)
Water bamboo	*Zizania latifolia*	Dietary fiber, vitamins	Cleansing masks, moisturizers	Cleansing, hydration, softens skin texture	Lipid-lowering, digestive health promotion	Lowers blood lipids, supports intestinal health	Wu et al. (2024), Yuan et al. (2024a, b), Zhang et al. (2024)

c, d). Similarly, water chestnuts are high in vitamins C and E along with a variety of antioxidants, which can effectively counteract the damage inflicted by free radicals (Li et al. 2024). In conclusion, these aquatic plants, characterized by their distinctive moisturizing and antioxidant capabilities, demonstrate considerable potential for applications in the beauty and skincare industries (Yuan et al. 2024a, b).

4.5.3 Applications in the Pharmaceutical Industry

The potential anti-aging and health benefits of aquatic plants represent a valuable resource for the pharmaceutical industry. However, rigorous scientific investigation and regulatory oversight are essential to guarantee their safe and effective use in human health applications. The primary components of water fern (*Azolla* sp.) and duckweed (*Lemna minor*) are flavonoids, phenolic acids, and polysaccharides. These compounds have been employed in traditional medicine to manage a range of health issues, particularly dermatological disorders (Saxena et al. 2021). Simultaneously, it may function as a means of eliminating antibiotic drug residues from aquatic ecosystems. (Maldonado et al. 2022) Spirulina, on the other hand, is rich in iron, proteins, and polysaccharides, which may contribute to enhanced immune function, increased hemoglobin levels, improved energy, overall health maintenance, and the mitigation of aging processes (Majewski et al. 2022). *Acorus calamus L.* and *Acorus tatarinowii* are characterized by the presence of several antioxidant constituents, including volatile oils, flavonoids, and polysaccharides. *Acorus calamus* L. has been shown to support cardiovascular health by inhibiting platelet aggregation, thereby mitigating the risk of cardiovascular diseases and decelerating the aging process (Sahu et al. 2024). In contrast, Acorus tatarinowii contributes to hepatic protection by reducing the serum levels of alanine aminotransferase and aspartate aminotransferase, thereby preserving normal liver metabolic function and delaying the aging process (Wang et al. 2023a, b, c, d). Due to the considerable diversity of aquatic plants, it is crucial for professionals to prioritize precise identification to reduce the potential negative consequences associated with differences in medicinal properties and toxicity, despite the morphological similarities that may exist among them (Ceschin et al. 2021). In addition, it is essential to examine the interactions between aquatic plants and other pharmacological agents to minimize the likelihood of adverse reactions and toxicological risks (Pawłowska and Biczak 2024).

4.6 Conclusions

The anti-aging properties of aquatic plants are of considerable significance across multiple domains. The advancement of novel anti-aging pharmaceuticals within the medical sector has the potential to enhance human health and overall quality of life, especially in the context of preventing and managing age-related ailments, including cardiovascular diseases, degenerative disorders, and cancer. A comprehensive exploration of the aging process, coupled with an investigation of the anti-aging mechanisms exhibited by aquatic plants, may yield valuable insights into the biological underpinnings of aging. These findings offer a novel perspective and potential avenue for further investigation of the characteristics of aging. In the context of personalized medicine, as research into the anti-aging properties of aquatic plants advances, there are variations in the anti-aging effectiveness of components derived from these plants across different individuals.

The formulation of natural skincare products utilizing the antioxidant, anti-inflammatory, and moisturizing properties inherent in aquatic plants presents a viable approach to address consumer preferences for safe and effective beauty solutions. By leveraging the benefits of these botanical sources, it is possible to diminish dependence on synthetically produced chemicals and mitigate the associated risks of adverse side effects. Maintaining equilibrium within ecological systems is essential for the overall health of the biosphere and plays a significant role in sustaining aquatic plant ecosystems. By employing sustainable practices for the utilization and conservation of the ecological environment, we can secure the long-term availability of aquatic plant resources. Furthermore, advancing the economic sector and commercializing products derived from aquatic plants can stimulate the growth of associated industries and generate economic value.

Competing Interests The authors have no relevant financial or non-financial interests to disclose.

References

Achmad AA, Tangdilintin F, Stephanie CK, Sulistiawati E, Rifai Y, Aliyah PAD, Manggau MA (2024) Development of dissolving microneedles loaded with fucoidan for enhanced anti-aging activity: an *in vivo* study in mice animal model. Eur J Pharm Biopharm 202:114362

Aguzzi J, Thomsen L, Flögel S, Robinson NJ, Picardi G, Chatzievangelou D, Bahamon N, Stefanni S, Grinyó J, Fanelli E, Corinaldesi C, Del Rio Fernandez J, Calisti M, Mienis F, Chatzidouros E, Costa C, Violino S, Tangherlini M, Danovaro R (2024) New technologies for monitoring and upscaling marine ecosystem restoration in deep-sea environments. Engineering 34:195–211

Al-Thani RF, Yasseen BT (2021) Perspectives of future water sources in Qatar by phytoremediation: biodiversity at ponds and modern approach. Int J Phytoremediation 23(8):866–889

Anetor GO, Nwobi NL, Igharo GO, Sonuga OO, Anetor JI (2022) Environmental pollutants and oxidative stress in terrestrial and aquatic organisms: examination of the total picture and implications for human health. Front Physiol 13:931386

Arivendan A, Winowlin J, Siva I, Adamkhan, Chrish B (2022) Extraction and characterization of natural aquatic plant fiber, powder and ash from water hyacinth (*Eichhornia crassipes*) as reinforcement of fiber, powder, and ash reinforced polymer composite. J Nat Fibers 19(14):9589–9599

Asahi H, Sawada K (2024) High abundances of degraded triterpenoids in *Miocene turbiditic* sediments of south-central Hokkaido, Japan. Org Geochem 188:104671

Bae U-J, Jang H-N, Lee S-H, Kim J-Y, Kim G-C (2022) *Oenanthe javanica* ethanolic extract alleviates inflammation and modifies gut microbiota in mice with DSS-induced colitis. Antioxidants 11(12):2429

Bao J, Liu D, Sun J, Su X, Cheng H, Qi L, Zhang Y, Lv Y, Ye Z, Yu X, Wei Q, Qiu Y, Su J, Li L (2022) Pancreatic cancer-associated diabetes mellitus is characterized by reduced β-cell secretory capacity, rather than insulin resistance. Diabetes Res Clin Pract 185:109223

Bayram B, Samur FG (2025) Are carotenoids like lutein and zeaxanthin an antiglycation agents? The potential cardiometabolic health effects and antiglycation mechanisms of action: a comprehensive review. Food Rev Int 42(2):1–26

Braga-Souto RN, Bürck M, Nakamoto MM, Braga ARC (2025) Cracking spirulina flavor: compounds, sensory evaluations, and solutions. Trends Food Sci Technol 156:104847

Buecker S, Sanders JM, Winget P, Leeb E, Grossmann L, Gibis M, Weiss J (2025) Uncovering the light absorption mechanism of the blue natural colorant allophycocyanin from *Arthrospira platensis* using molecular dynamics. Food Chem 466:141834

Ceschin S, Bellini A, Scalici M (2021) Aquatic plants and ecotoxicological assessment in freshwater ecosystems: a review. Environ Sci Pollut Res Int 28(5):4975–4988

Chen H-W, Yang M-Y, Hung T-W, Chang Y-C, Wang C-J (2019) *Nelumbo nucifera* leaves extract attenuate the pathological progression of diabetic nephropathy in high-fat diet-fed and streptozotocin-induced diabetic rats. J Food Drug Anal 27(3):736–748

Chen Q, Chen Y, Xiao L, Li Y, Zou S, Han D (2022) Co-production of lutein, zeaxanthin, and β-carotene by utilization of a mutant of the green alga *Chromochloris zofingiensis*. Algal Res 68:102882

Chodkowska KA, Barszcz M, Tuśnio A (2024) MicroRNA expression and oxidative stress markers in pectoral muscle of broiler chickens fed diets supplemented with phytobiotics composition. Sci Rep 14(1):4413

Dávid CZ, Hohmann J, Vasas A (2021) Chemistry and pharmacology of *Cyperaceae stilbenoids*: a review. Molecules 26(9):2794

Feng S, Luan D, Ning K, Shao P, Sun P (2019) Ultrafiltration isolation, hypoglycemic activity analysis and structural characterization of polysaccharides from *Brasenia schreberi*. Int J Biol Macromol 135:141–151

Ferdous UT, Balia Yusof ZN (2021) Insight into potential anticancer activity of algal flavonoids: current status and challenges. Molecules 26(22):6844

Figueroa V, Aguilera FMJM (2023) Seaweeds as novel foods and source of culinary flavors. Food Rev Int 39(1):1–26

Firmin A, Selosse M-A, Dunand C, Elger A (2022) Mixotrophy in aquatic plants, an overlooked ability. Trends Plant Sci 27(2):147–157

Forsch KO, Hahn-Woernle L, Sherrell RM, Roccanova VJ, Bu K, Burdige D, Vernet M, Barbeau KA (2021) Seasonal dispersal of fjord meltwaters as an important source of iron and manganese to coastal Antarctic phytoplankton. Biogeosciences 18(23):6349–6375

Galgali P, Supriya P, Arindam A, Rajkumar P, Routh J (2023) Remediation of potentially toxic elements-containing wastewaters using water hyacinth-a review. Int J Phytoremediation 25(2):172–186

Gao J, Liu L, Ma N, Yang J, Dong Z, Zhang J, Zhang J, Cai M (2020) Effect of ammonia stress on carbon metabolism in tolerant aquatic plant-*Myriophyllum aquaticum*. Environ Pollut 263:114412

Gębalski J, Filip G, Załuski D (2022) Paving the way towards effective plant-based inhibitors of hyaluronidase and tyrosinase: a critical review on a structure-activity relationship. J Enzyme Inhib Med Chem 37(1):1120–1195

Giuliani ME, Bigossi G, Lai G, Marcozzi S, Brunetti D, Malavolta M (2024) Marine compounds and age-related diseases: the path from pre-clinical research to approved drugs for the treatment of cardiovascular diseases and diabetes. Mar Drugs 22(5):210

González Fernández LA, Castillo Ramos V, Sánchez Polo M, Medellín Castillo NA (2023) Fundamentals in applications of algae biomass: a review. J Environ Manag 338:117830

Gravandi MM, Abdian S, Tahvilian M, Iranpanah A, Moradi SZ, Fakhri S, Echeverría J (2023) Therapeutic targeting of Ras/Raf/MAPK pathway by natural products: a systematic and mechanistic approach for neurodegeneration. Phytomedicine 115:154821

Han Y, Kim DH, Pack SP (2024) Marine-derived bioactive ingredients in functional foods for aging: nutritional and therapeutic perspectives. Mar Drugs 22(11):496

Hong J, Gruda NS (2020) The potential of introduction of Asian vegetables in Europe. Horticulturae 6(3):38

Hosseini SF, Masoud R, McClements DJ (2022) Bioactive functional ingredients from aquatic origin: a review of recent progress in marine-derived nutraceuticals. Crit Rev Food Sci Nutr 62(5):1242–1269

Hu X, Ma W, Zhang D, Tian Z, Yang Y, Huang Y, Hong Y (2025) Application of natural antioxidants as feed additives in aquaculture: a review. Biology (Basel) 14(1):87

Jewell MD, Bell G (2023) Eco-evolutionary contributions to community trait change in floating aquatic plants. Ecol 104(8):e4117

Ji H, Li J, Gang D, Yu H, Jia H, Hu C, Qu J (2024) Spatiotemporal dynamics of reactive oxygen species and its effect on beta-blockers' degradation in aquatic plants' rhizosphere. J Hazard Mater 476:135146

Jiang Q, Liu J, Huang S, Wang X-Y, Chen X, Liu G-H, Ye K, Song W, Masters CL, Wang J, Wang Y-J (2025) Antiageing strategy for neurodegenerative diseases: from mechanisms to clinical advances. Sig Transduct Target Ther 10(1):76

John OD, Surugau N, Kansedo J, Panchal SK, Brown L (2025) Plant-based functional foods from borneo. Nutrients 17(2):200

Karimi Alavijeh M, Safi S, Zarei A (2023) An efficient method for economic micropropagation of three aquatic plant species (*Lobelia cardinalis*, *Staurogyne repens*, and *Alternanthera reineckii*). Aquac Int 31(3):1623–1636

Kassis A, Fichot M-C, Horcajada M-N, Horstman AM, Duncan P, Bergonzelli G, Preitner N, Zimmermann D, Bosco N, Vidal K (2023) Nutritional and lifestyle management of the aging journey: a narrative review. Front Nutr 9:1087505

Koga H, Ikematsu S, Kimura S (2024) Diving into the water: amphibious plants as a model for investigating plant adaptations to aquatic environments. Annu Rev Plant Biol 75:579–604

Koley A, Mukhopadhyay P, Gupta N, Singh A, Ghosh A, Show BK, GhoshThakur R, Chaudhury S, Hazra AK, Balachandran S (2023) Biogas production potential of aquatic weeds as the next-generation feedstock for bioenergy production: a review. Environ Sci Pollut Res 30(52):111802–111832

Kristanti RA, Hadibarata T (2023) Phytoremediation of contaminated water using aquatic plants, its mechanism and enhancement. Curr Opin Environ Sci Health 32:100451

Kumar S, Huang X, Ji Q, Qayyum A, Zhou K, Ke W, Zhu H, Zhu G (2022) Influence of blanching on the gene expression profile of phenylpropanoid, flavonoid and vitamin biosynthesis, and their accumulation in *Oenanthe javanica*. Antioxidants 11(3):470

Kumar P, Banik SP, Ohia SE, Hiroyoshi M, Sanjoy C, Chin-Kun W, Sang SY, Apurva G, Manashi B, Bagchi D (2024) Current insights on the photoprotective mechanism of the macular carotenoids, lutein and zeaxanthin: safety, efficacy and bio-delivery. J Am Nutr Assoc 43(6):505–518

Li Z, Zhou W, Wang P, Chen Y, Huo S, Wang J, Tian D, Niu J, Zhao Y, Song X (2021) Transcriptome analysis reveals the senescence process controlling the flower opening and closure rhythm in the waterlilies (*nymphaea* L.). Front. Plant Sci 12:701633

Li S, He Y, Zhong S, Li Y, Di Y, Wang Q, Ren D, Liu S, Li D, Cao F (2023a) Antioxidant and anti-aging properties of polyphenol–polysaccharide complex extract from *Hizikia fusiforme*. Foods 12(20):3725

Li J, Kang Z, Yu H, Feng Y, Zhang X, Zhao Y, Dong L, Zhang L, Dong J, Li Y, Ma S (2023b) Potent insecticidal activity of *Eleocharis dulcis* (Burm. f.) Trin peel extract and its main components against aphids. Pest Manag Sci 79(4):1295–1304

Li F, Li R, Lu F, Xu L, Gan L, Chu W, Yan M, Gong H (2023c) Adverse effects of silver nanoparticles on aquatic plants and zooplankton: a review. Chemosphere 338:139459

Li Y, Song Y, Zhang J, Wan Y (2023d) Phytoremediation competence of composite heavy-metal-contaminated sediments by intercropping *Myriophyllum spicatum* L. with two species of plants. Int J Environ Res Public Health 20(4):3185

Li S, Liu Y, Wu Y, Ren L, Lu Y, Yamaguchi S, Lu Q, Hu C, Li D, Jiang N (2024) An outlook on platinum-based active ingredients for dermatologic and skincare applications. Nano 14(15):1303

Li R, Chen G, Liao W, Yuchi Y, Yang X, Zhang Z, Liu X, Mao Z, Li L, Zhao J, Li H, Huo W, Guo Y, Li S, Wu W, Wang C, Hou J (2025) The role of telomere shortening in ambient ozone exposure-related insulin resistance. J Hazard Mater 484:136768

Liang F, Liu L, Li C, Liu Y, Han S, Yang H, Li S, Hui W, Liu L, Yang C (2024) Systematic identification and functional characterization of the CFEM proteins in fishscale bamboo rhombic-spot pathogen *Neostagonosporella sichuanensis*. Front Plant Sci 15:1396273

Liu X-y, Liu D, Lin G-p, Wu Y-j, Gao L-y, Ai C, Huang Y-f, Wang M-f, El-Seedi HR, Chen X, Zhao C (2019) Anti-ageing and antioxidant effects of sulfate oligosaccharides from green algae *Ulva lactuca* and *Enteromorpha prolifera* in SAMP8 mice. Int J Biol Macromol 139:342–351

Liu J, Chang Y, Sun L, Du F, Cui J, Liu X, Li N, Wang W, Li J, Yao D (2021) Abundant allelochemicals and the inhibitory mechanism of the phenolic acids in water dropwort for the control of *Microcystis aeruginosa* blooms. Plants 10(12):2653

Liu D, Ouyang Y, Chen R, Wang M, Ai C, El-Seedi HR, Sarker MMR, Chen X, Zhao C (2022) Nutraceutical potentials of algal ulvan for healthy aging. Int J Biol Macromol 194:422–434

Liu P, Hu J, Wang Q, Tan J, Wei J, Yang H, Tang S, Huang H, Zou Y, Huang Z (2023) Physicochemical characterization and cosmetic application of kelp blanching water polysaccharides. Int J Biol Macromol 248:125981

Lü H, Zuo Y, Meng X, Ding X, Jian T, Liu Y, Ren B, Chen J, Li V (2022) Phenolic profiles, antioxidant activity and inhibition of digestive enzymes of water caltrop pericarps. J Sci Food Agric 102(6):2342–2351

Lu Y, Zhao S, Jia C, Xu Y, Zhang B, Niu M (2022) Textural properties of Chinese water chestnut (*Eleocharis dulcis*) during steam heating treatment. Foods 11(9):1175

Majewski M, Mercedes K-M, Verdasco-Martín CM, Cristina O, Ferrer M (2022) *Spirulina* extract improves age-induced vascular dysfunction. Pharm Biol 60(1):627–637

Maldonado I, Moreno EG, Terrazas VFZ (2022) Application of duckweed (*Lemna* sp.) and water fern (*Azolla* sp.) in the removal of pharmaceutical residues in water: state of art focus on antibiotics. Sci Total Environ 838:156565

Mannick JB, Lamming DW (2023) Targeting the biology of aging with mTOR inhibitors. Nat Aging 3(6):642–660

Mansour AT, Ashour M, Alprol AE, Alsaqufi AS (2022) Aquatic plants and aquatic animals in the context of sustainability: cultivation techniques, integration, and blue revolution. Sustainability 14(6):3257

Manzoor MF, Afraz MT, Yılmaz BB, Adil M, Arshad N, Goksen G, Ali M, Zeng X-A (2024) Recent progress in natural seaweed pigments: green extraction, health-promoting activities, techno-functional properties and role in intelligent food packaging. J Agric Food Res 15:100991

Mapelli-Brahm P, Gómez-Villegas P, Gonda ML, León-Vaz A, León R, Mildenberger J, Rebours C, Saravia V, Vero S, Vila E (2023) Microalgae, seaweeds and aquatic bacteria, archaea, and yeasts: sources of carotenoids with potential antioxidant and anti-inflammatory health-promoting actions in the sustainability era. Mar Drugs 21(6):340

Marafie SK, Al-Mulla F, Abubaker J (2024) mTOR: its critical role in metabolic diseases, cancer, and the aging process. Int J Mol Sci 25(11):6141

Matovic V, Ljujic B, Radojevic I, Đelic G, Kovacevic MM, Zivanovic S, Papic M, Milivojevic N, Petrovic I, Gazdic Jankovic M (2025) *Trapa natans* L. extract attenuates inflammation and oxidative damage in cisplatin-induced cardiotoxicity in rats by promoting M2 macrophage polarization. Mediat Inflamm 2025(1):6587305

Melini F, Melini V, Luziatelli F, Ficca AG, Ruzzi M (2019) Health-promoting components in fermented foods: an up-to-date systematic review. Nutrients 11(5):1189

Michalak M (2023) Plant extracts as skin care and therapeutic agents. Int J Mol Sci 24(20):15444

Michalak I, Ruchi T, Manish D, Mahmoud A, Farag MR, Sharun K, Emran TB, Dhama K (2022) Antioxidant effects of seaweeds and their active compounds on animal health and production-a review. Vet Q 42(1):48–67

Moon SH, Kim E, Kim HI, Kim SY, Seo HH, Lee JH, Lee MS, Lee SK, Moh SH, Bae S (2023) Skin-whitening effect of a callus extract of *Nelumbo nucifera* isolate Haman. Plants 12(23):3923

Naseem S, Bhat SU, Gani A, Bhat FA (2024) Starch exploration in *Nelumbo nucifera* and *Trapa natans*: understanding physicochemical and functional variations for future perspectives. Int J Biol Macromol 274:133077

Nguyen TQ, Sesin V, Kisiala A, Emery RJN (2021) Phytohormonal roles in plant responses to heavy metal stress: implications for using macrophytes in phytoremediation of aquatic ecosystems. Environ Toxicol Chem 40(1):7–22

Nguyen TM, Chen HH, Yang CC, Tsai YP, Kuo MY, Liao YT, Chang YC, Chen KF (2024) The application of water bamboo (*Zizania latifolia*) husk-derived biochar for copper-contaminated irrigation water treatment and soil amendment. Environ Sci-Water Res Tech 10(9):2048–2061

Nie J, Fu X, Wang L, Xu J, Gao X (2023) Impact of *Monascus purpureus* fermentation on antioxidant activity, free amino acid profiles and flavor properties of kelp (*Saccharina japonica)*. Food Chem 400:133990

Nova M, Citterio S, Martegani E, Colombo S (2024) Unraveling the anti-aging properties of phycocyanin from the cyanobacterium spirulina (*Arthrospira platensis*). Int J Mol Sci 25(8):4215

Nutho B, Tungmunnithum D (2024) Anti-aging potential of the two major flavonoids occurring in asian water lily using *in vitro* and in silico molecular modeling assessments. Antioxidants 13(5):601

Park MS, Son SU, Kim TE, Shim SH, Jang B-K, Park S, Shin KS (2025) Polysaccharide fraction isolated from *saccharin japonica* exhibits anti-cancer effects through immunostimulating activities. Mar Drugs 23(1):38

Pawłowska B, Biczak R (2024) Drugs in the environment – impact on plants: a review. Environ Toxicol Pharmacol 111:104557

Peng X, Lin Q, Liu Q, Huang S, Yan W, Zhang L, Ge F, Zhang Y, Wu Z (2022) Effect of submerged plant coverage on phytoplankton community dynamics and photosynthetic activity in situ. J Environ Manag 301:113822

Peng X, Xiao Y, Tang W, Dai Z, Zhang D, Zheng Y, Lin L (2025) Abscisic acid improves the phytoremediation ability of the aquatic accumulator plant *Nasturtium officinale* R. Br. In cadmium-contaminated soil. Environ Prog Sustain 44(2):e14542

Pereira L (2021) Macroalgae. Encyclopedia 1:177–188

Pereira L (2022) Therapeutic and nutritional uses of marine algae: a pharmacy in the ocean. Trad Med Res 7(3):19–30

Pérez DJ, Lombardero LR, Doucette WJ (2023) Influence of exposure time, physicochemical properties, and plant transpiration on the uptake dynamics and translocation of pharmaceutical and personal care products in the aquatic macrophyte *Typha latifolia*. Sci Total Environ 896:165107

Plaskova A, Mlcek J (2023) New insights of the application of water or ethanol-water plant extract rich in active compounds in food. Front Nutr 10:1118761

Polechońska L, Klink A (2021) Validation of *Hydrocharis morsus*-ranae as a possible bioindicator of trace element pollution in freshwaters using *Ceratophyllum demersum* as a reference species. Environ Pollut 269:116145

Pramanic A, Sharma S, Dhanorkar M, Prakash O, Singh P (2023) Endophytic microbiota of floating aquatic plants: recent developments and environmental prospects. World J Microbiol Biotechnol 39(4):96

Rehman MU, Sehar N, Dar NJ, Khan A, Arafah A, Rashid S, Rashid SM, Ganaie MA (2023) Mitochondrial dysfunctions, oxidative stress and neuroinflammation as therapeutic targets for neurodegenerative diseases: an update on current advances and impediments. Neurosci Biobehav Rev 144:104961

Rehman AU, Khan AU, Sohaib M, Rehman H (2024) Comparative analysis of nutritional properties, phytochemical profile, and antioxidant activities between red and green water chestnut (*Trapa natans*) fruits. Foods 13(12):1883

Ritu JR, Ambati RR, Ravishankar GA, Shahjahan M, Khan S (2023) Utilization of astaxanthin from microalgae and carotenoid rich algal biomass as a feed supplement in aquaculture and poultry industry: an overview. J Appl Phycol 35(1):145–171

Sahu T, Vedantam S, Vetrivel U (2024) *In silico* exploration of bioactive compounds from *Acorus calamus* L. for targeted treatment of ischemic heart disease: molecular insights and therapeutic prospects. J Drug Res Ayur Sci 9(3):150–157

Saxena MK, Neerja S, Kumar S, Dobhal MP, Soumana D (2021) Review potent pharmaceutical products from aquatic plants-review. Asian J Pharm Clin Res 14(1):48–63

Seferli M, Kotanidou C, Lefkaki M, Adamantidi T, Panoutsopoulou E, Finos MA, Krey G, Kamidis N, Stamatis N, Anastasiado N (2024) Bioactives of the freshwater aquatic plants, *Nelumbo nucifera* and *Lemna minor*, for functional foods, cosmetics and pharmaceutical applications, with antioxidant, anti-inflammatory and antithrombotic health promoting properties. Appl Sci 14(15):6634

Selvaraj D, Velvizhi G (2021) Sustainable ecological engineering systems for the treatment of domestic wastewater using emerging, floating and submerged macrophytes. J Environ Manag 286:112253

Shi H, Ernst E, Heinzel N, McCorkle S, Rolletschek H, Borisjuk L, Ortleb S, Martienssen R, Shanklin J, Schwender J (2023) Mechanisms of metabolic adaptation in the duckweed *Lemna gibba*: an integrated metabolic, transcriptomic and flux analysis. BMC Plant Biol 23(1):458

Singh AA, Rajeswari G, Nirmal LA, Jacob S (2021) Synthesis and extraction routes of allelochemicals from plants and microbes: a review. Rev Anal Chem 40(1):293–311

Song C, Sun C, Liu B, Xu P (2023) Oxidative stress in aquatic organisms. Antioxidants 12(6):1223

Sopawong P, Warodomwichit D, Srichamnong W, Methacanon P, Tangsuphoom N (2022) Effect of physical and enzymatic modifications on composition, properties and *in vitro* starch digestibility of sacred lotus (*Nelumbo nucifera*) seed flour. Foods 11(16):2473

Staszek P, Krasuska U, Ciacka K, Gniazdowska A (2021) ROS metabolism perturbation as an element of mode of action of allelochemicals. Antioxidants 10(11):1648

Sultanova Z, Shen A, Hencel K, Carlsson H, Crighton Z, Clifton D, Akay A, Maklakov AA (2025) Optimising age-specific insulin signalling to slow down reproductive ageing increases fitness in different nutritional environments. Aging Cell:e14481

Sun M, Zhang Y, Gao W, He Y, Wang Y, Sun Y, Kuang H (2024) Polysaccharides from *Porphyra haitanensis*: a review of their extraction, modification, structures, and bioactivities. Molecules 29(13):3105

Suraiya S, Ria SJ, Tanzim Riya MU, Ritu FY, Sumona AA, Rodela AB, Akter L, Uddin MS, Hasan MN (2024) Nutritional and biofunctional characterizations of four novel edible aquatic plants of Bangladesh. Heliyon 10(15):e35538

Tanpichai S, Thongdeelerd C, Chantaramanee T, Boonmahitthisud A (2023) Property enhancement of epoxidized natural rubber nanocomposites with water hyacinth-extracted cellulose nanofibers. Int J Biol Macromol 234:123741

Wan J, Yu X, Liu J, Li J, Ai T, Yin C, Liu H, Qin R (2023) A special polysaccharide hydrogel coated on *Brasenia schreberi*: preventive effects against ulcerative colitis *via* modulation of gut microbiota. Food Funct 14(8):3564–3575

Wang X, Ding G, Liu B, Wang Q (2020) Flavonoids and antioxidant activity of rare and endangered fern: *Isoetes sinensis*. PLoS One 15(5):e0232185

Wang X, Zhang Z, Zhang S, Yang F, Yang M, Zhou J, Hu Z, Xu X, Mao G, Chen G, Xiang W, Sun X, Xu N (2021) Antiaging compounds from marine organisms. Food Res Int 143:110313

Wang K, Liu H, Hu Q, Wang L, Liu J, Zheng Z, Zhang W, Ren J, Zhu F, Liu G-H (2022) Epigenetic regulation of aging: implications for interventions of aging and diseases. Signal Transduct Target Ther 7(1):374

Wang H, Chen J, Yi Y, Wang L, Hou W, Ai Y, Wang H, Min T (2023a) Regulation and mechanism of ethylene treatment on storage quality of fresh-cut lotus (*Nelumbo nucifera* Gaertn.) root slices. Sci Hortic 313:111900

Wang X, Li B, Sun S, Liu Q, Zhu J, Zhou X, Zhang H, Wu Q, Wang L (2023b) Analysis of proanthocyanidins and flavonols in the seedpods of Chinese Antique lotus: a rich source of antioxidants. Food Chem 415:135756

Wang M, Tang HP, Wang S, Hu WJ, Li JY, Yu Q, Bai QX, Yang BY, Kuang HX (2023c) *Acorus tatarinowii* Schott: a review of its botany, traditional uses, phytochemistry, and pharmacology. Molecules 28(11):4525

Wang Z, Yong L, Dandan M, Maomao Z, Zhaojun W, Fang Q, Jie C, Mark C, He Z (2023d) Alkaloids from lotus (*Nelumbo nucifera*): recent advances in biosynthesis, pharmacokinetics, bioactivity, safety, and industrial applications. Crit Rev Food Sci Nutr 63(21):4867–4900

Wang W, Wang H, Zang S (2025) Purification mechanism of emergent aquatic plants on polluted water: a review. J Environ Manag 375:124198

Wilawan BS, Chan S, Ling TC, Show PL, Ng E-P, Jonglertjunya W, Phadungbut P, Khoo KS (2024) Advancement of carotenogenesis of astaxanthin from *Haematococcus pluvialis*: recent insight and way forward. Mol Biotechnol 66(3):402–423

Wu W, Han Y, Niu B, Yang B, Liu R, Fang X, Chen H, Xiao S, Farag MA, Zheng S (2024) Recent advances in *Zizania latifolia*: a comprehensive review on phytochemical, health benefits and applications that maximize its value. Crit Rev Food Sci Nutr 64(21):7535–7549

Yanagido A, Ueno M, Jiang Z, Cho K, Yamaguchi K, Kim D, Oda T (2018) Increase in anti-inflammatory activities of radical-degraded porphyrans isolated from discolored nori (*Pyropia yezoensis*). Int J Biol Macromol 117:78–86

Yang H, He S, Feng Q, Liu Z, Xia S, Zhou Q, Wu Z, Zhang Y (2024a) Lotus (*Nelumbo nucifera*): a multidisciplinary review of its cultural, ecological, and nutraceutical significance. Bioresour Bioprocess 11(1):18

Yang Y, Lin H, Fu X (2024b) Fermentation of *Pyropia* spp. seaweed: a comprehensive review on processing conditions, biological activities and potential applications in the food industry. Crit Rev Food Sci Nutr:1–16

Yasuda M, Ikeoka M, Kondo S-I (2021) Skin-related enzyme inhibitory activity by hydrolyzable polyphenols in water chestnut (*Trapa natans*) husk. Biosci Biotechnol Biochem 85(3):666–674

Yu H, Zhang X, Hu J, Peng J, Qu J (2020) Ecotoxicity of polystyrene microplastics to submerged carnivorous *Utricularia vulgaris* plants in freshwater ecosystems. Environ Pollut 265:114830

Yuan M, Wang J, Geng L, Wu N, Yang Y, Zhang Q (2024a) A review: structure, bioactivity and potential application of algal polysaccharides in skin aging care and therapy. Int J Biol Macromol 272:132846

Yuan S, Zhang M, Yao Z, Liu J, Li X, Zhang Z, Li D (2024b) Isolation, structural characterization, and bioactivities of neutral polysaccharides from *Zizania latifolia*. Int J Biol Macromol 254:127679

Zhang B, Li W, Guo Y, Zhang Z, Shi W, Cui F, Lens PNL, Tay JH (2020) Microalgal-bacterial consortia: from interspecies interactions to biotechnological applications. Renew Sust Energ Rev 118:109563

Zhang Y, Nie R, Liu W, Dong S, Yang J, Wang X, Wang Y, Zheng L (2024) Sulfation on polysaccharides from *Zizania latifolia* extracted using ultrasound: characterization, antioxidant and anti-non-small cell lung cancer activities. Ultrason Sonochem 103:106803

Zhang J, Xing L, Meng W, Zhang X, Li J, Dong P (2025) Molecular weight distribution and structure analysis of phlorotannins in Sanhai kelp (*Saccharina japonica*) and evaluation of their antioxidant activities. Food Chem 469:142569

Zhao X, Yang J, Li G, Sun Z, Hu S, Chen Y, Guo W, Hou H (2021) Genome-wide identification and comparative analysis of the WRKY gene family in aquatic plants and their response to abiotic stresses in giant duckweed (*Spirodela polyrhiza*). Genomics 113(4):1761–1777

Zhao J, Liu C, Li H, Liu J, Jiang T, Yan D, Tong J, Dong L (2022) Review on ecological response of aquatic plants to balanced harvesting. Sustain 14(19):12451

Zhao Y, Simon M, Seluanov A, Gorbunova V (2023) DNA damage and repair in age-related inflammation. Nat Rev Immunol 23(2):75–89

Zhao J, Han Z, Ding L, Wang P, He X, Lin L (2024a) The molecular mechanism of aging and the role in neurodegenerative diseases. Heliyon 10(2)

Zhao R, Peng K, Yi Y, Sun Y, Jiang X, Wang H (2024b) Heating alters the nutritional and antioxidant characteristics of lotus root. LWT 210:116848

Zhao H, Zuo Z, Yang L, Zhang L, Lv T, Yu D, Wang Z (2024c) Similarities and differences in the physiological adaptation to water salinity between two life forms of aquatic plants in alpine and arid wetlands. Sci Total Environ 908:168449

Zhi M, Zhao Y, Zeng X, Maddela NR, Xiao Y, Chen Y, Prasad R, Zhou Z (2023) Filamentous cyanobacteria and hydrophobic protein in extracellular polymeric substances facilitate algae-bacteria aggregation during partial nitrification. Int J Biol Macromol 251:126379

Induction of Somatic Embryos, Their Maturation, and Regeneration in Anti-aging Medicinal Plants

5

Santiago-Hernández Iliana
and Bello-Bello Jericó Jabín

Abstract

Somatic embryogenesis is a type of plant tissue culture that regenerates plants by developing embryos from somatic or asexual plant embryos in vitro. There are two types of somatic embryogenesis: direct somatic embryogenesis (DSE) and indirect somatic embryogenesis (ISE). In DSE, embryos form directly on the surface of the original explant, whereas ISE requires the initial formation of a callus from which somatic embryos will be regenerated. This chapter examines various factors involved in embryogenesis and describes the effects of multiple plant growth regulators and additives on the induction, histodifferentiation, maturation, germination, and conversion into plantlets of somatic embryos, whether via DSE or ISE. Additionally, we describe examples of cellular-level changes that occur during somatic embryogenesis in various medicinal plants with potential anti-aging properties. Scientific research demonstrates that somatic embryogenesis is a valuable tool in plant biotechnology, serving as a plant regeneration technique with applications in mass micropropagation, synthetic seed production, germplasm conservation, genetic transformation, protoplast fusion, in vitro mutagenesis, and the breeding of medicinal plants. Furthermore, somatic embryogenesis represents a sustainable alternative to maximize the potential of medicinal plants.

Keywords

Cell differentiation · Conversion · Germination · Growth regulators · Histodifferentiation · Somatic embryogenesis

---

S.-H. Iliana
Postgraduate College-Campus Veracruz, Veracruz, Mexico

B.-B. J. Jabín (✉)
SECIHTI-Postgraduate College-Campus Cordoba, Veracruz, Mexico

S. Gantait et al. (eds.), *Medicinal Plants with Anti-Ageing Properties*,
https://doi.org/10.1007/978-981-95-8064-4_5

5.1 Introduction

Medicinal plants are valuable sources of bioactive metabolites, including polyphenols and antioxidants, with anti-aging activity (Duque et al. 2017; Iweala et al. 2024). Thus, their propagation and conservation are crucial to maintaining these phytogenetic resources. Plant biotechnology is one of the scientific disciplines that has shown the most development in recent years. It offers alternative solutions to a range of challenges related to the sustainable and efficient use of plants by humans. This discipline represents a valuable tool for increasing both the quantity and quality of plant-based foods and for developing new medicinal plant-based products. Plant biotechnology is based on plant tissue culture (PTC), a series of techniques that allow the culture and manipulation of plant cells, tissues, or organs under controlled, artificial conditions (Tarraf and De Carlo 2024). PTC can be applied to plant health, the in vitro production of secondary metabolites, genetic improvement, germplasm conservation, and micropropagation-based plant regeneration.

In vitro, plants can be regenerated via three different mechanisms: budding, organogenesis, and somatic embryogenesis. Budding derives from previously formed meristems and yields one or more complete plants. On the other hand, organogenesis and somatic embryogenesis can derive from any tissue and encompass a cell dedifferentiation and redifferentiation process to yield adventitious shoots or somatic embryos. During budding, new plants originate from multicellular structures. However, during organogenesis and somatic embryogenesis, new plants can originate from a single cell or from a group of cells (Bello-Bello et al. 2025; Yan et al. 2025). The tissue's response to in vitro culture depends on the interaction between the explant and the culture medium and conditions. This response can result in tissue development or more complex processes, such as morphogenesis, which can then lead to organogenesis (the formation of adventitious organs like roots or shoots), somatic embryogenesis (the formation of embryos from somatic cells), or the formation of callus tissue. This chapter reveals the effects of multiple plant growth regulators and additives on the induction and regeneration of somatic embryos either via a direct approach or via a callus-mediated approach in several anti-aging plants.

5.2 Types of Somatic Embryogenesis

One of the most notable characteristics of plant tissue culture is that certain cells, under specific in vitro culture conditions, can form embryos following a process very similar to zygotic embryogenesis. This process, called somatic embryogenesis, is one of the clearest examples of cell totipotency. Somatic embryos, as well as zygotic embryos, can form new plants following a germination and conversion process. The main difference between these types of embryogenesis is that somatic embryogenesis is asexual. Thus, the new plant will be identical to the initial donor plant (Aronen et al. 2025).

A plant cell can only undergo somatic embryogenesis if it is "determined" or "induced". Cells capable of developing somatic embryos are called embryogenic determined cells (EDC) or proembryogenic cells (Gulzar et al. 2020).

There are two types of in vitro somatic embryogenesis: direct (DSE) and indirect (ISE). DSE entails the formation of embryos directly on the original explant. On the other hand, ISE first requires the formation of a callus tissue or an embryogenic cell suspension from which somatic embryos can differentiate. This is a much more complex process compared to DSE. DSE occurs when the original explant contains proembryogenic cells, eliminating the need for an induction process. Therefore, this type of embryogenesis only requires optimal in vitro tissue conditions for differentiation. However, during ISE, the original explant lacks proembryogenic cells. Therefore, it is necessary to first provide an induction environment that favors proembryogenic cell development, followed by one that enhances embryo differentiation (Yang et al. 2024). The cells of very young tissues, such as immature zygotic embryos, possess proembryogenic cells. If that is not the case, these cells are very easy to induce. On the other hand, it is difficult for differentiated cells or adult tissues to yield proembryogenic cells. This is why somatic embryogenesis is only feasible in some tissues.

5.3 Overview of Stages of Somatic Embryogenesis

Somatic embryogenesis consists of various stages (Fig. 5.1). Each requires different stimuli and yields different phenomena. These stages are described below.

5.3.1 Induction

This process involves converting somatic cells into proembryogenic cells. The determining factors of this process include: (1) Genotype. Different genotypes, even those from the same species, vary significantly in their ability to activate the key elements required to initiate somatic embryogenesis; (2) differentiation degree of explant cells; and (3) plant growth regulators. Auxins are a determining factor in the induction stage. The most effective auxins for somatic embryogenesis induction include synthetic auxins, such as phenoxyacetic acids (2,4-dichlorophenoxyacetic acid and 2,4,5-trichlorophenoxyacetic acid) and other regulators, such as thidiazuron (TDZ) (Yu et al. 2024).

5.3.2 Histodifferentiation

During this stage, groups of proembryogenic cells differentiate into somatic embryos following a simultaneous cellular division and differentiation. To stop multiplication and induce differentiation, it is essential to first eliminate exogenous auxins. One of the first phenomena to occur during this stage is the establishment of

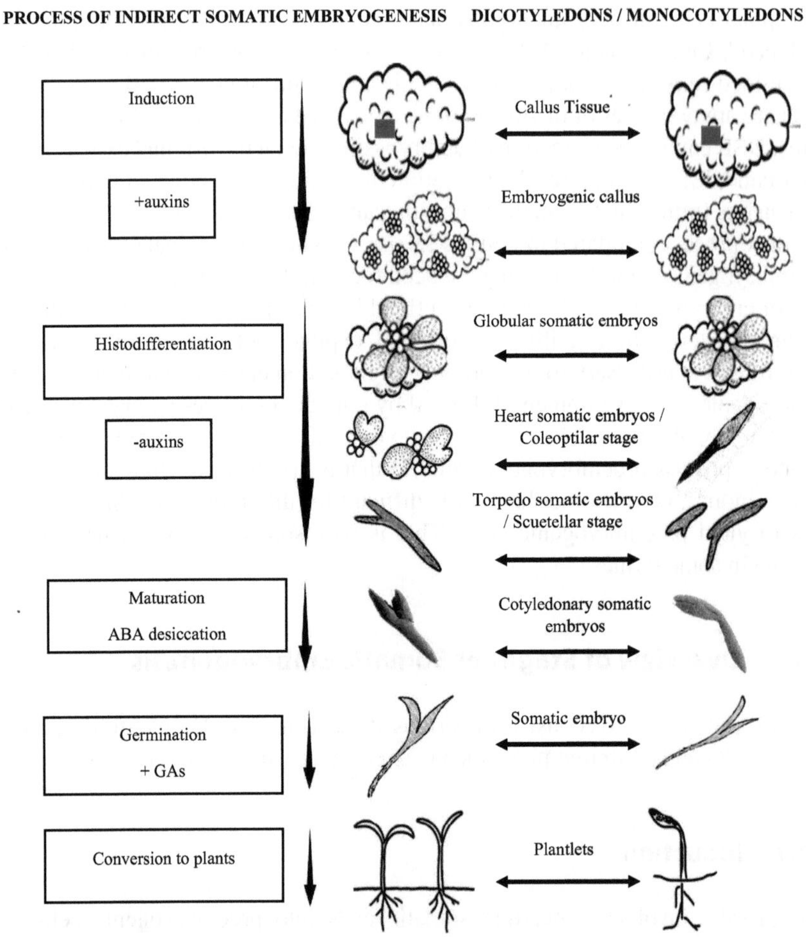

Fig. 5.1 Process of indirect and direct somatic embryogenesis. *ABA:* abscisic acid and *GAs:* gibberellins

polarity in proembryogenic cell masses. This polarity is maintained throughout embryonic development. During histodifferentiation, somatic embryos undergo a series of intermediate stages similar to those of zygotic embryogenesis. In dicots, these stages are the globular, heart, and torpedo stages (Fig. 5.2a). For monocots, these stages are globular, coleoptilar, and scutellar.

5.3.3 Maturation

Usually, a somatic embryo at the torpedo intermediate histodifferentiation stage cannot germinate. For germination to occur, the embryo must undergo a maturation stage characterized by cellular elongation without cell division. For most species,

Fig. 5.2 Plantlet regeneration via somatic embryogenesis in coffee (*Coffea* sp.). (**a**) Development of somatic embryos, globular (gl), heart (ht), torpedo (tor), cotyledonary (cot), (**b**) conversion to plantlets, and (**c**) acclimatized plantlets in greenhouse

the stimuli that induce maturation are abscisic acid and desiccation. In dicots, the mature somatic embryo is called the cotyledonary stage.

5.3.4 Germination

This stage entails the elongation and metabolic activation of a mature somatic embryo to become a plantlet. This process requires stimuli such as light, gibberellic acid, or cytokinins. Because somatic embryos lack storage tissues, germination can occur only in vitro, where the culture medium provides nutrients, or when artificial nutrient storage (artificial seeds) is provided.

5.3.5 Conversion

During conversion, a germinated somatic embryo with normal morphology transforms into a complete plantlet capable of continuing its development under non-axenic conditions (Fig. 5.2b, c) (Von Arnold et al. 2002).

5.4 Somatic Embryogenesis in Medicinal Plants with Anti-aging Effects

Somatic embryogenesis is an important technique used for the regeneration of species with medicinal use, such as *Ginkgo Biloba* (Laurain et al. 1996), (Von Arnold et al. 2002), *Centella asiatica* (Joshee et al. 2007), *Withania somnifera* (Jhankare et al. 2011), *Brassica oleracea* var. *Itálica* (Pavlović et al. 2013), *Phyllanthus emblica* (Isabel et al. 2014), *Salvia rosmarinus* (Aman and Afrasiab 2014), *Hibiscus sabdariffa* (Konar et al. 2018), *Bacopa monnieri* (Ali et al. 2020), *Phyllanthus emblica* (Gantait et al. 2021), *Panax ginseng* (Lee et al. 2023), and *Aloe vera* and *Cinnamomum zeylanicum* (Liu et al. 2024). The response to somatic embryogenesis in each species depends on the type of explant and the type and concentration of the growth regulator. Table 5.1 shows some examples of somatic embryogenesis in medicinal plants with anti-aging effects.

5.5 Application of Somatic Embryogenesis

SE is a regeneration technique used in micropropagation, synthetic seed formation, and cryopreservation.

5.5.1 Micropropagation

Micropropagation refers to the asexual propagation of plants using in vitro tissue culture techniques. It is one of the most used techniques in plant biotechnology. This is due to its significant productivity compared to traditional plant propagation techniques. Furthermore, micropropagation is important for the mass regeneration of genetically improved plants. The micropropagation of any plant species consists of five stages. Stage 0: stock plant selection and preparation, stage I: establishment, stage II: multiplication, stage III: elongation and rooting, and stage IV: acclimatization.

To date, micropropagation has great importance worldwide, having displaced traditional seedling production systems. As for the future of micropropagation, developed countries are moving toward automation and greater specialization of laboratories in the production of one or a few species. Furthermore, current work is underway to develop larger-scale somatic embryogenesis systems, including synthetic seed production systems.

Table 5.1 Somatic embryogenesis in medicinal plants with anti-aging effects

Specie	Type of explant	Growth regulators induction	Types of somatic embryogenesis	Author
Horse chestnut (*Aesculus hippocastanum*)	Flower	2,4-D + Kin	Direct	Radojevic (1988)
Jatamansi (*Nardostachys jatamansi*)	Leaf	IAA	Direct	Mathur (1993)
Ginkgo (*Ginkgo biloba*)	Leaf and seed	IAA + BA	Direct	Laurain et al. (1996)
Aloe vera (*Aloe vera*)	Leaf	2,4-D + Kin	Indirect	Von Arnold et al. (2002)
Ginger (*Zingiber officinale*)	Leaf and root	2,4-D + 6-BA + BAP	Indirect and direct	Guo and Zhang (2005)
Centella asiatica (*Centella asiatica*)	Seed	2,4-D	Indirect	Joshee et al. (2007)
Turmeric (*Phyllanthus emblica*)	Root and leaf	2,4-D + 6-BA	Direct	Gour and Kant (2010); Isabel et al. (2014)
Arjuna (*Terminalia arjuna*)	Zygotic embryo	2,4-D + Kin	Indirect	Anjaneyulu and Giri (2011)
Ashwagandha (*Withania somnifera*)	Leaf	2,4-D + Kin	Indirect	Jhankare et al. (2011)
Guduchi (*Tinospora cordifolia*)	Leaf	2,4-D	Direct	Khalilsaraie et al. (2011)
Broccoli (*Brassica oleracea*)	Root, cotyledon, hypocotyl, and stigmas	2,4-D + 6-BA	Direct	Pavlović et al. (2013)
Rosemary (*Salvia rosmarinus*)	Leaf	2, 4- D	Indirect	Aman and Afrasiab (2014)
Jamaica (*Hibiscus sabdariffa*)	Root	2,4-D + Kn	Indirect	Konar et al. (2018)
Bacopa (*Bacopa monnieri*)	Leaf	2,4-D + NAA	Indirect	Ali et al. (2020)
Indian Gooseberry (*Phyllanthus emblica*)	*Leaf*	2,4-D + BAP	Indirect	Gantait et al. (2021)
Ginseng (*Panax ginseng*)	Seeds	2,4-D	Indirect	Lee et al. (2023)
Cinnamon (*Cinnamomum eylanicum*)	Leaf	2,4-D + IAA	Indirect	Liu et al. (2024)
Liquorice (*Glycyrrhiza glabra*)	Root	2,4-D	Direct	Sarkar (2024)

2,4-D Dichlorophenoxyacetic acid, *NAA* α-Naphthalene acetic acid, *BAP N6* N6-Benzyladenin, *ABA* abscisic acid, *KIN* kinetin, *6-BA* 6-benzylaminopurine

5.5.2 Synthetic Seed Formation

Somatic embryogenesis sets the foundation for the production of synthetic seeds. Synthetic seeds are somatic embryos, buds, or vegetative parts artificially encapsulated that can develop into a new plant ex vitro (Bordolui 2024).

5.5.3 Cryopreservation

Cryopreservation involves maintaining tissues at extremely low temperatures (−196 °C) using liquid nitrogen for an indefinite time. Embryos are selected and prepared with osmo- and cryoprotectant solutions containing dimethyl sulfoxide, ethylene glycol, glycerol, sorbitol, and sucrose. Prepared embryos are then subjected to a dehydration process through desiccation, encapsulation-desiccation, or vitrification, followed by cooling and storage in liquid nitrogen to stop metabolic activity (Mikuła et al. 2011; Pramanik et al. 2023). For regeneration, post-cryopreservation embryos are rapidly thawed in water, washed to eliminate cryoprotectants, and transferred into a specific culture medium for germination (Popova et al. 2023; Cabero-Moreno et al. 2025).

5.6 Considerations and Future Prospects

Plant regeneration via somatic embryogenesis has been successful in only a few species compared to organogenesis, largely because of its more complex process. However, when successful, it yields a greater number of plants. To establish a somatic embryogenesis system, the main variables to consider are the type of explant and the type and concentration of the growth regulator used to induce the proembryogenic tissue.

Somatic embryos and the adventitious shoots obtained via organogenesis have key differences that allow their identification when their origin is unclear. These differences are described below.

- Somatic embryos are polar structures characterized by a visible apex and radicle. On the other hand, organogenesis structures are unipolar, consisting of only an apex (adventitious shoot) or a root.
- Somatic embryos lack a vascular connection with the original explant, which makes them relatively independent structures. Whereas adventitious shoots generally remain connected to the vascular system of the initial explant during direct organogenesis.
- At a molecular level, somatic embryos express a series of proteins related to the zygotic embryogenesis process that are not observed in adventitious shoots. These proteins are considered molecular markers of embryogenesis.

5.7 Conclusion

Somatic embryogenesis is a biotechnological tool used in the propagation and conservation of medicinal plants with anti-aging potential. The commercial cultivation of plants derived from embryogenesis represents an alternative to reduce the over-exploitation of wild populations and to sustainably manage species of importance to the pharmaceutical industry.

Competing Interests The authors have no relevant financial or non-financial interests to disclose.

References

Ali D, Alarifi S, Pandian A (2020) Somatic embryogenesis and in vitro plant regeneration of *Bacopa monnieri* (Linn.) Wettst., a potential medicinal water hyssop plant. Saudi J Biol Sci 28:353–359

Aman N, Afrasiab H (2014) Primary and secondary somatic embryogenesis from leaf explants of rosemary (*Rosmarinus officinalis* L.-Lamiaceae). Pak J Bot 46:903–909

Anjaneyulu C, Giri CC (2011) Direct somatic embryogenesis from mature zygotic embryo and conversion to plants in medicinal tree *Terminalia chebula* Retz. Tree For Sci Biotechnol 5:32–37

Aronen T, Varis S, Tikkinen M (2025) Somatic embryogenesis: concept, principles, and applications. In: Biotechnology applications in forestry, Volume 4 in Forest microbiology. Academic Press, pp 373–388

Bello-Bello JJ, Mancilla-Álvarez E, Spinoso-Castillo JL (2025) Scaling-up procedures and factors for mass micropropagation using temporary immersion systems. In Vitro Cell Dev Biol-Plant 61:321–332

Bordolui SK (2024) Prospects and limitations of synthetic seeds. Res Rev J Env Sci 6:6–16

Cabero-Moreno L, Landeras-López I, Bravo-Navas M, Sánchez-Romero C (2025) Induction of somatic embryogenesis and cryopreservation of *Abies pinsapo*. Boiss Front Plant Sci 16:1535113

Duque L, Bravo K, Osorio E (2017) A holistic anti-aging approach applied in selected cultivated medicinal plants: a view of photoprotection of the skin by different mechanisms. Ind Crop Prod 97:431–439

Gantait S, Mahanta M, Bera S, Verma S (2021) Advances in biotechnology of *Emblica officinalis* Gaertn syn *Phyllanthus emblica* L: a nutraceuticals-rich fruit tree with multifaceted ethnomedicinal uses. 3 Biotech 11:62

Gour VS, Kant T (2010) *In vitro* regeneration in *Emblica officinalis* from juvenile root-derived callus. J Indian Bot Soc 89:34–36

Gulzar B, Mujib A, Malik MQ, Sayeed R, Mamgain J, Ejaz B (2020) Genes, proteins and other networks regulating somatic embryogenesis in plants. J Genetic Eng Biotechnol 18:31

Guo Y, Zhang Z (2005) Establishment and plant regeneration of somatic embryogenic cell suspension cultures of the *Zingiber officinale* Rosc. Sci Hortic 107:90–96

Isabel Z, Fonnegra M, Inés A, Trujillo U, Morales A (2014) Organogénesis y embriogénesis en *Curcuma longa* L. a partir de capas delgadas de células, segmentos y bases de hoja *Curcuma longa* L. embryogenesis and organogenesis from thin cell layers, leaf segments and bases of leaf. Colombian. J Biotechnol 16:74

Iweala EJ, Adurosakin OE, Innocent U, Omonhinmin CA, Dania OE, Ugbogu EA (2024) Anti-aging potential of bioactive phytoconstituents found in edible medicinal plants: a review. Science 6:36

Jhankare A, Tripathi MK, Tiwari G, Pandey GN, Baghel BS, Tiwari S (2011) Establishment of embryogenic cell suspension cultures and plantlet regeneration in ashwagandha (*Withania somnifera* L.). Indian J Plant Physiol 16:174

Joshee N, Biswas B, Yadav A (2007) Somatic embryogenesis and plant development in *Centella asiatica* L., a highly prized medicinal plant of the tropics. HortSci 42:633–637

Khalilsaraie MF, Saima N, Meti NT, Bhadekar RK, Nerkar DP (2011) Cytological study and antimicrobial activity of embryogenic callus induced from leaf cultures of *Tinospora cordifolia* (Willd.) Miers. J Med Plant Res 5:3002–3006

Konar S, Karmakar J, Ray A, Adhikari S, Bandyopadhyay TK (2018) Regeneration of plantlets through somatic embryogenesis from root derived calli of *Hibiscus sabdariffa* L. (Roselle) and assessment of genetic stability by flow cytometry and ISSR analysis. PLoS One 13:202324

Laurain D, Chénieux JC, Tremouillaux-Guiller J (1996) Somatic embryogenesis from immature zygotic embryos of *Ginkgo biloba*. Plant Cell Tissue Organ Cult 44:19–24

Lee J, Kim J, Bang K, Kwon N, Kim Y, Jo I, Park Y (2023) Efficient somatic embryogenesis, regeneration and acclimatization of *Panax ginseng* Meyer: true-to-type conformity of plantlets as confirmed by ISSR analysis. Plants 12:1270

Liu Y, Guo H, Kang X, Wu R, Yuan M, Du L (2024) Study on somatic embryogenesis of *Cinnamomum camphora* based on transcriptome sequencing. Braz J Bot 47:325–342

Mathur J (1993) Somatic embryogenesis from callus cultures of *Nardostachys jatamansi*. Plant Cell Tissue Organ Cult 33:163–169

Mikuła A, Tomiczak K, Rybczyński JJ (2011) Cryopreservation enhances embryogenic capacity of *Gentiana cruciata* (L.) suspension culture and maintains (epi) genetic uniformity of regenerants. Plant Cell Rep 30:565–574

Pavlović S, Vinterhalter B, Zdravković-Korać S, Vinterhalter D, Zdravković J, Cvikić D, Mitić N (2013) Recurrent somatic embryogenesis and plant regeneration from immature zygotic embryos of cabbage (*Brassica oleracea* var. capitata) and cauliflower (*Brassica oleracea* var. botrytis). Plant Cell Tissue Organ Cult 113:397–406

Popova E, Kulichenko I, Kim HH (2023) Critical role of regrowth conditions in post-cryopreservation of in vitro plant germplasm. Biology 12:542

Pramanik A, Maheshwari S, Acharya N (2023) Medicinal plants, antioxidant potential, and applications to aging. In: Dhara AK, Mandal SC (eds) Role of herbal medicines. Springer, Singapore, pp 403–413

Radojevic L (1988) Plant regeneration of *Aesculus hippocastanum* L.(horse chestnut) through somatic embryogenesis. J Plant Physiol 132:322–326

Sarkar M (2024) Biochemical changes in embryogenic and nonembryonic callus of *Glycyrrhiza glabra* L. during somatic embryogenesis. Environ Ecol 42:437–445

Tarraf W, De Carlo A (2024) In vitro biotechnology for conservation and sustainable use of plant genetic resources. Plants 13:1897

Von Arnold S, Sabala I, Bozhkov P, Dyachok J, Filonova L (2002) Developmental pathways of somatic embryogenesis. Plant Cell Tissue Organ Cult 69:233–249

Yan BH, Yang MH, Chen Y, Li Q, Wang LB, Yu XL, Wang T (2025) A multi-stage protocol for indirect somatic embryogenesis and plantlet regeneration from leaf explants of *Quercus chungii*. Ind Crop Prod 236:121842

Yang G, Huang LJ, Jiang D, Huang J, Cui C, Li N (2024) Development of indirect somatic embryogenesis and plant regeneration system with immature embryos of the cultivated traditional Chinese medicinal herb *Polygonatum cyrtonema*. Ind Crop Prod 214:118557

Yu Y, Yang B, Ma D, Guo S, Liao F, Li Z (2024) Study on the induction of somatic embryogenesis and morphological structural changes during the development of *Handeliodendron bodinieri* L. Plant Cell Tissue Organ Cult 156:62

Enhancing Anti-aging Metabolite and Biomass Production via Cell Suspension, Elicitation and Bioreactor Strategies

6

Nguyen Hoang Loc, Le Thi Ha Thanh, and Nguyen Hoang Tue

Abstract

This review explores advanced plant cell culture (PCC) strategies to sustainably enhance the production of anti-aging secondary metabolites (e.g., polyphenols, terpenoids) and biomass, addressing the unsustainable reliance on wild-plant resources. The core methodology involves establishing highly productive cell suspension cultures from friable callus, which has proven effective for compounds from species like *Centella asiatica*, *Curcuma*, and *Panax ginseng*. Crucially, metabolite yield is significantly boosted through elicitation strategies, using both biotic (e.g., methyl jasmonate, yeast extract) and abiotic factors to mimic stress and activate key biosynthetic pathways. For industrial scalability, the paper highlights the indispensable role of bioreactor systems, particularly temporary immersion systems (TIS) and bubble column bioreactors (BCR), which provide controlled environments, optimize mass transfer, and ensure high, consistent product quality while minimizing shear stress. The robust integration of suspension culture, elicitation optimization, and advanced bioreactors offers a clear and viable pathway for the industrial production of high-quality, uniform plant-derived biopharmaceuticals for the cosmetic and pharmaceutical sectors.

Keywords

Biopharmaceuticals · Callus · Flavonoids · Medicinal plants · Phenolics · Terpenoids

N. H. Loc (✉) · L. T. H. Thanh · N. H. Tue
Institute of Bioactive Compounds and Department of Biotechnology, University of Sciences, Hue University, Hue, Vietnam
e-mail: nhloc@hueuni.edu.vn; nhloc@husc.edu.vn

S. Gantait et al. (eds.), *Medicinal Plants with Anti-Ageing Properties*,
https://doi.org/10.1007/978-981-95-8064-4_6

6.1 Introduction

Aging is defined as a complex biological process associated with the progressive decline of cellular and tissue function, leading to increased susceptibility to age-related diseases (Li et al. 2024). Mechanistically, this decline is largely driven by the accumulation of cellular damage and a concurrent decrease in the body's repair efficiency (López-Otín et al. 2023). Bioactive compounds from plants, particularly secondary metabolites such as phenolic compounds (including flavonoids, the most important subclass of polyphenol) and terpenoids, have shown significant potential in combating the aging process, thanks to their potent antioxidant and anti-inflammatory properties (Sun and Shahrajabian 2023; Câmara et al. 2024; Păcularu-Burada et al. 2024). However, the surging demand for natural products and extracts across the food, cosmetic, and pharmaceutical industries can lead to the over-exploitation of wild plant resources. This situation poses a serious threat to the sustainability of biodiversity (Chemat et al. 2019). Therefore, exploring alternative, biotechnology-based platforms is crucial to meet the increasing demand for high-quality bioactive compounds.

Plant cell culture (PCC) emerges as a promising biotechnological platform to overcome these drawbacks. PCC allows for the sustainable and large-scale production of these metabolites in a controlled environment (Rao and Ravishankar 2002; Ochoa-Villarreal et al. 2016). The potential of in vitro cell cultures for the production of therapeutically valuable compounds and their large-scale production in bioreactors has received significant attention in recent years. In vitro culture system further provides a distinct advantage for studying various cellular and molecular processes leading to secondary metabolite accumulation and their regulation (Singh et al. 2017). This chapter focuses on advanced strategies in PCC, ranging from establishing cell suspension culture, optimization using elicitors, to the application of sophisticated bioreactor systems to enhance the production of both anti-aging metabolites and biomass, opening up new avenues for the cosmetic and pharmaceutical industries.

6.2 Callus Induction

The initiation of cell suspension cultures fundamentally relies on the successful induction of friable calli, a mass of undifferentiated plant cells (Smith 2013). This initial step requires the meticulous selection of explants (such as young leaves or shoot apices) from the source medicinal plant, prioritizing tissues with high cellular competence and subjecting them to stringent surface sterilization to eliminate microbial contaminants. Callus formation is governed primarily by the composition of the culture medium, typically based on the Murashige and Skoog (1962) formulation. Critically, the medium must be optimized with a precise concentration and ratio of phytohormones (also known as plant growth regulators). Generally speaking, an intermediate ratio of auxin and cytokinin promotes callus induction, while a high ratio of auxin-to-cytokinin or cytokinin-to-auxin induces root and shoot regeneration, respectively (Skoog and Miller 1957). Successful induction yields callus tissue that is soft, pale yellow, and highly friable, qualities essential for subsequent mechanical dispersion (Fig. 6.1) to establish cell suspension culture. These cultures are typically maintained in dark or dim light conditions of around 500 lux at a controlled temperature of around 25 °C to favor undifferentiated growth.

Callus tissue serves as the ideal starting material for establishing cell suspension culture, as its highly friable nature is suitable for mechanical cell dispersion. This system provides the cell lines necessary for the rapid, large-scale production of high-value phytochemicals, thus addressing the growing needs of the pharmaceutical and cosmeceutical industries. Therefore, before proceeding to cell suspension culture, researchers must first determine the presence and quantify the content of the target compound within the established callus. Numerous studies highlight the potential of callus in generating compounds with significant anti-aging activities:

Triterpenoids from *Centella asiatica*, such as asiaticoside and madecassoside, are known for their distinct anti-aging mechanism. They stimulate collagen (types I & III) synthesis and dermal fibroblast proliferation, thereby enhancing skin elasticity. These compounds also neutralize free radicals due to their potent antioxidant activity, boost hydration by promoting hyaluronic acid, and protect the extracellular matrix by inhibiting degrading enzymes (Lee et al. 2006; Ratz-Łyko et al. 2016; Sun et al. 2020; Bandopadhyay et al. 2023). Buranasudja et al. (2021) successfully induced *C. asiatica* callus and obtained callus extracts exhibiting antioxidant and anti-aging properties.

Ferulic acid, a polyphenol, possesses potent antioxidant capabilities, crucial for neutralizing free radicals and mitigating oxidative stress, a major contributor to aging (Purushothaman and Rizwanullah 2024). This compound is also known to fortify the stability and efficacy of synergistic agents (e.g., vitamins C and E), thus offering optimal defense against UV damage and photoaging (Lin et al. 2005). Hoa et al. (2025) successfully produced ferulic acid from *Angelica sinensis* callus suspension culture, and the callus extract exhibited high antioxidant activities. Similarly, the callus extract from *Woodfordia fruticosa* (WSC-3), an Indian endemic plant,

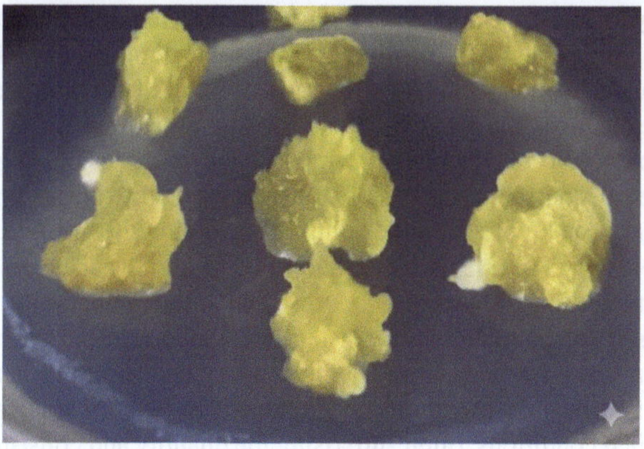

Fig. 6.1 Callus culture of *Polyscias fruticosa* (L.) Harms. (*Source*: Unpublished photograph of authors)

was confirmed by LC-MS analysis to be rich in polyphenols (including gallic acid and glucogallin) (Chalageri et al. 2019). WSC-3 demonstrated anti-aging effects by promoting collagen and elastin gene expression in human dermal fibroblasts (HDF) and effectively inhibiting H_2O_2-induced cellular senescence. Sagharyan et al. (2020) established an optimal procedure for callus induction and the production of phenolic compounds, including flavonoids, in the restricted medicinal plant *Nepeta binaloudensis*. Cho et al. (2020) evaluated the efficacy of *Leontopodium alpinum* callus extract (LACCE). LACCE showed strong in vitro antioxidant activity against UVB and suppressed inflammation and wrinkling while enhancing hydration. A subsequent clinical in vivo test confirmed that continuous application of LACCE significantly improved periorbital wrinkles, skin elasticity, dermal density, and skin thickness, suggesting its potential as a safe and effective candidate for anti-aging cosmetics. Amla (*Phyllanthus emblica*, syn. *Emblica officinalis*) is a rich source of phenolic compounds, vitamin C, and tannins (such as emblicanin). These provide potent antioxidant capacity that helps protect cells from aging factors (Variya et al. 2016; Chaikul et al. 2021; Gul et al. 2022). Studies on amla callus culture focus on the production of antioxidant compounds like phenolics and tannins. Antioxidant activity tests showed that digallic acid was the major compound accumulated in the 6-week-old in vitro callus, exhibiting the strongest antioxidant activity. Additionally, the compound putranjivain-A demonstrated significant antibacterial and antifungal activity. However, maintaining a high concentration of vitamin C under in vitro conditions remains a technical challenge that needs to be addressed to maximize the anti-aging efficacy of the cultured Amla cells (Singh and Sharma 2014).

6.3 Development and Characterization of Cell Suspension Cultures

Once established, friable callus aggregates are transferred from the solidified medium to a liquid medium which has typically similar composition (or sometimes it is necessary to adjust the amount of phytohormone), marking the transition to cell suspension culture (Fig. 6.2). Continuous agitation on an orbital shaker (around 80–130 rpm) is mandatory to ensure homogenous cell dispersion, maximize nutrient and gas exchange, and prevent cellular clumping. However, due to the natural tendency of plant cells to adhere together, it is not always possible to grow a suspension of single cells only (Bhatia 2015). Industrial scalability hinges on the choice of culture system. Batch culture is the simplest, involving growth in a fixed volume until the depletion of nutrients halts the cycle. More advanced systems, such as continuous culture (e.g., chemostat) or fed-batch culture (Bhatia 2015; Babu et al. 2025), are used for maintaining a constant metabolic state and achieving higher volumetric productivity (Fig. 6.3). The industrial application of plant cell suspension culture is driven by the ability to produce high-value secondary metabolites under controlled conditions, often surpassing the quantity and consistency achieved through traditional field cultivation. Table 6.1 presents a summary of various bioactive compounds—ranging from isothiocyanates and saponins to polyphenols and

Fig. 6.2 Cell suspension culture of *Angelica sinensis* (Oliv.) Diels in a shaking flask. (Source: Unpublished photograph of authors)

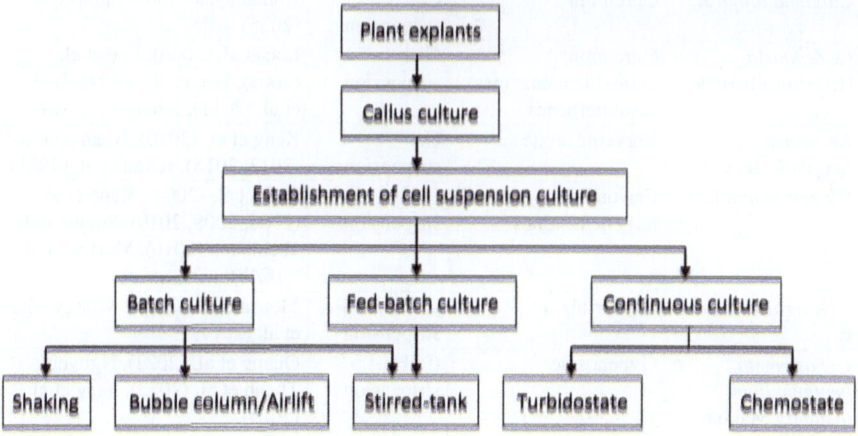

Fig. 6.3 Types of plant cell suspension culture. (Source: Unpublished diagram of authors)

anthocyanins—that have been successfully produced using this technique across several valuable medicinal plant species.

Characterization is formalized through cell growth kinetics, which involves monitoring biomass (usually dry weight) to plot the standard growth curve (lag, exponential, stationary, and decline/death phases) (Fig. 6.4) (Chisti and Moo-Young 2003). Reporting the specific growth rate (μ) and doubling time (td) is crucial for defining the optimal harvest window, which frequently coincides with the stationary phase when secondary metabolite biosynthesis peaks (Mehrara et al. 2007). Due to the fact that individual plant cells in a population of cultured cells invariably show cytogenetical and metabolic variations depending upon the stage of the growth cycle and culture conditions (Larkin and Scowcroft 1981; Maurya and Bahadur

Table 6.1 Some anti-aging bioactive compounds produced from callus and suspension cell cultures of valuable medicinal plants

Plant species	Product	Type of culture	References
Aesculus hippocastanum L.	Aescin	Callus	Profumo et al. (1992)
Aloe vera L.	Glucomannan, phenolics	Callus	Kim et al. (2012)
Bacopa monnieri (L.) Pennell	Bacoside	Cell suspension	Rahman et al. (2002), Kharde et al. (2018), Seth et al. (2020)
Brassica oleracea L.	Sulforaphane	Callus, cell suspension	Salama et al. (2021), Alwan et al. (2023)
Centella asiatica (L.) Urban	Centelloside (asiaticoside, madecassoside)	Callus, cell suspension	Nath and Buragohaine (2005), Loc and An (2010), Bonfill et al. (2011), Loc and Giang (2012), Loc and Nhat (2013), Loc et al. (2016), Loc et al. (2017)
Cinnamomum cassia (L.) J.Presl	Tanin	Callus, cell suspension	Yazaki and Okuda (1990)
Curcuma longa L.	Curcumin	Cell suspension	Kumanayaka and Bandupriya (2025)
C. Zedoaria (Christm.) Roscoe	Curcumin, antioxidant enzymes, sesquiterpenes	Cell suspension	Loc et al. (2006), Loc et al. (2008), Loc et al. (2009), Tuan et al. (2011), Lan et al. (2019)
Eurycoma longifolia Jack	Eurycomanone	Cell suspension	Keng et al. (2010), Nhan and Loc (2017, 2018), Kwan et al. (2021)
Ginkgo biloba L.	Flavonoid glycosides, terpene lactones	Cell suspension	Park et al. (2004), Kang et al. (2006, 2009, 2010), Sukito and Tachibana (2016), Murthy et al. (2024b)
Glycyrrhiza glabra L.	Glycyrrhizin	Callus, cell suspension	Mousa et al. (2007), Wongwicha et al. (2014)
Gynostemma pentaphyllum (Thunb.) Makino	Gypenoside	Cell suspension	Quang et al. (2022), Nguyen-Thanh et al. (2023), Sang et al. (2025)
Hibiscus sabdariffa L.	Anthocyanins and phenolics	Cell suspension	de Dios-Lopez et al. (2011)
Hibiscus syriacus L.	Extract	Cell suspension	Di Martino et al. (2017)
Impatiens balsamina L.	Phenolics and flavonoids	Callus	Kasem (2018)
Lithospermum erythrorhizon Siebold & Zucc.	Shikonin and derivatives	Cell suspension	Fujita et al. (1981a, b) Fujita et al. (1983), Tani et al. (1992, 1993), Gaisser and Heide (1996)
Mertensia maritima (L.) Gray	Rosmarinic acid, allantoin	Callus culture	Fedoreyev et al. (2012)
Nardostachys jatamansi (D.Don) DC.	Actinidine and glaziovine, polyphenol	Callus and cell suspension	Krishnan et al. (2024), Rawat et al. (2019)

(continued)

Table 6.1 (continued)

Plant species	Product	Type of culture	References
Ocimum basilicum L.	Phenolics, terpenoids, rosmarinic acid	Callus, cell suspension	Kintzios et al. (2003), Makri and Kintzios (2008), Açıkgöz (2020)
Olea europaea L.	Phenolics (verbascoside/acteoside), triterpene acids	Cell suspension	Saimaru et al. (2007), Saimaru and Orihara (2010), Orihara and Ebizuka (2010), Blando et al. (2024)
Panax ginseng C.A.Mey.	Ginsenosides (e.g., Rb1, Rg1, Rg3, Rd)	Callus, cell suspension	Lu et al. (2001), Bonfill et al. (2002), Lian et al. (2002), Hu et al. (2003), Thanh et al. (2004), Thanh et al. (2005), Thanh et al. (2014), Le et al. (2019)
Phoenix dactylifera L.	Phenolics, flavonoids	Cell suspension	Naik and Al-Khayri (2018), Al-Khayri and Naik (2020, 2022)
Phyllanthus emblica L.	Antioxidant	Callus	Singh and Sharma (2014)
Polyscias fruticosa (L.) Harms	Oleanolic acid, ladyginoside A	Cell suspension	Kim et al. (2021)
Rubus idaeus L.	Anthocyanins, phenolics, raspberry ketone	Cell suspension	Berejsza-Wysecki and Hrazdin (1994), Häkkinen et al. (2015), Tito et al. (2015)
Rosmarinus officinalis L.	Rosmarinic acid, carnosic acid	Cell suspension	Kuhlmann and Röhl (2006), Hassani (2011), Jordán et al. (2012)
Terminalia arjuna (Roxb.) Wight & Arn.	Luteolin, kaempferol, quercetin	Callus	Sharma (2014)
Tinospora cordifolia (Thunb.) Miers	Protoberberine	Cell suspension	Chintalwar et al. (2003), Srivastava and Chaturvedi (2025)
Vitis vinifera L.	Flavonoids, stilbenes, phenolics, anthocyanin, polyphenol	Cell suspension	Larronde et al. (1998), Krisa et al. (1999), Andi et al. (2019), Hassanen (2020), Sák et al. (2021)
Withania somnifera (L.) Dunal	Withanolides	Cell suspension	Ciddi (2006), Mirjalili et al. (2009), Sivanandhan et al. (2014), Ahmad et al. (2024)
Zingiber officinale Roscoe	Flavonoid, phenolics	Callus	Widyastuti et al. (2024), El-Nabarawy et al. (2015)

2022), the assessment of cellular totipotency through regeneration experiments (e.g., inducing somatic embryogenesis or organogenesis) confirms the genetic stability and viability of the cultured line (Bhatia 2015).

The principal bioactive compounds in *Curcuma* species are curcuminoids, with curcumin being the main polyphenol in turmeric (*C. longa*) and zedoary (*C. zedoaria*). They are renowned for their potent anti-inflammatory and antioxidant

Fig. 6.4 Plant cell growth profile in batch culture, including the four main phases of lag, exponential, stationary, and decline/death, and transitional phases of acceleration and deceleration. (Source: Unpublished diagram of authors)

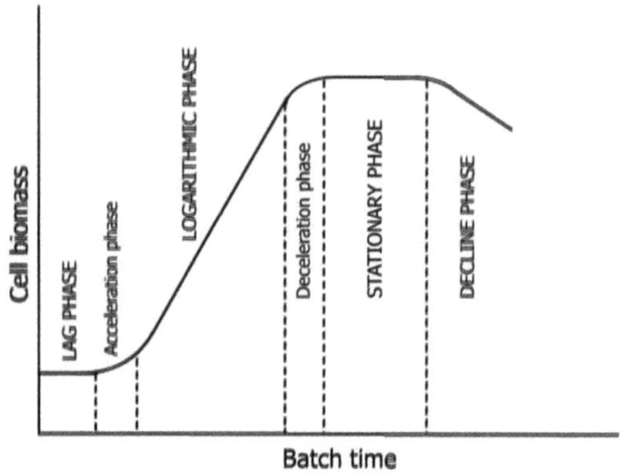

properties. Regarding anti-aging mechanisms, curcumin exhibits broad effects, including the modulation of longevity-related cellular signaling pathways such as AMPK and AKT/mTOR (Bielak-Zmijewska et al. 2019; Papadopoli et al. 2019; Yang et al. 2022). Some studies have successfully established cell lines of *Curcuma* species (*C. longa, C. aromatica, and C. zedoaria*) capable of producing curcuminoids and antioxidant enzymes (Loc et al. 2006, 2008; Tuan et al. 2011; Murthy et al. 2024a; Kumanayaka and Bandupriya 2025). Furthermore, the application of various elicitors, such as methyl jasmonate, salicylic acid or yeast extract, has been shown to significantly enhance curcumin synthesis in cell suspension cultures, resulting in yields many times higher than those obtained from in planta rhizome (Lan et al. 2019; Murthy et al. 2024a).

Centella asiatica is a rich source of triterpenoid saponins, including asiaticoside and madecassoside (James and Dubery 2009; Bylka et al. 2014; Loc et al. 2017; Sun et al. 2020). These compounds are key to its anti-aging properties through their capacity to improve collagen synthesis, promote wound healing, and offer skin protection (Maquart et al. 1990; Witkowska et al. 2024). In vitro cell suspension cultures of *C. asiatica* have shown superior efficacy, enabling rapid biomass production with adjustable or higher triterpenoid content (such as asiaticoside and madecassoside) compared to field-grown plants (Nath and Buragohaine 2005; Loc and An 2010; Bonfill et al. 2011; Loc and Giang 2012; Loc and Nhat 2013; Loc et al. 2016). Consequently, in vitro cell culture represents a sustainable method for producing these crucial anti-aging bioactive compounds for the cosmetic and pharmaceutical industries. The main bioactive compounds of ashwagandha plant (*Withania somnifera* L. Dunal.) are withanolides, including withaferin A, which act as critical adaptogens that protect cells from stress and inflammation, major causes of skin damage and aging (Davis and Kuttan 2002; Singh et al. 2011; Ambiye et al. 2013; Bungau et al. 2021; Chaudry et al. 2025). By mitigating oxidative stress and promoting cellular repair mechanisms, they indirectly support collagen synthesis. Furthermore, they exhibit cytoprotective (antioxidant) effects and specific effects on the skin and extracellular matrix (ECM), including collagen (Sikandan et al. 2018;

Narra et al. 2023; Balkrishna et al. 2023). Cell suspension culture has proven to be an effective method for generating biomass with high concentrations of withanolides, particularly withaferin A, a compound linked to skin structure support (Ciddi 2006; Mirjalili et al. 2009; Ahmad et al. 2024). Cell suspension culture can be optimized by combining elicitors and precursors to achieve significantly higher biosynthesis of withanolides (including withaferin A) compared to field-grown plants (Sivanandhan et al. 2014). This in vitro system ensures a stable supply of high-quality anti-aging bioactives, overcoming the limitations of conventional farming methods (Verpoorte et al. 2002; Ramachandra Rao and Ravishankar 2002; Smetanska 2008).

Ginseng (*Panax ginseng*), with its key active compounds being the ginsenoside group (saponins/triterpenoid saponins), is renowned for its potent anti-aging properties by enhancing collagen synthesis and protecting the skin against damage from ultraviolet (UV) radiation (Lim et al. 2015; Kang et al. 2024). Ginseng cell suspension culture has become a popular industrial technique for producing specific ginsenosides, and optimizing the bioreactor is a crucial factor in achieving commercial concentrations, ensuring the stable quantity and quality of this antioxidant bioactive (Furuya et al. 1983; Thanh et al. 2014; Xu et al. 2023; Xue et al. 2025). Ginsenosides combat aging through three main mechanisms: (1) Antioxidant action (neutralizing reactive oxygen species (ROS), protecting DNA, lipids, and proteins); (2) Anti-inflammatory response (reducing chronic inflammatory reactions that cause tissue degeneration and premature aging); and (3) Structural protection (stimulating collagen/elastin production while stabilizing telomeres to prolong the healthy lifespan of skin cells) (Hwang et al. 2017; Cong et al. 2023). Specifically, ginsenoside Rg1 (PPT group) is highly important for anti-aging skin benefits and cell protection (Wang et al. 2023); ginsenosides Rb1 & Rd (PPD group) focus on cell and nervous system protection, supporting recovery; and ginsenoside Rg3 and compound K (rare metabolites), often enhanced through processing or biotransformation culture from Rb1, are highly valued for their ability to fight inflammation, protect DNA, and prevent cellular aging (cellular senescence) (Liu et al. 2022).

Broccoli (*Brassica oleracea*) supplies the bioactive sulforaphane, a potent isothiocyanate known for its ability to activate cellular detoxification pathways and exhibit pronounced anti-aging activity (Yagishita et al. 2019; Petkovic et al. 2021). Callus or cell suspension culture allows for tight control over the production of glucosinolate precursors, which are subsequently hydrolyzed into sulforaphane (Salama et al. 2021; Alwan et al. 2023). This cell-derived product is optimized for antioxidant activity, proving valuable for the anti-aging cosmetic and functional food industries. The principal anti-aging bioactives of ginkgo (*Ginkgo biloba*) are flavonoid glycosides and terpene lactones (ginkgolides, bilobalide), which confer neuroprotective antioxidant properties and improve blood circulation (Chen et al. 2013; Zuo et al. 2017). Although ginkgo cell culture has been commercialized in some countries, production yield remains a challenge (Park et al. 2004; Murthy et al. 2024b). Therefore, enhancement strategies like precursor feeding and the use of elicitors are research priorities aimed at improving the anti-aging and antioxidant activity of cultured cells (Kang et al. 2006, 2009, 2010; Sabater-Jara et al. 2013; Sukito and Tachibana 2016). Licorice (*Glycyrrhiza glabra*) contains the bioactive glycyrrhizin, which exhibits anti-inflammatory, antioxidant, and skin-restoring and soothing properties, playing a vital role in anti-aging formulations (Schröfelbauer et al. 2009; Darvishi et al. 2022). Furthermore, cell suspension culture has been successfully applied to produce glycyrrhizin, providing a controlled supply of this anti-aging bioactive with consistent quality (Wu et al. 1974; Henry et al. 1991; Mousa et al. 2007). Oleanolic acid, a triterpenoid derived from *Polyscias fruticosa*, exhibits significant anti-aging activity in the skin. This activity is demonstrated by its ability to protect epidermal cells and fibroblasts from environmental aggressors like fine particulate matter (PM10), while simultaneously inhibiting aging-related enzymes such as Matrix Metalloproteinase-1 (MMP-1) (Zhang et al. 2018; Kim et al. 2021). However, the number of published studies focusing on the suspension cell culture of *P. fruticosa* specifically for oleanolic acid production remains limited, with primary reports attributed to Kim et al. (2021) and Titova et al. (2023). Eurycomanone, a major quassinoid isolated from Tongkat Ali (*Eurycoma longifolia*), possesses intrinsic antioxidant properties, but its most significant anti-aging mechanism stems from hormonal modulation. Eurycomanone acts primarily by inhibiting the aromatase enzyme, which is responsible for converting testosterone into estrogen. This inhibition helps to maintain or enhance natural testosterone levels, a key mechanism linked to improving age-related health, vitality, and physical function in men (Low et al. 2013a, 2013b; Leisegang et al. 2022). Furthermore, some studies have been conducted to utilize *E. longifolia* suspension cell culture as a biotechnological method for the production of eurycomanone (Nhan and Loc 2017, 2018; Kwan et al. 2021; Keng et al. 2010). Rosemary (*Rosmarinus officinalis*) is rich in rosmarinic acid (RA) and carnosic acid (CA), both recognized as extremely potent antioxidants. Cell suspension culture has demonstrated the capacity to enhance RA production at high concentrations through the use of elicitors (e.g., chitosan), thereby optimizing the antioxidant potential of the resulting cell biomass for cosmetic and pharmaceutical applications (Kuhlmann and Röhl 2006; Hassani 2011; Jordán et al. 2012). These studies confirm that the suspension cell culture of *R. officinalis* is an effective biotechnological method. Specifically, the application of elicitors, such as chitosan, is key to significantly boosting the accumulation of these potent antioxidants. This strategy effectively converts the cell biomass into a highly promising source of raw material for industrial applications.

6.4 Elicitation Strategies for Enhanced Metabolite Production

Elicitation is an effective strategy to significantly enhance the yield of secondary metabolites by mimicking stress conditions or pathogen attacks (Ramirez-Estrada et al. 2016; Selwal et al. 2025).

6.4.1 Mechanisms of Elicitor Action

This section explores the molecular mechanisms by which biotic and abiotic elicitors influence metabolic pathways.

The mechanism involves elicitors activating membrane receptors, leading to the release of secondary messengers (e.g., Ca^{2+}, Reactive Oxygen Species—ROS), ultimately triggering the gene expression of secondary metabolite synthesis (e.g., enzymes in the phenylpropanoid pathway for flavonoids/phenolics or curcuminoid, or the isoprenoid pathway—including the MEP/MVA pathway—for terpenoids) (Loc et al. 2016; Ramirez-Estrada et al. 2016; Lan et al. 2018; Bajwa et al. 2021).

Biotic elicitors (e.g., chitin, yeast extract—YE, methyl jasmonate—MeJA) often mimic pathogenic factors, while abiotic elicitors are physical or chemical factors (e.g., UV light, heavy metal salts like Ag^+, growth regulators) (Largia et al. 2022; Naik and Al–Khayri 2016).

6.4.2 Types of Elicitors and Their Application in Cell Culture Systems

This section classifies elicitors and reviews the diverse strategies used to maximize the production of target compounds (Table 6.2). Elicitors are broadly categorized into biotic and abiotic factors, each employing distinct signaling pathways to enhance metabolite yield.

6.4.3 Biotic Elicitors

These elicitors mimic components of pathogen attack or stress signaling molecules.

Microbial Derivatives Components derived from microorganisms, such as chitosan (from fungal cell walls) and YE, are frequently used to trigger the phenylpropanoid pathway. For example, YE has been shown to successfully induce the accumulation of defense-related phenolics and flavonoids in various cultures, often by activating receptors that perceive Pathogen-Associated Molecular Patterns (PAMPs) (Bavi et al. 2022; Zaman et al. 2022; Elbouzidi et al. 2025; Roy Chowdhury et al. 2025).

Signaling Molecules MeJA, an ester of the plant hormone jasmonic acid, is a powerful signaling molecule (Wasternack and Hause 2013). MeJA-mediated signaling is crucial for defense gene activation and is highly effective in promoting the biosynthesis of complex compounds like shikonin (naphthoquinone) and gypenosides (saponins) in respective cell cultures (Wang et al. 2014; Quang et al. 2022). Salicylic acid (SA), another key signaling molecule, often promotes the accumulation of defense-related phenolics (Dong et al. 2010; Yin et al. 2024).

Table 6.2 Classification of common elicitors used in plant cell culture

Elicitor class	Origin/nature	Common elicitors	Mechanism of action
Biotic elicitors	Microbial derivatives (components from microbes or pathogens)	Chitosan YE Chitin Peptidoglycans Glucans Cell wall fragments	Mimic pathogen-associated molecular patterns (PAMPs). Activate membrane receptors → release of secondary messengers (Ca^{2+}, ROS) → trigger gene expression for secondary metabolite synthesis
	Signaling molecules (endogenous plant signaling molecules)	MeJA SA JA	MeJA/JA: Activate the Jasmonate-mediated signaling pathway. SA: Activate signaling pathways related to systemic acquired resistance (SAR).
Abiotic elicitors	Heavy metal salts (inorganic chemical agents that induce stress)	$AgNO_3$ $CdCl_2$ $CuSO_4$ $HgCl_2$ $ZnSO_4$	Induce controlled, mild oxidative stress → activate defense response and upregulate related enzymes.
	Physical factors (non-biological environmental factors)	UV light Osmotic stress Ultrasound Cold shock	Alter cell membrane permeability and enzyme activity → trigger general stress response mechanism → shift metabolism.
	Growth regulators (exogenous)	JA SA Brassinosteroids ABA Polyamines	Regulate endogenous plant development and stress response pathways, activating specific metabolic routes.
	Other inorganic chemicals (non-heavy metal)	H_2O_2 Ozone NaCl (as salt stress)	H_2O_2 acts as a signaling molecule → enhances controlled oxidative stress. NaCl causes osmotic and ionic stress.
Integrated strategies	Combined/sequential use (combination of different elicitor types)	$AgNO_3$ followed by MeJA, YE and MeJA combined	Exploit synergistic effects by using different entry points and activation mechanisms → finely tune metabolic pathways.

6.4.4 Abiotic Elicitors

These are non-biological factors that impose controlled stress on the cells, activating general stress response mechanisms.

Heavy Metal Salts Salts such as silver nitrate ($AgNO_3$), cadmium chloride ($CdCl_2$), or copper sulfate ($CuSO_4$) are utilized to induce mild oxidative stress. $AgNO_3$, for instance, has been effectively used to enhance the yield of phenolic compounds in date palm cell cultures by upregulating specific defense enzymes (Al-Khayri and Naik 2020).

Physical Factors Controlled exposure to UV light or the application of osmotic stress (e.g., high sugar concentration) can alter cell permeability and enzyme activity, shifting metabolism towards secondary product formation (Guru et al. 2022).

Hormonal Elicitors Some phytohormones constitute a widely utilized group of factors as elicitors for enhancing the content of bioactive compounds in a broad array of in vitro cultured medicinal plants. Commonly studied plant hormones employed in elicitation strategies include jasmonates, salicylic acid, brassinosteroids, abscisic acid (ABA), and gibberellic acid (Jamwal et al. 2018).

Integrated and Sequential Elicitation Strategies Optimal strategies rarely rely on a single factor. Combined or sequential use of elicitors (e.g., applying $AgNO_3$ followed by MeJA) is often employed, as synergistic effects can lead to significantly higher yields compared to single elicitor application. This multi-factor approach allows researchers to finely tune the complex metabolic pathways, exploiting the different entry points and activation mechanisms of various elicitors to achieve maximum productivity (Largia et al. 2022).

6.5 Bioreactor Systems for Biomass Production

To produce biomass and metabolites at an industrial scale, the use of advanced bioreactor systems is indispensable.

6.5.1 Principles and Utilization of Temporary Immersion Systems

Temporary immersion systems (TISs) are sophisticated, semi- or fully-automated bioreactors specifically engineered for the mass propagation of plant tissues, embryos, organs, and even cell cultures, as well as for the optimization of secondary metabolite production. The operating principle relies on a precisely controlled cycle alternating between two phases: the immersion phase (briefly submerging the material in liquid nutrient medium for uniform nutrient uptake) and the ventilation phase (draining the medium to supply oxygen (O2) and facilitate the removal of volatile, potentially toxic gases like ethylene). This regulated cycling offers significant advantages over conventional culture, including enhanced growth rates, greater biomass uniformity, minimization of both shear stress and the risk of hyperhydricity

Table 6.3 Temporary immersion systems (TISs) and bubble column bioreactor

Feature	Temporary immersion systems (TISs)	Bubble column bioreactor
Configuration	Two-tank/flask type or single-tank type connected to a gas pumping/vacuum system and a sterile filter unit. Culture material is placed on a support rack.	Typically a vertical column or cylinder with a large height-to-diameter ratio, containing no internal mechanical stirring parts. Equipped with a gas sparger at the bottom.
Operating principle	Intermittent/recirculating operation. Culture material is temporarily submerged in liquid medium (usually for a few minutes) using gas pressure or a vacuum pump, and then the medium is drained to expose the material to air.	Continuous operation. Gas is continuously sparged from the bottom to create bubbles, which mix the medium and supply O_2. Mixing intensity depends on bubble size and velocity.
Mixing/mass transfer mechanism	No mechanical agitation. Nutrient/gas mass transfer relies on the cyclic liquid immersion and gas exposure, optimizing O_2 and CO_2 exchange at the culture material surface.	Pneumatic agitation. Mixing and mass transfer are driven by gas bubble circulation.
Primary applications	Culture of organized structures: Mass propagation of tissues, shoot clusters, plantlets (whole plants), hairy roots, and somatic embryos.	Culture of unorganized structures: Cell suspension cultures and hairy root cultures.
Advantages	Reduced vitrification (glassy appearance); optimized mass transfer; reduced physical damage to tissues/organs.	Simple and easy to operate; lower cost; less shear stress compared to stirred-tank bioreactors (suitable for sensitive cell suspensions).
Disadvantages	More complex and costly to set up and operate; not ideal for cell cultures requiring strong mixing.	O_2 transfer efficiency can be lower; risk of cell settling at the bottom of the column.

(tissue waterlogging), and lower labor costs. The enhanced control over nutrient and gas exchange allows for the precise timing of elicitor and precursor application to specialized organ cultures (e.g., hairy roots), thereby maximizing target metabolite yields (Table 6.3). Typical TIS models commonly used in plant biotechnology include RITA (recipient for automated temporary immersion) and TIB (temporary immersion bioreactor) (Etienne and Berthouly 2002; Ziv 2005; Mirzabe et al. 2022; De Carlo et al. 2021).

6.5.2 Bubble Bioreactors for Suspension Cell and Biomass

Bubble column reactors (BCRs) are pneumatically mixed reactors considered an ideal, cost-effective, and scalable choice for suspension cell culture, hairy root culture, and large-scale biomass production in plant biotechnology, especially where a low-shear environment is critical. A BCR is a cylindrical container typically featuring an aspect ratio of 2:1 or 3:1, occasionally reaching 6:1 for specific applications. The core principle of operation is elegantly simple: compressed air, O_2, or an air-gas mixture is continuously or intermittently sparged through a porous element (sparger) at the bottom of the culture vessel, generating bubbles that serve a dual function (Paek et al. 2005).

Firstly, these bubbles supply essential oxygen (O_2) for cell respiration. Secondly, the rising bubbles create a continuous, gentle mixing action within the liquid medium, effectively ensuring the uniform distribution of cells, nutrients, and dissolved gases throughout the vessel. This uniform environment minimizes concentration gradients and promotes consistent growth. The air-mixing bioreactor—of which the bubble-type is a prime example—performs both aeration and mixing of the cell suspension using compressed air. These bioreactors are typically shaped as a vertical tank equipped with a gas distribution device or sparger installed at the bottom. This design is relatively simple, features no rubbing or moving parts, and is thus highly functional and reliable. This type of bioreactor was used in the first experiments on scaling up the cultivation of plant cell suspensions and has been effectively exploited for decades since (Table 6.3). Bubble-type bioreactors were employed for the experimental cultivation of various plant cell lines, such as *Taxus cuspidata* and *Eurycoma longifolia* cells. However, several studies demonstrated that bubble-type bioreactors have relatively low mass transfer characteristics, and they are, therefore, not recommended for cell suspensions with a high viscosity or high final cell biomass concentration (Titova et al. 2024).

Key operational parameters that require careful optimization include the aeration rate (volume of gas per volume of medium per minute, or *vvm*) and the oxygen mass transfer coefficient k_La, which directly dictates the efficiency of oxygen delivery to the cells. Applications are broad, notably including the production of high-value suspended cell biomass from medicinal plants (e.g., anti-aging compounds), where the goal is maximizing cellular yield. The integration with elicitation strategy is highly effective in BCRs; by allowing precise control over the culture environment, researchers can introduce elicitors (biotic or abiotic stress compounds) and precursors at optimal growth phases to induce or enhance the biosynthetic pathways, thereby achieving both a high biomass yield and maximal secondary metabolite content simultaneously. BCRs are preferred over more mechanically complex systems due to their simple design, low maintenance, and minimal risk of mechanical shear stress to fragile plant cells (Dong et al. 2013; Georgiev and Weber 2014).

6.6 Conclusion and Future Perspectives

This chapter has elucidated the potential of various plant cell culture strategies to enhance the production of anti-aging metabolites and biomass. The robust combination of suspension culture techniques and elicitation optimization with the application of advanced bioreactor systems—such as Temporary Immersion Systems (TIS) and bubble column reactors—represents a clear pathway toward the industrial production of high-quality, safe, and uniform plant-derived biopharmaceuticals. Future research should prioritize gene editing to optimize critical biosynthetic pathways and focus on developing next-generation bioreactor systems equipped with sophisticated online monitoring and control capabilities for complex culture parameters.

Competing Interests The authors have no relevant financial or non-financial interests to disclose.

References

Açıkgöz MA (2020) Establishment of cell suspension cultures of *Ocimum basilicum* L. and enhanced production of pharmaceutical active ingredients. Ind Crops Prod 148:112278

Ahmad Z, Shareen, Ganie IB, Firdaus F, Ramakrishnan M, Shahzad A, Ding Y (2024) Enhancing withanolide production in the *Withania* species: advances in *in vitro* culture and synthetic biology approaches. Plants 13:2171

Al-Khayri JM, Naik PM (2020) Elicitor-induced production of biomass and pharmaceutical phenolic compounds in cell suspension culture of date palm (*Phoenix dactylifera* L.). Molecules 25:4669

Al-Khayri JM, Naik PM (2022) Influence of 2iP and 2,4-D concentrations on accumulation of biomass, phenolics, flavonoids and radical scavenging activity in date palm (*Phoenix dactylifera* L.) cell suspension culture. Horticulturae 8:683

Alwan AM, Ibrahem MM, Nadir DS (2023) The role of iron oxide nanoparticles and the cultivation of cell suspension cultures of broccoli plant *Brassica oleracea* and detection of sulforaphane compound *ex vivo*. AIP Conf Proc 2839:070004

Ambiye VR, Langade D, Dongre S, Aptikar P, Kulkarni M, Dongre A (2013) Clinical evaluation of the spermatogenic activity of the root extract of ashwagandha (*Withania somnifera*) in oligospermic males: A pilot study. Evid Based Complement Alternat Med 2013:571420

Andi SA, Gholami M, Ford CM, Maskani F (2019) The effect of light, phenylalanine and methyl jasmonate, alone or in combination, on growth and secondary metabolism in cell suspension cultures of *Vitis vinifera*. J Photochem Photobiol B 199:111625

Babu R, Veeramani M, Aadinath W, Muthuvijayan V, Singh S, Srivastava S (2025) Model-based fed-batch cultivation of *Viola odorata* plant cells exhibiting antimalarial and anticancer activity. Front Bioeng Biotechnol 13:1528570

Bajwa MN, Bibi A, Idrees MZ (2021) Elicitation, a mechanistic approach to change the metabolic pathway of plants to produce pharmacological important compounds in *in-vitro* cell cultures. Glob J Eng Sci 8:000678

Balkrishna A, Tomar M, Bhattacharya K, Varshney A (2023) Withania somnifera-derived carbon dots protect human epidermal cells against UVB-induced cell death and support growth factor-mediated wound healing. Nanoscale Adv 5:1331–1344

Bandopadhyay S, Mandal S, Ghorai M, Jha NK, Kumar M, Radha, Ghosh A, Proćków J, Pérez de la Lastra JM, Dey A (2023) Therapeutic properties and pharmacological activities of asiaticoside and madecassoside: A review. J Cell Mol Med 27:593–608

Bavi K, Khavari-Nejad RA, Najafi F, Ghanati F (2022) Phenolics and terpenoids change in response to yeast extract and chitosan elicitation in *Zataria multiflora* cell suspension culture. 3 Biotech 12:163

Berejsza-Wysecki W, Hrazdin G (1994) Establishment of callus and cell suspension cultures of raspberry (*Rubus idaeus* cv. Royalty). Plant Cell Tissue Organ Cult 37:213–216

Bhatia S (2015) Plant tissue culture. In: Bhatia S, Sharma K, Dahiya R, Bera T (eds) Modern applications of plant biotechnology in pharmaceutical sciences. Academic Press, pp 31–107

Bielak-Zmijewska A, Grabowska W, Ciolko A, Bojko A, Mosieniak G, Bijoch Ł, Sikora E (2019) The role of curcumin in the modulation of ageing. Int J Mol Sci 20:1239

Blando F, Mita G, Di Sansebastiano GP, Nicoletti I, Donati E (2024) Highly efficient verbascoside production from olive (*Olea europea* L. var. Cellina di Nardò) *in vitro* cell cultures. J Agric Food Chem 72:1162–1169

Bonfill M, Cusidó RM, Palazón J, Piñol MT, Morales C (2002) Influence of auxins on organogenesis and ginsenoside production in *Panax ginseng* calluses. Plant Cell Tissue Organ Cult 68:73–78

Bonfill M, Mangas S, Moyano E, Cusido RM, Palazón J (2011) Production of centellosides and phytosterols in cell suspension cultures of *Centella asiatica*. Plant Cell Tissue Organ Cult 104:61–67

Bungau S, Vesa CM, Abid A, Behl T, Tit DM, Purza AL, Pasca B, Todan LM, Endres L (2021) Withaferin A—a promising phytochemical compound with multiple results in dermatological diseases. Molecules 26:2407

Buranasudja V, Rani D, Malla A, Kobtrakul K, Vimolmangkang S (2021) Insights into antioxidant activities and anti-skin-aging potential of callus extract from *Centella asiatica* (L.). Sci Rep 11:13459

Bylka W, Znajdek-Awiżeń P, Studzińska-Sroka E, Dańczak-Pazdrowska A, Brzezińska M (2014) *Centella asiatica* in dermatology: An overview. Phytother Res 28:1117–1124

Câmara JS, Perestrelo R, Ferreira R, Berenguer CV, Pereira JAM, Castilho PC (2024) Plant—derived terpenoids: A plethora of bioactive compounds with several health functions and industrial applications—a comprehensive overview. Molecules 29:3861

Chaikul P, Kanlayavattanakul M, Somkumnerd J, Lourith N (2021) *Phyllanthus emblica* L. (Amla) branch: A safe and effective ingredient against skin aging. J Tradit Complement Med 11:390–399

Chalageri G, Dhananjaya SP, Raghavendra P, Sharath Kumar LM, Babu UV, Varma SR (2019) Substituting plant vegetative parts with callus cell extracts: Case study with *Woodfordia fruticosa* Kurz.—A potent ingredient in skin care formulations. S Afr J Bot 123:351–360

Chaudry A, Han H, Fischer DL, Rozenberg SS (2025) The therapeutic potential of *Withania somnifera* (Ashwagandha) in non-malignant dermatological conditions. SKIN J Cutaneous Med 9:2417–2425

Chemat F, Abert-Vian M, Fabiano-Tixier AS, Strube J, Uhlenbrock L, Gunjevic V, Cravotto G (2019) Green extraction of natural products. Origins, current status, and future challenges. TrAC Trends Anal Chem 118:248–263

Chen F, Li L, Xu F, Sun Y, Du F, Ma X, Zhong C, Li X, Wang F, Zhang N, Li C (2013) Systemic and cerebral exposure to and pharmacokinetics of flavonols and terpene lactones after dosing standardized *Ginkgo biloba* leaf extracts to rats via different routes of administration. Br J Pharmacol 170:440–457

Chintalwar GJ, Gupta S, Gopalakrishnan R, Bapat VA (2003) Protoberberine alkaloids from callus and cell suspension cultures of *Tinospora cordifolia*. Pharm Biol 41:81–86

Chisti Y, Moo-Young M (2003) Bioreactors. In: Meyers RA (ed) Encyclopedia of physical science and technology, 3rd edn. Academic, pp 247–271

Cho WK, Kim HI, Kim SY, Seo HH, Song J, Kim J, Shin DS, Jo Y, Choi H, Lee JH, Moh SH (2020) Anti-aging effects of *Leontopodium alpinum* (edelweiss) callus culture extract through transcriptome profiling. Genes (Basel) 11:230

Ciddi V (2006) Withaferin A from cell cultures of *Withania somnifera*. Indian J Pharm Sci 68:490–492

Cong L, Ma J, Zhang Y, Zhou Y, Cong X, Hao M (2023) Effect of anti-skin disorders of ginsenosides: A systematic review. J Ginseng Res 47:605–614

Darvishi M, Shamsaie Mehrgan M, Khajehrahimi AE (2022) Effect of licorice (*Glycyrrhiza glabra*) extract as an immunostimulant on serum and skin mucus immune parameters, transcriptomic responses of immune-related gene, and disease resistance against *Yersinia ruckeri* in rainbow trout (*Oncorhynchus mykiss*). Front Vet Sci 9:811684

Davis L, Kuttan G (2002) Effect of *Withania somnifera* on cell mediated immune responses in mice. J Exp Clin Cancer Res 21:585–590
De Carlo A, Tarraf W, Lambardi M, Benelli C (2021) Temporary immersion system for production of biomass and bioactive compounds from medicinal plants. Agronomy 11:2414
de Dios-Lopez A, Montalvo-Gonzalez E, Andrade-Gonzalez I, Gomez-Leyva JF (2011) Induction of anthocyanins and phenolic compounds in cell cultures of Roselle (*Hibiscus sabdariffa* L.) *in vitro*. Rev Chapingo Ser Hortic 17:77–87
Di Martino O, Tito A, De Lucia A, Cimmino A, Cicotti F, Apone F, Colucci G, Calabrò V (2017) *Hibiscus syriacus* extract from an established cell culture stimulates skin wound healing. Biomed Res Int 2017:7932019
Dong J, Wan G, Liang Z (2010) Accumulation of salicylic acid-induced phenolic compounds and raised activities of secondary metabolic and antioxidative enzymes in *Salvia miltiorrhiza* cell culture. J Biotechnol 148:99–104
Dong Y, Gao WY, Man S, Zuo B, Wang J, Huang L, Xiao P (2013) Effect of bioreactor angle and aeration rate on growth and hydromechanics parameters in bioreactor culture of ginseng suspension cells. Acta Physiol Plant 35:1497–1501
Elbouzidi A, Taibi M, Baraich A, Haddou M, Mothana RA, Alsufyani SA, Darwish HW, Molinié R, Fontaine J-X, Fliniaux O, Mesnard F, Addi M (2025) Elicitor-driven enhancement of phenolic compounds in geranium callus cultures: phytochemical profiling via LC-MS/MS and biological activities. Front Chem 13:1537877
El-Nabarawy MA, El-Kafafi SH, Hamza MA, Omar MA (2015) The effect of some factors on stimulating the growth and production of active substances in *Zingiber officinale* callus cultures. Ann Agric Sci 60:1–9
Etienne H, Berthouly M (2002) Temporary immersion systems in plant micropropagation. Plant Cell Tissue Organ Cult 69:215–231
Fedoreyev SA, Inyushkina YV, Bulgakov VP, Veselova MV, Tchernoded GK, Gerasimenko AV, Zhuravlev YN (2012) Production of allantoin, rabdosiin and rosmarinic acid in callus cultures of the seacoastal plant *Mertensia maritima* (Boraginaceae). Plant Cell Tissue Organ Cult 110:183–188
Fujita Y, Hara Y, Ogino T, Suga C (1981a) Production of shikonin derivatives by cell suspension cultures of *Lithospermum etythrorhizon*. I. Effects of nitrogen sources on the production of shikonin derivatives. Plant Cell Rep 1:59–60
Fujita Y, Hara Y, Suga C, Morimoto T (1981b) Production of shikonin derivatives by cell suspension cultures of *Lithospermum erythrorhizon*. II. A new medium for the production of shikonin derivatives. Plant Cell Rep 1(2):61–63
Fujita Y, Maeda Y, Suga C, Morimoto T (1983) Production of shikonin derivatives by cell suspension cultures of *Lithospermum erythrorhizon*. III. Comparison of shikonin derivatives of cultured cells and Ko-shikon. Plant Cell Rep 2(4):192–193
Furuya T, Yoshikawa T, Orihara Y, Oda H (1983) Saponin production in cell suspension cultures of *Panax ginseng*. Planta Med 48:83–87
Gaisser S, Heide L (1996) Inhibition and regulation of shikonin biosynthesis in suspension cultures of *Lithospermum*. Phytochem 41:1065–1072
Georgiev MI, Weber J (2014) Bioreactors for plant cells: hardware configuration and internal environment optimization as tools for wider commercialization. Biotechnol Lett 36:1359–1367
Gul M, Liu ZW, Iahtisham-Ul-Haq, Rabail R, Faheem F, Walayat N, Nawaz A, Shabbir MA, Munekata PES, Lorenzo JM, Aadil RM (2022) Functional and nutraceutical significance of amla (*Phyllanthus emblica* L.): a review. Antioxidants (Basel) 11:816
Guru A, Dwivedi P, Kaur P, Pandey DK (2022) Exploring the role of elicitors in enhancing medicinal values of plants under *in vitro* condition. S Afr J Bot 149:1029–1043
Häkkinen ST, Seppänen-Laakso T, Oksman-Caldentey KM, Rischer H (2015) Bioconversion to raspberry ketone is achieved by several non-related plant cell cultures. Front Plant Sci 6:1035
Hassanen SA (2020) Production of anthocyanin in cell suspension culture of grape (*Vitis vinifera* L.). IOSR J Biotech Biochem 6:23–31

Hassani HH (2011) Cytotoxicty and DNA fragmentation in cancer cell lines, Hep-2 and Amn-3, induced by a novel maroon pigment from mutant strain of *Rugamonas rubra* Rr62. J Al-Nahrain Univ 14:110–114

Henry M, Edy AM, Desmarest P, Manoir JD (1991) *Glycyrrhiza glabra* L. (Licorice): cell culture, regeneration, and the production of glycyrrhizin. In: Bajaj YPS (ed) Medicinal and aromatic plants III. Biotechnology in agriculture and forestry, vol 15. Springer, Berlin/Heidelberg

Hoa PTB, Phuc NH, Phu HN, Nam NVH, Thu LNM, Thanh LTH, Loc NH (2025) *In vitro* plant regeneration, callus suspension culture, and production of bioactive metabolites in dong quai (*Angelica sinensis*). S Afr J Bot 183:193–201

Hu X, Neill S, Cai W, Tang Z (2003) Hydrogen peroxide and jasmonic acid mediate oligogalacturonic acid-induced saponin accumulation in suspension-cultured cells of *Panax ginseng*. Physiol Plant 118:414–421

Hwang E, Park SY, Yin CS, Kim HT, Kim YM, Yi TH (2017) Antiaging effects of the mixture of *Panax ginseng* and *Crataegus pinnatifida* in human dermal fibroblasts and healthy human skin. J Ginseng Res 41:69–77

James JT, Dubery IA (2009) Pentacyclic triterpenoids from the medicinal herb, *Centella asiatica* (L.) Urban. Molecules 14:3922–3941

Jamwal K, Bhattacharya S, Puri S (2018) Plant growth regulator mediated consequences of secondary metabolites in medicinal plants. J Appl Res Med Aromat Plants 9:26–38

Jordán MJ, Lax V, Rota MC, Lorán S, Sotomayor JA (2012) Relevance of carnosic acid, carnosol, and rosmarinic acid concentrations in the *in vitro* antioxidant and antimicrobial activities of *Rosmarinus officinalis* (L.) methanolic extracts. J Agric Food Chem 60:9603–9608

Kang SM, Min JY, Kim YD, Karigar CS, Kim SW, Goo GH, Choi MS (2006) Effects of methyl jasmonate and salicylic acid on the production of bilobalide and ginkgolides in cell cultures of *Ginkgo biloba*. *In Vitro* Cell Dev Biol Plant 42:44–49

Kang SM, Min JY, Kim YD, Karigar CS, Kim SW, Goo GH, Choi MS (2009) Effect of biotic elicitors on the accumulation of bilobalide and ginkgolides in *Ginkgo biloba* cell cultures. J Biotechnol 139:84–88

Kang SM, Min JY, Park DJ, Jeong MJ, Song HJ, Heo CM, Choi MS (2010) Potassium chloride elicits enhancement of bilobalide and ginkgolides production by *Ginkgo biloba* cell cultures. Forest Sci Technol 6:49–54

Kang M, Park S, Son SR, Noh Y, Jang DS, Lee S (2024) Anti-aging and anti-inflammatory effects of compounds from fresh *Panax ginseng* roots: a study on TNF-α/IFN-γ-induced skin cell damage. Molecules 29:5479

Kasem M (2018) Callus production and suspension elicitation of *Impatiens balsamina* L., plant for enhancing accumulation of phenolics and flavonoids content. J Plant Prod 9:241–248

Keng CL, Wei AS, Bhatt A (2010) Elicitation effect on cell biomass and production of alkaloids in cell suspension culture of the tropical tree *Eurycoma longifolia*. UNED Res J 2:239–244

Kharde A, Kore S, Khetmalas MB (2018) Elicitation of bacoside content using plant growth regulators in cell suspension culture of *Bacopa monnieri* (L.) Wettst. Plant Tissue Cult Biotechnol 28:191–199

Kim MU, Cho YJ, Lee SY (2012) Production and optimization of extracellular polysaccharide by suspension cultivation of *Aloe vera* L. callus. Food Eng Prog 16:233–241

Kim YJ, Lee JE, Jang HS, Hong SY, Lee JB, Park SY, Hwang JS (2021) Oleanolic acid protects the skin from particulate matter-induced aging. Biomol Ther (Seoul) 29:220–226

Kintzios S, Makri O, Panagiotopoulos E, Scapeti M (2003) *In vitro* rosmarinic acid accumulation in sweet basil (*Ocimum basilicum* L.). Biotechnol Lett 25:405–408

Krisa S, Larronde F, Budzinski H, Decendit A, Deffieux G, Mérillon J-M (1999) Stilbene production by *Vitis vinifera* cell suspension cultures: methyl jasmonate induction and ^{13}C biolabeling. J Nat Prod 62:1688–1690

Krishnan N, Singh PK, Devadasan V, Mariappanadar V, Gopinath SCB, Chinni SV, Raman P (2024) Enhanced production of actinidine and glaziovine alkaloids from *Nardostachys jatamansi* (D. Don) DC. through cell suspension culture with elicitors treatment. Process Biochem 138:139–149

Kuhlmann A, Röhl C (2006) Phenolic antioxidant compounds produced by *in vitro*. Cultures of rosemary (*Rosmarinus officinalis*) and their anti-inflammatory effect on lipopolysaccharide-activated microglia. Pharm Biol 44:401–410

Kumanayaka DS, Bandupriya HDD (2025) Initiating cell suspension cultures of *Curcuma longa* L. (turmeric). In: Proceedings of the 13th YSF Symposium—2025, pp 165–169

Kwan LS, Tan SY, Hirata Y, Chan LK, Nagaoka Y, Uesato S, Boey PL (2021) Biotic elicitation at different feeding time in cell suspension cultures of *Eurycoma longifolia* Jack, a valuable medicinal plant, for enhancement of cytotoxic activity of bioactive compounds against human colon cancer cell line. *In Vitro* Cell Dev Biol Plant 58:15–27

Lan TTP, Huy ND, Luong NN, Nghi NV, Tan TH, Quan LV, Loc NH (2018) Identification and characterization of genes in the curcuminoid pathway of *Curcuma zedoaria* Roscoe. Curr Pharm Biotechnol 19:839–846

Lan TTP, Huy ND, Luong NN, Quang HT, Tan TH, Thu LTA, Huy NX, Loc NH (2019) Effect of salicylic acid and yeast extract on curcuminoids biosynthesis gene expression and curcumin accumulation in cells of *Curcuma zedoaria*. J Plant Biotechnol 46:172–179

Largia MJV, Shilpha J, Satish L, Swamy MK, Ramesh M (2022) Elicitation: an efficient strategy for enriched production of plant secondary metabolites. In: Swamy MK, Kumar A (eds) Phytochemical genomics. Springer, Singapore, pp 477–497

Larkin PJ, Scowcroft WR (1981) Somaclonal variation—a novel source of variability from cell cultures for plant improvement. Theor Appl Genet 60:197–214

Larronde F, Krisa S, Decendit A et al (1998) Regulation of polyphenol production in *Vitis vinifera* cell suspension cultures by sugars. Plant Cell Rep 17:946–950

Le KC, Jeong CS, Lee H, Paek KY, Park SY (2019) Ginsenoside accumulation profiles in long- and short-term cell suspension and adventitious root cultures in *Panax ginseng*. Hortic Environ Biotechnol 60:125–134

Lee J, Jung E, Kim Y, Park J, Park J, Hong S, Kim J, Hyun C, Kim YS, Park D (2006) Asiaticoside induces human collagen I synthesis through TGFbeta receptor I kinase (TbetaRI kinase)—independent Smad signaling. Planta Med 72:324–328

Leisegang K, Finelli R, Sikka SC, Panner Selvam MK (2022) *Eurycoma longifolia* (Jack) improves serum total testosterone in men: a systematic review and meta-analysis of clinical trials. Medicina (Kaunas) 58:1047

Li Y, Tian X, Luo J, Bao T, Wang S, Wu X (2024) Molecular mechanisms of aging and anti-aging strategies. Cell Commun Signal 22:285

Lian ML, Chakrabarty D, Paek KY (2002) Effect of plant growth regulators and medium composition on cell growth and saponin production during cell-suspension culture of mountain ginseng (*Panax ginseng* CA Mayer). J Plant Biol 45:201–206

Lim TG, Lee CC, Dong Z, Lee KW (2015) Ginsenosides and their metabolites: a review of their pharmacological activities in the skin. Arch Dermatol Res 307:397–403

Lin FH, Lin JY, Gupta RD, Tournas JA, Burch JA, Selim MA, Monteiro-Riviere NA, Grichnik JM, Zielinski J, Pinnell SR (2005) Ferulic acid stabilizes a solution of vitamins C and E and doubles its photoprotection of skin. J Invest Dermatol 125:826–832

Liu T, Zhu L, Wang L (2022) A narrative review of the pharmacology of ginsenoside compound K. Ann Transl Med 10:234

Loc NH, An NTT (2010) Asiaticoside production from centella (*Centella asiatica* L. Urban) cell culture. Biotechnol Bioproc E 15:1065–1070

Loc NH, Giang NT (2012) Effects of elicitors on the enhancement of asiaticoside biosynthesis in cell cultures of centella (*Centella asiatica* L. Urban). Chem Pap 66:642–648

Loc NH, Nhat NTD (2013) Production of asiaticoside from centella (*Centella asiatica* L. Urban) cells in bioreactor. Asian Pac J Trop Biomed 3:806–810

Loc NH, Ha TTT, Hirata Y (2006) Effect of several factors on cell biomass production of zedoary (*Curcuma zedoaria* Roscoe) in bioreactor. Vietnam J Biotechnol 4:213–220

Loc NH, Diem DTH, Binh DHN, Huong DT, Kim TG, Yang MS (2008) Isolation and characterization of antioxidation enzymes from cells of zedoary (*Curcuma zedoaria* Roscoe) cultured in a 5-L bioreactor. Mol Biotechnol 38:81–87

Loc NH, Tuan VC, Binh DHN, Phuong TTB, Kim TG, Yang MS (2009) Accumulation of sesquiterpenes and polysaccharides in cells of zedoary (*Curcuma zedoaria* roscoe) cultured in a 10 L bioreactor. Biotechnol Bioproc E 14:619–624

Loc NH, Giang NT, Huy ND (2016) Effect of salicylic acid on expression level of genes related with isoprenoid pathway in centella (*Centella asiatica* (L.) Urban) cells. 3 Biotech 6:1–7

Loc NH, Giang NT, Huy ND, Lan TTP (2017) Accumulation of madecassoside—a major component of centelloside—in centella (*Centella asiatica* (L.) urban) cells elicited by salicylic acid. Period Biol 119:81–85

López-Otín C, Blasco MA, Partridge L, Serrano M, Kroemer G (2023) Hallmarks of aging: An expanding universe. Cell 86:243–278

Low BS, Choi SB, Abdul Wahab H, Das PK, Chan KL (2013a) Eurycomanone, the major quassinoid in *Eurycoma longifolia* root extract increases spermatogenesis by inhibiting the activity of phosphodiesterase and aromatase in steroidogenesis. J Ethnopharmacol 149:201–207

Low BS, Das PK, Chan KL (2013b) Standardized quassinoid-rich *Eurycoma longifolia* extract improved spermatogenesis and fertility in male rats via the hypothalamic-pituitary-gonadal axis. J Ethnopharmacol 145:706–714

Lu M, Wong H, Teng W (2001) Effects of elicitation on the production of saponin in cell culture of *Panax ginseng*. Plant Cell Rep 20:674–677

Makri O, Kintzios S (2008) *Ocimum* sp. (Basil): botany, cultivation, pharmaceutical properties, and biotechnology. J Herbs Spices Med Plants 13:123–150

Maquart FX, Bellon G, Gillery P, Wegrowski Y, Borel JP (1990) Stimulation of collagen synthesis in fibroblast cultures by a triterpene extracted from *Centella asiatica*. Connect Tissue Res 24:107–120

Maurya PK, Bahadur V (2022) Somoclones and somaclonal variants: a review. Environ Ecol 40:2532–2539

Mehrara E, Forssell-Aronsson E, Ahlman H, Bernhardt P (2007) Specific growth rate versus doubling time for quantitative characterization of tumor growth rate. Cancer Res 67:3970–3975

Mirjalili MH, Moyano E, Bonfill M, Cusido RM, Palazón J (2009) Steroidal lactones from *Withania somnifera*, an ancient plant for novel medicine. Molecules 14:2373–2393

Mirzabe AH, Hajiahmad A, Fadavi A, Rafiee S (2022) Temporary immersion systems (TISs): a comprehensive review. J Biotechnol 357:56–83

Mousa NA, Siaguru P, Wiryowidagdo S, Wagih ME (2007) Establishment of regenerative callus and cell suspension system of licorice (*Glycyrrhiza glabra*) for the production of the sweetener glycyrrhizin *in vitro*. Sugar Tech 9:72–82

Murashige T, Skoog F (1962) A revised medium for rapid growth and bio assays with tobacco tissue cultures. Plant Physiol 15:473–497

Murthy HN, Yadav GG, Paek KY, Park SY (2024a) Curcuminoid production from plant cells and organ cultures for application in food and pharmaceutical industries. Food Rev Int 41:399–417

Murthy HN, Yadav GG, Paek KY, Park SY (2024b) Production of terpene trilactones from cell and organ cultures of *Ginkgo biloba*. Plants (Basel) 13:2575

Naik PM, Al-Khayri JM (2018) Cell suspension culture as a means to produce polyphenols from date palm (*Phoenix dactylifera* L.). Ciência e Agrotecnol 42:464–473

Naik PM, Al–Khayri JM (2016) Abiotic and biotic elicitors–role in secondary metabolites production through in vitro culture of medicinal plants. In: Shanker AK, Shanker C (eds) Abiotic and biotic stress in plants—recent advances and future perspectives. IntechOpen, London, pp 247–277

Narra K, Naik SK, Ghatge AS (2023) A study of efficacy and safety of ashwagandha (*Withania somnifera*) lotion on facial skin in photoaged healthy adults. Cureus 15:e36168

Nath S, Buragohaine AK (2005) Establishment of callus and cell suspension cultures of *Centella asiatica*. Biol Plant 49:411–413

Nguyen-Thanh T, Dang-Ngoc S, Tran-Quoc D, Hoang-Tan Q (2023) Gypenosides production and spermatogenesis recovery potentials of extracts from cell suspension cultures of *Gynostemma pentaphyllum*. Avicenna J Med Biotechnol 15:216–222

Nhan NH, Loc NH (2017) Production of eurycomanone from cell suspension culture of *Eurycoma longifolia*. Pharm Biol 55:2234–2239

Nhan NH, Loc NH (2018) Enhancement of eurycomanone biosynthesis in cell culture of long jack (*Eurycoma longifolia*) by elicitor treatment. J Plant Biotechnol 45:340–346

Ochoa-Villarreal M, Howat S, Hong S, Jang MO, Jin YW, Lee EK, Loake GJ (2016) Plant cell culture strategies for the production of natural products. BMB Rep 49:149–158

Orihara Y, Ebizuka Y (2010) Production of triterpene acids by cell-suspension cultures of *Olea europaea*. In: Preedy VR, Watson RR (eds) Olives and olive oil in health and disease prevention. Academic Press, pp 341–347

Păcularu-Burada B, Cîrîc AI, Begea M (2024) Anti-aging effects of flavonoids from plant extracts. Foods 13:2441

Paek KY, Chakrabarty D, Hahn EJ (2005) Application of bioreactor systems for large scale production of horticultural and medicinal plants. Plant Cell Tissue Organ Cult 81:287–300

Papadopoli D, Boulay K, Kazak L, Pollak M, Mallette FA, Topisirovic I, Hulea L (2019) mTOR as a central regulator of lifespan and aging. F1000Res 8:F1000 Faculty Rev-998

Park YG, Kim SJ, Kang YM, Jung HY, Prasad DT, Kim SW, Chung YG, Choi MS (2004) Production of ginkgolides and bilobalide from optimized the *Ginkgo biloba* cell culture. Biotechnol Bioproc E 9:41–46

Petkovic M, Leal EC, Alves I, Bose C, Palade PT, Singh P, Awasthi S, Børsheim E, Dalgaard LT, Singh SP, Carvalho E (2021) Dietary supplementation with sulforaphane ameliorates skin aging through activation of the Keap1-Nrf2 pathway. J Nutr Biochem 98:108817

Profumo P, Caviglia AM, Gastaldo P (1992) Formation of aescin glucosides by callus tissue from cotyledonary explants of *Aesculus hippocastanum* L. Plant Sci 85:161–164

Purushothaman JR, Rizwanullah M (2024) Ferulic acid: A comprehensive review. Cureus 16:e68063

Quang HT, Thi PTD, Sang DN, Tram TTN, Huy ND, Dung TQ, The QTT (2022) Effects of plant elicitors on growth and gypenosides biosynthesis in cell culture of *Giao co lam* (*Gynostemma pentaphyllum*). Molecules 27:2972

Rahman L, Verma PC, Singh D, Gupta MM (2002) Bacoside production by suspension cultures of *Bacopa monnieri* (L.) Pennell. Biotechnol Lett 24:1427–1429

Ramachandra Rao S, Ravishankar GA (2002) Plant cell cultures: chemical factories of secondary metabolites. Biotechnol Adv 20:101–153

Ramirez-Estrada K, Vidal-Limon H, Hidalgo D, Moyano E, Golenioswki M, Cusidó RM, Palazon J (2016) Elicitation, an effective strategy for the biotechnological production of bioactive high-added value compounds in plant cell factories. Molecules 21:182

Rao SR, Ravishankar GA (2002) Plant cell cultures: chemical factories of secondary metabolites. Biotechnol Adv 20:101–153

Ratz-Łyko A, Arct J, Pytkowska K (2016) Moisturizing and antiinflammatory properties of cosmetic formulations containing *Centella asiatica* extract. Indian J Pharm Sci 78:27–33

Rawat V, Ghildiyal A, Singh L, Jugran AK, Bhatt ID, Nandi SK, Pande V (2019) Methyl jasmonate induced polyphenols and antioxidant production in callus suspension culture of *Nardostachys jatamansi*. Plant Biosyst 154:851–859

Roy Chowdhury M, Mehmet M, Mukherjee J, Debnath AJ, Ražná K (2025) Chitosan as an elicitor in plant tissue cultures: methodological challenges. Molecules 30:3476

Sabater-Jara AB, Souliman-Youssef S, Novo-Uzal E, Almagro L, Belchí-Navarro S, Pedreño MA (2013) Biotechnological approaches to enhance the biosynthesis of ginkgolides and bilobalide in *Ginkgo biloba*. Phytochem Rev 12:191–205

Sagharyan M, Ganjeali A, Cheniany M, Mousavi Kouhi SM (2020) Optimization of callus induction with enhancing production of phenolic compounds production and antioxidants activity in callus cultures of *Nepeta binaloudensis* Jamzad (Lamiaceae). Iran J Biotechnol 18:e2621

Saimaru H, Orihara Y (2010) Biosynthesis of acteoside in cultured cells of *Olea europaea*. J Nat Med 64:139–145

Saimaru H, Orihara Y, Tansakul P, Kang YH, Shibuya M, Ebizuka Y (2007) Production of triterpene acids by cell suspension cultures of *Olea europaea*. Chem Pharm Bull (Tokyo) 55:784–788

Sák M, Dokupilová I, Kaňuková Š, Mrkvová M, Mihálik D, Hauptvogel P, Kraic J (2021) Biotic and abiotic elicitors of stilbenes production in *Vitis vinifera* L. cell culture. Plants 10:490

Salama A, Abdelhameed MF, Mostafa S, Nada SA, Taha HS, Amer AA (2021) Influence of extract derived cell cultures of broccoli against osteoporosis in ovariectomized rats. Egypt J Chem 64:3521–3539

Sang DN, Thi PTD, Thao TT, Dung TQ, Quang HT (2025) Study on biomass and gypenoside production of *Gynostemma pentaphyllum* via suspenssion cell culture. Hue Univ J Agric Res Dev 134:19–29

Schröfelbauer B, Raffetseder J, Hauner M, Wolkerstorfer A, Ernst W, Szolar OH (2009) Glycyrrhizin, the main active compound in liquorice, attenuates pro-inflammatory responses by interfering with membrane-dependent receptor signalling. Biochem J 421:473–482

Selwal N, Supriadi K, Rahayu F, Sukmadjaja D, Khamidah A, Budiaarto K, Antarlina SS, Tripatmasari M, Wani AK (2025) Elicitation strategies for enhanced secondary metabolite synthesis in plant cell cultures and its role in plant defense mechanism. Plant Gene 41:100485

Seth B, Sahoo KK, Aravind KR, Sahu BB, Singh VR, Patra N (2020) Statistical optimization of bacoside A biosynthesis in plant cell suspension cultures using response surface methodology. 3 Biotech 10:264

Sharma P (2014) Production of flavonoids from *Terminalia arjuna* (ROXB) *in vivo* and *in vitro* tissue culture. Int J ChemTech Res 6:881–885

Sikandan A, Shinomiya T, Nagahara Y (2018) Ashwagandha root extract exerts anti-inflammatory effects in HaCaT cells by inhibiting the MAPK/NF-κB pathways and by regulating cytokines. Int J Mol Med 42:425–434

Singh B, Sharma RA (2014) Antioxidant and antimicrobial activities of callus culture and fruits of *Phyllanthus emblica* L. J Herbs Spices Med Plants 21:230–242

Singh N, Bhalla M, de Jager P, Gilca M (2011) An overview on ashwagandha: a rasayana (rejuvenator) of ayurveda. Afr J Tradit Complement Altern Med 8:208–213

Singh P, Guleri R, Angurala A, Kaur K, Kaur K, Kaul SC, Wadhwa R, Pati PK (2017) Addressing challenges to enhance the bioactives of *Withania somnifera* through organ, tissue, and cell culture based approaches. Biomed Res Int 2017:3278494

Sivanandhan G, Selvaraj N, Ganapathi A, Manickavasagam M (2014) Enhanced biosynthesis of withanolides by elicitation and precursor feeding in cell suspension culture of *Withania somnifera* (L.) Dunal in shake-flask culture and bioreactor. PLoS One 9:e104005

Skoog F, Miller CO (1957) Chemical regulation of growth and organ formation in plant tissues cultured *in vitro*. Symp Soc Exp Biol 11:118–130

Smetanska I (2008) Production of secondary metabolites using plant cell cultures. Adv Biochem Eng Biotechnol 111:187–228

Smith RH (2013) Callus induction. In: Smith RH (ed) Plant tissue culture, 3rd edn. Academic Press, pp 63–79

Srivastava V, Chaturvedi R (2025) Chemoprofiling of *in vitro* cell cultures of *Tinospora cordifolia* for protoberberine alkaloid content analysis and to understand their therapeutic potential against *Staphylococcus aureus*. Ind Crop Prod 234:121573

Sukito A, Tachibana S (2016) Effect of methyl jasmonate and salycilic acid synergism on enhancement of bilobalide and ginkgolide production by immobilized cell cultures of *Ginkgo biloba*. Bioresour Bioprocess 3:24

Sun W, Shahrajabian MH (2023) Therapeutic potential of phenolic compounds in medicinal plants—natural health products for human health. Molecules 28:1845

Sun B, Wu L, Wu Y, Zhang C, Qin L, Hayashi M, Kudo M, Gao M, Liu T (2020) Therapeutic potential of *Centella asiatica* and its triterpenes: a review. Front Pharmacol 11:568032

Tani M, Fukui H, Shimomura M, Tabata M (1992) Structure of endogenous oligogalacturonides inducing shikonin biosynthesis in *Lithospermum* cell cultures. Phytochem 31:2719–2723

Tani M, Takeda K, Yazaki K, Tabata M (1993) Effects of oligogalacturonides on biosynthesis of shikonin in *Lithospermum* cell cultures. Phytochem 34:1285–1290

Thanh NT, Murthy HN, Yu KW, Jeong CS, Hahn EJ, Paek KY (2004) Effect of inoculum size on biomass accumulation and ginsenoside production by large-scale cell suspension cultures of *Panax ginseng*. J Plant Biotechnol 6:265–268

Thanh N, Murthy H, Yu K, Hahn E, Paek K (2005) Methyl jasmonate elicitation enhanced synthesis of ginsenoside by cell suspension cultures of *Panax ginseng* in 5-l balloon type bubble bioreactors. Appl Microbiol Biotechnol 67:197–201

Thanh NT, Murthy HN, Paek KY (2014) Optimization of ginseng cell culture in airlift bioreactors and developing the large-scale production system. Ind Crop Prod 60:343–348

Tito A, Bimonte M, Carola A, De Lucia A, Barbulova A, Tortora A, Colucci G, Apone F (2015) An oil-soluble extract of *Rubus idaeus* cells enhances hydration and water homeostasis in skin cells. Int J Cosmet Sci 37:588–594

Titova MV, Kochkin DV, Sukhanova ES, Gorshkova EN, Tyurina TM, Ivanov IM, Lunkova MK, Tsvetkova EV, Orlova A, Popova EV, Nosov AM (2023) Suspension cell culture of *Polyscias fruticosa* (L.) Harms in bubble-type bioreactors—growth characteristics, triterpene glycosides accumulation and biological activity. Plants 12:3641

Titova M, Popova E, Nosov A (2024) Bioreactor systems for plant cell cultivation at the Institute of Plant Physiology of the Russian Academy of Sciences: 50 years of technology evolution from laboratory to industrial implications. Plants 13:430

Tuan VC, Hoang VD, Loc NH (2011) Cell suspension culture of zedoary (*Curcuma zedoaria* Roscoe). VNU J Sci: Nat Sci Technol 27:64–70

Variya BC, Bakrania AK, Patel SS (2016) *Emblica officinalis* (Amla): a review for its phytochemistry, ethnomedicinal uses and medicinal potentials with respect to molecular mechanisms. Pharmacol Res 111:180–200

Verpoorte R, Contin A, Memelink J (2002) Biotechnology for the production of plant secondary metabolites. Phytochem Rev 1:13–25

Wang S, Guo LP, Xie T, Liu Y, Sun H, Wang Z, Lu Y (2014) Different secondary metabolic responses to MeJA treatment in shikonin-proficient and shikonin-deficient cell lines from *Arnebia euchroma* (Royle) Johnst. Plant Cell Tissue Organ Cult 119:587–598

Wang S, Qiao J, Jiang C, Pan D, Yu S, Chen J, Liu S, Zhang P, Zhao D, Liu M (2023) Ginsenoside Rg1 delays chronological aging in a yeast model via CDC19- and SDH2-mediated cellular metabolism. Antioxidants (Basel) 12:296

Wasternack C, Hause B (2013) Jasmonates: biosynthesis, perception, signal transduction and action in plant stress response, growth and development. Ann Bot 111:1021–1058

Widyastuti DA, Santosa D, Nuringtyas TR, Rohman A (2024) Callus induction of red ginger (*Zingiber officinale* var. *rubrum*) with 2,4-D and kinetin combination to enhance total phenolic and flavonoid content. IOP Conf Ser Earth Environ Sci 1364:012051

Witkowska K, Paczkowska-Walendowska M, Garbiec E, Cielecka-Piontek J (2024) Topical application of *Centella asiatica* in wound healing: recent insights into mechanisms and clinical efficacy. Pharmaceutics 16:1252

Wongwicha W, Tanaka H, Shoyama Y, Tuvshintogtokh I (2014) Production of glycyrrhizin in callus cultures of licorice. Z Naturforsch C 63:413–417

Wu CH, Zabawa EM, Townsley PM (1974) The single cell suspension culture of the licorice plant, *Glycyrrhiza glabra*. Can Inst Food Sci Tech J 7:105–109

Xu F, Valappil AK, Mathiyalagan R, Tran TNA, Ramadhania ZM, Awais M, Yang DC (2023) *In vitro* cultivation and ginsenosides accumulation in *Panax ginseng*: a review. Plants (Basel) 12:3165

Xue Y, Zhang R, Li T, Deng Q, Luo W, Chang R, Zeng D, Tan J, Sun T, Liu YG, Xiang Y, Zhu Q, Chai N (2025) Sustainable production of ginsenosides: advances in biosynthesis and metabolic engineering. Plants (Basel) 14:2821

Yagishita Y, Fahey JW, Dinkova-Kostova AT, Kensler TW (2019) Broccoli or sulforaphane: is it the source or dose that matters? Molecules 24:3593

Yang L, Shi J, Wang X, Zhang R (2022) Curcumin alleviates D-galactose—induced cardiomyocyte senescence by promoting autophagy via the SIRT1/AMPK/mTOR pathway. Evid Based Complement Alternat Med 2022:2990843

Yazaki K, Okuda T (1990) Condensed tannin production in callus and suspension cultures of *Cinnamomum cassia*. Phytochemistry 29:1559–1562

Yin Y, Hu M, Yang Z, Zhu J, Fang W (2024) Salicylic acid promotes phenolic acid biosynthesis for the production of phenol acid-rich barley sprouts. J Sci Food Agric 104:5350–5359

Zaman G, Farooq U, Bajwa MN, Hassan I, Ahmed MB (2022) Effects of yeast extract on the production of phenylpropanoid metabolites in callus culture of purple basil (*Ocimum Basilicum* L. var *purpurascens*) and their *in-vitro* evaluation for antioxidant potential. Plant Cell Tissue Organ Cult 150:543–553

Zhang W, Feng J, Cheng B, Lu Q, Chen X (2018) Oleanolic acid protects against oxidative stress-induced human umbilical vein endothelial cell injury by activating AKT/eNOS signaling. Mol Med Rep 18:3641–3648

Ziv M (2005) Simple bioreactors for mass propagation of plants. Plant Cell Tissue Organ Cult 81:277–285

Zuo W, Yan F, Zhang B, Li J, Mei D (2017) Advances in the studies of *Ginkgo biloba* leaves extract on aging-related diseases. Aging Dis 8:812–826

Light-Emitting Diode-Induced Enhancement of Anti-aging Secondary Metabolites Under In Vitro Conditions

7

Ghada M. Abbas, Abdullah Haikal, Amira E. Shaaban, and Ahmed Zayed

Abstract

The rising demand for natural anti-aging therapeutics has accelerated research into the controlled production of secondary metabolites with antioxidant, anti-inflammatory, and collagen-boosting effects, which are vital for counteracting age-related degeneration. Among the latest biotechnological innovations, light-emitting diodes (LEDs) have emerged as efficient elicitors offering durability, cool temperature, long life, diverse wavelengths, and small diode size, resulting in improved yield and quality of bioactive compounds in vitro cultures and intact plants. This chapter explores how specific LED wavelengths, particularly red, blue, and combined spectra, stimulate biosynthetic pathways in medicinal plants known for their anti-aging benefits. Mechanistically, LEDs offer durability, cool temperature, long life, diverse wavelengths, and small diode size, which influence key genes and enzymes involved in the phenylpropanoid, flavonoid, and carotenoid biosynthesis pathways, modulate reactive oxygen species (ROS) signaling, and enhance phytohormonal regulation. This chapter will shed light on how LEDs are beneficial for the accumulation of different secondary metabolites in multiple shoots, callus, and/or cell suspension culture. Integrating plant photophysiology with secondary metabolism has highlighted how LED-based light manipulation supports sustainable and intensified production of high-value phytochemicals targeting age-related degeneration.

G. M. Abbas · A. E. Shaaban
Department of Pharmacognosy, Horus University—Egypt, New Damietta, Egypt

A. Haikal
Department of Pharmacognosy, Mansoura University, Mansoura, Egypt

A. Zayed (✉)
Pharmacognosy Department, Tanta University, Tanta, Egypt
e-mail: ahmed.zayed1@pharm.tanta.edu.eg

S. Gantait et al. (eds.), *Medicinal Plants with Anti-Ageing Properties*,
https://doi.org/10.1007/978-981-95-8064-4_7

Keywords

Cosmetics and nutraceuticals · Gene expression · In vitro culture · Phenylpropanoid pathway · Photophysiology · Reactive oxygen species

7.1 Introduction

Aging is a complex biological process associated with progressive decline in physiological integrity and function (Maldonado et al. 2023). Recently, there has been growing interest in delaying or mitigating aging-related deterioration, not only for beauty-related reasons but also to prevent chronic degenerative diseases such as cardiovascular disorders, neurodegeneration, and skin damage (Li et al. 2021; Martín Giménez et al. 2022; Tenchov et al. 2024). This interest has encouraged a global market shift toward natural, plant-based instead of synthetic anti-aging products, attributed to consumer preference for safe, biocompatible, and sustainable products (Goyal and Jerold 2023; Nhani et al. 2024).

The global shift toward sustainable and environmentally conscious beauty solutions has prompted potential growth in the cosmetics sector. Recent projections estimate that the sustainable cosmetics market will reach approximately USD 937.13 million by 2030, expanding at a compound annual growth rate (CAGR) of 7.7% between 2024 and 2030 (Nhani et al. 2024). This trend reflects increasing consumer demand for products derived from natural resources, including plant and animal by-products, which are rich in bioactive compounds such as antioxidants, collagen-stimulating peptides, and anti-inflammatory agents. In addition to addressing efficacy and safety, the incorporation of natural products into cosmetic formulations aligns with global sustainability objectives, especially those outlined in the United Nations 2030 Agenda for Sustainable Development (Bom et al. 2019; Rocca et al. 2022). Among these goals are climate action (SDG 13), responsible consumption and production (SDG 12), and the protection of terrestrial ecosystems (SDG 15).

Polyphenols, flavonoids, alkaloids, terpenoids, and saponins are examples of secondary metabolites that are essential for reducing cellular and biochemical damage associated with aging (Riaz et al. 2023; Wang et al. 2023; Dixit et al. 2025). Antioxidant activity via scavenging reactive oxygen species and preventing lipid peroxidation, modulation of inflammatory pathways (e.g., suppression of NF-κB), and enhancement of extracellular matrix stability (e.g., stimulation of collagen synthesis and inhibition of matrix metalloproteinases) are some mechanisms in which these metabolites produce their anti-aging effects (Xia et al. 2016; Shin et al. 2019; Hussen et al. 2025). For example, asiaticoside and madecassoside from *Centella asiatica* promote collagen synthesis and fibroblast proliferation, aiding in skin regeneration (Kwon et al. 2012; Bylka et al. 2013). In addition, curcumin from *Curcuma longa* is a known inhibitor of oxidative stress and inflammatory cytokines (Pulido-Moran et al. 2016; Hatipoglu and Keskin 2022), whereas ginsenosides from *Panax ginseng* have immunomodulatory and neuroprotective qualities

(Zheng et al. 2018). Therefore, these metabolites are good candidates for preventative and therapeutic strategies that focus on both extrinsic and intrinsic aging processes.

However, the biosynthetic pathways of previously mentioned products are greatly influenced in medicinal plants by environmental factors, including temperature variations, soil composition, water availability, and exposure to biotic or abiotic stressors (Reshi et al. 2023; Shiade et al. 2024). While geographic origin, including altitude, latitude, and local climate, further contributes to chemical variability in wild or field-grown plant materials, seasonal variations can result in noticeable changes in the concentration and profile of secondary metabolites (Ogwu et al. 2025). The standardization and large-scale manufacturing of phytochemicals for medical or cosmetic uses are both hampered by this intrinsic inconsistency, and therefore, controlled in vitro culture systems provide a dependable substitute for generating bioactive compounds with greater reproducibility in order to get around these restrictions (Espinosa-Leal et al. 2018; Mohaddab et al. 2022). Within these systems, elicitation techniques, such as the application of biotic (e.g., microbial extracts) or abiotic (e.g., light, temperature, or osmotic stress) factors, can be employed to stimulate and enhance secondary metabolite accumulation (Fazili et al. 2022; Han and Miao 2024).

Particularly, light-emitting diode (Sobhy et al. 2019) elicitation is a cutting-edge and environmentally friendly method of boosting secondary metabolite biosynthesis in both plant tissue cultures and intact plants (Alrifai et al. 2019; Martínez-Chávez et al. 2024). LEDs are superior to conventional light sources in many ways, including low heat emission, low energy consumption, long operating life, and precise wavelength tuning, which makes them perfect for use in sterile and confined in vitro settings (Nowakowska et al. 2023).

Because red and blue LEDs have different physiological effects, they have been studied the most. Blue light affects stomatal opening, chlorophyll synthesis, and flavonoid production (Matthews et al. 2020; Le et al. 2021), whereas red light is frequently linked to increased biomass accumulation, flowering, and root development (Karimi et al. 2022). It has been demonstrated that their combined use enhances the biosynthesis of valuable phytochemicals in a synergistic manner (Alallaq et al. 2020; Wu et al. 2022). For instance, blue LED illumination dramatically increased the accumulation of eugenol in *Ocimum basilicum* callus cultures to 0.273 mg/g dry weight (DW), about a 2.25-fold increase, and rosmarinic acid reached about 96.0 mg/g DW, more than doubling its concentration compared to white light controls (Nadeem et al. 2019). Meanwhile, biosynthetic genes like β-amyrin synthase (β-AS) and squalene synthase (SQS) were upregulated when blue LEDs were used in *Centella asiatica*, increasing the total triterpenoid content by about 70% (Kamol et al. 2024).

Hence, this chapter aims to investigate the use of particular LED wavelengths as abiotic elicitors to promote the biosynthesis of secondary metabolites produced by in vitro cultures exhibiting anti-aging effects. It draws attention to the physiological and molecular processes, such as pathway activation and gene expression modulation, that underlie the metabolite production induced by LEDs. Case studies on medicinal plants will illustrate practical uses of LED integration in plant biotechnology, along with a discussion of its scalability and sustainability. The main goal is to demonstrate LED-based elicitation as a viable method for creating valuable bioactive compounds for anti-aging purposes.

7.2 LED Technology in Plant Cultivation

Light plays a vital role in plant growth and development; it is not merely an energy source for photosynthesis, but a crucial environmental signal that regulates morphogenesis, metabolism, and stress responses. In controlled cultivation systems, such as plant factories, vertical farming units, and tissue culture laboratories, artificial lighting becomes crucial for enhancing plant growth and maximizing the production of secondary metabolites (Gawande et al. 2023; Livadariu et al. 2023; Sena et al. 2024).

7.2.1 Properties and Advantages of LED Lighting

LEDs are solid-state semiconductor devices that convert electrical energy into light in narrow spectral bands. In recent years, LEDs have emerged as the dominant lighting technology for controlled-environment plant production systems such as greenhouses and vertical farms, as well as for in vitro culture applications (Abdelkader and Tammam 2021). In this section, the properties and advantages of LED lighting will be discussed.

Physical/Technical Properties LEDs are highly efficient, durable light sources that emit within the optimal 450–670 nm range for plant growth, offering superior photosynthetic photon efficacy (>2.5 μmol·J^{-1}) compared to traditional lamps (Kusuma et al. 2022; Wong and Zhou 2024). Their low heat output prevents tissue damage, enabling proximity to cultures, and their long lifespan (over 50,000 h) ensures stable, cost-effective performance in controlled plant cultivation systems (Kusuma et al. 2022; Sena et al. 2024; Wong and Zhou 2024).

Spectral Specificity (Narrow-Band Emission and Tunability) LEDs emit narrow wavelength bands (±10–20 nm), enabling precise targeting of plant photoreceptors and efficient energy use. Customized “light recipes” combining red, blue, green, and far-red spectra can fine-tune growth and metabolism. In *Castanea sativa*, multispectral LEDs Red:Blue (RB), Red:Blue:Far-red (RBFr), and R:Blue:Green:Far-red (RBGFr) significantly improved shoot and root development during in vitro propagation, achieving rates (from 3.9 to 4.3 shoots/explant) (Marino et al. 2025).

Energy Efficiency, Lifetime, and Reduced Heat Emission LEDs convert a high fraction of electrical energy into photons rather than heat. Their minimal convection heating and low infrared output enable close mounting to cultures without inducing medium evaporation, detrimental heat stress, or tissue desiccation. Their lifespans are long, often exceeding several tens of thousands of hours, thereby minimizing replacement frequency and reducing maintenance costs in sterile culture environments. Recent studies continue to confirm that LED arrays reduce energy and cooling demands. Spectral arrangement and intensity are shown to influence growth and metabolic performance while reducing waste heat in growth systems (Sena et al. 2024).

Adjustable Intensity, Photoperiod, and Dynamic Control LEDs can be fine-tuned to regulate wavelength, intensity, photo period (duration of light/dark), and duty cycle (continuous versus pulsed). These controls are important to optimize morphogenesis and metabolite biosynthesis in vitro (Cavallaro and Muleo 2022). The green and far-red LED combinations improved the induction of shoots and roots in *Castanea sativa* (Marino et al. 2025). Equally, in *Cunninghamia lanceolata*, a 16-h mixed-spectrum regime enhanced balanced shoot-root growth, whereas pulsed LEDs at 0.5–1 kHz in baby-leaf lettuce enhanced biomass, pigment content, and photosynthetic efficiency. Blue LEDs in *Operculina turpethum* callus had a great effect of enhancing biomass and metabolite buildup (Xu et al. 2020; Miliauskienė et al. 2021; Biswal et al. 2022).

Form Factor, Spatial Arrangement, and Uniformity LEDs are small and versatile, making them suitable for shelf setups, inter-canopy placement, side illumination, and multi-directional lighting applications. Uniformity is easier to attain, reducing "hot-spots" or shading (which can lead to non-uniform metabolite production). Several recent experiments in closed soilless systems (e.g., red lettuce under mixed LED spectra) demonstrate that uniform spectra supplemented with white light result in enhanced phenolic and antioxidant distribution across leaves, as well as improved total antioxidant capacity (Sawatdee et al. 2023).

Reliability, Scalability, and Low Maintenance LEDs are solid-state devices; they are robust, have fewer failure modes compared to older high-pressure lamps, and require less frequent maintenance. In in vitro laboratories, the reduction of disturbance from fixture replacement is valuable (Lee et al. 2022).

7.3 Photoreceptors and Wavelength Specificity in Plants

Light is not only utilized by plants as a source of energy to promote photosynthesis, but also as an important environmental cue in controlling germination, growth, flowering, and senescence. Specialized photoreceptors mediate these responses to wavelengths ranging between the UV-B and far-red light (Kong and Okajima 2016).

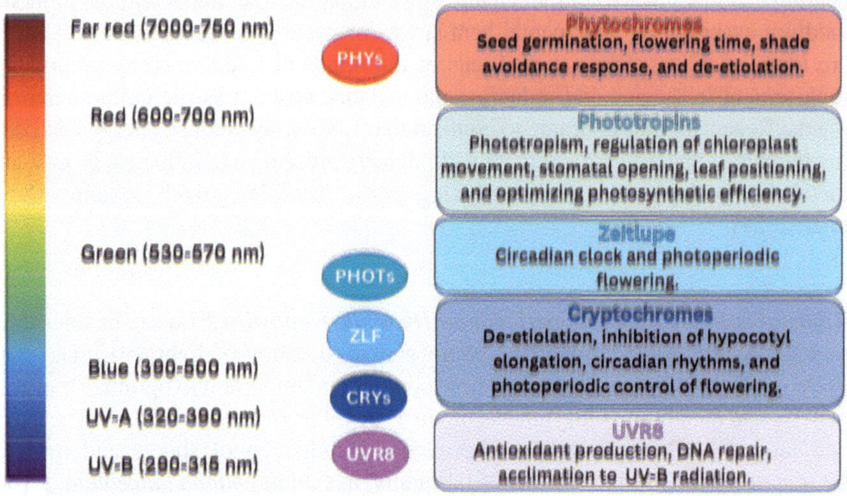

Fig. 7.1 Spectral wavelength-specificity in plant photoreceptors to regulate multiple physiological pathways, phytochromes (PHYs), phototropins (PHOTs), Zeitlupe family proteins (ZLF), cryptochromes (CRYs), and UV resistance Locus 8 (UVR8). (Source: Diagram prepared by the authors)

Plants have used special types of photoreceptors to sense and react to light at a variety of different wavelengths across the electromagnetic spectrum. All photoreceptor families are sensitive to certain spectral regions, which allows plants to optimize their development and physiological responses to varying light conditions (Paik and Huq 2019). Outlined below are the different photoreceptor families identified in plants and summarized in Fig. 7.1.

Phytochromes: Red and Far-Red Light Perception Phytochromes are dimeric specialized photoreceptors, which absorb red (660–680 nm) and far-red (730–750 nm) light, in two reversible states: the red-absorbing Pr and the far-red-absorbing Pfr. When Pr is exposed to red light, it is converted to the active form Pfr and causes signaling cascades that control expression of genes and development. This photo-reversible process allows plants to detect and adjust to alterations of light quality and intensity and regulate such physiological events as germination, flowering, de-etiolation, and shade avoidance (Paik and Huq 2019; Cai and Huq 2024) (Fig. 7.1).

Cryptochromes (Cry): Blue and UV-A Light-Sensing Cryptochromes are flavoproteins that absorb blue (450–495 nm) and UV-A (315–400 nm) light. They mediate inhibition of hypocotyl elongation, circadian rhythms, and photoperiodic control of flowering. They form signaling complexes to regulate transcription in response to blue light. They play crucial roles in regulating circadian rhythms, photomorphogenesis, and photoperiodic control of flowering. They form signaling complexes to regulate transcription in response to blue light. Recent research has demonstrated the wavelength-specific and photoreceptor-dependent effects of cryptochromes in protecting plants from UV-induced stress (Serrano et al. 2021; Zeng et al. 2025) (Fig. 7.1).

Phototropins: Blue Light and Phototropism Phototropins are blue-light-dependent serine/threonine kinase receptors that mediate phototropism, the process by which plants grow directionally toward a light source. They also regulate chloroplast movement, stomatal opening, and leaf positioning, optimizing photosynthetic efficiency. Phototropins are sensitive to blue light and contribute significantly to the adjustment of the plant architecture for optimal light interception (Hart and Gardner 2021) (Fig. 7.1).

UVR8: UV-B Light Perception UV-B receptors use tryptophan residues as chromophores. The UV-B photoreceptor UVR8 specifically absorbs UV-B light (280–315 nm and mediates responses to UV-B-induced stress. UV-B absorption causes UVR8 homodimers to split into active monomers, which interact with COP1, regulate gene expression, and enhance UV tolerance. Upon UV-B exposure, UVR8 undergoes a conformational change that activates downstream signaling pathways, leading to the expression of genes involved in antioxidant production, DNA repair, and other protective mechanisms (Jenkins 2017; Li et al. 2020).

Zeitlupe Family Proteins: Blue Light and Circadian Regulation ZEITLUPE (ZTL)/FLAVIN-BINDING, KELCH REPEAT, F-BOX 1 (FKF1)/LOV kelch protein2 (LKP2) is another class of blue light receptor. The Zeitlupe (ZTL) family of proteins, including FKF1 and LKP2, is involved in the regulation of the circadian clock and photoperiodic flowering. These proteins are sensitive to blue light and modulate the timing of flowering in response to changes in light quality and duration. Their activity is crucial for synchronizing plant development with environmental light cycles (Paik and Huq 2019) (Fig. 7.1).

7.4 Influence of LEDs on Plant Development and Stress Response

LEDs have revolutionized controlled-environment agriculture by allowing precise control over light spectrum, intensity, and photoperiod, enabling targeted regulation of plant growth and metabolism. This technology influences photosynthesis, morphogenesis, and stress responses, enhancing both productivity and resilience. Overall, LEDs serve as an efficient, sustainable tool for improving crop yield and quality (Azizi et al. 2025).

7.4.1 Impact of LED Light Spectra on Plant Development

The spectral quality of LED light profoundly influences plant growth and development. Blue light (400–500 nm) promotes chlorophyll formation, stomatal opening, and root elongation, enhancing photosynthetic efficiency and structural strength. Red light (600–700 nm) boosts photosynthesis and regulates key transitions like flowering and fruiting, while its combination with blue light creates a synergistic effect that supports vigorous, balanced growth. Green light, though less explored, penetrates deeper into plant canopies, complementing red–blue spectra to improve biomass accumulation. Light intensity also plays a vital role—optimal levels promote healthy growth, whereas excessive exposure can cause photoinhibition, and insufficient light limits productivity (Yaoyuan et al. 2025).

7.4.2 LED Lighting and Stress Response Mechanisms

LED light not only regulates plant growth but also enhances stress tolerance. Blue–red light combinations boost antioxidant enzymes such as SOD, catalase, and peroxidases, helping plants counter oxidative damage from drought, salinity, or heat (Lazzarin et al. 2021; Roeber et al. 2021). Moreover, LEDs promote the accumulation of protective metabolites, anthocyanins, flavonoids, and phenolics that act as UV shields and reactive oxygen species (ROS) scavengers, strengthening plants' defense systems (Sena et al. 2024).

Studies on strawberry, basil, and lemon balm demonstrate that combined blue–red LED lighting enhances plant tolerance to abiotic stress. In strawberries and basil, this spectrum reduced sodium buildup while improving potassium and magnesium uptake, maintaining ionic balance and water regulation under salinity stress (Rafeie et al. 2022; Malekzadeh et al. 2024). Similarly, in lemon balm, blue–red LED "priming" improved drought resilience by maintaining higher leaf water content, chlorophyll stability, and antioxidant activity. These results suggest that LEDs can precondition plants to better withstand future stress by enhancing physiological and metabolic adaptation (Ahmadi et al. 2025).

7.4.3 Practical Applications in Agriculture

The ability to tailor LED lighting conditions offers significant advantages in agricultural practices. By adjusting light spectrum and intensity, cultivators can optimize plant growth, improve stress resilience, and enhance crop yield and quality. For instance, the use of specific LED wavelengths can promote the accumulation of secondary metabolites, such as antioxidants and flavonoids, which are beneficial for both plant health and human nutrition. Additionally, LED lighting can be utilized to mitigate the effects of environmental stressors. For example, the application of supplemental LED light has been shown to improve the growth and development of plants under salinity and alkalinity stress conditions (Gai et al. 2023).

Beyond field cultivation and greenhouse, LEDs have become a powerful tool in in vitro plant culture systems. For instance, in *Scrophularia kakudensis*, exposure to red and blue LEDs promoted shoot elongation, root growth, and biomass accumulation, while significantly enhancing phenolic and flavonoid contents compared to conventional light sources (Manivannan et al. 2020). Similarly, in *Cichorium* spp. (chicory), Red-blue LED treatments stimulated a higher accumulation of flavonoids and phenolics in callus cultures, with red light exhibiting the strongest antifungal activity (Zhiponova et al. 2025). A recent review of over 25 studies further highlighted that combined red and blue spectra are consistently the most effective for improving phenolic content and antioxidant activity in medicinal plants grown in vitro (Zauli et al. 2022). These findings illustrate how LEDs can be harnessed not only for propagation and conservation but also for the biotechnological production of bioactive compounds under controlled culture conditions.

7.5 LEDs-Stimulated Anti-aging Metabolite Production under In Vitro Systems

7.5.1 Plant Tissue Cultures (Callus, Shoot, and Hairy Root Systems)

The use of LED lighting in in vitro plant culture is an effective method of increasing anti-aging metabolites, including phenolics, flavonoids, iridoids, anthocyanins, and carotenoids. The spectral quality, intensity, and exposure period are fine-tuned by using LEDs to direct metabolic processes to antioxidant, anti-inflammatory, and neuroprotective compounds.

The different LED spectra (white, red, blue, red-blue mixes, and UV-supplemented) had a significant influence on the synthesis of steviol glycoside and phenolic compounds in the cultures of *Stevia rebaudiana* 50% of green, 35% red, and 15% blue increased biomass, whereas 50% far-red, 35% red, and 15% blue increased stevioside (1.62 mg/g DW) and 50% UV, 35% red, and 15% blue increased rebaudioside A (3.15 mg/g DW). The identical far-red mixture activated the production of five phenolic acids, namely neochlorogenic, chlorogenic, isochlorogenic A, 4-feruloylquinic, and rosmarinic acids, which have been shown to have antioxidant, collagen-protective, and anti-pigmentation properties (Ptak et al. 2024). The same improvement was also observed with quercitrin, which is a flavonoid that possesses photoprotective and anti-wrinkle properties. In *Swertia chirayita*, the red lights were the ones that gave the highest biomass (3.56 ± 0.04 g/L), whereas the blue lights produced the highest amarogentin level (7.025 ± 0.04 mg/mg DW), total phenolics (22.33 ± 1.05 mg GA/g DW), flavonoids (29 ± 1.03 mg QE/g DW), and DPPH scavenging activity (50.4 ± 0.16%) relative to other light conditions (Disasa et al. 2020; Gupta and Sood 2023). Taken together, these findings establish the fact that LED spectral modulation is a successful tool in the enhancement of in vitro biosynthesis of potent anti-aging metabolites, leading to the enhancement of both plant development and the production of bioactive compounds, as summarized in Fig. 7.2. Table 7.1 summarizes how LED spectral treatments affect in vitro anti-aging secondary metabolites.

7.5.2 Intact Plant In Vitro Cultures (Seedlings, Synseeds, and Controlled Growth Chambers)

In vitro cultivation of intact plantlets and seedlings under controlled growth chambers has provided compelling evidence for LED-driven enhancement of anti-aging metabolites. Blue light, in particular, has shown strong efficacy in stimulating phenolic and flavonoid accumulation across multiple species. For example, cherry tomato seedlings (*Solanum lycopersicum* L.) subjected to blue LED illumination (456 nm) demonstrated higher accumulation of phenolic compounds and total flavonoids. The antioxidant capacity of seedlings grown under blue LEDs was 2.1-fold higher than that of seedlings grown under green LEDs, strengthening their potential use in nutraceutical formulations (Kim et al. 2014). *Lactuca sativa* seedlings, cultivated in a solar greenhouse, exposed to blue LEDs (peak at 437 nm), exhibited a 2.07-fold increase in total flavonoids (15.70 ± 1.05 mg/g DW) and vitamin C (137.69 ± 2.36 mg/kg FW) (Martinez et al. 2024). Applying red:blue LED treatment during in vitro propagation of *Anoectochilus roxburghii*, a medicinal orchid, enhanced flavonoid (1.81 mg/g) compared to other light sources (Gam et al. 2020). Table 7.2, demonstrates different examples of LED-induced anti-aging secondary metabolites derived from in vitro intact plants.

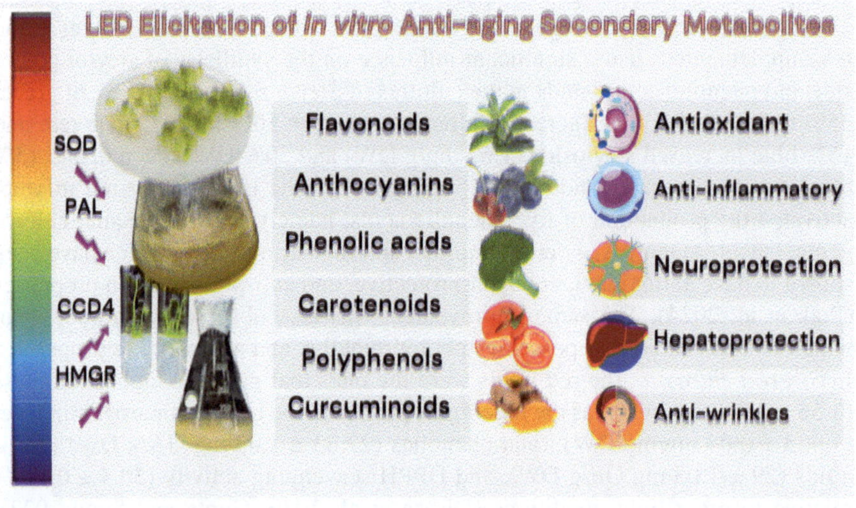

Fig. 7.2 LED elicitation on in vitro poroduction of secondary metabolites with anti-aging and other relevant properties. (Source: Diagram prepared by the authors)

Table 7.1 LED-induced anti-aging secondary metabolites derived from in vitro tissue culture

Plant species (culture)	LED/light treatment	Enhanced metabolites	Anti-aging relevance	References
Cichoriumendive Supp. *Pumillum* (callus cultures)	Red LED light	Myricetin (1648 μg/mL) and kamapherol 61.42 μg/mL	Phenolics improve resistance against environmental stressors, reduce oxidative damage to cells, and maintain skin elasticity through collagen stabilization	Abdelhamid et al. (2025)
Curcuma longa (shoot culture)	(50% red (peak at 660 nm) + 50% blue (420–460 nm))	Flavonoids, polyphenols, reducing sugars, and curcumin	Compounds central to antioxidant and anti-inflammatory activity	Marchant et al. (2022)
Withania somnifera (callus culture)	Red LED light (610–715 nm)	Withaferin A (24.3 ± 1.8 mg. g^{-1} DW) chlorogenic acid (11.6 ± 0.4)	A potent bioactive with adaptogenic and anti-aging effects	Adil et al. (2019a, b)
Salvia bulleyana (transformed shoot cultures)	(70% red light and 30% blue light)	Polyphenols (24.3 mg/g of DW), a two-fold increase compared to control treatments. And rosmarinic acid (20.0 mg/g of DW), three-fold increases compared to control treatments	Rosmarinic acid suppresses matrix metalloproteinases (MMPs) and reduces hyperpigmentation. It also contributes to systemic anti-aging effects through neuroprotection, cardiovascular protection, and modulation of inflammatory pathways that drive age-related diseases	Lu et al. (2017), Hassanzadeh-Taheri et al. (2021), Krzemińska et al. (2023), and Girgenti et al. (2025)
Artemisia annua (suspension culture)	Red LED light	Artemisinin	Artemisinin and related sesquiterpenes exhibit strong antioxidant and anti-inflammatory properties, and have been linked to lifespan extension in model organisms	Wang et al. (2024a, b) and Lim et al. (2025)
Eclipta alba (callus culture)	Red LEDs (660 nm, 24-h)	Phenolics (57.8 mg/g) and flavonoid content (11.1 mg/g) wedelolactone (32.54 mg/g), dimethyl wedelolactone (23.67 ± 1.04 mg/g), coumarin (1.26 ± 0.04 mg/g), and eclalbatin (5.00 ± 0.83 mg/g)	Antioxidant, hepatoprotective, and anti-inflammatory effects, thereby contributing to anti-aging potential	Khurshid et al. (2020)

(continued)

Table 7.1 (continued)

Plant species (culture)	LED/light treatment	Enhanced metabolites	Anti-aging relevance	References
Myrtus communis (shoot culture)	Red LEDs with 35 μmol m^{-2} s^{-1} PPFD	Flavonoids Gallic acid (224.05 ± 0.36 mg 100 g^{-1} DW)	Gallic acid contributes to dermal anti-aging by inhibiting MMPs, reducing hyperpigmentation, and enhancing collagen stability, in addition to its systemic benefits in neuroprotection and cardiovascular protection	Cioć et al. (2018), Wu et al. (2021), and Shan et al. (2022)
Ocimum basilicum (callus culture)	Blue light (460 nm) White (400–700 nm) LED light (660 nm)	Rosmarinic acid (96.0 mg/g DW) and eugenol (0.273 mg/g DW), 2.46 and 2.25 times greater than control, respectively Chicoric acid (81.40 mg/g DW) Peonidin (0.127 mg/g DW) Cyanidin (0.1216 mg/g DW)	Both eugenol and chicoric acid have antioxidant and anti-inflammatory activitiesAnthocyanins have potential roles in age-related diseases, including lifespan extension, neurodegenerative diseases, skeletal diseases, cardiovascular diseases, cancer, and metabolic syndrome.	Tsai et al. (2017), Nadeem et al. (2019), Damasceno et al. (2024), Wang et al. (2024a, b), and Ma et al. (2025)
Cnidium officinale Makin (callus culture)	(50% red: 50% blue LED light)	Total phenolic and flavonoid contents	Antioxidant and anti-inflammatory activity	Adil et al. (2019a, b)
Dendrobium hybrid orchid (protocorn-like body culture)	Blue: Red (1:1) LED light (peak at 440 and 660 nm, respectively)	Total flavonoid contents (16.79 μmol/g fresh weight)	Antioxidant and anti-inflammatory activity	Yeow et al. (2020)
Silybum marianum (callus culture)	Red LED light (660 nm)	Silymarin (17.67 mg/g DW) (twofold)	Hepatoprotective	Seo et al. (2015) and Younas et al. (2018)

Table 7.2 LED-induced anti-aging secondary metabolites derived from in vitro intact plants

Plant species (culture)	LED/light treatment	Enhanced metabolites	Anti-aging relevance	References
Curculigo orchioides CGC (leaves of synseed-derived seedlings)	Blue LED light (470)	Flavonoids 9.48 mg RE/g leaf fresh weight, and phenolics	Antioxidant and anti-inflammatory activity	Dutta Gupta et al. (2019)
Rehmannia glutinosa In vitro shoot culture CGC (root and leaf)	Blue LED light (450 nm) and blue: violet LED light (420–440 nm)	*p*-OH-cinnamic acid, chlorogenic, and cichoric	Cinnamic acid derivatives exhibit antioxidant, anti-inflammatory, and UV protection properties. Clorogenic acid activates the insulin/ IGF-1 signaling pathway, enhancing DAF-16 activity.	Gunia-Krzyżak et al. (2018), Taulavuori et al. (2018), and Yang et al. (2023)
Triticum aestivium (CGC seedling)	Blue LED light (470 nm) Red LED light (660 nm)	Gallic acid and quercetin Ferulic acid	Antioxidant and anti-inflammatory activity	Cuong et al. (2019)
Fagopyrum sp. (CGC seedling)	(Red: Blue LED ratio of 7:3) White light	Total phenolics (56.46 mg/g DW) Vitexin, rutin, and total flavonoid Anthocyanins (7.85 mg/g DW) Chlorogenic acid (5.62 mg/g DW)	Anthocyanins have antioxidant, anti-inflammatory, and metabolic regulatory properties	Ma et al. (2025)

CGC controlled growth chambers, *RE* rutin equivalents

7.6 Molecular and Biochemical Insights

7.6.1 Gene Expression Modulation under LED Spectra

By altering gene expression through the modulation of biosynthetic enzymes and transcription factors, in vitro LED treatments, particularly blue, red, or mixed wavelengths, significantly enhance the production of anti-aging secondary metabolites such as carotenoids, polyphenols, and curcuminoids. This response has been consistently observed across a variety of plant species, including Brassica, dill, basil relatives, turmeric, myrtle, and other medicinal herbs, suggesting that LED lighting is a powerful tool for controlling biosynthesis of beneficial phytochemicals, as shown in Table 7.3 and Fig. 7.3.

Table 7.3 LED-induced modulation of anti-aging metabolites

Plant	LED spectrum	Enhanced metabolites	Gene/TF modulation	References
Panax ginseng	Red and blue	Ginsenosides (e.g., Rg1, Rb1)	Increase *HMGR*, *SS*, *SE*, *DS* in ginsenoside pathway	Di et al. (2023)
Withania somnifera	Red, violet	Withanolides (Withaferin A), phenolics, flavonoids	Increase biomass and metabolites	Adil et al. (2019a, b)
Ocimum basilicum (Basil)	Supplemental 638 nm; 665 nm; B455 + R638	Increase Phenolics, α-tocopherol, ascorbic acid, DPPH scavenging capacity	Light-dose and wavelength-dependent antioxidant responses	Jung et al. (2021)
Petroselinum crispum (Parsley)	Supplemental 638 nm/665 nm red + blue baseline	Increase *β*-carotene, increase DPPH scavenging capacity	Similar photoprotective antioxidant activation	Samuolienė et al. (2016)
Brassica	Blue + white	Carotenoids	Increase *PSY*, *βLCY*, *βOHASE1*, *CCD4*	Frede et al. (2023)
Anethum graveolens	Red:Blue (50:50), blue	Chlorogenic and rosmarinic acids, monoterpenes	Increase phenylpropanoid and monoterpene genes	Goswami and Mitra (2023)
Agastache rugosa	White	Rosmarinic acid, tilianin	Increase *C4H*, *TAT*, *CHI*.	Jung et al. (2021)
Sarcandra glabra	Blue, red, white	Flavonoids, coumarins	Increase R2R3-MYB, bHLH TFs	Xie et al. (2020)
Myrtus communis	Red	Catechin, EGCG, myricetin, gallic acid	Increase phenolic synthesis	Cioć et al. (2018)
Peucedanum japonicum	3R:3B:3IR etc.	Chlorogenic acid	Regulation during somatic embryogenesis	Chen et al. (2016)
Curcuma longa	Red/blue combinations	Curcuminoids, polyphenols, antioxidants	General phytochemical enhancement	Marchant et al. (2022)

Blue LEDs are known to upregulate genes involved in secondary metabolism, stabilize the HY5 transcription factor, and activate cryptochrome receptors. Red LEDs influence the phenylpropanoid pathway and modulate phytochrome-interacting factors (PIFs) via phytochromes. When used together, red and blue lights often act synergistically to enhance antioxidant levels.

In *Panax ginseng*, exposure to red and blue LEDs significantly alters the expression of genes related to ginsenoside biosynthesis. Red light upregulates key biosynthetic genes, leading to increased ginsenoside accumulation. Ginsenoside Re content

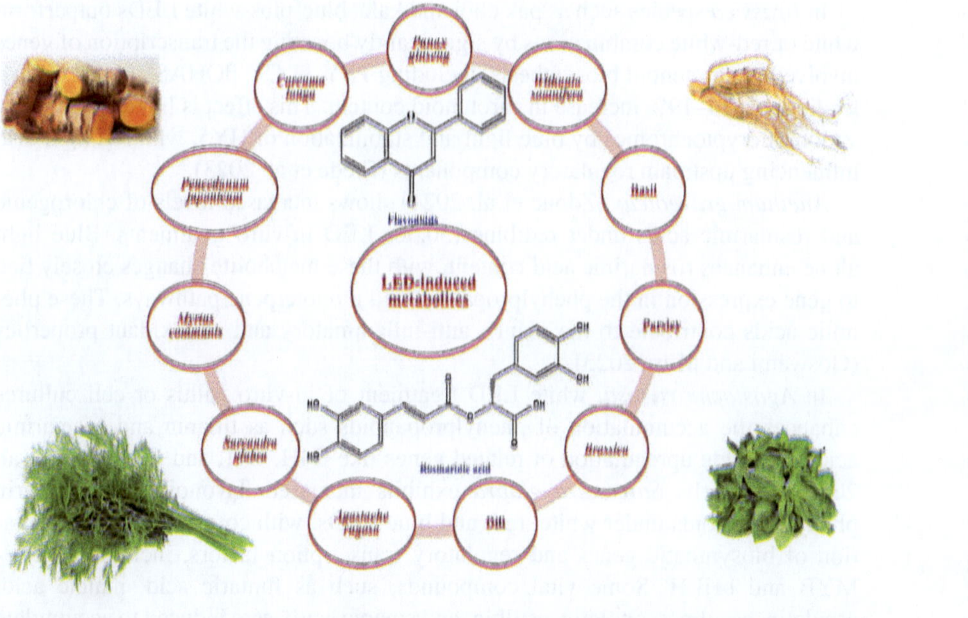

Fig. 7.3 LED-induced secondary metabolites in various plants. (Source: Diagram prepared by the authors)

was considerably elevated by blue- and red-light treatments, which were 14.85% and 7.01% greater than the CK treatment, respectively, and 30.19% and 21.31% higher than the natural light treatment (Di et al. 2023). In *Withania somnifera* (Ashwagandha), although gene-level data are limited, callus cultures exhibit enhanced biomass and withanolide production (e.g., withaferin A) under red light, while violet light increases flavonoid and phenol content. The quantity of total phenol and flavonoids in violet-light-grown callus was larger than that of other light treatments (42 and 2.5 mg/g DW, respectively). With phenolic and flavonoid concentrations of 39.6 mg/g DW and 2.08 mg/g DW, respectively, the red-light-treated callus comes in second. Interestingly, callus treated with white light showed decreased levels of flavonoids (0.57 mg/g DW) and phenolics (18.8 mg/g DW). These substances are crucial for reducing plant stress, and the amount of each varies according to the level of stress brought on by various monochromatic lighting (Adil et al. 2019a, b).

Ocimum basilicum (basil) exposed to red LEDs (638 and 665 nm) shows a decrease in lutein and β-carotene, but a marked increase in phenolics, α-tocopherol, ascorbic acid, and overall antioxidant activity via activation of the phenylpropanoid pathway (Jung et al. 2021). *Petroselinum crispum* (parsley) demonstrates improved DPPH scavenging activity and elevated β-carotene levels with red light exposure, suggesting a light-dose-dependent antioxidant enhancement. While only a substantial rise in the total phenolic content (41.5%) was seen in the parsley extract, HPS lams combined with R638 LEDs considerably caused accumulation of total phenols (36.7%), α-tocopherols (18.6%), lutein (48.8%), and β-carotene (47.9%) in basil extracts (Samuolienė et al. 2016).

In *Brassica* species such as pak choi and kale, blue-plus-white LEDs outperform white or red-white combinations by significantly boosting the transcription of genes involved in carotenoid biosynthesis, including PSY, βLCY, βOHASE1, and CCD4, leading to a 14–19% increase in carotenoid content. This effect is linked to the activation of cryptochromes by blue light and stabilization of HY5, with red light also influencing upstream regulatory components (Frede et al. 2023).

Anethum graveolens (Zdouc et al. 2024) shows increased levels of chlorogenic and rosmarinic acids under red:blue (50,50) LED in-vitro treatments. Blue light alone enhances rosmarinic acid content, with these metabolite changes closely tied to gene expression in the phenylpropanoid and monoterpene pathways. These phenolic acids contribute to the plant's anti-inflammatory and antioxidant properties (Goswami and Mitra 2023).

In *Agastache rugosa*, white LED treatment of in-vitro callus or cell cultures enhances the accumulation of phenylpropanoids such as tilianin and rosmarinic acid, alongside upregulation of related genes like C4H, TAT, and CHI (Jung et al. 2021). Similarly, *Sarcandra glabra* exhibits increased flavonoid and coumarin phenylpropanoids under white, red, and blue LEDs, with corresponding upregulation of biosynthetic genes and regulatory transcription factors, including R2R3-MYB and bHLH. Some vital compounds, such as fumaric acid, maleic acid, esculetin, esculin, scopoletin, astilbin, and sinapic acid, were induced to accumulate 1.24–1.77 times by red light (Xie et al. 2020).

Myrtus communis (common myrtle) produces peak levels of leaf flavonoids (e.g., catechin, epigallocatechin gallate, and myricetin) and phenolic acids such as gallic acid when cultured under red LEDs. In this setting, total secondary metabolite content reached approximately 1045 mg/100 g DW, compared to 674 mg under blue light (Cioć et al. 2018). *Peucedanum japonicum* somatic embryos and callus cultures also respond to specific LED spectra, producing significant amounts of chlorogenic acid ranging from 3.44 to 10.5 mg/g DW, with gene expression modulated throughout development (Chen et al. 2016).

Finally, *Curcuma longa* (turmeric) benefits from combined blue and red LED treatments, which enhance the accumulation of curcuminoids and polyphenols in vitro. The phytochemical concentration and antioxidant activity are, however, highly dependent on the specific spectral quality of the applied LEDs. Phytochemicals such as polyphenols, flavonoids, and reducing sugars were also more abundant in plants exposed to red:blue light, which promoted the production of curcumin during the acclimation period (Marchant et al. 2022).

7.6.2 Enzyme Activity and ROS Signaling

ROS as Signaling Molecules and Their Role Under LED Under the light spectrum, plants release reactive oxygen species (ROS) such as hydrogen peroxide (H_2O_2) and superoxide anion ($O_2^{\bullet-}$), which act as secondary messengers. These oxidants modify proteins, including transcription factors, through cysteine oxidation, thereby linking light perception to subsequent metabolic reactions. Narrow-spectrum LEDs, such as blue and red light, function as moderate stressors that increase ROS levels, which in turn activate secondary metabolite pathways like phenylpropanoid and flavonoid biosynthesis, as well as antioxidant enzymes (Yeow et al. 2020).

Enzyme Activity Changes Under LED Exposure Key enzymes of the phenylpropanoid pathway are upregulated under LED light treatments; for example, PAL, C4H, and 4CL show increased activity in *Agastache rugosa* when exposed to blue/red LEDs, leading to higher levels of tilianin and rosmarinic acid as shown in Table 7.4 (Park et al. 2020; Zhang et al. 2021). Similarly, CHI and TAT enzymes are activated by comparable LED conditions and play important roles in flavonoid production (Jung et al. 2021). Antioxidant enzymes that mitigate reactive oxygen species (ROS) are also influenced by LED exposure; activities of SOD, CAT, APX, and GR vary in red lettuce micro-shoots depending on the LED spectrum, reflecting a dynamic response to ROS (Sawatdee et al. 2023). Interestingly, red LED light quadrupled SOD activity and enhanced phenolic and flavonoid contents in *Silybum marianum* callus cultures compared to controls as presented in Table 7.4 (Hashim et al. 2021).

Linking Enzyme Activity to Secondary Metabolite Accumulation Flavonoids and phenolics serve dual roles as metabolic products and ROS scavengers. For instance, *Ocimum basilicum* callus exposed to blue/red LEDs shows increased levels of flavonoids such as cyanidin, chicoric acid, and rosmarinic acid (Hashim et al. 2021). Similarly, blue/red LED treatments enhance PAL enzyme activity in *Dendrobium*

Table 7.4 Effect of LED light spectra on enzyme activity and secondary metabolite accumulation in plant cultures

Plant/culture	Enzymes	Metabolites	LED Light	References
Agastache rugosa	PAL, C4H, 4CL, CHI, TAT	Rosmarinic acid, tilianin	White/blue/red LED	Park et al. (2020) and Zhang et al. (2021)
Silybum marianum	SOD	Silymarin compounds	Red LED	Hashim et al. (2021)
Ocimum basilicum	Phenylpropanoid pathway	Rosmarinic acid, cyanidin	Blue/red LED	Hashim et al. (2021)
Red lettuce	SOD, CAT, APX, GR	Phenolics, antioxidant capacity	Blue-white/red LED	Sawatdee et al. (2023)
Dendrobium hybrid PLBs	PAL, CHS	Total phenolics, flavonoids	Blue/red LED after pre-culture	Yeow et al. (2020)
Myrtus communis and *Panax ginseng*	Various antioxidant and biosynthetic enzymes	Diverse phenolic acids, flavonoids, saponins	Red, blue, green LEDs	Hashim et al. (2021)

protocorms, resulting in elevated production of flavonoids and phenolics (Yeow et al. 2020). The oxidative stress induced by LED exposure suggests a metabolic shift favoring the synthesis of antioxidant compounds (Yeow et al. 2020).

7.6.3 Synergistic Role of Light and Hormones

Light–Hormone Crosstalk in Plant Tissue Culture Hormones as integrators of light signals: Endogenous levels of auxin, cytokinins (CK), gibberellins (GA), ethylene, brassinosteroids (BR), and abscisic acid (ABA) are influenced by LED light, particularly blue and red, and help to coordinate stress signaling, metabolite production, and growth. Hormone-related genes such as auxin transporters (PINs), CK biosynthesis (e.g., IPT genes), and ethylene-related factors are regulated by HY5, which is stabilized by red and blue light photoreceptors, including phytochromes and cryptochromes, deactivating COP1 (Trivellini et al. 2023).

Auxin and Cytokinins Interaction with Light Red light influences auxin homeostasis by increasing the expression of PIN1, PIN3, and PIN4 transporters, thereby enhancing auxin movement toward the roots, which supports nutrient uptake and root development. In terms of CK regulation, light, particularly red light (RL) and white light (WL), promotes shoot proliferation under high CK conditions by upregulating genes responsible for CK biosynthesis, such as IPTs and LOG, while simultaneously downregulating CK degradation genes like CKX. Moreover, the combination of red LED light and benzylaminopurine (Zdouc et al. 2024) has been shown to synergistically enhance shoot proliferation and the accumulation of secondary metabolites, including flavonoids and phenolic acids, in species such as *Myrtus communis* and *Pinus radiate* as shown in Table 7.5 (Cavallaro et al. 2022).

Table 7.5 Effects of LED light treatments and hormonal interactions on plant growth and secondary metabolite production

Plant/system	LED treatment	Hormones involved	References
M. communis	100% red + BAP	CK + light → shoot proliferation and increase flavonoids (catechin, myricetin)	Hashim et al. (2021) and Cavallaro et al. (2022)
Lavandula, *M. communis* and *Spirea*	RL + BA/putrescine	Shoot induction via cytokinin-light synergy	Cavallaro et al. (2022)
Scrophularia kakudensis	Blue/red LED	ROS (via light) → antioxidant enzymes and PAL → phenolics	Manivannan et al. (2021)
Melissa officinalis	Red/blue mix	Light + increased gene expression (tyrosine aminotransferase, rosmarinic acid synthase, 3-deoxy-D-arabino-heptulosonate synthase) → rosmarinic acid and amino acids	Ahmadi et al. (2022)

Ethylene and Other Hormones Affected by Light The interaction between ethylene and light plays a significant role in organogenic responses, as ethylene signaling is influenced by light conditions. For example, the combination of ethylene, benzyladenine (BA), and specific LED spectra has been shown to enhance shoot development in *Spirea* (Cavallaro et al. 2022). Additionally, jasmonic acid (JA) and ROS appear to share similar signaling pathways, particularly in response to stress such as wounding, suggesting a coordinated mechanism where hormones and ROS act together to regulate phenylpropanoid-related enzymes like PAL, even though this specific connection is less emphasized in studies focusing on LED exposure (Torres-Contreras et al. 2023).

Redox Enzymes and Secondary Metabolite Biosynthesis ROS buildup is caused by LED-activated hormone signaling, which enhances the synthesis of secondary metabolites (phenolics, flavonoids), activates phenylpropanoid pathway enzymes (e.g., PAL), and stimulates antioxidant enzyme activity (e.g., SOD, CAT, APX, and GPX). Blue and red LEDs raised the activities of SOD, CAT, APX, and PAL in *Scrophularia kakudensis*, which resulted in higher metabolite levels, as shown in Table 7.5 (Manivannan et al. 2021).

7.7 LEDs in Scalable Culture Platforms

The incorporation of LEDs into scalable culture platforms provides major advantages over conventional lighting, including superior energy efficiency, reduced heat generation, and lower operational costs. Their long lifespan minimizes maintenance and ensures system stability. Most importantly, LEDs allow precise spectral tuning, enabling targeted wavelength emission to optimize photosynthesis, regulate secondary metabolite production, and tailor growth conditions to the specific physiological requirements of cultured organisms (Lee and Kim 2010; Dutta Gupta and Jatothu 2013; Mahari et al. 2024) as depicted in Fig. 7.4.

Recent research has shown that LED lighting can greatly improve the production of secondary metabolites that possess potent antioxidant and possible anti-aging effects. This enhancement occurs by adjusting the biosynthesis of specialized metabolites via targeted light wavelengths, which stimulate photoreceptors and affect gene expression as well as biosynthetic pathways. Utilizing LED technology in plant cultures has demonstrated encouraging outcomes in elevating the levels of diverse bioactive compounds with antioxidant characteristics, essential for anti-aging purposes.

In the callus cultures of *Moringa oleifera*, sustained exposure to white light led to a rise in the levels of kaempferol, neochlorogenic acid, and quercetin. Conversely, blue light was particularly effective in promoting the synthesis of chlorogenic acid, all of which are well-established antioxidants (Bajwa et al. 2023). In a similar manner, the application of blue LED light to the adventitious roots of *Panax ginseng* resulted in elevated concentrations of phenolic compounds, including ferulic acid,

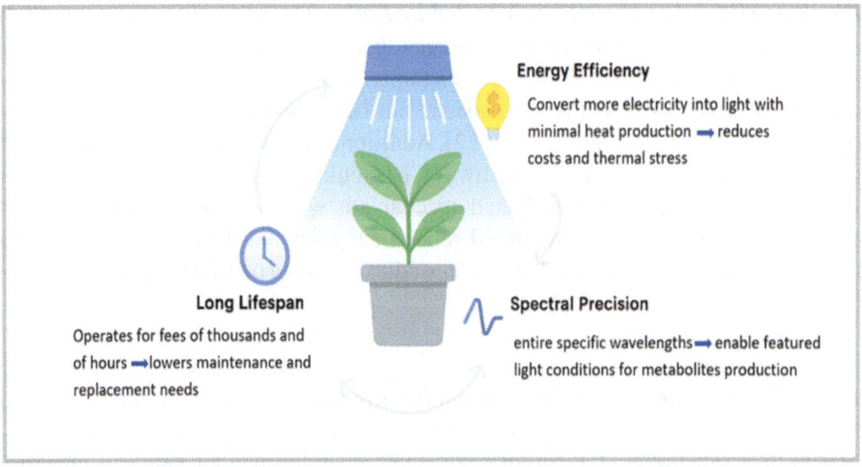

Fig. 7.4 Benefits and effects of LED integration in scalable culture platforms. (Source: Diagram prepared by the authors)

p-coumaric acid, and sinapic acid. Furthermore, this treatment also led to an increase in total tocopherol and sterol levels, thereby improving the overall antioxidant capacity of the tissue (Ghimire et al. 2017). In *Achilleae millefolii* L. (yarrow), in vitro cultures subjected to controlled lighting conditions revealed substantial amounts of flavonoids, phenolic acids, and antioxidant vitamins, i.e., C and E. Among these are bioactive compounds such as caffeic acid and apigenin, emphasizing their importance in geroprotective and anti-aging properties (Asyakina et al. 2021). Furthermore, these findings emphasize the importance of LED-mediated light quality in regulating phytochemical pathways, Fig. 7.5, suggesting a beneficial strategy for augmenting the nutraceutical and therapeutic benefits of medicinal plants.

Phenolic compounds, which encompass flavonoids, tannins, and phenolic acids, exhibit a strong correlation with the antioxidant capacity found in fruits and vegetables. The synthesis of these compounds is enhanced by LED lighting, especially blue light, which promotes pathways such as PAL, CHS, and DFR. Under blue LED treatment, flavonoids like isoflavones in soybean sprouts, as well as compounds like kaempferol and quercetin in tea plants, are significantly increased, highlighting the importance of wavelength-specific gene expression in the regulation of metabolism (Nam et al. 2018; Taulavuori et al. 2018). Anthocyanins, which are a subgroup of flavonoids, provide pigmentation, possess significant antioxidant properties, and offer various health advantages. Their accumulation is significantly influenced by light, with blue and red wavelengths typically promoting synthesis, whereas green light may reduce the accumulation induced by blue light (Giliberto et al. 2005; Li and Kubota 2009; Lee et al. 2016). Research on tomato, lettuce, kale, broccoli, and Arabidopsis validates the complex interplay of light signaling in the biosynthesis of anthocyanins (Stutte et al. 2009; Pinho et al. 2017) (Fig. 7.3).

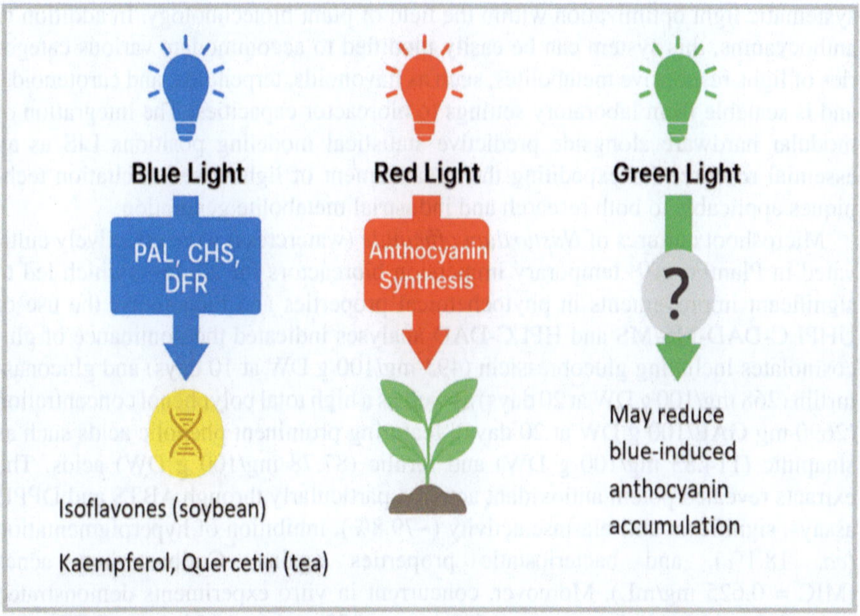

Fig. 7.5 LED light effects on biosynthetic pathways and antioxidant production. (Source: Diagram prepared by the authors)

7.7.1 Immersion Bioreactors

Immersion bioreactors that incorporate LED technology have become prominent as efficient platforms for increasing the yield of anti-aging secondary metabolites. This phenomenon highlights the merging of two crucial biotechnological innovations: temporary immersion systems (TIS) for plant tissue culture and LED technology for controlled illumination. Such integrated systems fulfil the rising demand for cost-effective and scalable techniques in plant micropropagation and the production of secondary metabolites (De Carlo et al. 2021; Mirzabe et al. 2022). The evolution of technology has arisen as a response to the challenges posed by standard tissue culture practices, characterized by increased labor expenses, potential contamination, and inadequate resource efficiency (Barretto et al. 2016; Monja-Mio et al. 2016).

Beuel et al. (2021) introduced LEDitSHAKE (LiS), a modular LED-based system designed to enhance secondary metabolite production in plant cell suspension cultures. The setup employs 3D-printed flasks, allowing 12 independent light treatments within a standard incubator. Using a design-of-experiments (Zdouc et al. 2024) approach on grapevine cells, they tested 24 combinations of red, green, blue, and UV light. Results revealed that blue and UV wavelengths promoted anthocyanin synthesis, while red light suppressed pigment formation and green light showed compound-specific effects. Optimized conditions increased peonidin- and cyanidin-glucosides up to 1.8- and 2.3-fold, respectively (Beuel et al. 2021). The research emphasizes the promise of LiS as a versatile and economic framework for systematic light optimization within the field of plant biotechnology. In addition to anthocyanins, this system can be easily modified to accommodate various categories of light-responsive metabolites, such as flavonoids, terpenoids, and carotenoids, and is scalable from laboratory settings to bioreactor capacities. The integration of modular hardware alongside predictive statistical modeling positions LiS as an essential resource for expediting the advancement of light-based elicitation techniques applicable to both research and industrial metabolite generation.

Microshoot cultures of *Nasturtium officinale* (watercress) were effectively cultivated in PlantForm™ temporary immersion bioreactors for 20 days, which led to significant improvements in phytochemical properties and bioactivity; the use of UHPLC-DAD-MS/MS and HPLC-DAD analyses indicated the dominance of glucosinolates including glucobrassicin (493 mg/100 g DW at 10 days) and gluconasturtiin (268 mg/100 g DW at 20 days) as well as a high total polyphenol concentration (2690 mg GAE/100 g DW at 20 days), featuring prominent phenolic acids such as sinapinic (114.83 mg/100 g DW) and ferulic (87.78 mg/100 g DW) acids. The extracts revealed potent antioxidant activity, particularly through ABTS and DPPH assays, significant anti-elastase activity (~79.8%), inhibition of hyperpigmentation (*ca.* 18.1%), and bacteriostatic properties against Cutibacterium acnes (MIC = 0.625 mg/mL). Moreover, concurrent in vitro experiments demonstrated that a defined LED lighting pattern 50% green, 35% red, and 15% blue (RGB) considerably enhanced both biomass accumulation (Gi ≈ 11.8) and glucosinolate production (up to 5.8 times higher than fluorescent controls), while ensuring similar profiles of phenolics, photosynthetic pigments, and antioxidant capacity (Klimek-Szczykutowicz et al. 2022, 2025a, b) (Fig. 7.6).

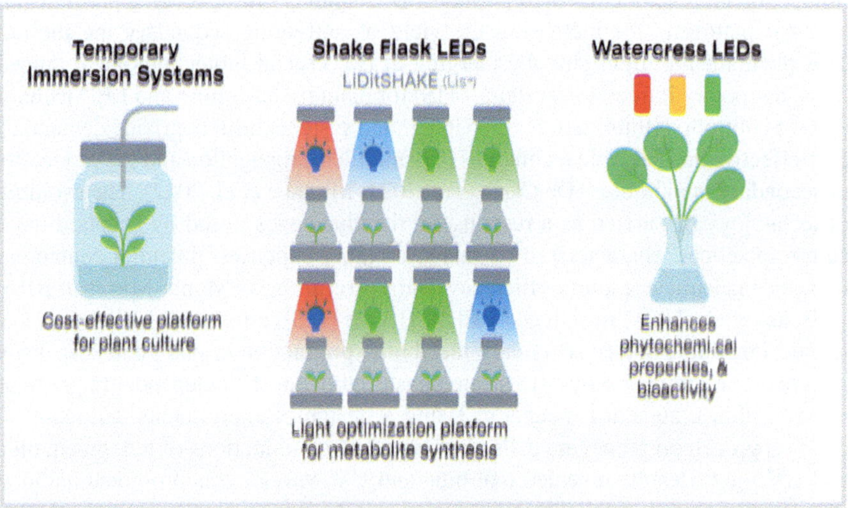

Fig. 7.6 LED-integrated immersion bioreactors. (Source: Diagram prepared by the authors)

7.7.2 Integration with Automation and Smart Farming

The incorporation of automation and intelligent farming into LED scalable platforms is essential for boosting agricultural productivity and sustainability. By utilizing Internet of Things (Boussaoudi et al. 2023) technologies, farmers are able to optimize resource management, enhance crop yields, and minimize environmental impacts. This integration allows for the collection and analysis of real-time data, which supports informed decision-making and effective farm management. LED lighting systems play a crucial role in contemporary controlled-environment agriculture, attributed to their exceptional energy efficiency, minimal thermal output, adjustable spectrum, and extended operational lifespan. Effective LED drivers are crucial as they manage the power supply and safeguard LEDs against voltage variations, thereby providing stable lighting that is vital for plant growth (Lozano-Castellanos et al. 2025).

Within smart farming applications, LED drivers facilitate the adjustable control of light intensity and spectrum, promoting enhanced photosynthesis, elevated crop quality, and efficient use of resources. LED drivers can be broadly classified into passive and switched-mode designs. Passive drivers utilize simple circuits that incorporate reactive components, ensuring reliability and extended operational life. However, they are prone to significant power losses, which diminishes their suitability for applications that demand high efficiency (Chen et al. 2019). On the other hand, switched-mode drivers, including buck, boost, buck-boost, Ćuk, and phase-shifted converters, work at high frequencies, which helps to minimize power loss and reduce size, while also delivering accurate current and voltage control with low ripple and high efficiency (Yadlapalli et al. 2020).

Recent innovations concentrate on single-stage LED driver architectures that merge power-factor correction (PFC) with conversion in a compact format, resulting in decreased costs and complexity. Illustrative cases consist of resonant DC–DC converters with switched capacitors, quadratic boost converters, and buck–boost systems that apply zero-voltage-switching strategies, achieving efficiencies of about 94.8% (Vishnuram et al. 2022). Multistring drivers that utilize single-inductor multiple-output (SIMO) architectures have been suggested, successfully attaining balanced current regulation and adjustable PWM dimming across various LED strings, thereby enhancing performance and minimizing power loss (Tahan and Hu 2017). The regulation of LED drivers within smart farming environments is largely contingent upon modulation strategies and automation. PWM (pulse-width modulation) is chosen over amplitude modulation for its superior linear control of brightness and its capacity to maintain spectral integrity (Hadj Abdelkader et al. 2023).

The use of microcontrollers in distributed control architectures, like Arduino and Raspberry Pi, governs dimming and environmental control Arduino is recognized for its low-power precision capabilities, while Raspberry Pi offers advanced analytics and integration with the Internet of Things (Lozano-Castellanos et al. 2025). In extensive indoor agricultural facilities, microcontrollers linked through 6LoWPAN networks and employing RPL routing protocols facilitate low-latency, energy-efficient control of lighting across multiple chambers (Hadj Abdelkader et al. 2023).

The performance of LED drivers is enhanced by intelligent control via advanced algorithms. Adaptive light regulation is achieved through PID controllers, often used in conjunction with clustering techniques like K-Means++ and FPKM, which balance energy usage with the needs of crops (Lozano-Castellanos et al. 2025). More advanced techniques such as Normal Boundary Intersection (NBI) for the optimization of light setpoints, genetic algorithms combined with digital twin simulations, and RBF neural networks enhanced through quantum genetic algorithms have been utilized for optical configuration and predictive control in smart agriculture (Lozano-Castellanos et al. 2025). Anticipating future developments, the integration of wide-bandgap semiconductors, Internet of Things (Boussaoudi et al. 2023) enabled smart drivers, and programmable lighting configurations are becoming prominent trends in the realm of energy-efficient, automated LED systems (Yadlapalli et al. 2020).

7.7.3 Economic and Environmental Perspectives

The transition from conventional lighting systems to LED technologies demonstrates clear economic and environmental advantages. It was showed that, although LEDs have higher upfront costs than fluorescent lamps, their longer lifespan, lower energy consumption, and reduced maintenance requirements make them more cost-effective over time, emphasizing the importance of lifecycle cost analysis (Pereira and Carvelli 2018), Similarly, replacing grid-powered high-pressure sodium (HPS) streetlights with solar-PV-powered LEDs could yield savings exceeding 70 million PKR within 25 years while enhancing municipal energy independence and budget stability (Touqeer et al. 2016).

7.8 Future Perspectives and Applications

7.8.1 Use of LEDs in Green Cosmetics and Nutraceuticals

Although there have been significant developments in plant biotechnology, there are very limited applications of LED-induced metabolites in large-scale industrial applications in cosmetics and nutraceuticals. Recent studies are mainly focused on improving the yield of secondary metabolites in vitro, and this has great potential in the future. With a tight control of spectral quality, intensity, and photoperiod, the LEDs are capable of triggering photoreceptors that control the production of phenolics, flavonoids, terpenoids, and anthocyanins, which are the most important antioxidant, anti-aging, and photoprotective compounds. Various in vitro studies point to the possibility of LEDs increasing the production of anti-aging metabolites. Blue light enhanced rosmarinic acid and eugenol levels in *Ocimum basilicum* by 2.3 and 2.1-fold, respectively (Nadeem et al. 2019). Red LEDs almost doubled the silymarin of *Silybum marianum*, increasing its nutraceutical value (Younas et al. 2018), and red-blue LEDs increased the wedelolactone of *Eclipta alba* by approximately

200%, which is associated with anti-inflammatory and collagen-protective benefits (Khurshid et al. 2020). However, these encouraging results are mostly pre-industrial since LED elicitation is still applied to optimize metabolites on experimental scales and not at large-scale production.

To further boost LED-induced metabolite accumulation, recent studies have integrated dual elicitation systems, combining light modulation with physical or biochemical stimuli. For example, *Dendrobium officinale* plantlets cultured in vitro under a high red: blue LED ratio (9:1) showed increased total phenolic content when supplemented with salicylic acid (1.5 mg/L) or yeast extract (1 g/L); the combined treatment outperformed LED alone both in phenolic yield (6.24 mg GAE/g DW, 1.57-times higher than the control) and in plantlet growth metrics (Huong et al. 2022). Similarly, in *Vitis vinifera* cell suspensions, exposure to red LED (1.34 μE/m^2/s, peak wave length 625 ± 10 nm) plus methyl jasmonate (10 μM dissolved in 100% *v/v* ethanol) elicitation produced significantly higher stilbene (90%), anthocyanin (52%, at day 9) and catechins levels, while red light alone had more modest effects; the combination also accelerated the time to maximum metabolite accumulation, indicating synergy between photoreceptor activation and hormone/elicitor signaling pathways (Tassoni et al. 2012). These results suggest that LED light can prime or enhance elicitor responsiveness, for instance by increasing expression of PAL, other phenylpropanoid-pathway enzymes, or antioxidant enzymes, thus leading to higher yields of bioactives relevant to antiwrinkle, UV protection, or skin repair formulations (Artés-Hernández et al. 2022).

Roadmap for Future Research The gap areas that hamper the use of LED in improving the production of anti-aging metabolites in vitro should be filled in future studies. The use of systematic studies in poorly studied species such as *Phyllanthus emblica*, which contain vitamin C and antioxidants, should be used to experiment with monochromatic versus mixed LED spectra in enhancing metabolites (Gautam et al. 2021). On the same note, *Bacopa monnieri*, which has well-established regeneration conditions, demonstrates that the LED wavelength, intensity and photoperiod, and especially ratios of red:blue can be adjusted to enhance growth and bacoside biosynthesis, indicating the possibility of a scalable production of cognitive-health nutraceuticals (Aasim 2018).

White and blue LED treatment of *Centella asiatica* enhanced centelloside levels by up to 73% and neuroprotective phenolics, which suggests that maximum spectral regimes with blue, red, far-red, or green LEDs could be used to enhance asiaticoside and madecassoside production further (Kamol et al. 2024). In the same way, the study of *Ginkgo biloba* propagation and seedlings demonstrates that the red and blue light ratios alter the growth and flavonol structure, which may suggest that appropriate ratios of red: blue LED could be adopted to enhance flavonol glycosides related to vascular and neuroprotective anti-aging effects (Wang et al. 2022).

Future work should combine the application of multi-omics tools, including transcriptomics, metabolomics, proteomics, in conjunction with LED treatments to identify genetic and metabolic mechanisms underlying light-regulated biosynthetic processes. To determine yield, reproducibility, and efficiency, it is necessary to scale production in bioreactors. Testing should go as far as anti-aging, collagen preservation, and delayed senescence. Advances will be based on interdisciplinary cooperation and the creation of standard spectral databases.

7.8.2 Opportunities for Precision Agriculture and Personalized Plant-Based Therapies

Benefits in the Realm of Precision Agriculture LED technology plays a pivotal role in precision agriculture by enhancing plant growth, nutritional quality, and bioactive compound accumulation through precise spectral control. It was demonstrated that red-blue LED supplementation in crops like pak choi, tomato, and lettuce significantly improved light-use efficiency up to 115% compared to fluorescent lighting, while promoting higher biomass, larger leaf area, reduced nitrate levels, and increased antioxidant content. These improvements stem from LEDs' spectral precision, energy efficiency, and durability (Bian et al. 2017). Furthermore, the integration of customized spectra with intelligent control systems enables real-time optimization of plant growth conditions *via* automation and AI, transforming LEDs into adaptive tools for sustainable, resource-efficient, and economically viable controlled environment agriculture (Al Murad et al. 2021; Arcel et al. 2021).

Livadariu and his research team conducted a comprehensive synthesis of 95 studies examining the impact of LED lighting on plant growth and bioactive compound accumulation. Their analysis demonstrates that spectrally tuned LED illumination in controlled environments significantly improves crop nutritional quality by enhancing the production of phenols, flavonoids, carotenoids, terpenes, and glucosinolates. Specific light wavelengths were found to activate photoreceptor-mediated pathways, boosting enzyme activity and secondary metabolite biosynthesis. Moreover, LEDs proved effective in both pre- and postharvest applications due to their precise wavelength control, low heat output, and compatibility with closed cultivation systems. Overall, the study highlights LEDs as energy-efficient tools for maximizing bioactive compound content and maintaining superior crop quality (Livadariu et al. 2023).

Impact on Plant-Based Therapies About 95 studies demonstrate that LED lighting in controlled environments enhances plant growth and bioactive compound synthesis, notably increasing phenols, flavonoids, carotenoids, terpenes, and glucosinolates through photoreceptor-activated metabolic pathways, while improving postharvest quality with minimal energy use (Livadariu et al. 2023). Complementarily, another work emphasized the synergy between LED technology and intelligent control systems in plant factories, highlighting how adjustable light spectra and low heat emission optimize photomorphogenesis, yield, and quality. This investigation also indicated the global adoption of automated, sensor-based, and AI-driven LED systems that enable precise, energy-efficient, and sustainable crop production in modern agriculture (Arcel et al. 2021).

7.9 Conclusion

Enhancing the production of anti-aging secondary metabolites in plant in vitro cells by LED-based elicitation is a viable and environmentally friendly approach. By altering certain wavelengths, especially red and blue light, LEDs can promote gene expression and metabolic pathways associated with beneficial phytochemicals. This method provides a scalable and repeatable platform for producing high-quality bioactives, facilitating potential uses in nutraceuticals and sustainable cosmetics. Across a variety of plant species, LED light spectrums, particularly red and blue wavelengths, effectively increase gene expression involved in the production of important secondary metabolites, including flavonoids, polyphenols, and carotenoids. These lights enhance the activity of antioxidant enzymes by activating transcription factors and photoreceptors, which in turn initiate metabolic pathways and increase ROS signaling. LED light also affects the manufacturing and signaling of hormones, including auxin, cytokinin, and ethylene, which promote the growth of roots and shoots as well as the storage of metabolites. With important implications in metabolic engineering and therapeutic plant biotechnology, the precise and long-lasting method of maximizing plant growth and secondary metabolism in vitro is made possible by the synergistic interplay of light and hormones.

Acknowledgment Ahmed Zayed is thankful to the State Government of Rhineland-Palatinate (Germany) within the framework of the Biotech Fellow International Research for funding the project "Intensification of the Production of Biologically Active Substances: Plant Cell Cultures as Production Systems".

Competing Interests The authors have no relevant financial or non-financial interests to disclose.

References

Aasim M (2018) Efficacy of light emitting diodes (LEDs) lighting system for in vitro shoot regeneration of medicinal water hyssop (*Bacopa monnieri* L. PENNEL). Romanian Biotechnol Lett 23:13197–13204

Abdelhamid SA, Marzouk AI, Asker MS, El Shabrawi HM (2025) RED light promotes flavonoid and phenolic accumulation in *Cichorium* spp. callus culture as anti-candida agent. Sci Rep 15:2194

Abdelkader MM, Tammam AK (2021) Organic-inorganic hybrid lead halide perovskites for optoelectronic and electronic applications. In: Khan I, khan a, khan MMA et al (eds) Hybrid perovskite composite materials. Woodhead publishing, pp 267–289

Adil M, Abbasi BH, ul Haq I (2019a) Red light controlled callus morphogenetic patterns and secondary metabolites production in *Withania somnifera* L. Biotechnol Rep 24:e00380

Adil M, Ren X, Jeong BR (2019b) Light elicited growth, antioxidant enzymes activities and production of medicinal compounds in callus culture of Cnidium officinale Makino. J Photochem Photobiol B: Biol 196:111509

Ahmadi T, Shabani L, Sabzalian MR (2022) Light emitting diodes improved the metabolism of rosmarinic acid and amino acids at the transcriptional level in two genotypes of *Melissa officinalis* L. Funct Plant Biol 49:1055–1069

Ahmadi T, Shabani L, Sabzalian MR, Hassannejad S (2025) Comparative analysis of LED priming effects on two medicinal lemon balm genotypes one and three weeks post-drought stress. BMC Plant Biol 25(1):266

Al Murad M, Razi BR, Jeong PM, Samy A, Muneer S (2021) Light emitting diodes (LEDs) as agricultural lighting: impact and its potential on improving physiology, flowering, and secondary metabolites of crops. Sustainability 13(4):1985

Alallaq S, Ranjan A, Brunoni F, Novák O, Lakehal A, Bellini C (2020) Red light controls adventitious root regeneration by modulating hormone homeostasis in *Picea abies* seedlings. Front Plant Sci 11:586140

Alrifai O, Hao X, Marcone MF, Tsao R (2019) Current review of the modulatory effects of lED lghts on photosynthesis of secondary metabolites and future perspectives of microgreen vegetables. J Agric Food Chem 67(22):6075–6090

Arcel MM, Lin X, Huang J, Wu J, Zheng S (2021) The application of LED illumination and intelligent control in plant factory, a new direction for modern agriculture: a review. J Phys Conf Ser. 1732 1, 012178. IOP Publishing

Artés-Hernández F, Castillejo N, Martínez-Zamora L (2022) UV and visible Spectrum LED lighting as abiotic elicitors of bioactive compounds in sprouts, microgreens, and baby leaves-a comprehensive review including their mode of action. Foods 11(3):265

Asyakina LK, Fotina NV, Izgarysheva NV, Slavyanskiy AA, Neverova OA (2021) Geroprotective potential of in vitro bioactive compounds isolated from yarrow (*Achilleae millefolii* L.) cell cultures. Foods and Raw Materials 9(1):126–134

Azizi S, Aliniaeifard S, Zarbakhsh S, Esmaeili S, Baghalian K, Gruda NS (2025) Photobiology, photosynthesis, and plant responses under artificial lighting in controlled environment agriculture. Sci Hortic 349:114248

Bajwa MN, Khanum M, Zaman G, Ullah MA, Farooq U, Waqas M, Ahmad N, Hano C, Abbasi BH (2023) Effect of wide-spectrum monochromatic lights on growth, phytochemistry, nutraceuticals, and antioxidant potential of in vitro callus cultures of *Moringa oleifera.* Molecules 28(3):1497

Barretto S, Michoux F, Nixon PJ (2016) Temporary immersion bioreactors for the contained production of recombinant proteins in transplastomic plants. In: Recombinant Proteins from Plants: Methods and Protocols. Springer, pp 149–160

Beuel A-K, Jablonka N, Heesel J, Severin K, Spiegel H, Rasche S (2021) LEDitSHAKE: a lighting system to optimize the secondary metabolite content of plant cell suspension cultures. Sci Rep 11(1):23353

Bian Z, Jiang N, Grundy S, Lu C (2017) Uncovering LED light effects on plant growth: new angles and perspectives-LED light for improving plant growth, nutrition and energy-use efficiency. In: International symposium on new technologies for environment control, energy-saving and crop production in greenhouse and plant, p 1227

Biswal B, Jena B, Giri AK, Acharya L (2022) Monochromatic light elicited biomass accumulation, antioxidant activity, and secondary metabolite production in callus culture of *Operculina turpethum* (L.). Plant Cell Tissue Organ Cult 149(1):123–134

Bom S, Jorge J, Ribeiro H, Marto J (2019) A step forward on sustainability in the cosmetics industry: a review. J Clean Produc 225:270–290

Boussaoudi I, Mtil N, Lamarti A, Cherroud S, Ainane A, Ainane T, Biotechnology A (2023) Antimicrobial and antioxidant activities of bay laurel (*Laurus nobilis* L.) essential oil from the region of Khenifra (Morocco). J Anal Sci Appl Biotechnol 5(2):116–124

Bylka W, Znajdek-Awiżeń P, Studzińska-Sroka E, Brzezińska M (2013) *Centella asiatica* in cosmetology. Postepy Dermatol Alergol 30(1):46–49

Cai X, Huq E (2024) Shining light on plant growth: recent insights into phytochrome-interacting factors. J Exp Bot 76(3):646–663

Cavallaro V, Muleo R (2022) The effects of LED light spectra and intensities on plant growth. Plants 11:1911

Cavallaro V, Pellegrino A, Muleo R, Forgione I (2022) Light and plant growth regulators on in vitro proliferation. Plants 11(7):844

Chen C-C, Agrawal DC, Lee M-R, Lee R-J, Kuo C-L, Wu C-R, Tsay H-S, Chang H-C (2016) Influence of LED light spectra on in vitro somatic embryogenesis and LC–MS analysis of chlorogenic acid and rutin in *Peucedanum japonicum* Thunb.: a medicinal herb. Bot Stud 57(1):9

Chen W, Cheng KWE, Shao J (2019) Circuit topology analysis for led lighting and its formulation development. Energies 12(21):4203

Cioć M, Szewczyk A, Żupnik M, Kalisz A, Pawłowska B (2018) LED lighting affects plant growth, morphogenesis and phytochemical contents of *Myrtus communis* L. in vitro. Plant Cell Tissue Organ Cult 132(3):433–447

Cuong DM, Ha TW, Park CH, Kim NS, Yeo HJ, Chun SW, Kim C, Park SU (2019) Effects of LED lights on expression of genes involved in phenylpropanoid biosynthesis and accumulation of phenylpropanoids in Wheat Sprout. Agronomy 9(6):307

Damasceno ROS, Pinheiro JLS, Rodrigues LHM, Gomes Duarte ABS, Emídio JJ, Diniz LRL, de Sousa DP (2024) Anti-Inflammatory and antioxidant activities of eugenol: an update. Pharmaceuticals 17(11):1505

De Carlo A, Tarraf W, Lambardi W, Benelli C (2021) Temporary immersion system for production of biomass and bioactive compounds from medicinal plants. Agronomy 11(12):2414

Di P, Sun Z, Cheng L, Han M, Yang L, Yang L (2023) LED light irradiations differentially affect the physiological characteristics, ginsenoside content, and expressions of ginsenoside biosynthetic pathway genes in panax ginseng. Agriculture 13(4):807

Disasa D, Cheng L, Manzoor M, Liu Q, Wang Y, Xiang L, Qi J (2020) Amarogentin from *Gentiana rigescens* franch exhibits antiaging and neuroprotective effects through antioxidative stress. Oxidative Med Cell Longev 2020:3184019

Dixit V, Chaubey KK, Dayal D, Chole PB, Bachheti MBTRK, Gupta PC, Bachheti A, Worku LA (2025) Natural products in aging: cellular mechanisms and emerging therapeutics for age-related disorders. J Aging Res 2025(1):6868732

Dutta Gupta S, Jatothu B (2013) Fundamentals and applications of light-emitting diodes (LEDs) in in vitro plant growth and morphogenesis. Plant Biotechnol Rep 7(3):211–220

Dutta Gupta S, Kumar A, Agarwal A (2019) Impact of light-emitting diodes (LEDs) on the growth and morphogenesis of encapsulated shoot buds of *Curculigo orchioides* Gaertn., an endangered medicinal herb. Acta Physiol Plant 41(4):50

Espinosa-Leal CA, Puente-Garza CA, García-Lara S (2018) In vitro plant tissue culture: means for production of biological active compounds. Planta 248(1):1–18

Fazili MA, Bashir I, Ahmad M, Yaqoob U, Geelani SN (2022) In vitro strategies for the enhancement of secondary metabolite production in plants: a review. Bull Natl Res Cent 46(1):35

Frede K, Winkelmann S, Busse L, Baldermann S (2023) The effect of LED light quality on the carotenoid metabolism and related gene expression in the genus *brassica*. BMC Plant Biol 23(1):328

Gai Q-Y, Feng X, Jiao J, Xu X-J, Fu J-X, He X-J, Fu Y-J (2023) Blue LED light promoting the growth, accumulation of high-value isoflavonoids and astragalosides, antioxidant response, and biosynthesis gene expression in *Astragalus membranaceus* (Fisch.) Bunge hairy root cultures. Plant Cell Tissue Organ Cult 153(3):511–523

Gam DT, Khoi PH, Ngoc PB, Linh LK, Hung NK, Anh PTL, Thu NT, Hien NTT, Khanh TD, Ha CH (2020) Blue LED light promoting the growth, accumulation of high-value isoflavonoids and astragalosides, antioxidant response, and biosynthesis gene expression in *Astragalus membranaceus* (Fisch.) Bunge hairy root cultures. Plant Cell Tissue Organ Cult 9(10):1344

Gautam 1A, Pant M, Pant G, Pant K, Yadav MRJW (2021) In vitro propagation of *Phyllanthus emblica* (L.) through different explants. Webology 18(1):1488–1502

Gawande V, Raut S, Beese PB, Agnihotri N (2023) Artificial light spectra and its impact on plant physiological processes and secondary metabolism. Int J Plant Soil Sci 35(18):2060–2070

Ghimire BK, Lee JG, Yoo JH, Kim JK, Yu CY (2017) The influence of light-emitting diodes (LEDs) on the growth, antioxidant activities, and metabolites in adventitious root of Panax ginseng CA Meyer. In: Light emitting diodes for agriculture: smart lighting. Springer, pp 259–272

Giliberto L, Perrotta G, Pallara P, Weller JL, Fraser PD, Bramley PM, Fiore A, Tavazza M, Giuliano G (2005) Manipulation of the blue light photoreceptor cryptochrome 2 in tomato affects vegetative development, flowering time, and fruit antioxidant content. Plant Physiol 137(1):199–208

Girgenti A, Picone P, Buttacavoli M, Palumbo L, Naselli F, Presti EL, Pecora D, Cipollina C, Annunziata F, Pinto A, Tamborini L, Nuzzo D (2025) Mitochondria-targeted rosmarinic acid: its role against oxidative damage. Biomed Pharmacother 187:118114

Goswami A, Mitra A (2023) Light spectra manipulation stimulates growth, specialized metabolites and nutritional quality in *Anethum graveolens*. J Photochem Photobiol B Biol 249:112812

Goyal N, Jerold F (2023) Biocosmetics: technological advances and future outlook. Environ Sci Pollut Res 30(10):25148–25169

Gunia-Krzyżak A, Słoczyńska K, Popiół J, Koczurkiewicz P, Marona H, Pękala E (2018) Cinnamic acid derivatives in cosmetics - current use and future prospects. Int J Cosmet Sci 40(4):356–366

Gupta R, Sood H (2023) Emerging technologies for the production of in vitro raised quality rich Swertia chirayita by using LED lights. Plants 15(2):1714

Hadj Abdelkader O, Bouzebiba H, Pena D, Aguiar AP (2023) Energy-efficient IoT-based light control system in smart indoor agriculture. Sensors 23(18):7670

Han T, Miao G (2024) Strategies, achievements, and potential challenges of plant and microbial chassis in the biosynthesis of plant secondary metabolites. Molecules 29(9):2106

Hart JE, Gardner KH (2021) Lighting the way: recent insights into the structure and regulation of phototropin blue light receptors. J Biol Chem 296:100594

Hashim M, Ahmad B, Drouet S, Hano C, Abbasi BH, Anjum S (2021) Comparative effects of different light sources on the production of key secondary metabolites in plants in vitro cultures. Plants 10(8):1521

Hassanzadeh-Taheri M, Ahmadi-Zohan A, Mohammadifard M, Hosseini M (2021) Rosmarinic acid attenuates lipopolysaccharide-induced neuroinflammation and cognitive impairment in rats. J Chem Neuroanat 117:102008

Hatipoglu D, Keskin E (2022) The effect of curcumin on some cytokines, antioxidants and liver function tests in rats induced by aflatoxin B1. Heliyon 8(7):e09890

Huong T, Trang H, Giap D, Trần T-T, Hoang T, Khai H, Nguyễn M, Duong TN (2022) Influence of LED light, salicylic acid and yeast extract on growth and phenolic content of *Dendrobium officinale* plantlets. Asian J Plant Sci 22:13–24

Hussen NHA, Abdulla SK, Ali NM, Ahmed VA, Hasan AH, Qadir EE (2025) Role of antioxidants in skin aging and the molecular mechanism of ROS: a comprehensive review. Aspects Mol Med 5:100063

Jenkins GI (2017) Photomorphogenic responses to ultraviolet-B light. Plant Cell Environ 40(11):2544–2557

Jung W-S, Chung I-M, Hwang MH, Kim S-H, Yu CY, Ghimire BK (2021) Application of light-emitting diodes for improving the nutritional quality and bioactive compound levels of some crops and medicinal plants. Molecules 26(5):1477

Kamol P, Nukool W, Pumjaroen S, Inthima P, Kongbangkerd A, Suphrom A, Buddhachat K (2024) Harnessing postharvest light emitting diode (LED) technology of *Centella asiatica* (L) Urb to improve centelloside content by up-regulating gene expressions in the triterpenoid pathway. Heliyon 10(1):e23639

Karimi M, Ahmadi N, Ebrahimi M (2022) Red LED light promotes biomass, flowering and secondary metabolites accumulation in hydroponically grown *Hypericum perforatum* L. (cv. Topas). Ind crops PRO 175:114239

Khurshid R, Ullah MA, Tungmunnithum D, Drouet S, Shah M, Zaeem A, Hameed S, Hano C, Abbasi BH (2020) Lights triggered differential accumulation of antioxidant and antidiabetic secondary metabolites in callus culture of *Eclipta alba* L. PLoS One 15(6):e0233963

Kim E-Y, Park S-A, Park B-J, Lee Y, Oh M-M (2014) Growth and antioxidant phenolic compounds in cherry tomato seedlings grown under monochromatic light-emitting diodes. Hortic Environ Biotechnol 55(6):506–513

Klimek-Szczykutowicz M, Prokopiuk B, Dziurka K, Pawłowska B, Ekiert H, Szopa A (2022) The influence of different wavelengths of LED light on the production of glucosinolates and phenolic compounds and the antioxidant potential in in vitro cultures of *Nasturtium officinale* (watercress). Plant Cell Tissue Organ Cult 149(1):113–122

Klimek-Szczykutowicz M, Ekiert H, Szopa A (2025a) Increasing of biologically active metabolite production using plant biotechnology techniques in *Nasturtium officinale* (watercress) micro-shoot cultures. In: Phytochemicals for health. Elsevier, Amsterdam, pp 193–223

Klimek-Szczykutowicz M, Malinowska MA, Gałka A, Blažević I, Đulović A, Paprocka P, Wrzosek M, Szopa A (2025b) Nasturtium officinale microshoot culture multiplied in PlantForm bioreactor—phytochemical profiling and biological activity. Molecules 30(4):936

Kong SG, Okajima K (2016) Diverse photoreceptors and light responses in plants. J Plant Res 129(2):111–114

Krzemińska M, Hnatuszko-Konka K, Weremczuk-Jeżyna I, Owczarek-Januszkiewicz A, Ejsmont W, Olszewska MA, Grzegorczyk-Karolak I (2023) Effect of light conditions on polyphenol production in transformed shoot culture of *salvia bulleyana* Diels. Molecules 28(12):4603

Kusuma P, Pattison PM, Bugbee B (2022) Chapter 7 – Photon efficacy in horticulture: turning LED packages into LED luminaires. In: Kozai T, Niu G, Masabni J (eds) Plant Factory Basics, Applications and Advances. Academic Press, pp 115–128

Kwon MC, Choi WY, Seo YC, Kim JS, Yoon CS, Lim HW, Kim HS, Ahn J, Lee HY (2012) Enhancement of the skin-protective activities of *Centella asiatica* L. urban by a Nano-encapsulation process. J Biotechnol 157(1):100–106

Lazzarin M, Meisenburg M, Meijer D, van Ieperen W, Marcelis LFM, Kappers IF, van der Krol AR, van Loon JJA, Dicke M (2021) LEDs make it resilient: effects on plant growth and defense. Trends Plant Sci 26(5):496–508

Le AT, Phan NH, Do TK (2021) Effect of blue light on the photosynthesis and flavonoid accumulation in leaves of *Hedyotis corymbosa* (L.) Lam. Plant Sci Today 8(4):955–962

Lee C-G, Kim Z-H (2010) Photobioreactor for large-scale culture of microalgal. Patents, Google

Lee M-J, Son K-H, Oh M-M (2016) Increase in biomass and bioactive compounds in lettuce under various ratios of red to far-red LED light supplemented with blue LED light. Hortic Environ Biotechnol 57(2):139–147

Lee SWR, Jeffery CCL, Mian T, Huaiyu Y (2022) LEDs for solid-state lighting. In: From LED to Solid State Lighting: Principles, Materials, Packaging, Characterization, and Applications. Wiley, pp 1–13

Li Q, Kubota C (2009) Effects of supplemental light quality on growth and phytochemicals of baby leaf lettuce. Environ Exp Bot 67(1):59–64

Li X, Liu Z, Ren H, Kundu M, Wang L, Gao J, Zhong D (2020) Dynamics and mechanism of light harvesting in UV photoreceptor UVR8††electronic supplementary information (ESI) available: methods. Chem Sci 11(46):12553–12569

Li Z, Zhang Z, Ren Y, Wang Y, Fang J, Yue H, Ma S, Guan F (2021) Aging and age-related diseases: from mechanisms to therapeutic strategies. Biogerontology 22(2):165–187

Lim WH, Khaw ML, Yungeree O, Hew WH, Parab AR, Chew BL, Wahyuni DK, Subramaniam S (2025) Effects of LEDs, macronutrients and culture conditions on biomass and artemisinin production using *Artemisia annua* L. suspension cultures. Biotechnol Prog 41:e70041

Livadariu O, Maximilian C, Rahmanifar B, Cornea CP (2023) LED technology applied to plant development for promoting the accumulation of bioactive compounds: a review. 12(5):1075

Lozano-Castellanos LF, Navas-Gracia LM, Lozano-Castellanos IC, Correa-Guimaraes A (2025) Technologies applied to artificial lighting in indoor agriculture: a review. Sustainability 17(7):3196

Lu C, Zou Y, Liu Y, Niu Y (2017) Rosmarinic acid counteracts activation of hepatic stellate cells via inhibiting the ROS-dependent MMP-2 activity: involvement of Nrf2 antioxidant system. Toxicol Appl Pharmacol 318:69–78

Ma X, Jin Z, Rao Z, Zheng L (2025) Health benefits of anthocyanins against age-related diseases. Front Nutr 12:1618072

Mahari WAW, Razali WAW, Waiho K, Wong KY, Foo SS, Kamaruzzan AS, Derek CJC, Ma NL, Chang J-S, Dong C-D (2024) Light-emitting diodes (LEDs) for culturing microalgae and cyanobacteria. Chem Eng J 485:149619

Maldonado E, Morales-Pison S, Urbina F, Solari A (2023) Aging hallmarks and the role of oxidative stress. Antioxidants 12(3):651

Malekzadeh MR, Roosta HR, Kalaji HM (2024) Enhancing strawberry resilience to saline, alkaline, and combined stresses with light spectra: impacts on growth, enzymatic activity, nutrient uptake, and osmotic regulation. BMC Plant Biol 24(1):1038

Manivannan A, Soundararajan P, Park YG, Jeong BR (2020) Physiological and proteomic insights into red and blue light-mediated enhancement of in vitro growth in *Scrophularia kakudensis*-a potential medicinal plant. Front Plant Sci 11:607007

Marchant MJ, Molina P, Montecinos M, Guzmán L, Balada C, Castro M (2022) Effects of LED light spectra on the development, phytochemical profile, and antioxidant activity of *Curcuma longa* from Easter Island. Plants 11(20):2701

Marino LA, Ruffa P, Mozzanini E, Patono DL, Sereno A, Pavese V (2025) LEDs in plant tissue culture: boosting micropropagation of *Castanea sativa* cultivars. J Plant Growth Regul. https://doi.org/10.1007/s00344-025-11812-6

Martín Giménez VM, de las Heras N, Lahera V, Tresguerres JAF, Reiter RJ, Manucha W (2022) Melatonin as an anti-aging therapy for age-related cardiovascular and neurodegenerative diseases. Front Aging Neurosci 14(14):888292

Martinez AA, Bonilla JO, Aguilera LJ, Callegari EA, Paez MD, Kurina-Sanz M, Orden AA (2024) Light-mediated elicitation of *Tessaria absinthioides* calli for the production of anthocyanins: correlation with protein expression. Plant Cell Tissue Organ Cult 159(1):20

Martínez-Chávez LA, Hernández-Ramírez MY, Feregrino-Pérez AA, Esquivel Escalante K (2024) Cutting-edge strategies to enhance bioactive compound production in plants: potential value of integration of elicitation, metabolic engineering, and green nanotechnology. Agronomy 14(12):2822

Matthews JSA, Vialet-Chabrand S, Lawson T (2020) Role of blue and red light in stomatal dynamic behaviour. J Exp Bot 71(7):2253–2269

Miliauskienė J, Karlicek RF Jr, Kolmos E (2021) Effect of multispectral pulsed light-emitting diodes on the growth, photosynthetic and antioxidant response of baby leaf lettuce (*Lactuca sativa* L.). Plants 10(4):762

Mirzabe AH, Hajiahmad A, Fadavi A, Rafiee S (2022) Temporary immersion systems (TISs): a comprehensive review. J Biotech 357:56–83

Mohaddab M, Goumi YE, Gallo M, Montesano D, Zengin G, Bouyahya A, Fakiri M (2022) Biotechnology and in vitro vulture as an alternative system for secondary metabolite production. Molecules 27(22):8093

Monja-Mio KM, Herrera-Alamillo MÁ, Robert ML (2016) Somatic embryogenesis in temporary immersion bioreactors. In: Somatic embryogenesis: fundamental aspects and applications. Springer, pp 435–454

Nadeem M, Abbasi BH, Younas M, Ahmad W, Zahir A, Hano C (2019) LED-enhanced biosynthesis of biologically active ingredients in callus cultures of *Ocimum basilicum*. J Photochem Photobiol B 190:172–178

Nam TG, Kim D-O, Eom SH (2018) Effects of light sources on major flavonoids and antioxidant activity in common buckwheat sprouts. Food Sci Biotech 27(1):169–176

Nhani GBB, Di Filippo LD, de Paula GA, Mantovanelli VR, da Fonseca PP, Tashiro FM, Monteiro DC, Fonseca-Santos B, Duarte JL, Chorilli M (2024) High-tech sustainable beauty: exploring nanotechnology for the development of cosmetics using plant and animal by-products. Cosmetics 11(4):112

Nowakowska K, Kowalczyk K, Pawełczak A, Gajc-Wolska J (2023) Energy efficiency of LEDs during micropropagation of Helleborus 'Molly's white. Agriculture 13(6):1265

Ogwu MC, Izah SC, Joshua MT (2025) Ecological and environmental determinants of phytochemical variability in forest trees. Phytochem Rev. https://doi.org/10.1007/s11101-025-10066-0

Paik I, Huq E (2019) Plant photoreceptors: multi-functional sensory proteins and their signaling networks. Semin Cell Dev Biol 92:114–121

Park WT, Yeo SK, Sathasivam R, Park JS, Kim JK, Park SU (2020) Influence of light-emitting diodes on phenylpropanoid biosynthetic gene expression and phenylpropanoid accumulation in *Agastache rugosa*. Appl Biol Chem 63(1):25

Pereira YT, Carvelli E (2018) Vantagens econômicas, ambientais e de durabilidade da lâmpada de LED em comparação às lâmpadas fluorescentes tubulares. Uningá Rev 33(2):180–190

Pinho P, Jokinen K, Halonen L (2017) The influence of the LED light spectrum on the growth and nutrient uptake of hydroponically grown lettuce. Light Res Technol 49(7):866–881

Ptak A, Szewczyk A, Simlat M, Pawłowska B, Warchoł M (2024) LED light improves shoot multiplication, steviol glycosides and phenolic compounds biosynthesis in *Stevia rebaudiana* Bertoni in vitro culture. Sci Rep 14(1):30860

Pulido-Moran M, Moreno-Fernandez J, Ramirez-Tortosa C, Ramirez-Tortosa M (2016) Curcumin and health. Molecules 21(3):264

Rafeie M, Shabani L, Sabzalian MR, Gharibi S (2022) Pretreatment with LEDs regulates antioxidant capacity and polyphenolic profile in two genotypes of basil under salinity stress. Protoplasma 259(6):1567–1583

Reshi ZA, Ahmad W, Lukatkin AS, Javed SB (2023) From nature to lab: a review of secondary metabolite biosynthetic pathways, environmental influences, and in vitro approaches. Meta 13(8):895

Riaz M, Khalid R, Afzal M, Anjum F, Fatima H, Zia S, Rasool G, Egbuna C, Mtewa AG, Uche CZ, Aslam MA (2023) Phytobioactive compounds as therapeutic agents for human diseases: a review. Food Sci Nutr 11(6):2500–2529

Rocca R, Acerbi F, Fumagalli L, Taisch M (2022) Sustainability paradigm in the cosmetics industry: state of the art. Cleaner Waste Syst 3:100057

Roeber VM, Bajaj I, Rohde M, Schmülling T, Cortleven A (2021) Light acts as a stressor and influences abiotic and biotic stress responses in plants. Plant Cell Environ 44(3):645–664

Samuolienė G, Brazaitytė A, Viršilė A, Jankauskienė J, Sakalauskienė S, Duchovskis P (2016) Red light-dose or wavelength-dependent photoresponse of antioxidants in herb microgreens. PLoS One 11(9):e0163405

Sawatdee S, Jarunglumlert T, Pavasant P, Sakihama Y, Flood AE, Prommuak C (2023) Effect of mixed light emitting diode spectrum on antioxidants content and antioxidant activity of red lettuce grown in a closed soilless system. BMC Plant Biol 23(1):351

Sena S, Kumari S, Kumar V, Husen A (2024) Light emitting diode (LED) lights for the improvement of plant performance and production: a comprehensive review. Curr Res Biotechnol 7:100184

Seo J-M, Arasu MV, Kim Y-B, Park SU, Kim SU (2015) Phenylalanine and LED lights enhance phenolic compound production in Tartary buckwheat sprouts. Food Chem 177:204–213

Serrano A, Vanhaelewyn L, Vandenbussche F, Boccalandro H, Maldonado B, Van Der Straeten D, Ballaré C, Arana M (2021) Cryptochromes are the dominant photoreceptors mediating heliotropic responses of Arabidopsis inflorescences. Plant Cell Environ 44:3246–3256

Shan H, Geng L, Jiang X, Song M, Wang J, Liu Z, Zhuo X, Wu Z, Hu J, Ji Z, Wang S, Chan P, Qu J, Zhang W, Liu GH (2022) Large-scale chemical screen identifies gallic acid as a geroprotector for human stem cells. Protein Cell 13(7):532–539

Shiade SRG, Zand-Silakhoor A, Fathi A, Rahimi R, Minkina T, Rajput VD, Zulfiqar U, Chaudhary T (2024) Plant metabolites and signaling pathways in response to biotic and abiotic stresses: exploring bio stimulant applications. Plant Stress 12:100454

Shin J-W, Kwon S-H, Choi J-Y, Na J-I, Huh C-H, Choi H-R, Park K-C (2019) Molecular mechanisms of dermal aging and antiaging approaches. Int J Mol Sci 20(9):2126

Sobhy MK, Mowafy S, Lasheen DS, Farag NA, Abouzid KAM (2019) 3D-QSAR pharmacophore modelling, virtual screening and docking studies for lead discovery of a novel scaffold for VEGFR 2 inhibitors: design, synthesis and biological evaluation. Bioorg Chem 89:102988

Stutte GW, Edney S, Skerritt T (2009) Photoregulation of bioprotectant content of red leaf lettuce with light-emitting diodes. Hort Science 44(1):79–82

Tahan M, Hu T (2017) Multiple string LED driver with flexible and high-performance PWM dimming control. IEEE Trans Power Electron 32(12):9293–9306

Tassoni A, Durante L, Ferri M (2012) Combined elicitation of methyl-jasmonate and red light on stilbene and anthocyanin biosynthesis. J Plant Physiol 169(8):775–781

Taulavuori K, Pyysalo A, Taulavuori E, Julkunen-Tiitto R (2018) Responses of phenolic acid and flavonoid synthesis to blue and blue-violet light depends on plant species. Environ Exp Bot 150:183–187

Tenchov R, Sasso JM, Wang X, Zhou QA (2024) Aging hallmarks and progression and age-related diseases: a landscape view of research advancement. ACS Chem Neurosci 15(1):1–30

Torres-Contreras AM, Nair V, Senés-Guerrero C, Pacheco A, González-Agüero M, Ramos-Parra PA, Cisneros-Zevallos L, Jacobo-Velázquez DA (2023) Cross-talk and physiological role of jasmonic acid, ethylene, and reactive oxygen species in wound-induced phenolic biosynthesis in broccoli. Plants 12(7):1434

Touqeer J, Memon H, Soomro S, Tunio N (2016) Economic and environmental analysis of converting grid supplied HPS lights to solar PV powered LEDs in street lighting at Khairpur Mirs' Pakistan. Indian J Sci Technol 9(47):1–6

Trivellini A, Toscano S, Romano D, Ferrante a (2023) the role of blue and red light in the orchestration of secondary metabolites, nutrient transport and plant quality. Plants 12(10):2026

Tsai KL, Kao CL, Hung CH, Cheng YH, Lin HC, Chu PM (2017) Chicoric acid is a potent anti-atherosclerotic ingredient by anti-oxidant action and anti-inflammation capacity. Oncotarget 8(18):29600–29612

Vishnuram P, Narayanamoorthi R, Vijayakumar K, Bajaj M, Shouran M, Abdul Samad B, Kamel S (2022) A comprehensive review on various non-isolated power converter topologies for a light-emitting diode driver. Front Energy Res 10:1055258

Wang G, Zhang L, Wang G, Cao F (2022) Growth and flavonol accumulation of *Ginkgo biloba* leaves affected by red and blue light. Ind Crop Prod 187:115488

Wang X, Ma Y, Xu Q, Shikov AN, Pozharitskaya ON, Flisyuk EV, Liu M, Li H, Vargas-Murga L, Duez P (2023) Flavonoids and saponins: what have we got or missed? Phytomedicine 109:154580

Wang B, Tang X, Mao B, Zhang Q, Tian F, Zhao J, Cui S, Chen W (2024a) Anti-aging effects and mechanisms of anthocyanins and their intestinal microflora metabolites. Crit Rev Food Sci Nutr 64(8):2358–2374

Wang B, Zhu S, Guo M, Ma RD, Tang YL, Nie YX, Gu HF (2024b) Artemisinin ameliorates cognitive decline by inhibiting hippocampal neuronal ferroptosis via Nrf2 activation in T2DM mice. Mol Med 30(1):35

Wong TI, Zhou X (2024) Photon efficacy and cost investigation of LEDs at different wavelengths and color temperatures for horticulture. Discov Agric 2(1):125

Wu YZ, Tsai YY, Chang LS, Chen YJ (2021) Evaluation of Gallic acid-coated gold nanoparticles as an anti-aging ingredient. Pharmaceuticals 14(11):1071

Wu D, Liu M, Yu W, Cui M, Huang X, Ning F, Chingin K, Luo L (2022) Red:Blue LED light proportion affects biomass accumulation and polyamine metabolism in *Anoectochilus roxburghii* studied by nano-electrospray ionization mass spectrometry. Ind Crop Prod 188:115636

Xia S, Zhang X, Zheng S, Khanabdali R, Kalionis B, Wu J, Wan W, Tai X (2016) an update on Inflamm-aging: mechanisms, prevention, and treatment. J Immunol Res 2016(1):8426874

Xie D, Chen L, Zhou C, Tarin MWK, Yang D, Ren K, He T, Rong J, Zheng Y (2020) Transcriptomic and metabolomic profiling reveals the effect of LED light quality on morphological traits, and phenylpropanoid-derived compounds accumulation in *Sarcandra glabra* seedlings. BMC Plant Biol 20(1):476

Xu Y, Yang M, Cheng F, Liu S, Liang Y (2020) Effects of LED photoperiods and light qualities on in vitro growth and chlorophyll fluorescence of *Cunninghamia lanceolata*. BMC Plant Biol 20(1):269

Yadlapalli RT, Narasipuram RP, Kotapati A (2020) An overview of energy efficient solid state LED driver topologies. Int J Energy Res 44(2):612–630

Yang Y, Chen X, Ye D, Wei C, Pang X, Kong C, Fang Y, Yang H, Zhang Y, Liu Y (2023) Exploring the anti-aging effects of chlorogenic acid and the underlying mechanisms based on a *Caenorhabditis elegans* model. J Tradit Chin Med Sci 10(2):208–217

Yaoyuan Z, Lau N-S, Subramaniam S (2025) Illuminating the harvest: the regulatory effects of LEDs on pigment accumulation in various food crops. Physiol Mol Biol Plants 31(5):693–708

Yeow LC, Chew B, Sreeramanan S (2020) Elevation of secondary metabolites production through light-emitting diodes (LEDs) illumination in protocorm-like bodies (PLBs) of *dendrobium* hybrid orchid rich in phytochemicals with therapeutic effects. Biotechnol Rep 27:e00497

Younas M, Drouet S, Nadeem M, Giglioli-Guivarc'h N, Hano C, Abbasi BH (2018) Differential accumulation of silymarin induced by exposure of *Silybum marianum* L. callus cultures to several spectres of monochromatic lights. Photochem Photobiol B Biol 184:61–70

Zauli I, Appolloni E, Carotti L, Paucek I, Quaini S, Pennisi G, Orsini F, Gianquinto G (2022) Effects of different LEDs wavelengths on secondary metabolites accumulation in medicinal plants cultivated in vitro: a review of recent literature. Acta Hortic 1337:233–240

Zdouc M M, Blin K, Louwen Nico L L et al (2024) MIBiG 4.0: advancing biosynthetic gene cluster curation through global collaboration. Nucleic Acids Res 53(D1):D678–D690

Zeng D, Lv J, Li X, Liu H (2025) The Arabidopsis blue-light photoreceptor CRY2 is active in darkness to inhibit root growth. Cell 188(1):60–76.e20

Zhang S, Zhang L, Zou H, Qiu L, Zheng Y, Yang D, Wang Y (2021) Effects of light on secondary metabolite biosynthesis in medicinal plants. Front Plant Sci 12:781236

Zheng M, Xin Y, Li Y, Xu F, Xi F, Guo H, Cui X, Cao H, Zhang X, Han C (2018) Ginsenosides: a potential neuroprotective agent. Biomed Res Int 2018(1):8174345

Zhiponova M, Zehirov G, Rusanov K, Rusanova M, Stefanova M, Ganeva T, Paunov M, Ganeva V, Mishev K, Dobrev PI, Vaculíková R, Motyka V, Yordanova Z, Chaneva G, Vassileva V (2025) Blue–red LED light modulates morphophysiological and metabolic responses in the medicinal plant *Nepeta nuda*. Plants 14(15):2285

Enhancement of Anti-aging Properties of Medicinal Plants Through Induced Auto-polyploidization

8

Rohit Bharati and Eloy Fernández-Cusimamani

Abstract

Aging is a natural biological process that causes gradual deterioration of physiological functions, oxidative stress, and cellular damage. It can be influenced by genetic factors as well as environmental stressors. Although aging is irreversible, its progression can be modulated by various interventions, including the use of plant-based bioactive compounds. Natural compounds with anti-aging potential have recently gained considerable traction due to their safety, low toxicity, broad biological activities, and cost-effectiveness. Medicinal plants are a vast inventory of these compounds, offering a wide range of phytochemicals such as polyphenols, flavonoids, terpenoids, and alkaloids, which are known for their ability to combat oxidative stress, inflammation, and other aging-related processes. Enhancing the production of these compounds within the plants themselves offers a sustainable and efficient approach to enhancing the anti-aging properties of these plants. Induced autopolyploidization has emerged as a promising biotechnological tool that alters gene dosage and cellular metabolism, significantly enhancing morphophysiological traits of plants. Polyploid genotypes often show elevated biosynthesis of phytochemicals by many folds, including those with anti-aging properties, compared to their diploid counterparts, making it a valuable strategy in medicinal plant improvement. Hence, this chapter details the molecular and physiological mechanisms underlying metabolite enrichment,

R. Bharati (✉)
Plant Virus and Vector Interactions, Czech Agrifood Research Center, Prague, Czech Republic

Faculty of Economics and Management, Department of Economic Theories, Czech University of Life Sciences Prague, Prague 6, Czech Republic
e-mail: rohit.bharati@carc.cz

E. Fernández-Cusimamani
The Faculty of Tropical AgriSciences, Department of Crop Sciences and Agroforestry, Czech University of Life Sciences Prague, Prague 6, Suchdol, Czech Republic

S. Gantait et al. (eds.), *Medicinal Plants with Anti-Ageing Properties*,
https://doi.org/10.1007/978-981-95-8064-4_8

reviews effective antimitotic agents used for induction, and outlines methodologies for polyploid identification and screening. Furthermore, it highlights key medicinal species demonstrating improved anti-aging efficacy upon polyploidization and discusses future opportunities for combining polyploidy with advanced biotechnologies to optimize therapeutic outcomes. Overall, this approach offers a sustainable and innovative strategy to develop superior medicinal plants that mitigate multiple hallmarks of cellular aging, supporting natural solutions for healthy aging and longevity.

Keywords

Anti-mitotic agents · Artificial polyploidization · Bioactive compounds · Biotechnology · Medicinal plants · Metabolites

8.1 Introduction

Aging is an inevitable, multifactorial biological process characterized by the decline of cellular and physiological integrity over time. It arises from the cumulative effects of molecular damage, metabolic dysregulation, and impaired cellular homeostasis, ultimately leading to functional deterioration and disease susceptibility. Recent advances in biogerontology categorize aging as an interplay of 12 interconnected hallmarks, classified into three major categories, including primary, antagonistic, and integrative hallmarks, that collectively result in aging phenotypes (López-Otín et al. 2023).

Primary hallmarks include genomic instability, telomere attrition, epigenetic alterations, and loss of proteostasis, which represent the fundamental sources of molecular damage that initiate the aging process. These alterations compromise genome integrity, chromatin architecture, and protein homeostasis, thereby undermining cellular stability and function. Antagonistic hallmarks, including disabled macroautophagy, deregulated nutrient sensing, mitochondrial dysfunction, and cellular senescence, are adaptive responses that initially counteract damage but become detrimental when chronically activated, resulting in metabolic imbalance and oxidative stress. Lastly, integrative hallmarks include stem cell exhaustion, altered intercellular communication, chronic inflammation, and dysbiosis, which result in organism-wide dysfunctions that integrate and amplify the effects of earlier molecular and cellular disturbances, consolidating the physiological decline associated with aging (López-Otín et al. 2023).

While aging is a multifactorial and irreversible process, several of its hallmarks can be modulated or mitigated through bioactive natural metabolites. Compounds such as polyphenols, flavonoids, terpenoids, and alkaloids act on multiple levels of this network, for instance, they can reduce genomic instability and telomere attrition by enhancing DNA repair and telomerase activity (Matsuno et al. 2020; Kapetanou et al. 2024; Liu et al. 2024), attenuate epigenetic drift through regulation of sirtuins and histone-modifying enzymes (Malireddy et al. 2012; Rahnasto-Rilla et al. 2018; Arora et al. 2020; Rogina and Tissenbaum 2024), and help maintain proteostasis by stimulating chaperone activity and autophagic clearance (Yessenkyzy et al. 2020; García-Aguilar et al. 2021; Jafari et al. 2024). Likewise, metabolites like resveratrol, quercetin, curcumin, and rosmarinic acid restore mitochondrial efficiency, activate macroautophagy, and rebalance nutrient-sensing pathways such as AMPK and mTOR, thereby alleviating oxidative stress and cellular senescence (Ungvari et al. 2011; Gibellini et al. 2015; Kim et al. 2024). At the systemic level, many plant-derived antioxidants suppress chronic inflammation, preserve stem cell function, and contribute to a healthy microbiome composition (Yahfoufi et al. 2018; Rodríguez-Vera et al. 2022; Anghel et al. 2024; Pereira et al. 2024). Thus, natural metabolites serve as pleiotropic regulators that can counteract several hallmarks simultaneously, offering a biologically harmonious strategy for promoting longevity and functional health.

Medicinal and aromatic plants are rich reservoirs of these bioactive compounds such as polyphenols, flavonoids, alkaloids, and terpenoids that can scavenge free radicals, reduce inflammation, and enhance cellular repair mechanisms. However, the natural abundance of these bioactive compounds often remains suboptimal for large-scale therapeutic or nutraceutical applications, necessitating novel strategies to sustainably enhance their production within plant system. In recent years, induced autopolyploidization, which involves the artificial doubling of the chromosome set using a suitable anti-mitotic agent, has emerged as a promising biotechnological approach to boost secondary metabolite synthesis. It has been employed to several medicinal plants to elevate the anti-aging metabolites and properties (Table 8.1). Polyploid induction often works by altering gene dosage, cellular metabolism, and epigenetic regulation, and can significantly enhance the biosynthetic capacity of medicinal plants without introducing genetic material from external sources (Chen 2007). This chapter examines the role of induced autopolyploidization in enhancing the anti-aging properties of medicinal and aromatic plants. It discusses the molecular and physiological mechanisms underpinning this enhancement, outlines the methodological toolkit for polyploid induction and screening, and highlights the potential of integrating polyploidy with modern biotechnological tools to optimize therapeutic efficacy and productivity.

8.2 Artificial Polyploidization: Mechanism and Relevance in Enhancing Anti-aging Properties

The anti-aging potential of many medicinal plants is strongly linked to their secondary metabolite profile, including flavonoids, phenolics, alkaloids, terpenoids, and antioxidants that mitigate oxidative stress and cellular senescence. Artificial polyploidization can reprogram plant metabolism by altering gene dosage, epigenetic and transcriptional networks regulation, ultimately enhancing the biosynthesis of bioactive compounds with anti-aging efficacy. Figure 8.1 illustrates three panels: the first panel highlights a simplified process of induced autopolyploidization, showing the key stages and factors involved in chromosome doubling; the second panel depicts the molecular modulation mechanisms leading to enhanced

Table 8.1 Medicinal plants with enhanced anti-aging properties through induced auto-polyploidization: key metabolites, mechanisms, and impact of polyploidy

Plant species	Key metabolite/s	Anti-aging mechanism	Impact of polyploidy on anti-aging properties and associated metabolites	References
Zingiber officinale	6-Gingerol	Antioxidant, anti-inflammatory, neuroprotective via gingerols and shogaols; prevents oxidative stress, reduces reactive oxygen species (ROS), enhances antioxidant enzyme activity (SOD, GPx, CAT); protects against age-related degenerative diseases	In vitro induction of stable tetraploids by colchicine treatment increased 6-gingerol content significantly (498.36 μg/g) compared to controls (375.20 μg/g). Tetraploids showed superior morphological features, increased photosynthesis, and molecular genetic variation confirming successful polyploid induction and enhanced metabolite production	Mohd Sahardi and Makpol (2019) and Kumar et al. (2024)
Panax ginseng C.A. Meyer	Ginsenosides (especially Rg1)	Antioxidant, anti-inflammatory, neuroprotective via ginsenosides (Rg1, Rg3, Rb1); promotes DNA damage repair, reverses cell cycle arrest, enhances collagen synthesis, regulates inflammatory cytokines (IL-6, TNF-α); improves skin firmness and elasticity	Colchicine-induced octoploidy increased fresh and dry biomass significantly compared to natural tetraploids. Total and Rb-group ginsenoside content decreased, but Rg-group ginsenosides, notably Rg1, increased by 22%, which has neuroprotective and anti-aging activities. Polyploid induction was successful and beneficial for selective metabolite enhancement	Kim et al. (2004, 2025) and Kan et al. (2023)

Talinum paniculatum (Jacq.) Gaertn.	Various metabolites including mascaroside, undecylenic acid, tetracosatetraenoic acid	Antioxidant, free radical scavenger via flavonoids and saponins; protects skin from oxidative damage, enhances skin vitality, reduces visible aging signs; supports antioxidant defense mechanisms	Tetraploids induced by colchicine (optimal 1–3 mg/mL for 48 h) showed significant phenotypic changes (e.g., larger leaves, fruits, pollen) but reduced aboveground biomass. Metabolomic analysis revealed 22 metabolite concentrations significantly increased (e.g., 27.99% increase in mascaroside, 15.19% in undecylenic acid) and 74 significantly decreased. Polyploid induction was successful, and the metabolite profile was altered in favor of some medicinal compounds	Emelda et al. (2022) and Liu et al. (2025)
Bacopa monnieri L. Wettst	Bacosides (A, Bacopaside II, Bacopaside X, Bacopasaponin C)	Neuroprotective, antioxidant via bacosides A and B, alkaloids, and flavonoids; prevents neuronal apoptosis, inhibits acetylcholinesterase, repairs damaged neurons through kinase activity stimulation; reduces oxidative stress markers and pro-inflammatory cytokines (TNF-α, IL-6)	Colchicine treatment (optimal 0.05% for 48 h) on node and leaf explants effectively induced stable tetraploids with higher biomass and greater multiple shoot regeneration. Tetraploid plants showed significantly increased bacoside-A3, bacoside-II, and bacoside-C content compared to diploid progenitors. The BmoOSC gene expression, linked to triterpenoid biosynthesis, was significantly upregulated in tetraploids, correlating with increased metabolite production. Mixoploids showed intermediate changes	Sujipuli et al. (2021) and Govindarasu and Murugesh (2025)
			Colchicine treatment (0.5% for 48 h) induced autotetraploidy with up to 14.6% efficiency. Autotetraploid clones showed significantly increased fresh weight (up to 2.8-fold), dry weight (up to 2.0-fold), and total bacoside content (up to 2.3-fold compared to diploids). Polyploidy altered morphological traits and boosted metabolite production, enhancing the medicinal value of the plant	Inthima and Sujipuli (2019)

(continued)

Table 8.1 (continued)

Plant species	Key metabolite/s	Anti-aging mechanism	Impact of polyploidy on anti-aging properties and associated metabolites	References
Glycyrrhiza glabra var. glandulifera	Glycyrrhizic acid, anthocyanin	Antioxidant, anti-inflammatory, collagen-boosting via glycyrrhizin, flavonoids, and isoflavones; brightens skin, inhibits melanin production, reduces fine lines, provides UV protection; scavenges DPPH free radicals	Seeds treated with colchicine (0.03–0.1% for 24 h) resulted in increased cellular DNA content, indicating polyploidy (tetraploids and mixoploids). Treated plantlets showed increased stomata size, thicker roots, shoots, and leaves, though shorter length and slower growth rate. Callus cultures from these plantlets showed a significant increase in anthocyanin (up to 3.5-fold at 0.08% colchicine) and glycyrrhizic acid (up to 2.3-fold compared to control), especially in 0.1% and 0.03% treatments. This confirms colchicine-induced polyploidy effectively enhances secondary metabolite production linked to anti-aging	Bernard et al. (2012) and Wahab et al. (2021)
Centella asiatica L. Urban	Triterpenoids: Asiaticoside, Madecassoside, Asiatic acid, Madecassic acid	Collagen-stimulating, antioxidant, anti-inflammatory via triterpenes (madecassoside, asiaticoside, asiatic acid, madecassic acid); promotes wound healing, enhances type I collagen synthesis via TGF-β/Smad signaling pathway; strengthens skin barrier function and reduces transepidermal water loss	Tetraploids induced with 0.05–0.2% colchicine showed larger stomata and increased stomatal index. Metabolite content showed an 11% increase in total triterpenoids vs diploids, although statistically not significant	Kaensaksiri et al. (2011), Wong et al. (2021), and Bikiaris et al. (2025)
			Tetraploid lines showed significantly greater leaf size, petiole length, and biomass, including 70% higher total triterpenoids after 4 months' cultivation. Quantitative analysis confirmed higher yields of asiaticoside, madecassoside, and overall triterpenoids, supporting polyploidy for metabolite enhancement	Thong-on et al. (2014)
			Tetraploids were induced by treating shoot tips with 0.05% colchicine via cotton plug method. Tetraploid plants showed significant morphological changes including increased stolon length, leaf size, petiole length, and biomass yield. Phytochemical analysis revealed that metabolite content per unit biomass (asiaticoside, madecassoside, asiatic acid, madecassic acid) was similar between diploid and tetraploid plants, but due to increased biomass, the total metabolite yield per hectare increased substantially (asiaticoside yield increased by ~71.7% compared to diploid)	Rohini et al. (2024)

Wedelia chinensis (Osbeck) Merrill	Wedelolactone, demethylwedelolactone, luteolin, apigenin	Antioxidant, cholinesterase-inhibitory, neuroprotective via apigenin and flavonoids; prevents neurodegeneration, protects against oxidative stress and lipid peroxidation; exhibits DPPH radical scavenging superior to standard catechin	Colchicine treatment (0.025–0.2%) on in vitro apical shoots and nodal segments induced stable triploid and tetraploid plants without obvious morphological changes except larger stomata. Metabolite profiling via UPLC-MS/MS showed fluctuating patterns: some polyploid plants had significantly increased demethylwedelolactone (up to ~1.5-fold), wedelolactone (up to 22.74 μg/g vs 18.19 μg/g in WT), and apigenin (up to ~1.4-fold), while others showed decreases. Luteolin generally decreased or remained similar. No single polyploid plant consistently increased all four metabolites	Islam et al. (2021) and Tsai et al. (2021)
Salvia officinalis	Rosmarinic acid, catechin, caffeic acid, ferulic acid, chlorogenic acid, other phenolics	Antioxidant, anti-aging enzyme inhibitor via phenolics, flavonoids, and rosmarinic acid; inhibits collagenase, elastase, hyaluronidase (up to 50% inhibition of aging enzymes); reduces wrinkle scores, improves skin firmness and elasticity	Polyploidy was induced by colchicine treatment producing autotetraploid genotypes. Diploids had higher total phenolic content and rosmarinic acid than autotetraploids. Antioxidant activities varied by method; diploids generally had higher antiradical and antioxidant activities except for the b-carotene/linoleic acid bleaching assay where autotetraploids performed better. Differing metabolite synthesis patterns suggest complex regulation of secondary metabolite accumulation in polyploids	Ceramella et al. (2025) and Tuğlu et al. (2025)
			In vitro and in vivo colchicine treatments (0.05–0.2%, 24–48 h) induced tetraploids (highest efficiency 19.05%). HPLC showed a significant decrease in ursolic acid and rosmarinic acid content in polyploids compared to diploids. However, betulinic acid content significantly increased in both in vitro and in vivo polyploids. Expression of triterpene biosynthesis genes (squalene epoxidase SQE and lupeol synthase LUS) was significantly elevated in polyploids (up to 2.9-fold in vitro tetraploids). Overall, polyploidization randomly alters metabolite profiles, enhancing some compounds like betulinic acid correlated with gene upregulation, but reducing others like ursolic and rosmarinic acid	Tavan et al. (2021)

(continued)

Table 8.1 (continued)

Plant species	Key metabolite/s	Anti-aging mechanism	Impact of polyploidy on anti-aging properties and associated metabolites	References
Mentha spicata L.	Carvone, limonene	Antioxidant, cognitive-enhancing via rosmarinic acid and polyphenols; improves working memory, reduces protein carbonyls in hippocampus, protects against oxidative byproducts; enhances sleep quality and mood	Polyploid plants (hexaploids and mixoploids) were induced from in vitro nodal segments treated with oryzalin at 20, 40, and 60 μM for 24 and 48 h, yielding 8% polyploid induction at 40 μM/48 h. Essential oil yield was increased by up to 48.85% compared to triploid controls (2.59% vs 1.74%). GC-MS analysis revealed significant increases in major oil components, carvone and limonene, in specific genotypes	Bharati et al. (2023a) and Silva et al. (2023)
Melissa officinalis L.	Rosmarinic acid, geranial, neral, citronellal	Antioxidant, neuroprotective, cognitive-enhancing via rosmarinic acid and caffeic acid; anti-inflammatory, improves cognitive function, treats Alzheimer's disease symptoms; reduces oxidative damage in neural tissues	Oryzalin treatment (20, 40, 60 μM for 24–48 h) on nodal segments induced stable tetraploids confirmed by flow cytometry and chromosome counting ($2n = 4x = 64$). Peltate trichome size, responsible for oil secretion, increased by 45%, leading to a 75% increase in essential oil yield. GC-MS revealed increased geranial (+11.06%) and neral (+9.49%) content in tetraploids, while citronellal remained unchanged	Bharati et al. (2023b) and Awlqadr et al. (2025)
Lippia integrifolia (Gris) Hier.	Essential oils: mainly monoterpenes (sabinene, terpinen-4-ol, p-cymene, limonene, terpinene) and sesquiterpenes	Antioxidant, antimicrobial, anti-inflammatory; protects skin from environmental damage and premature aging through polyphenolic compound activity	Success in generating autotetraploid plants via colchicine (0.01% for 15 days) in vitro resulting in morphological changes: larger leaves, inflorescences, stomata, trichomes, and pollen grains compared to diploid mother plant. Essential oil yield in tetraploids increased by an average of 35%, in some up to 60%, compared to mother plant. Quantitative changes noted in essential oil composition, particularly enhanced monoterpenes such as sabinene (up to 11.6% from 0.8%) and other bioactives, suggesting improved medicinal and aromatic properties	Marcial et al. (2014) and Iannicelli et al. (2016)

Thymus persicus	Betulinic acid, oleanolic acid, ursolic acid	Antioxidant, anti-inflammatory via phenolics; reduces oxidative stress markers and free radical formation; contributes to healthy aging through polyphenol-mediated antioxidant defense	Induction of tetraploids using colchicine (0.3% for 12–24 h) resulted in significant increases in all three pentacyclic triterpenoids in aerial parts. Tetraploid plants showed marked increases of 69.7% (betulinic acid), 42.8% (oleanolic acid), and 140.7% (ursolic acid) compared to diploid controls. Colchicine was the anti-mitotic agent used	Tavan et al. (2015) and Abed et al. (2022)
Rubus persicus, Rubus caesius, Rubus sanctus	Anthocyanins, phenolic substances, flavonoids, antioxidants, vitamin C	Antioxidant, anti-inflammatory; rich in polyphenols and tannins; protects against oxidative damage and age-related inflammation; supports skin health through free radical scavenging	Polyploid induction via colchicine and oryzalin increased fruit size, flower size, drupelet count, anthocyanin content, antioxidant activity, and nectar sugar levels. Octaploid *Rubus persicus* showed the largest increases, including highest antioxidant activity (85.7%). Polyploidy altered phytohormone levels positively affecting fruit and flower traits. While polyploidy reduced pollen number/viability and seed germination, it enhanced metabolite production beneficial for anti-aging	Cuevas-Rodríguez et al. (2010) and Sabooni and Gharaghani (2022)
Salvia multicaulis	Ursolic acid, betulinic acid, oleanolic acid, rosmarinic acid, caffeic acid	Anti-inflammatory, antioxidant, antimicrobial properties via triterpenic acids and phenolic acids	Polyploidy induced in vitro using colchicine (0.05–0.2%, most effective 0.1% for 48 h) produced hexaploid plants. In hexaploids, oleanolic acid content increased 1.33-fold, and rosmarinic and caffeic acid contents increased around 1.5-fold compared to tetraploids. However, ursolic acid and betulinic acid decreased. Gene expression analysis showed upregulation of squalene synthase (SQS) and β-amyrin synthase (BAS) genes, correlating with increased oleanolic acid. Morphological and anatomical changes confirmed polyploidy. Polyploidy effects on metabolites were selective and complex	Tavan et al. (2022) and Hamed et al. (2024)

(continued)

Table 8.1 (continued)

Plant species	Key metabolite/s	Anti-aging mechanism	Impact of polyploidy on anti-aging properties and associated metabolites	References
Satureja khuzistanica	Carvacrol, thymol, p-cymene, sabinene, limonene, linalool, others	Antioxidant, antimicrobial, anti-inflammatory effects via phenolic compounds and essential oils	Polyploid induction using colchicine (0.05% for 6 h) significantly increased biomass, leaf size, stomatal size, and secondary metabolites, including essential oils (35% increase) and phenolic compounds (150% increase). Flow cytometry and chromosome counting confirmed tetraploid formation. Morphological, physiological, and phytochemical traits were positively affected, suggesting improved medicinal and commercial potential of tetraploid plants	Shariat and Sefidkon (2021) and Kavoosi et al. (2025)
Ficus carica	Phenolic compounds, total soluble sugars, total soluble proteins, phytohormones (zeatin, indole-3-acetic acid, gibberellic acid, salicylic acid, jasmonic acid)	Anti-oxidant, phytohormonal regulation of growth, stress tolerance via phenolic compounds, soluble sugars, soluble proteins	Tetraploidy induced in two cultivars (Sabz and Torsh) via colchicine treatment (312–625 μM, 24–72 h) confirmed by flow cytometry and chromosome counting. Tetraploids showed enhanced growth traits (stem height, leaf size), increased chlorophyll content, higher photosynthesis rates, and significant increases in phenolic compounds, total soluble sugars, and proteins. Growth stimulator phytohormones (ZR, IAA, GA3, SA, JA) were elevated, while growth inhibitor ABA decreased. Antioxidant activity decreased slightly. Tetraploids exhibited beneficial physiological and phytochemical changes attributed to polyploid induction with colchicine	Ali et al. (2012), Abdolinejad et al. (2021), and Yahiaoui et al. (2024)
Thymus vulgaris	Thymol, carvacrol, essential oils	Antibacterial, anti-spasmolytic, antitumor, antiviral due to thymol and carvacrol terpenoids	Polyploidy induced in vivo via colchicine (0.3% for 6 h at 4-leaf stage) successfully doubled chromosome number from 30 to 60, confirmed by flow cytometry and chromosome counting. Tetraploids showed larger leaves, taller and thicker stems, denser branching, fewer but larger stomata, larger trichomes, and increased number of chloroplasts and mitochondria by 1.66× and 1.63× respectively. Expression of cytochrome P450 genes CYP71D178, CYP71D180, and CYP71D181 increased 3.27, 7.39, and 2.15 times respectively, linked to increased essential oil content by 64.7%, thymol by 40.9%, and carvacrol by 18.6%. The anti-mitotic agent used was colchicine	Habashy et al. (2018), Pandur et al. (2022), and Mohammadi et al. (2023)

Fagopyrum tataricum	Flavonoids, seed proteins	Antioxidant, nutritional, medicinal properties via flavonoids, seed proteins	Tetraploidy was successfully induced using colchicine (0.25% for 48–96 h on seedling apical meristems). Induced tetraploids had significantly larger leaves, flowers, seeds, higher seed protein (up to ~39.6% increase), and higher seed flavonoid content (~80% increase in some cultivars) compared to diploids	Wang et al. (2017)
Rhinacanthus nasutus	Rhinacanthins (rhinacanthin-C, -D, -N), phenolics, flavonoids	Antioxidant, anti-inflammatory, skin disease treatment	In vitro polyploidization induced on calli using colchicine at 0.05% for 48 h resulted in tetraploid calli with doubled DNA content confirmed by flow cytometry. Tetraploid calli produced significantly higher total phenolic content (81.28 mg GAE/g extract) and total flavonoid content (35.33 mg QE/g extract), nearly double those of diploid calli. Antioxidant activities (FRAP, DPPH, ABTS assays) were markedly increased in tetraploid calli compared to diploid and mixoploid calli. Polyploidy induction via colchicine effectively enhanced secondary metabolite production and antioxidant potential, supporting medicinal value enhancement	Siriwatanametanon et al. (2010) and Yaowachai et al. (2025)
Stevia rebaudiana	Rebaudioside-A, stevioside	Natural non-calorie sweetener; anti-diabetic, antioxidant, and anti-inflammatory effects attributed to steviol glycosides	Polyploidy was induced by seed treatment with colchicine (0.05%, 0.1%, 0.2%) for 12–48 h. Tetraploids showed significantly increased stomata size, leaf length, and rebaudioside-A content (highest 2.84% DW at 0.2% colchicine for 24 h). These findings demonstrate that colchicine-induced polyploidy can enhance anti-aging metabolite production and morphological traits in Stevia rebaudiana	Talei et al. (2020) and Papaefthimiou et al. (2023)
Echinacea purpurea	Cichoric acid, chlorogenic acid	Antioxidant, antibacterial, antifungal, immunostimulant via caffeic acid derivatives including cichoric acid and chlorogenic acid	Tetraploidy was induced using colchicine (0.25%) treatment for 24, 48, and 72 h on seedling root tips. Chlorophyll a, b and carotenoids were increased, while stomatal density, leaf index, plant height, and photosystem II efficiency decreased compared to diploids. Tetraploids had 45% more cichoric acid and 71% more chlorogenic acid than diploids, indicating enhanced secondary metabolite production	Abdoli et al. (2013) and Vieira et al. (2023)

(continued)

Table 8.1 (continued)

Plant species	Key metabolite/s	Anti-aging mechanism	Impact of polyploidy on anti-aging properties and associated metabolites	References
Stachys byzantina	Linalool, α-cadinol, cubenol, α-terpineol, menthone	Antiviral, antifungal, antibacterial, antioxidant via essential oils rich in monoterpenes	Tetraploidy was induced in vitro using colchicine (0.1%, 0.2%, 0.4% for 12, 24, 48 h) on the apical meristem. Optimal induction efficiency (~18%) at 0.2% for 12 h. Chlorophyll a, b, and total chlorophyll increased, carotenoids unchanged. Significant increase in essential oil components: linalool nearly doubled, α-cadinol and cubenol doubled, α-terpineol and menthone also increased	Hamarashid et al. (2022) and Lima et al. (2024)
Clausena lansium	Flavonoids, alkaloids, phenolic acids	Antioxidant, antimicrobial, anti-inflammatory, antitumor activities via flavonoids, alkaloids, and phenolic acids	Polyploidy was induced by colchicine treatment and hybridization in three cultivars, yielding 13 stable polyploid lines. Metabolomic profiling revealed significantly higher accumulation of flavonoids, alkaloids, and phenolic acids in tetraploids compared to diploids; triploids showed moderate increases	Adebajo et al. (2009) and Ding et al. (2025)
Allium cepa L.	Flavonoids (quercetin, fisetin), phenolic acids, anthocyanins, sulfur compounds	Antioxidant, anti-inflammatory, skin, and cardiovascular protective effects due to flavonoids, phenolic acids, sulfur compounds	Polyploidy was induced with colchicine at 0.05%, 0.1%, and 0.2% for 24 h. The antioxidant activity (DPPH scavenging) was highest in tetraploid bulbs (72.58%). Higher dissolved sugars were recorded in tetraploids (refractive index 1.369)	Marefati et al. (2021) and Abubakar et al. (2025)
Cichorium intybus	Total phenolic compounds, caffeic acid derivatives (chlorogenic acid, cichoric acid), flavonoids	Antioxidant, anti-inflammatory, hepatoprotective, and anticancer effects	Tetraploidy was induced using colchicine treatment on seeds and shoot apical meristems; shoot apical meristem treatment was more efficient for tetraploid induction. Total phenolics nearly doubled (67.89 mg/g DW vs 36.59 mg/g DW in diploids), and CGA content increased about ten-fold in tetraploids	Ghotbi Ravandi et al. (2013) and Chatterjee et al. (2025)
Cichorium intybus			Colchicine-induced tetraploids (0.25% for 24–72 h) exhibited larger stomata, pollen grains, and leaves, with increased chlorophyll a, b, and carotenoids. Tetraploids showed doubled phenolic acids (notably chlorogenic acid increased approximately 2.4-fold compared to diploids)	Mahdy et al. (2024)

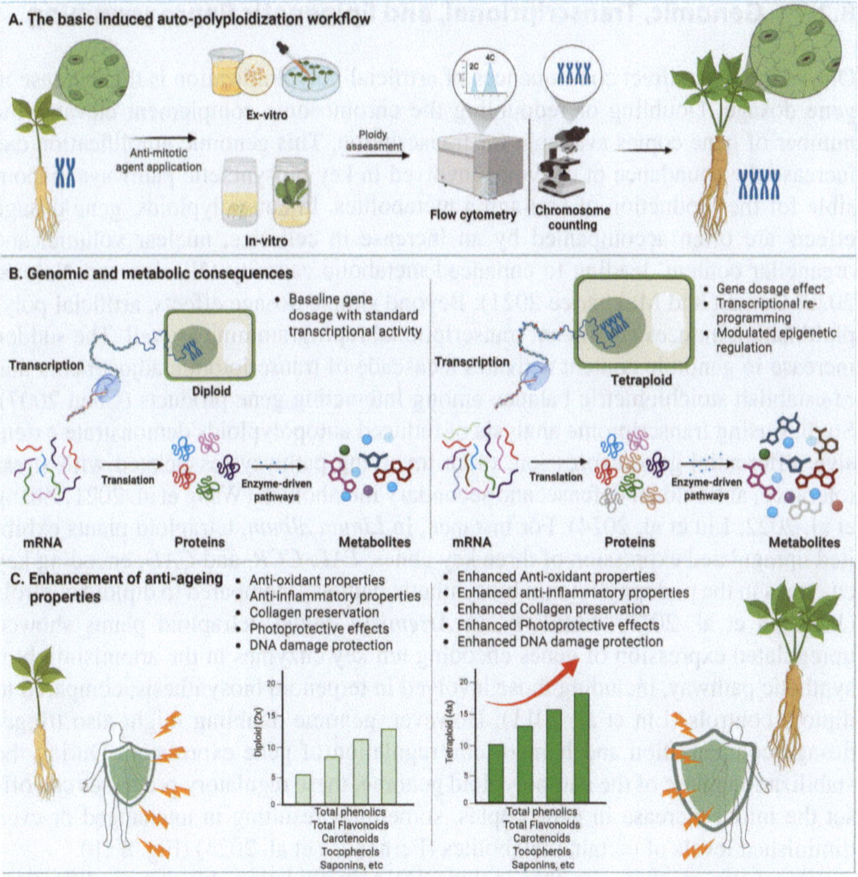

Fig 8.1 Illustration of induced autopolyploidization and its effects on anti-aging traits in plants. (**a**) Workflow of autopolyploidization using anti-mitotic agents via in vitro and ex vitro methods, followed by ploidy verification through flow cytometry and chromosome counting. (**b**) Comparison of diploid and tetraploid states showing genomic and metabolic outcomes such as gene dosage effects, transcriptional shifts, and altered protein and metabolite levels. (**c**) Overview of enhanced anti-aging activities in polyploids—greater antioxidant and anti-inflammatory capacity, improved collagen maintenance, photoprotection, and DNA damage resistance—illustrated by comparative metabolite bar graphs. (Created in BioRender. Bharati, R. (2026) https://BioRender.com/sjb577b)

metabolite biosynthesis caused by polyploidy; and the third panel presents the improved anti-aging properties observed in medicinal plants as a consequence of autopolyploidization (Fig. 8.1).

8.2.1 Genomic, Transcriptional, and Epigenetic Reprogramming

One of the most direct consequences of artificial polyploidization is the increase in gene dosage. Doubling or redoubling the chromosome complement elevates the number of gene copies available for transcription. This genomic amplification can increase the abundance of enzymes involved in key biosynthetic pathways responsible for the production of anti-aging metabolites. In autopolyploids, gene dosage effects are often accompanied by an increase in cell size, nuclear volume, and organellar content, leading to enhanced metabolic capacity (Niazian and Nalousi 2020; Gantait and Mukherjee 2021). Beyond simple dosage effects, artificial polyploidization induces significant transcriptional reprogramming as well. The sudden increase in genomic content activates a cascade of transcriptomic adjustments that re-establish stoichiometric balance among interacting gene products (Chen 2007). Studies using transcriptome analysis of induced autopolyploids demonstrate extensive differential gene expression, often involving pathways associated with stress tolerance, antioxidant defense, and secondary metabolism (Wang et al. 2021; Zhang et al. 2022; Liu et al. 2024). For instance, in *Linum album*, tetraploid plants exhibited upregulated expression of three key genes, *PAL, CCR,* and *CAD*, encoding key enzymes in the podophyllotoxin biosynthetic pathway, compared to diploid controls (Javadian et al. 2017). Similarly, in *Artemisia annua*, tetraploid plants showed upregulated expression of genes encoding ten key enzymes in the artemisinin biosynthetic pathway, including those involved in terpenoid biosynthesis, compared to diploid controls (Lin et al. 2011). However, genome doubling might also trigger dosage compensation and homeostatic regulation of gene expression. During the stabilization phase of the autopolyploid genome, these regulatory processes can offset the initial increase in gene copies, sometimes resulting in unchanged or even diminished levels of certain metabolites (Fernandes et al. 2023) (Fig. 8.1b).

Epigenetic mechanisms that include DNA methylation, histone modification, and small RNA-mediated silencing may also play critical roles in stabilizing gene expression following genome doubling (Rao et al. 2023). Artificial polyploidization may induce epigenetic relaxation or remodeling, allowing the reactivation of previously silenced loci and the reconfiguration of metabolic gene networks (Song and Chen 2015). This can lead to de-repression of key biosynthetic genes, enhancing the accumulation of metabolites with anti-aging potential such as resveratrol, quercetin, or rosmarinic acid. Moreover, polyploidization often enhances stress-responsive transcription factors that coordinate the expression of antioxidant enzyme genes (Wang et al. 2021, 2023; Tossi et al. 2022; Zhang et al. 2022). The reprogramming of transcriptional and epigenetic landscapes thus can integrate metabolic and redox regulation, contributing to improved cellular longevity and reduced oxidative damage, contributing to enhanced anti-aging effects.

8.2.2 Implication on Metabolite Production and Its Anti-aging Properties

The primary objective of artificial polyploidization in medicinal plants is to enhance the biosynthesis and accumulation of metabolites that confer medicinal properties, such as anti-aging effects. The synergistic impact of increased gene dosage and transcriptional reprogramming often results in upregulated flux through secondary metabolic pathways such as the phenylpropanoid, terpenoid, and alkaloid biosynthetic routes (Madani et al. 2021). The resultant effect is a substantial accumulation of specific compounds that directly target and mitigate the core hallmarks of aging, thereby boosting the plant's overall medicinal efficacy including anti-aging properties. For instance, induced polyploid plants of *Foeniculum vulgare* (fennel) displayed significantly increased phenolic and flavonoid contents as well as antioxidant activity, indicating that polyploid induction can enhance plant's anti-aging properties by boosting its natural antioxidant potential (Hariwal et al. 2025). Similarly, in *Zingiber officinale*, the induction of stable tetraploids resulted in a 32.8% increase in 6-gingerol content, a potent compound known for its antioxidant and anti-inflammatory properties, which protect against age-related degenerative diseases by reducing reactive oxygen species (ROS) and enhancing endogenous antioxidant enzymes (Kumar et al. 2024).

Polyploidy also influences the neuroprotective potential of plants. In *Bacopa monnieri*, tetraploid plants not only showed significantly increased bacoside content but also a marked upregulation of the *BmoOSC* gene, which is pivotal for triterpenoid biosynthesis. This genetic enhancement correlates with a higher capacity to prevent neuronal apoptosis and repair damaged neurons, directly countering cognitive decline (Sujipuli et al. 2021). The augmentation of anti-aging benefits by polyploidy also extends to skin health and cellular protection. Tetraploid lines of *Thymus persicus* exhibited remarkable increases in pentacyclic triterpenoids, with ursolic acid content surging by 140.7%, a compound renowned for its ability to reduce oxidative stress markers and inflammation, thereby contributing to healthy aging (Tavan et al. 2015).

The biochemical enhancements in polyploids, whether measured as elevated antioxidant capacity, increased key phytochemicals, or superior anti-inflammatory activity, directly translate into a more potent ability to mitigate oxidative stress, protect against cellular damage, and slow the progression of age-related physiological decline. Therefore, induced autopolyploidy serves as a powerful biotechnological strategy to genetically fortify medicinal plants, effectively turning them into more concentrated and effective sources of natural anti-aging compounds for nutraceutical and pharmaceutical applications (Fig. 8.1c). The enhanced anti-aging properties of medicinal plants induced by autopolyploidization, along with their key metabolites and mechanisms, are summarized in Table 8.1, illustrating the significant impact of polyploidy on metabolite production and therapeutic potential.

8.3 Induction to Screening: A Methodological Toolkit

8.3.1 Induction Approaches: In Vivo vs In Vitro

Artificial polyploidy induction can be implemented using mainly by in vivo (often termed ex vitro) or in vitro methodologies. The in vivo approach involves treating intact plants or organs, such as seeds or buds, with antimitotic agents directly in a non-sterile, experimental environment. Common application techniques in this method include soaking, immersion, or the use of droplet cultures (Niazian and Nalousi 2020). This strategy has successfully generated polyploids in various species, including grapevine, common thyme, and cannabis (Mansouri and Bagheri 2017; Kara and Doğan 2023; Sanaei-Hoveida et al. 2024). The in vivo approach is relatively cheaper and does not require sophisticated equipment or specialized technical expertise to be employed. Despite its successes, the in vivo method is generally characterized by lower effectiveness and a relatively poor yield of stable polyploids.

In contrast, the in vitro approach involves the application of antimitotic agents to plants growing in a sterile, controlled environment on a defined culture medium. Treatments vary in concentration and duration and can be administered by incorporating the agent directly into the growth medium or by applying it as a solution directly to the explants (Zahumenická et al. 2018; Homaidan Shmeit et al. 2020). The higher control over environmental conditions and direct exposure to the targeted explant tissue, the in vitro approach results in a significantly higher polyploid induction rate (Bharati and Severová 2025). An additional advantage of the in vitro system is its tendency to produce fewer chimeric plants compared to in vivo treatments, leading to more genetically uniform polyploid populations (Gao et al. 2022). Consequently, the in vitro polyploidization has emerged as a preferred technique, particularly for economically important species like medicinal plants.

8.3.2 Antimitotic Agents

Antimitotic agents play a pivotal role in the induction of auto-polyploids by interfering with the microtubule dynamics responsible for chromosome movement during cell division, particularly during spindle formation and chromosome segregation. The failure in the spindle apparatus functionality prevents the replicated chromosomes from being divided into daughter cells, resulting in the doubling of the chromosome number within a single nucleus. This chromosomal duplication is the principal phenomenon behind the polyploid plants, which often exhibit enhanced vigor, metabolic activity, and improved secondary metabolite profiles, features that influence their potential anti-aging and other medicinal properties. Choosing an appropriate antimitotic agent is crucial, as the efficiency of chromosome doubling and the stability of the resulting polyploid plants depend on both the chemical nature of the compound and its mode of interaction with plant cells. Over the years, several compounds with anti-mitotic activity have been explored for inducing polyploidy in medicinal plants.

Among the commonly used anti-mitotic agents, colchicine remains the most commonly and extensively applied in medicinal and aromatic plants. It is an alkaloid derived from the autumn crocus (*Colchicum autumnale* L.) and disrupts microtubule polymerization by binding to tubulin proteins, thereby preventing spindle fiber formation. This results in metaphase arrest and subsequent chromosome doubling (Eng and Ho 2019). The popularity of colchicine for polyploid induction in plants stems from its effectiveness, water solubility, and thermostability (Bharati et al. 2023a). Despite its advantages, colchicine has several demerits, like a narrow effective concentration range, requires a relatively higher concentration and can be toxic to both plants and handlers if not used carefully (Niazian and Nalousi 2020). Consequently, high concentrations (millimolar levels) are required to achieve effective chromosome doubling in plants, which can lead to increased cytotoxicity, malformation, and plant mortality (Neenu et al. 2024).

As safer and more plant-specific alternatives, dinitroaniline herbicides such as oryzalin, trifluralin, pendimethalin, ethalfluralin, and butralin have gained increasing attention in recent years (Touchell et al. 2020; Neenu et al. 2024). These compounds interfere with microtubule formation by binding specifically to plant α-tubulin, thus inhibiting microtubule polymerization during mitosis (Touchell et al. 2020). This mechanism offers greater selectivity toward plant cells compared to colchicine, with high affinity for plant tubulins and no affinity for animal tubulins, thereby reducing cytotoxic effects to handlers (Niazian and Nalousi 2020). Oryzalin, in particular, has been widely used in both medicinal and ornamental plants, where it demonstrates high chromosome-doubling efficiency at concentrations 50–250 times lower than colchicine (10–100 μM compared to millimolar levels for colchicine) (Talebi et al. 2017; Madani et al. 2021; Bharati et al. 2023a; Neenu et al. 2024). Oryzalin's effectiveness has been linked to its strong binding affinity to α-tubulin, which disrupts spindle fiber formation and prevents chromosome segregation while increasing plant survival rates during treatment (Touchell et al. 2020).

Another alternative is amiprophos-methyl (APM), a phosphorothioamidate herbicide that binds to the same site on α-tubulin as oryzalin. However, it has the advantage of increased water solubility, thereby reducing the need for additional solvents, such as Dimethyl sulfoxide (DMSO), which is frequently added to antimitotic solutions at concentrations of 1.0–2.0% to enhance permeability and penetration into plant cells (Neenu et al. 2024). Overall, these dinitroaniline and alternative agents provide more controlled, efficient, and safer approaches to induce polyploidy compared to colchicine, advancing the breeding of elite medicinal plant varieties with improved agronomic and phytochemical profiles.

8.3.3 Polyploid Identification Approaches

The identification and confirmation of polyploid individuals following artificial induction are among the most crucial stages in any polyploidization experiment. The effectiveness of induction treatments can only be evaluated once the resulting plants are accurately characterized as true polyploids rather than mixoploids or diploid escapes. Selection can be broadly divided into direct and indirect methods, depending on whether the ploidy level is measured through cytological or morphological characteristics.

Direct methods involve physically examining chromosomes within dividing cells. The classical approach is chromosome counting, usually performed in mitotic cells of root tips or meristematic tissues after pretreatment with spindle inhibitors, followed by fixation, staining, and microscopic analysis (Bharati et al. 2024). Although labor-intensive, chromosome counting provides clear evidence of chromosome doubling. In advanced systems, fluorescence in situ hybridization (FISH) techniques are employed to visualize specific chromosome sets or repetitive DNA sequences, enabling high-resolution confirmation of genome duplication and detection of structural rearrangements that may accompany polyploidization. FISH employs species-specific or repetitive sequence probes, predominantly ribosomal DNA (rDNA) sequences including 5S, 18S, 45S, and 25S rDNA, which hybridize to complementary genomic sequences on chromosomes and generate fluorescent signals at loci of probe-target DNA annealing (Mekonen and Ali 2023). Following polyploid induction, the copy number of rDNA loci characteristically increases proportionally with ploidy level. For example, in colchicine-induced *Lippia alba* (lemon verbena), a medicinal aromatic plant, FISH mapping using 45S rDNA probes revealed a dose-dependent increase in ribosomal signal sites corresponding to genome duplication: diploid plants ($2n = 30$) displayed six terminal 45S rDNA sites, while induced triploid plants ($2n = 45$) displayed nine 45S rDNA sites, and induced tetraploid plants ($2n = 60$) displayed 12 45S rDNA sites, directly reflecting the proportional doubling and tripling of chromosome sets (Julião et al. 2020).

Indirect methods, by contrast, offer rapid estimates of nuclear DNA content and cellular traits associated with polyploidy. Flow cytometry is now the most commonly employed approach, as it offers an efficient and precise method for confirming genome duplication by quantifying nuclear DNA content in thousands of nuclei within seconds. This method utilizes fluorescent dyes, such as propidium iodide or DAPI, to stain nuclear DNA, with fluorescence intensity measured relative to a known diploid standard, providing a precise ploidy ratio. Flow cytometry can also detect mixoploidy, evaluate uniformity within populations, and monitor long-term ploidy stability across generations (Doležel et al. 2007).

Alongside flow cytometry, morpho-anatomical markers are often used as supplementary indicators. Polyploid plants generally exhibit enlarged stomata with reduced density, thicker leaves, larger flowers and seeds, and increased overall cell size due to augmented nuclear and cytoplasmic volume (Manzoor et al. 2019). Analysis of stomatal size and frequency, which can be measured via leaf epidermal impressions or microscopy, is particularly reliable as an early indicator of polyploidization, correlating strongly with genome size. Furthermore, pollen grain diameter, guard cell length, and seed size also serve as accessible diagnostic parameters for large-scale screening before cytological confirmation.

The most robust identification methods combine both direct and indirect techniques. For example, initial flow cytometry or stomatal analysis can quickly identify potential polyploid candidates, which are then confirmed through cytological chromosome counting or FISH to verify ploidy stability. Such integrated approaches ensure accuracy, efficiency, and reliability in identifying true autopolyploid individuals suitable for further characterization and breeding.

8.3.4 Choosing for Anti-aging Potential

Following polyploid confirmation, screening for genotypes with enhanced anti-aging potential is crucial to identify potential candidates suitable for further development and cultivation. The selection strategy can employ diverse approaches tailored to available resources, technical capabilities, and project scope. Indirect assessment methods, including morphological, anatomical, and some non-destructive physiological evaluations, offer accessible, rapid, and cost-effective options for preliminary candidate identification and pre-screening of large polyploid populations (Niazian and Nalousi 2020). Morphological and anatomical evaluation, including observation of stomatal size and density via leaf impressions, measurement of trichomes, and assessment of leaf thickness and cell volume, provides readily observable indicators of improvements that can be quickly surveyed across multiple genotypes. Non-destructive physiological characterization using SPAD and pulse-amplitude modulated (PAM) chlorophyll fluorometry enables quantification of photosynthetic efficiency parameters such as maximum quantum yield (Fv/Fm) and electron transport rate (ETR), which directly indicate the photosynthetic capacity and energy availability driving continuous synthesis of energy-demanding secondary metabolites (Bharati et al. 2024). For example, *Melissa officinalis* tetraploids induced with oryzalin exhibited significantly enhanced linear electron flow (LEF) in the morning (23.52 μmol electrons m^{-2} s^{-1} in tetraploids versus 7.19 in diploids) and afternoon (31.63 versus 14.0), demonstrating superior light-energy conversion and photosynthetic efficiency that correlates with reported elevated essential oil and metabolite accumulation (Bharati et al. 2023b). These indirect approaches, requiring minimal specialized equipment and investment, can efficiently screen numerous polyploid lines to identify those displaying superior morphophysiological vigor. By prioritizing candidates with enhanced chlorophyll content, superior Fv/Fm and LEF values, enlarged trichomes and stomata, researchers can substantially narrow candidate pools before proceeding to resource-intensive biochemical validation, thus optimizing screening efficiency and ensuring that expensive direct analyses are reserved for genuinely promising genotypes.

For candidates advancing from indirect pre-screening, a comprehensive biochemical and molecular assessment provides a definitive evaluation of anti-aging potential. Molecular assessment of biosynthetic gene expression via RT-qPCR represents the first tier of direct validation, quantifying transcript abundance of genes governing secondary metabolism such as *PAL* (*phenylalanine ammonia lyase*), *CHS* (*chalcone synthase*), and *GGPPS* (*geranylgeranyl pyrophosphate synthase*), which are central to the production of polyphenols and flavonoids with established anti-aging effects (Tavan et al. 2021).

Analytical validation using chromatographic platforms such as HPLC, GC-MS, or LC-MS/MS can then confirm the actual accumulation of anti-aging compounds like rosmarinic acid, resveratrol, quercetin, or ginsenosides (Bouzroud et al. 2023). Finally, functional bioassays provide the most direct evidence of anti-aging efficacy, assessing parameters such as antioxidant capacity, anti-inflammatory activity, anti-glycation effects, collagen preservation, and cellular protection under oxidative stress (Ondua et al. 2019; Hariwal et al. 2025).

Importantly, the choice of screening approach should remain context-dependent and flexible. Not all studies require every level of assessment. For example, researchers with limited resources may focus on molecular and metabolite profiling, while projects emphasizing biological efficacy may prioritize bioassay-based validation. Integrating both indirect and direct approaches in a tiered manner, beginning with broad, low-cost screening and progressing to targeted molecular and functional analyses, ensures scientific robustness. This modular strategy enables researchers to identify the most promising polyploid genotypes for further development and commercialization with confidence. Table 8.2 summarizes the key experimental details and results related to the induction of autopolyploidy in various medicinal plants, highlighting the methods, antimitotic agents used, and the corresponding effects on metabolite production and anti-aging properties.

8.4 Factors Influencing Auto-polyploidization

The efficiency and stability of auto-polyploidization are governed by a complex interplay of biological, physiological, environmental, and technical factors that collectively determine the success of chromosome doubling. The type of antimitotic agent, its concentration, and exposure duration are among the most decisive variables. For instance, colchicine, while traditionally the most widely used agent, often requires higher concentrations and longer exposure times compared to oryzalin or trifluralin, which can achieve chromosome doubling at micromolar levels with minimal cytotoxic effects. However, excessive exposure to any of these agents may result in tissue necrosis, stunted growth, or mixoploidy. Therefore, optimization of both dosage and duration is essential for maximizing induction efficiency while maintaining plant viability.

The developmental stage and physiological condition of the treated tissue also play a critical role. Actively dividing cells, such as those in meristematic regions, embryogenic calli, or young shoots, are more susceptible to spindle inhibition, yielding higher rates of stable chromosome doubling. Conversely, mature or differentiated tissues often exhibit reduced mitotic activity, leading to incomplete or uneven genome duplication. The type of explant used in vitro (e.g., leaf discs, nodal segments, or axillary buds) can influence both induction frequency and subsequent regeneration capacity (Niazian and Nalousi 2020). Explants with high totipotency tend to recover more rapidly from mitotic inhibition, generating uniform polyploid plantlets, while explants with lower totipotency, such as mature leaves or older stem tissues, usually show lower induction efficiency and higher rates of chimerism or mixoploidy, resulting in heterogeneous regenerated plants.

Table 8.2 Common biochemical, molecular, and physiological parameters for screening polyploid medicinal plants with enhanced anti-aging potential

Parameter	Type	Indicator of polyploidy	Measurement method	Relevance to anti-aging potential	References
Stomatal size and density	Morphological	Larger stomata and reduced density due to increased cell and nuclear size in polyploids	Leaf epidermal imprint or microscopy	Stomatal size and density in induced polyploids can be used as indirect markers for enhanced primary and secondary metabolism	Iannicelli et al. (2020)
Chlorophyll content (Chl a, Chl b, total) and photosynthetic efficiency (Fv/Fm, ETR, NPQ)	Physiological	Higher total chlorophyll content, elevated Chl a/b ratio, and improved photochemical efficiency in polyploids	SPAD readings, spectrophotometric analysis, and pulse-amplitude modulated (PAM) chlorophyll fluorometry (MultispeQ, FluorPen)	Enhanced chlorophyll in polyploids boosts photosynthesis by increasing light absorption and energy conversion in chloroplasts. This leads to more photosynthates, sugars and primary metabolites, which supply carbon and energy to secondary metabolite pathways like phenylpropanoids, raising levels of metabolites	Landi et al. (2020) and Wu et al. (2025)
Gene expression (*PAL, CHS, GGPPS*)	Molecular	Upregulation of genes involved in secondary metabolism and antioxidant defense pathways	RT-qPCR	Elevated transcript levels of biosynthetic and antioxidant genes reflect enhanced cellular capacity to produce protective metabolites and enzymes, directly strengthening the plant's anti-aging metabolic network	Sujipuli et al. (2021), Tavan et al. (2021, 2022)
Secondary metabolite profile (e.g., rosmarinic acid, resveratrol, quercetin, ginsenosides)	Biochemical	Quantitative increase in bioactive compounds with antioxidant or anti-inflammatory functions	HPLC, GC-MS, or UPLC, LC-MS/MS	Enhanced production of anti-aging phytochemicals directly improves therapeutic potential. These metabolites modulate oxidative signaling, DNA repair, mitochondrial function, and collagen synthesis, extending tissue longevity	Abdolinejad et al. (2021), Bharati et al. (2023a), and Bouzroud et al. (2023)

(continued)

Table 8.2 (continued)

Parameter	Type	Indicator of polyploidy	Measurement method	Relevance to anti-aging potential	References
Total antioxidant capacity (DPPH, FRAP, ABTS, ORAC)	Biochemical	Elevated radical-scavenging activity in polyploid extracts	Spectrophotometric antioxidant assays	Global antioxidant capacity provides an integrated measure of the plant's ability to counter oxidative stress, reflecting overall metabolic resilience and potential to slow cellular aging processes	Hariwal et al. (2025)
Total flavonoid content (TFC)	Biochemical	Elevated flavonoid accumulation in polyploid lines	Aluminum chloride colorimetric assay	Flavonoids protect cells from oxidative stress by directly neutralizing ROS and modulating longevity-related signaling pathways such as Nrf2, AMPK, and mTOR. Enhanced TFC thus reflects an improved intrinsic anti-aging defense system	Ulewicz-Magulska and Wesolowski (2019)
Total phenolic content (TPC)	Biochemical	Significant increase in total phenolics in polyploids	Folin–Ciocalteu assay	Phenolics are key redox regulators that scavenge reactive oxygen species (ROS) and chelate metal ions. Higher TPC indicates stronger antioxidant buffering, reducing oxidative damage to DNA, proteins, and lipids—Core molecular drivers of aging	Hariwal et al. (2025)
Anti-inflammatory enzyme inhibition (COX, LOX)	Biochemical	Polyploid genotypes exhibit stronger inhibition of inflammatory enzymes	Enzyme inhibition assay	Greater inhibition of COX/LOX reduces inflammation-related aging pathways	Ondua et al. (2019)

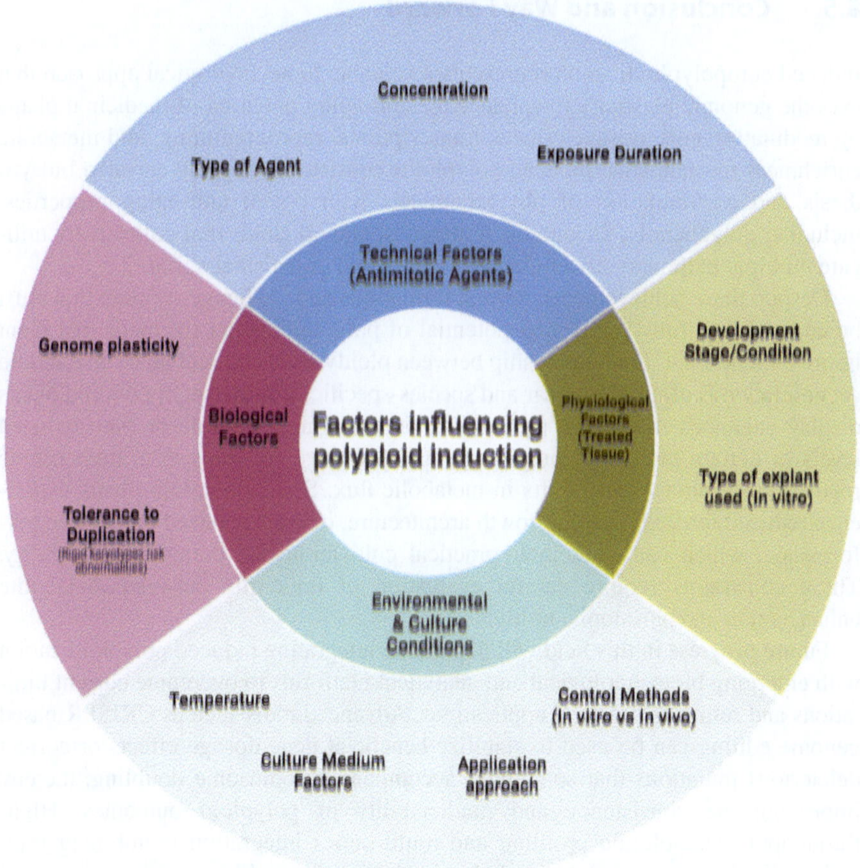

Fig. 8.2 Schematic representation of key factors influencing polyploid induction in plants. The diagram categorizes influences into four major domains: technical factors, physiological factors, biological factors, and environmental & culture conditions. These interconnected variables collectively determine the success and efficiency of artificial polyploidization methodologies

Environmental and culture conditions, including temperature, light intensity, and hormonal balance in the culture medium, further modulate the success of polyploidization. For instance, mild temperature stress during treatment can enhance spindle disruption, whereas excessive light intensity or osmotic stress may hinder cell recovery (Niazian and Nalousi 2020). In vitro conditions generally provide superior control over these parameters compared to in vivo approaches, leading to higher success rates and fewer chimeras (Bharati and Severová 2025). Moreover, the inherent genomic plasticity of the species determines its tolerance to genome duplication. Species with naturally variable chromosome numbers or a history of polyploidy are often more amenable to artificial polyploidization, whereas species with rigid karyotypes may experience developmental abnormalities or sterility following chromosome doubling (Leitch and Leitch 2008) (Fig. 8.2).

8.5 Conclusion and Way Forward

Induced autopolyploidization represents a valuable biotechnological approach that uses the genomic plasticity to enhance the anti-aging potential of medicinal plants by modulating gene dosage effects, transcriptional reprogramming, and metabolic enrichment mechanisms. Induced polyploids consistently display elevated biosynthesis and accumulation of phytochemicals with potent anti-aging properties, including polyphenols, flavonoids, terpenoids, and alkaloids that collectively mitigate multiple hallmarks of cellular senescence and organismal aging.

Despite these achievements, several limitations and challenges remain that must be addressed to fully realize the potential of polyploidization for medicinal plant improvement. First, the relationship between ploidy level and secondary metabolite accumulation is often non-linear and species-specific. While many polyploid plants display enhanced metabolite production, others exhibit variable or even reduced levels of certain bioactive compounds due to dosage compensation, homeostatic regulation, or unfavorable shifts in metabolic flux. Some polyploid plants experience reduced fertility, altered growth architecture, or compromised agronomic performance, which can limit their practical cultivation and commercial viability. These constraints require careful evaluation of trade-offs between metabolite enhancement and agronomic utility.

Future progress in this field will depend on integrating induced polyploidization with emerging biotechnological and analytical platforms to overcome current limitations and refine its practical applications. Advanced tools such as CRISPR-based genome editing can be used to stabilize beneficial gene dosage effects or correct deleterious mutations that sometimes accompany chromosome doubling, thereby improving the consistency and predictability of polyploid outcomes. High-throughput metabolomic profiling and multi-omics integration (combining transcriptomic, proteomic, and metabolomic data) will enable precise mapping of metabolic changes associated with different ploidy levels, facilitating the identification of key regulatory genes and pathways responsible for enhanced production of anti-aging phytochemicals. Moreover, the application of machine learning and AI-assisted phenotypic screening can accelerate the selection of elite polyploid genotypes by predicting metabolite yields and linking morphological traits to biochemical outputs. These approaches will make the selection process more efficient, reducing the reliance on time-consuming and resource-intensive empirical screening.

In parallel, coupling polyploidization with sustainable agricultural practices, precision cultivation systems, and biofortification strategies can ensure that enhanced polyploid lines are not only biochemically superior but also agronomically viable and economically accessible. Together, these advances will transform induced autopolyploidization from a largely experimental tool into a rational, data-driven strategy for the development of next-generation medicinal plants. Ultimately, this integrative framework will bridge traditional phytomedicinal wisdom with modern biogerontological science, paving the way for scalable, natural therapeutics that promote healthy aging and extended health span across global populations.

Acknowledgments This work was funded by the Faculty of Economics and Management, Czech University of Life Sciences in Prague, grant number 2025A1004. The authors also acknowledge the financial support from the Czech Agrifood Research Center, Prague Czech Republic.

Competing Interests The authors have no relevant financial or non-financial interests to disclose.

References

Abdoli M, Moieni A, Naghdi Badi H (2013) Morphological, physiological, cytological and phytochemical studies in diploid and colchicine-induced tetraploid plants of Echinacea purpurea (L.). Acta Physiol Plant 35:2075–2083

Abdolinejad R, Shekafandeh A, Jowkar A (2021) In vitro tetraploidy induction creates enhancements in morphological, physiological and phytochemical characteristics in the fig tree (Ficus Carica L.). Plant Physiol Biochem 166:191–202

Abed DZ, Sadeghian R, Mohammadi S, Akram M (2022) Thymus persicus (Ronniger ex Rech. F.) Jalas alleviates nociceptive and neuropathic pain behavior in mice: multiple mechanisms of action. J Ethnopharmacol 283:114695

Abubakar MA, Gennadievna NE, Mikhailovna SO, Mikhailovna KE (2025) Effect of induced polyploidy on morphology, antioxidant activity, and dissolved sugars in Allium cepa L. Horticulturae 11:154

Adebajo AC, Iwalewa EO, Obuotor EM, Ibikunle GF, Omisore NO, Adewunmi CO, Obaparusi OO, Klaes M, Adetogun GE, Schmidt TJ et al (2009) Pharmacological properties of the extract and some isolated compounds of Clausena lansium stem bark: anti-trichomonal, antidiabetic, anti-inflammatory, hepatoprotective and antioxidant effects. J Ethnopharmacol 122:10–19

Ali B, Mujeeb M, Aeri V, Mir SR, Faiyazuddin M, Shakeel F (2012) Anti-inflammatory and antioxidant activity of *Ficus carica* Linn. Leaves. Nat Prod Res 26:460–465

Anghel AC, Ţăranu I, Orţan A, Marcu Spinu S, Dragoi Cudalbeanu M, Rosu PM, Băbeanu NE (2024) Polyphenols and microbiota modulation: insights from swine and other animal models for human therapeutic strategies. Molecules 29:6026

Arora I, Sharma M, Sun LY, Tollefsbol TO (2020) The epigenetic link between polyphenols, aging and age-related diseases. Genes 11:1094

Awlqadr FH, Altemimi AB, Qadir SA, Mohammed OA, Saeed MN, Hesarinejad MA, Lakhssassi N (2025) Bioactive compounds, medicinal benefits, and contemporary extraction methods for lemon balm (*Melissa officinalis*). Food Sci Nutr 13:e70864

Bernard F, Moghbel N, Hassannejad S (2012) Treatment of licorice seeds with colchicine: changes in seedling DNA levels and anthocyanin and glycyrrhizic acid contents of derived callus cultures. Nat Prod Commun 7:1934578X1200701112

Bharati R, Severová L (2025) Artificial polyploidy as a tool for improving growth and stress resilience in tree species. Front For Glob Change 8:1569384

Bharati R, Fernández-Cusimamani E, Gupta A, Novy P, Moses O, Severová L, Svoboda R, Šrédl K (2023a) Oryzalin induces polyploids with superior morphology and increased levels of essential oil production in Mentha spicata L. Ind Crop Prod 198:116683

Bharati R, Gupta A, Novy P, Severová L, Šrédl K, Žiarovská J, Fernández-Cusimamani E (2023b) Synthetic polyploid induction influences morphological, physiological, and photosynthetic characteristics in Melissa officinalis L. Front Plant Sci 14:1332428

Bharati R, Shmeit YH, Šedivá JH, Cong TTN, Kundu JK, Severová L, Svoboda R, Fernández-Cusimamani E (2024) Comparative assessment of morphological, cytological, and photosynthetic characteristics of the induced octoploid and its tetraploid counterpart of Celosia argentea L. BMC Plant Biol 24:1227

Bikiaris RE, Paczkowska-Walendowska M, Koumentakou I, Niti A, Cielecka-Piontek J, Kyzas GZ (2025) Bioactivity and physicochemical characterization of *Centella asiatica* and Marigold extract serums: antioxidant, anti-inflammatory, anti-tyrosinase and skin barrier function insights. Int J Cosmet Sci 47:1070–1084. https://doi.org/10.1111/ics.70006

Bouzroud S, El Maaiden E, Sobeh M, Merghoub N, Boukcim H, Kouisni L, El Kharrassi Y (2023) Biotechnological approaches to producing natural antioxidants: anti-ageing and skin longevity prospects. IJMS 24:1397

Ceramella J, Chavarria D, Iacopetta D, Fernandes C, Marra M, Gattuso A, Loizzo MR, Sicari V, Tundis R, Alcaro S et al (2025) Salvia officinalis L. leaf extracts: chemical analysis and biological studies. J Ethnopharmacol 352:120193

Chatterjee A, Marjit S, Ghosh A, Mondal A, Banerjee B, Dey S, Pratihar S, Behera PC, Maity A, Khatun R et al (2025) Assessment of anti-inflammatory, antioxidant, antimicrobial activities of Cichorium intybus different extracts with identification of anti-inflammatory compound using computational and network pharmacological analysis. Discover Plants 2:45

Chen ZJ (2007) Genetic and epigenetic mechanisms for gene expression and phenotypic variation in plant Polyploids. Annu Rev Plant Biol 58:377–406

Cuevas-Rodríguez EO, Dia VP, Yousef GG, García-Saucedo PA, López-Medina J, Paredes-López O, Gonzalez De Mejia E, Lila MA (2010) Inhibition of pro-inflammatory responses and antioxidant capacity of Mexican blackberry (Rubus spp.) extracts. J Agric Food Chem 58:9542–9548

Ding Y, Wu L, Wei H, Zhang Z, Zhao J, Hu G, Qin Y, Zhang Z (2025) Polyploid induction enhances secondary metabolite biosynthesis in Clausena lansium: morphological and metabolomic insights. Agriculture 15:1566

Doležel J, Greilhuber J, Suda J (2007) Estimation of nuclear DNA content in plants using flow cytometry. Nat Protoc 2:2233–2244

Emelda A, Sukmawati, Dongke AL, Marzuki I (2022) Anti-inflamatory activity of water extract of Talinum paniculatum (Jact). Gaertn leaves wistar rat. IOP Conf Ser Earth Environ Sci 1041:012073

Eng W-H, Ho W-S (2019) Polyploidization using colchicine in horticultural plants: a review. Sci Hortic 246:604–617

Fernandes HP, Choi YH, Vrieling K, De Bresser M, Sewalt B, Tonolo F (2023) Cultivar-dependent phenotypic and chemotypic responses of drug-type Cannabis sativa L. to polyploidization. Front. Plant Sci 14:1233191

Gantait S, Mukherjee E (2021) Induced autopolyploidy—a promising approach for enhanced biosynthesis of plant secondary metabolites: an insight. J Genet Eng Biotechnol 19:4

Gao Y, Ma J, Chen J, Xu Q, Jia Y, Chen H, Li W, Lin L (2022) Establishing tetraploid embryogenic cell lines of Magnolia officinalis to facilitate tetraploid plantlet production and phenotyping. Front Plant Sci 13:900768

García-Aguilar A, Palomino O, Benito M, Guillén C (2021) Dietary polyphenols in metabolic and neurodegenerative diseases: molecular targets in autophagy and biological effects. Antioxidants 10:142

Ghotbi Ravandi E, Rezanejad F, Zolala J, Dehghan E (2013) The effects of chromosome-doubling on selected morphological and phytochemical characteristics of *Cichorium intybus* L. J Hortic Sci Biotechnol 88:701–709

Gibellini L, Bianchini E, De Biasi S, Nasi M, Cossarizza A, Pinti M (2015) Natural compounds modulating mitochondrial functions. Evid Based Complement Alternat Med 2015:1–13

Govindarasu S, Murugesh S (2025) Secondary metabolite based anti-inflammatory, antioxidant activities, drug like properties, molecular docking enzyme inhibitors of selected medicinal plant Bacopa monnieri (L)Wettst. The Microbe 8:100478

Habashy NH, Abu Serie MM, Attia WE, Abdelgaleil SAM (2018) Chemical characterization, antioxidant and anti-inflammatory properties of Greek Thymus vulgaris extracts and their possible synergism with Egyptian Chlorella vulgaris. J Funct Foods 40:317–328

Hamarashid SH, Khaledian Y, Soleimani F (2022) In vitro polyploidy-mediated enhancement of secondary metabolites content in Stachys byzantina L. Genet Resour Crop Evol 69:719–728

Hamed AR, Nabih HK, El-Rashedy AA, Mohamed TA, Mostafa OE, K. Ali S, Efferth T, Hegazy M-EF (2024) Salvimulticanol from Salvia multicaulis suppresses LPS-induced inflammation in RAW2647 macrophages: in vitro and in silico studies. 3 Biotech 14:144

Hariwal M, Verma S, Kumar S (2025) Comparative analysis of total phenolic, flavonoids and antioxidant profile in methanolic extracts of colchicine treated Foeniculum vulgare mill. Ind Crop Prod 236:121991

Homaidan Shmeit Y, Fernandez E, Novy P, Kloucek P, Orosz M, Kokoska L (2020) Autopolyploidy effect on morphological variation and essential oil content in Thymus vulgaris L. Sci Hortic 263:109095

Iannicelli J, Elechosa MA, Juárez MA, Martínez A, Bugallo V, Bandoni AL, Escandón AS, Van Baren CM (2016) Effect of polyploidization in the production of essential oils in Lippia integrifolia. Ind Crop Prod 81:20–29

Iannicelli J, Guariniello J, Tossi VE, Regalado JJ, Di Ciaccio L, Van Baren CM, Pitta Álvarez SI, Escandón AS (2020) The "polyploid effect" in the breeding of aromatic and medicinal species. Sci Hortic 260:108854

Inthima P, Sujipuli K (2019) Improvement of growth and bacoside production in *Bacopa monnieri* through induced autotetraploidy with colchicine. PeerJ 7:e7966

Islam MA, Zaman S, Biswas K, Al-Amin MY, Hasan MK, Alam AHMK, Tanaka T, Sadik G (2021) Evaluation of cholinesterase inhibitory and antioxidant activity of Wedelia chinensis and isolation of apigenin as an active compound. BMC Complement Med Ther 21:204

Jafari M, Macho-González A, Diaz A, Lindenau K, Santiago-Fernández O, Zeng M, Massey AC, De Cabo R, Kaushik S, Cuervo AM (2024) Calorie restriction and calorie-restriction mimetics activate chaperone-mediated autophagy. Proc Natl Acad Sci USA 121:e2317945121

Javadian N, Karimzadeh G, Sharifi M, Moieni A, Behmanesh M (2017) In vitro polyploidy induction: changes in morphology, podophyllotoxin biosynthesis, and expression of the related genes in Linum album (Linaceae). Planta 245:1165–1178

Julião SA, Ribeiro CDV, Lopes JML, Matos EMD, Reis AC, Peixoto PHP, Machado MA, Azevedo ALS, Grazul RM, Campos JMSD et al (2020) Induction of synthetic Polyploids and assessment of genomic stability in Lippia alba. Front Plant Sci 11:292

Kaensaksiri T, Soontornchainaksaeng P, Soonthornchareonnon N, Prathanturarug S (2011) In vitro induction of polyploidy in Centella asiatica (L.) urban. Plant Cell Tissue Organ Cult 107:187–194

Kan H, Zhang D, Chen W, Wang S, He Z, Pang S, Qu S, Wang Y (2023) Identification of anti-inflammatory components in Panax ginseng of Sijunzi decoction based on spectrum-effect relationship. Chin Herb Med 15:123–131

Kapetanou M, Athanasopoulou S, Goutas A, Makatsori D, Trachana V, Gonos E (2024) α-Terpineol induces Shelterin components TRF1 and TRF2 to mitigate senescence and telomere integrity loss via a telomerase-independent pathway. Antioxidants 13:1258

Kara Z, Doğan O (2023) Mutagenic effects of nitrogen protoxide and oryzalin on "41 B" and "Fercal" grapevine rootstocks seedlings. Breed Sci 73:355–364

Kavoosi G, Siahbalaei R, Salehi F, Mastinu A (2025) Bioactive properties and metabolite profiling of Satureja species: antioxidant potential, anti-amylase activity, and nutritional composition. Food Meas. https://doi.org/10.1007/s11694-025-03749-x

Kim Y-S, Hahn E-J, Murthy HN, Paek K-Y (2004) Effect of polyploidy induction on biomass and ginsenoside accumulations in adventitious roots of ginseng. J Plant Biol 47:356–360

Kim M-B, Lee J, Lee J-Y (2024) Targeting mitochondrial dysfunction for the prevention and treatment of metabolic disease by bioactive food components. J Lipid Atheroscler 13:306

Kim HJ, Cho JY, Kim M-Y (2025) A comprehensive review of the effects of Panax ginseng and its constituents against inflammatory diseases. J Ginseng Res 49:605–612

Kumar K, Thakur L, Thakur M, Bhardwaj V, Gupta M (2024) In vitro polyploidy induction in ginger (Zingiber officinale Rosc.) cv. Himgiri: morpho-physiological, biochemical, and molecular evaluation of regenerants. In Vitro Cell Dev Biol-Plant 60:693–710

Landi M, Zivcak M, Sytar O, Brestic M, Allakhverdiev SI (2020) Plasticity of photosynthetic processes and the accumulation of secondary metabolites in plants in response to monochromatic light environments: a review. Biochimica et Biophysica Acta (BBA) 1861:148131

Leitch AR, Leitch IJ (2008) Genomic plasticity and the diversity of polyploid plants. Science 320:481–483

Lima JADS, Leite VC, Silva JP, Ferrarez MA, Bahia GD, Rezende LVN, Guedes MCMR, Macedo GC, Silva NPD, Tavares GD et al (2024) *Stachys byzantina* K. Koch in the treatment of skin inflammation: a comprehensive evaluation of its therapeutic properties. ACS Omega 9:49899–49912

Lin X, Zhou Y, Zhang J, Lu X, Zhang F, Shen Q, Wu S, Chen Y, Wang T, Tang K (2011) Enhancement of artemisinin content in tetraploid *Artemisia annua* plants by modulating the expression of genes in artemisinin biosynthetic pathway. Biotech App Biochem 58:50–57

Liu Y, Fang M, Tu X, Mo X, Zhang L, Yang B, Wang F, Kim Y-B, Huang C, Chen L et al (2024) Dietary polyphenols as anti-aging agents: targeting the hallmarks of aging. Nutrients 16:3305

Liu Y, Huang X, Gao X, Zhang X, Huang H, Li W, Zhang Y (2025) Tetraploidization altered phenotypic traits and metabolite profile of Java ginseng (Talinum paniculatum (Jacq.) Gaertn.). Plants 14:480

López-Otín C, Blasco MA, Partridge L, Serrano M, Kroemer G (2023) Hallmarks of aging: an expanding universe. Cell 186:243–278

Madani H, Escrich A, Hosseini B, Sanchez-Muñoz R, Khojasteh A, Palazon J (2021) Effect of polyploidy induction on natural metabolite production in medicinal plants. Biomolecules 11:899

Mahdy D, Arafa A, Sarhan A, Hussien B (2024) Colchicine's impact on the growth and chemical composition of withania somnifera l. plant. Egypt J Chem 68: 123–128

Malireddy S, Kotha SR, Secor JD, Gurney TO, Abbott JL, Maulik G, Maddipati KR, Parinandi NL (2012) Phytochemical antioxidants modulate mammalian cellular Epigenome: implications in health and disease. Antioxid Redox Signal 17:327–339

Mansouri H, Bagheri M (2017) Induction of polyploidy and its effect on Cannabis sativa L. In: Chandra S, Lata H, ElSohly MA (eds) Cannabis sativa L.—botany and biotechnology. Springer, Cham, pp 365–383

Manzoor A, Ahmad T, Bashir M, Hafiz I, Silvestri C (2019) Studies on colchicine induced chromosome doubling for enhancement of quality traits in ornamental plants. Plants 8:194

Marcial G, Sendker J, Brandt S, De Lampasona MP, Catalán CAN, De Valdez GF, Hensel A (2014) Gastroprotection as an example: Antiadhesion against helicobacter pylori, anti-inflammatory and antioxidant activities of aqueous extracts from the aerial parts of Lippia integrifolia Hieron. J Ethnopharmacol 155:1125–1133

Marefati N, Ghorani V, Shakeri F, Boskabady M, Kianian F, Rezaee R, Boskabady MH (2021) A review of anti-inflammatory, antioxidant, and immunomodulatory effects of *Allium cepa* and its main constituents. Pharm Biol 59:285–300

Matsuno Y, Atsumi Y, Alauddin M, Rana MM, Fujimori H, Hyodo M, Shimizu A, Ikuta T, Tani H, Torigoe H et al (2020) Resveratrol and its related polyphenols contribute to the maintenance of genome stability. Sci Rep 10:5388

Mekonen A, Ali A (2023) A review on principles of FISH and GISH and its role in cytogenetic study. Glob Res Environ Sustain 1:15–26

Mohammadi V, Talebi S, Ahmadnasab M, Mollahassanzadeh H (2023) The effect of induced polyploidy on phytochemistry, cellular organelles and the expression of genes involved in thymol and carvacrol biosynthetic pathway in thyme (Thymus vulgaris). Front Plant Sci 14:1228844

Mohd Sahardi NFN, Makpol S (2019) Ginger (*Zingiber officinale* roscoe) in the prevention of ageing and degenerative diseases: review of current evidence. Evid Based Complement Alternat Med 2019:1–13

Neenu MG, Aswathi A, Prasath D (2024) Synthetic polyploidy in spice crops: a review. Crop Sci 64:2–23

Niazian M, Nalousi AM (2020) Artificial polyploidy induction for improvement of ornamental and medicinal plants. Plant Cell Tissue Organ Cult 142:447–469

Ondua M, Njoya EM, Abdalla MA, McGaw LJ (2019) Anti-inflammatory and antioxidant properties of leaf extracts of eleven south African medicinal plants used traditionally to treat inflammation. J Ethnopharmacol 234:27–35

Pandur E, Micalizzi G, Mondello L, Horváth A, Sipos K, Horváth G (2022) Antioxidant and anti-inflammatory effects of thyme (*Thymus vulgaris* L.) essential oils prepared at different plant phenophases on *Pseudomonas aeruginosa* LPS-activated THP-1 macrophages. Antioxidants 11:1330

Papaefthimiou M, Kontou PI, Bagos PG, Braliou GG (2023) Antioxidant activity of leaf extracts from *Stevia rebaudiana* Bertoni exerts attenuating effect on diseased experimental rats: a systematic review and meta-analysis. Nutrients 15:3325

Pereira QC, Fortunato IM, Oliveira FDS, Alvarez MC, Santos TWD, Ribeiro ML (2024) Polyphenolic compounds: orchestrating intestinal microbiota harmony during aging. Nutrients 16:1066

Rahnasto-Rilla M, Tyni J, Huovinen M, Jarho E, Kulikowicz T, Ravichandran S, Bohr VA, Ferrucci L, Lahtela-Kakkonen M, Moaddel R (2018) Natural polyphenols as sirtuin 6 modulators. Sci Rep 8:4163

Rao X, Ren J, Wang W, Chen R, Xie Q, Xu Y, Li D, Song Z, He Y, Cai D et al (2023) Comparative DNA-methylome and transcriptome analysis reveals heterosis- and polyploidy-associated epigenetic changes in rice. Crop J 11:427–437

Rodríguez-Vera D, Abad-García A, Vargas-Mendoza N, Pinto-Almazán R, Farfán-García E, Morales-González J, Soriano-Ursúa M (2022) Polyphenols as potential enhancers of stem cell therapy against neurodegeneration. Neural Regen Res 17:2093

Rogina B, Tissenbaum HA (2024) SIRT1, resveratrol and aging. Front Genet 15:1393181

Rohini MR, Rao VK, Rajendiran S (2024) Studies on the effect of induced polyploidy on yield and phytochemical content in *Centella asiatica* (L.) urban. J Hortic Sci. https://doi.org/10.24154/jhs.v19i2.2469

Sabooni N, Gharaghani A (2022) Induced polyploidy deeply influences reproductive life cycles, related phytochemical features, and phytohormonal activities in blackberry species. Front Plant Sci 13:938284

Sanaei-Hoveida Z, Mortazavian SMM, Norouzi M, Sadat-Noori SA (2024) Elevating morphology and essential oil in cumin genotypes through polyploidy induction. Sci Hortic 329:113031

Shariat A, Sefidkon F (2021) Tetraploid induction in savory (*Satureja khuzistanica*): cytological, morphological, phytochemical and physiological changes. Plant Cell Tissue Organ Cult 146:137–148

Silva BN, Cadavez V, Caleja C, Pereira E, Calhelha RC, Añibarro-Ortega M, Finimundy T, Kostić M, Soković M, Teixeira JA et al (2023) Phytochemical composition and bioactive potential of *Melissa officinalis* L., *Salvia officinalis* L and *Mentha spicata* L extracts. Foods 12:947

Siriwatanametanon N, Fiebich BL, Efferth T, Prieto JM, Heinrich M (2010) Traditionally used Thai medicinal plants: in vitro anti-inflammatory, anticancer and antioxidant activities. J Ethnopharmacol 130:196–207

Song Q, Chen ZJ (2015) Epigenetic and developmental regulation in plant polyploids. Curr Opinion Plant Biol 24:101–109

Sujipuli K, Inthima P, Yimtragool N, Warnnissorn N, Warnnissorn P, Prasarnpun S (2021) In vitro colchicine-induced polyploids from different explant segments of Bacopa monnieri. ScienceAsia 47:30

Talebi SF, Saharkhiz MJ, Kermani MJ, Sharafi Y, Raouf Fard F (2017) Effect of different antimitotic agents on polyploid induction of anise hyssop (*Agastache foeniculum* L.). Caryologia 70:184–193

Talei D, Nekouei MK, Mardi M, Kadkhodaei S (2020) Improving productivity of steviol glycosides in *Stevia rebaudiana* via induced polyploidy. J Crop Sci Biotechnol 23:301–309

Tavan M, Mirjalili MH, Karimzadeh G (2015) In vitro polyploidy induction: changes in morphological, anatomical and phytochemical characteristics of *Thymus persicus* (Lamiaceae). Plant Cell Tissue Organ Cult 122:573–583

Tavan M, Sarikhani H, Mirjalili MH, Rigano MM, Azizi A (2021) Triterpenic and phenolic acids production changed in Salvia officinalis via in vitro and in vivo polyploidization: a consequence of altered genes expression. Phytochemistry 189:112803

Tavan M, Azizi A, Sarikhani H, Mirjalili MH, Rigano MM (2022) Induced polyploidy and broad variation in phytochemical traits and altered gene expression in Salvia multicaulis. Sci Hortic 291:110592

Thong-on W, Arimatsu P, Pitiporn S, Soonthornchareonnon N, Prathanturarug S (2014) Field evaluation of in vitro-induced tetraploid and diploid *Centella asiatica* (L.) urban. J Nat Med 68:267–273

Tossi VE, Martínez Tosar LJ, Laino LE, Iannicelli J, Regalado JJ, Escandón AS, Baroli I, Causin HF, Pitta-Álvarez SI (2022) Impact of polyploidy on plant tolerance to abiotic and biotic stresses. Front Plant Sci 13:869423

Touchell DH, Palmer IE, Ranney TG (2020) In vitro ploidy manipulation for crop improvement. Front Plant Sci 11:722

Tsai Y-T, Chen P-Y, To K-Y (2021) Induction of polyploidy and metabolic profiling in the medicinal herb *Wedelia chinensis*. Plants 10:1232

Tuğlu Ü, Cessur A, Baydar NG, Baydar H (2025) Comparison of phenolic composition, antiradical and antioxidant activities of diploid and autotetraploid genotypes of *Salvia officinalis* L. S Afr J Bot 180:160–170

Ulewicz-Magulska B, Wesolowski M (2019) Total phenolic contents and antioxidant potential of herbs used for medical and culinary purposes. Plant Foods Hum Nutr 74:61–67

Ungvari Z, Sonntag WE, De Cabo R, Baur JA, Csiszar A (2011) Mitochondrial protection by resveratrol. Exerc Sport Sci Rev 39:128–132

Vieira SF, Gonçalves SM, Gonçalves VMF, Llaguno CP, Macías F, Tiritan ME, Cunha C, Carvalho A, Reis RL, Ferreira H et al (2023) *Echinacea purpurea* fractions represent promising plant-based anti-inflammatory formulations. Antioxidants 12:425

Wahab S, Annadurai S, Abullais SS, Das G, Ahmad W, Ahmad MF, Kandasamy G, Vasudevan R, Ali MS, Amir M (2021) *Glycyrrhiza glabra* (licorice): a comprehensive review on its phytochemistry, biological activities, clinical evidence and toxicology. Plants 10:2751

Wang L-J, Sheng M-Y, Wen P-C, Du J-Y (2017) Morphological, physiological, cytological and phytochemical studies in diploid and colchicine-induced tetraploid plants of *Fagopyrum tataricum* (L.) Gaertn. Bot Stud 58:2

Wang C, Wang T, Yin M, Eller F, Liu L, Brix H, Guo W (2021) Transcriptome analysis of tetraploid and octoploid common reed (*Phragmites australis*). Front Plant Sci 12:653183

Wang N, Wang Y, Wang C, Leng Z, Qi F, Wang S, Zhou Y, Meng W, Liu K, Zhang C et al (2023) Evaluating the differential response of transcription factors in diploid versus autotetraploid rice leaves subjected to diverse saline–alkali stresses. Genes 14:1151

Wong JH, Barron AM, Abdullah JM (2021) Mitoprotective effects of *Centella asiatica* (L.) Urb.: anti-inflammatory and neuroprotective opportunities in neurodegenerative disease. Front Pharmacol 12:687935

Wu W, Wu H, Liang R, Huang S, Meng L, Zhang M, Xie F, Zhu H (2025) Light regulates the synthesis and accumulation of plant secondary metabolites. Front Plant Sci 16:1644472

Yahfoufi N, Alsadi N, Jambi M, Matar C (2018) The immunomodulatory and anti-inflammatory role of polyphenols. Nutrients 10:1618

Yahiaoui S, Kati DE, Chaalal M, Ali LMA, El Cheikh K, Depaepe G, Morère A, Menut C, Bettache N, Bachir-Bey M (2024) Antioxidant, antiproliferative, anti-inflammatory, and enzyme inhibition potentials of *Ficus carica* wood bark and related bioactive phenolic metabolites. Wood Sci Technol 58:1051–1075

Yaowachai W, Luecha P, Taratima W (2025) Phytochemical production and antioxidant activity improvement of *Rhinacanthus nasutus* (L.) Kurz calli by *in vitro* polyploidization. PeerJ 13:e19160

Yessenkyzy A, Saliev T, Zhanaliyeva M, Masoud A-R, Umbayev B, Sergazy S, Krivykh E, Gulyayev A, Nurgozhin T (2020) Polyphenols as caloric-restriction mimetics and autophagy inducers in aging research. Nutrients 12:1344

Zahumenická P, Fernández E, Šedivá J, Žiarovská J, Ros-Santaella JL, Martínez-Fernández D, Russo D, Milella L (2018) Morphological, physiological and genomic comparisons between diploids and induced tetraploids in *Anemone sylvestris* L. Plant Cell Tissue Organ Cult 132:317–327

Zhang C, Meng W, Wang Y, Zhou Y, Wang S, Qi F, Wang N, Ma J (2022) Comparative analysis of physiological, hormonal and transcriptomic responses reveal mechanisms of saline-alkali tolerance in autotetraploid rice (*Oryza sativa* L.). IJMS 23:16146

Integrating Cryopreservation and Synthetic Seed Technologies for the Global Conservation and Exchange of Anti-aging Medicinal Plants

9

P. E. Rajasekharan, Marcos Edel Martínez Montero, and Harshid Pulparambil

Abstract

The rising global demand for anti-aging medicinal plants has intensified the need for advanced biotechnological approaches to ensure their long-term conservation, sustainable utilization, and equitable germplasm exchange. This chapter presents an integrative framework combining cryopreservation and synthetic seed technologies for the effective ex situ conservation of high-value species known for their therapeutic properties, including *Withania somnifera*, *Curcuma longa*, *Phyllanthus emblica*, *Centella asiatica*, *Panax ginseng*, and *Ginkgo biloba*. Emphasis is placed on optimized protocols, such as droplet vitrification, encapsulation-dehydration, and DMSO-based cryoprotection, for preserving critical tissues, including shoot tips, embryogenic calli, and pollen. In parallel, synthetic seed technology is discussed through the lens of encapsulation matrices and regeneration efficiency of somatic embryos and shoot buds. A bibliometric analysis of over 150 peer-reviewed publications (2000–2023), conducted using VOSviewer and Scopus/Web of Science datasets, identifies global research trends, collaborative networks, and underexplored taxonomic and geographic areas. Case studies from tropical and temperate bioregions illustrate practical

P. E. Rajasekharan (✉)
Division of Flower and Medicinal Crops, ICAR-Indian Institute of Horticultural Research (IIHR), Bengaluru, India

M. E. M. Montero
Universidad de Ciego de Ávila "Máximo Gómez Báez", Ciego de Ávila, Cuba

H. Pulparambil
KSCSTE–Malabar Botanical Garden and Institute for Plant Sciences, Kozhikode, Kerala, India

S. Gantait et al. (eds.), *Medicinal Plants with Anti-Ageing Properties*,
https://doi.org/10.1007/978-981-95-8064-4_9

applications, technology transfer pathways, and compliance with international treaties such as the Nagoya Protocol. The chapter concludes by outlining scalable strategies for commercial deployment, highlighting AI-assisted protocol optimization, public–private partnerships, and enabling policy frameworks. Together, these integrated approaches offer a climate-resilient and ethically grounded pathway for preserving the genetic diversity of medicinal plants threatened by overexploitation and environmental change.

Keywords

Alginate encapsulation · Germplasm conservation · Somatic embryogenesis · Bibliometric analysis · Germplasm exchange

9.1 Introduction

9.1.1 The Anti-aging Therapeutic Market and Its Botanical Foundations

The global anti-aging therapeutic market has increasingly embraced botanicals and phytochemicals as pivotal agents that address both the visible and systemic manifestations of aging. Driven by consumer demand for natural, multifunctional interventions, plant-derived compounds are now central to dermatological, nutraceutical, and neuroprotective product development. Medicinal plants rich in polyphenols, flavonoids, terpenoids, and alkaloids are recognized for their dual capacity to act as direct antioxidants and as modulators of cellular defense systems, influencing critical longevity-associated pathways (Argyropoulou et al. 2013; Mondal et al. 2014).

Bioactive compounds from *Curcuma longa*, *Camellia sinensis*, *Vitis vinifera*, *Ginkgo biloba*, *Panax ginseng*, *Aloe vera*, and *Centella asiatica* demonstrate potent anti-wrinkle, collagen-preserving, and anti-inflammatory effects, forming the scientific basis of modern cosmeceuticals (Mondal et al. 2014). Parallelly, neuroprotective botanicals such as *Withania somnifera*, *Bacopa monnieri*, and *Ginseng* have shown potential in mitigating oxidative and inflammatory cascades implicated in cognitive decline and age-related neurodegenerative disorders (Argyropoulou et al. 2013). This growing reliance on phytochemicals reflects a paradigm shift from single-target pharmacology to system-level modulation of aging processes, positioning botanicals as sustainable, safe, and clinically versatile anti-aging therapeutics.

Consequently, global market expansion in cosmeceuticals and nutraceuticals now intertwines with phyto-pharmacological innovation, highlighting the urgent need for conservation and cryopreservation of medicinal plant resources to secure the bioactive diversity underpinning the anti-aging sector.

9.1.2 Threats to Medicinal Plant Biodiversity: Overharvesting and Climate Change

The growing global reliance on botanicals for anti-aging and therapeutic applications has intensified pressures on wild medicinal plant populations, with overharvesting and climate change emerging as the two most critical threats to their survival. Unsustainable collection practices, often driven by escalating market demand for roots, bulbs, and whole-plant materials, have led to severe population declines and even local extinctions in several taxa (Chen et al. 2019; Lorite 2024; Mykhailenko et al. 2025). The impact is particularly acute for slow-growing, endemic, or habitat-restricted species, where removal of reproductive individuals impedes natural regeneration and disrupts ecological balance. These losses not only erode biodiversity but also threaten traditional healthcare systems and the livelihoods of communities dependent on herbal resources (van Wyk and Prinsloo 2018; Alum 2025). Concurrently, climate change exacerbates these vulnerabilities by altering temperature regimes, precipitation patterns, and the frequency of extreme climatic events, leading to habitat fragmentation, range shifts, and reduced habitat suitability (Applequist et al. 2019; Kougioumoutzis et al. 2024; Ngezahayo et al. 2025; Wani et al. 2024). Beyond distributional impacts, climatic stress also influences secondary metabolism, affecting the yield and quality of pharmacologically active compounds and thereby diminishing therapeutic efficacy (Cahyaningsih et al. 2021; Manish 2022; Dad and Rashid 2022). The interplay of overexploitation and climate-driven ecological shifts creates a feedback loop of decline, wherein depleted populations become less resilient to environmental stressors, while changing habitats further constrain sustainable harvesting and regeneration (Groner et al. 2022; Kougioumoutzis et al. 2024; Mykhailenko et al. 2025). Addressing these convergent threats demands integrated conservation strategies that combine sustainable utilization, ex situ preservation, habitat restoration, and adaptive management to safeguard the genetic and biochemical diversity that underpins the global anti-aging and medicinal plant sector (Applequist et al. 2019; Alum 2025).

9.1.3 Biotechnology in Germplasm Conservation: Emerging Paradigms

Amid escalating biodiversity loss from overexploitation and climate change, biotechnology has emerged as a cornerstone in modern plant germplasm conservation, offering precise, scalable, and long-term strategies to preserve genetic resources of medicinal and economically important species. While traditional seed banking remains fundamental, many anti-aging medicinal plants, often possessing recalcitrant seeds, complex reproductive systems, or restricted habitats, require alternative biotechnological interventions for effective conservation. In vitro culture and slow growth storage (SGS) techniques enable the safe multiplication and medium-term preservation of elite genotypes while minimizing somaclonal variation and maintenance costs (Cruz-Cruz et al. 2013; Coelho et al. 2020; Salgotra and Chauhan 2023).

However, cryopreservation stands out as the most reliable long-term ex situ conservation approach, ensuring indefinite storage of viable germplasm at ultra-low temperatures (−196 °C) without genetic alteration. Recent advancements, including vitrification, encapsulation-dehydration, and cryo-plate methods, have significantly enhanced recovery rates and species applicability, extending the scope of cryogenic preservation to recalcitrant and endangered taxa (Ruta et al. 2020: Hay and Sershen 2021; Malik et al. 2025). Complementary molecular tools, such as PCR-based markers, genomic sequencing, and barcoding, further strengthen these efforts by enabling genetic integrity monitoring, pathogen detection, and diversity assessment across cryo-stored accessions (Niazian 2019; Salgotra and Chauhan 2023). The integration of these technologies has shifted conservation from mere storage toward dynamic genetic resource management, aligning with global priorities for safeguarding phytochemical diversity and ensuring the sustainable supply of plant-based bioactive compounds critical to the anti-aging therapeutic sector (Coelho et al. 2020; Yadav et al. 2024b).

9.2 Anti-aging Medicinal Plants: Properties and Conservation Status

9.2.1 Mechanisms of Aging and Plant-Based Interventions

Aging represents a multifactorial biological process characterized by a progressive decline in physiological integrity, leading to impaired function and increased susceptibility to disease. It arises from the cumulative interaction of molecular and cellular dysfunctions, including oxidative stress, chronic inflammation, genomic instability, and altered nutrient sensing. These interconnected hallmarks, such as telomere attrition, mitochondrial dysfunction, loss of proteostasis, stem cell exhaustion, and deregulated intercellular signaling, collectively drive the decline in cellular resilience and tissue homeostasis (Yu et al. 2021; Guo et al. 2022; Maldonado et al. 2023; Li et al. 2024). Central to this process are imbalances in redox regulation and inflammatory signaling, which exacerbate molecular damage and metabolic dysregulation, ultimately accelerating aging-related pathologies.

Plant-based interventions have emerged as promising strategies to modulate these mechanisms and promote healthy longevity. Phytochemicals such as polyphenols, flavonoids, terpenoids, and alkaloids exert potent antioxidant and anti-inflammatory activities, mitigating oxidative stress and suppressing chronic inflammation, two pivotal accelerators of cellular aging (Salehi et al. 2020; Babich et al. 2022; Bjørklund et al. 2022; Das et al. 2024). Beyond direct free radical scavenging, compounds like resveratrol, curcumin, and catechins influence key longevity pathways, including AMPK, SIRT1, and mTOR, thereby modulating energy metabolism, autophagy, and stress resistance (Yu et al. 2021; Wang et al. 2024). Moreover, plant-derived metabolites promote proteostasis and autophagic clearance of damaged cellular components, enhance mitochondrial function, and protect neuronal networks from age-related decline (Phu et al. 2020; Babich et al. 2022).

Increasing evidence also highlights the role of plant-based diets in shaping a beneficial gut microbiome and reducing the formation of advanced glycation end-products (AGEs), further contributing to systemic anti-aging effects (Meccariello and D'Angelo 2021; Iweala et al. 2024). Collectively, these findings underscore the multifaceted potential of medicinal plants as modulators of aging biology, integrating antioxidant defense, metabolic regulation, and cellular maintenance into a cohesive framework for extending healthspan.

9.2.2 Key Species: Pharmacological Profiles and Commercial Value

Among anti-aging botanicals, six species, *Withania somnifera* (L.) Dunal, *Curcuma longa* L., *Phyllanthus emblica* L., *Centella asiatica* (L.) Urb., *Panax ginseng* C.A.Mey., and *Ginkgo biloba* L., are globally recognized for their potent pharmacological activities and commercial importance. Rich in bioactives such as withanolides, curcuminoids, ginsenosides, and flavonol glycosides, these plants exhibit antioxidant, anti-inflammatory, neuroprotective, adaptogenic, and immunomodulatory effects. Their efficacy against oxidative stress, cognitive decline, and skin aging has secured their prominence in the nutraceutical, pharmaceutical, and functional food industries.

Their broad market applications underscore their economic significance: *W. somnifera* is commercialized as capsules, powders, and extracts for adaptogenic and neuroprotective benefits (Tetali et al. 2021; Sharifi-Rad et al. 2021; Gaurav et al. 2023; Durg et al. 2020); *C. longa* dominates supplement, cosmetic, and functional food industries through curcumin-based products (Balasubramaniam et al. 2024); *P. emblica* is valued for its antioxidant and vitamin C-rich extracts used in nutraceuticals and skincare; *C. asiatica* is incorporated into cognitive and dermatological formulations (Moise et al. 2024); *P. ginseng* remains a high-value crop in Asia for tonics and energy supplements (Natarajan et al. 2022); and *G. biloba* retains global prominence in memory and cognitive health products (Malík and Tlustoš 2023). Given their therapeutic relevance, economic value, and growing demand, conserving these species through advanced germplasm preservation strategies, particularly cryopreservation, is vital for ensuring sustainable utilization and long-term genetic security (Table 9.1).

9.2.3 Conservation Priorities: Endangered Status and Regulatory Frameworks

The growing global demand for *Withania somnifera*, *Curcuma longa*, *Phyllanthus emblica*, *Centella asiatica*, *Panax ginseng*, and *Ginkgo biloba* underscores the urgency of conserving their genetic resources. These species are central to traditional and modern healthcare systems, yet face varying degrees of threat due to overharvesting, habitat degradation, and unsustainable commercial extraction.

Table 9.1 Bioactive constituents and anti-aging properties of selected medicinal plant species

Plant species	Family	Part used	Major bioactive compounds	Anti-aging properties	Notable applications	References
Withania somnifera	Solanaceae	Roots, leaves	Withanolides, sitoindosides	Antioxidant, stress-reducing, adaptogenic, neuroprotective, immunomodulatory	Cognitive enhancement, neurodegenerative disease	Kulkarni and Dhir (2008), Dar et al. (2015), Tsaltaki et al. (2019), Paul et al. (2021), Gaurav et al. (2023), Lerose et al. (2024), Yadav et al. (2024a), and Khalid et al. (2025), Singh et al. (2021)
Panax ginseng	Araliaceae	Root	Ginsenosides, panaxosides	Anti-wrinkle, enhances longevity, Adaptogenic, neuroprotective, immunomodulatory, anti-inflammatory	Stress release, cognitive function, Alzheimer's adjunct	Tsaltaki et al. (2019), Natarajan et al. (2022), Malík and Tlustoš (2023), Balasubramaniam et al. (2024), and Tripathi et al. (2024)
Curcuma longa	Zingiberaceae	Rhizome	Curcumin, turmerone	Anti-inflammatory, antioxidant, skin repair	Neuroprotection, Alzheimer's disease adjunct	Zheng et al. (2020), Verma et al. (2018), Balasubramaniam et al. (2024), and Moise et al. (2024)
Centella asiatica	Apiaceae	Whole plant	Asiaticoside, madecassoside	Promotes collagen synthesis, skin regeneration, neuroprotective, antioxidant, anti-inflammatory	Memory, cognitive enhancement	Seevaratnam et al. (2012), Buranasudja et al. (2021), Karsono et al. (2021), and Thong-on et al. (2024)
Phyllanthus emblica	Phyllanthaceae	Fruit	Ascorbic acid, ellagic acid	Antioxidant, protects DNA from oxidative stress, anti-inflammatory, hepatoprotective	General health, liver support	Pientaweeratch et al. (2016), Wu et al. (2022), and Malík and Tlustoš (2023)
Ginkgo biloba	Ginkgoaceae	Leaves, seeds	Ginkgolides	Neuroprotective, antioxidant, vasodilatory	Memory, cognitive disorders, Alzheimer's adjunct	Shi et al. (2010), Zuo et al. (2017), Tsaltaki et al. (2019), Barbalho et al. (2022), and Malík and Tlustoš (2023)

Conservation priorities thus emphasize sustainable harvesting, maintenance of genetic diversity, and the implementation of region-specific regulatory frameworks to ensure their long-term availability and ecological stability.

Among the six species, *Withania somnifera* faces concerns related to high demand, low seed viability, and poor germination, prompting the adoption of in vitro propagation and tissue culture techniques as conservation tools (Shasmita et al. 2018). *Curcuma longa* and *Phyllanthus emblica* are widely cultivated and not globally endangered, though ongoing conservation programs aim to preserve genetic diversity and promote sustainable agricultural practices (Malík and Tlustoš 2023). *Centella asiatica*, while not listed as endangered, experiences local population declines due to overexploitation, making controlled cultivation and habitat management essential (Seevaratnam et al. 2012; Shasmita et al. 2018). In contrast, *Panax ginseng* is among the most critically impacted, with wild populations in Asia and North America listed under CITES Appendix II due to extensive overharvesting; conservation strategies include developing disease-resistant cultivars, regulating wild harvests, and protecting natural habitats (Zhang et al. 2020; McGraw et al. 2013). *Ginkgo biloba* is categorized as endangered species according to IUCN criteria, even though it poses minimal conservation concern, as it is extensively cultivated worldwide and considered a "living fossil" species with remarkable resilience (Malík and Tlustoš 2023).

Regulatory frameworks reflect these differing conservation priorities. While *P. ginseng* requires CITES permits for international trade, other species, *W. somnifera*, *C. longa*, *P. emblica*, *C. asiatica*, and *G. biloba*, are primarily managed through national and regional guidelines that promote sustainable harvesting, cultivation, and quality control (Shasmita et al. 2018; Zhang et al. 2020). Broader conservation strategies incorporate in vitro propagation, germplasm storage, and habitat restoration as key interventions to mitigate genetic erosion and ensure continued resource availability.

9.3 Global Research Landscape: A Bibliometric Approach

9.3.1 Methodology: Database Selection, Search Strategy, and Analysis Tools

The bibliometric analysis employed data from *Scopus* and *Web of Science* to map global research trends on cryopreservation and synthetic seed technologies for conserving anti-aging medicinal plants. Publications from 2000 to 2025, including journal articles, reviews, and relevant proceedings, were retrieved using Boolean search strings combining terms related to cryopreservation, synthetic seeds, and medicinal or anti-aging plants. Duplicate records were removed, and metadata were standardized for authors, institutions, and keywords. Analytical tools such as VOS viewer were used to assess publication trends, collaboration networks, and thematic structures. The analysis followed transparent, reproducible steps such as data cleaning, normalization, and visualization, to ensure the robustness of results and highlight research gaps, collaborative patterns, and geographic imbalances in this emerging interdisciplinary field.

9.3.2 Publication Trends and Geographic Distribution (2000–2025)

Between 2000 and 2025, the global publication output on cryopreservation and synthetic seed technologies for medicinal and anti-aging plants showed a consistent upward trajectory, reflecting growing scientific and conservation interest. From fewer than 10 publications annually in the early 2000s, the output began to rise steadily after 2010, corresponding to increasing awareness of ex-situ conservation and the emergence of biotechnological applications for safeguarding phytomedicinal resources. A pronounced surge occurred after 2016, aligning with international commitments to biodiversity conservation, the implementation of the Nagoya Protocol, and rising demand for plant-derived bioactive compounds in pharmaceutical and nutraceutical industries (Fig. 9.1). The cumulative trend suggests a transition from descriptive and protocol-oriented studies to integrative research linking cryobiology, phytochemistry, and conservation policy.

Geographically, the research landscape is dominated by contributions from Asia, Europe, and North America, with notable regional variations. India and China together accounted for a significant proportion of total publications, reflecting their extensive medicinal plant diversity and strong academic emphasis on plant biotechnology and conservation. Japan and South Korea contributed a smaller but technologically advanced share, emphasizing cryogenic protocols and cellular-level studies. In Europe, Italy, Spain, and the UK emerged as key contributors, often

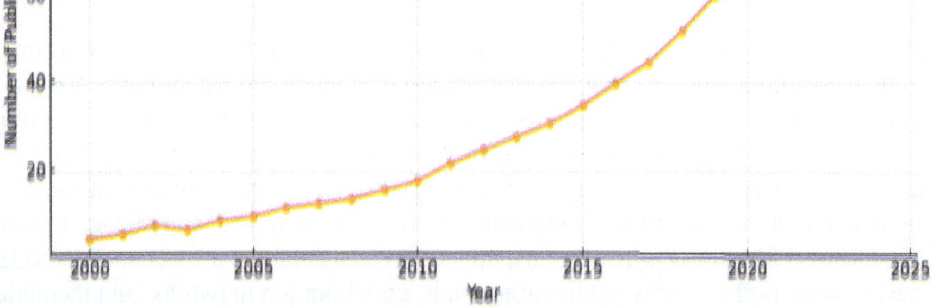

Fig. 9.1 Publication trend graph showing the steady increase in global research output on cryopreservation and synthetic seed technologies for medicinal and anti-aging plants from 2000 to 2025

focusing on germplasm conservation frameworks and comparative analyses between cryogenic and conventional preservation techniques. The United States and Canada exhibited sustained engagement through interdisciplinary collaborations linking academic institutions with botanical gardens and cryobanks.

In the Global South, emerging research hubs were observed in Brazil, South Africa, and Thailand, where conservation-focused institutions increasingly apply cryopreservation and synthetic seed approaches to endemic and high-value medicinal species. Despite this encouraging diversification, large regions, particularly Africa (outside South Africa) and tropical island systems, remain underrepresented, indicating a pressing need for capacity building, infrastructure investment, and cross-regional partnerships.

Temporal and geographic mapping revealed evolving patterns of international collaboration. Early studies were largely country-specific, but from 2015 onward, co-authorship networks expanded markedly, with multinational projects supported by global initiatives such as the *Global Plant Cryobank Network* and *International Society for Horticultural Science* working groups. Such networks have begun to bridge the gap between cryobiological innovation and field-level conservation, yet institutional collaboration remains concentrated among a few high-output centers, underlining the importance of equitable technology transfer.

Overall, the bibliometric patterns underscore that while scientific attention to cryopreservation and synthetic seed technologies for medicinal plants has increased exponentially, research intensity remains unevenly distributed. Sustained growth after 2020 indicates a maturing research ecosystem; however, the next phase must prioritize inclusion of biodiversity-rich but data-poor regions and promote coordinated global frameworks to ensure the conservation and ethical exchange of anti-aging medicinal plant germplasm.

9.3.3 Collaborative Networks and Thematic Clusters

The global research landscape on cryopreservation and synthetic seed technologies for medicinal and anti-aging plants is sustained by a compact but growing network of collaborations concentrated in Asia and Europe. Co-authorship mapping revealed three major clusters: an Asian cluster led by India and China, focusing on protocol development and applied plant biotechnology; a European cluster (notably Italy, Spain, and the UK) emphasizing germplasm conservation frameworks and cryogenic system optimization; and a trans-Pacific cluster linking North America and Japan, integrating cryobiology with phytochemical and metabolomic analyses. Emerging collaborations in Brazil, South Africa, and Thailand mark an encouraging rise in South–South cooperation, though regional participation remains uneven.

Keyword co-occurrence analysis identified five thematic foci: cryogenic protocol optimization, synthetic seed encapsulation, phytochemical stability, germplasm conservation, and policy integration under the Nagoya Protocol. A few institutions in India, China, and Italy serve as central nodes driving knowledge exchange and multi-institutional projects, yet overall network density remains moderate. These findings suggest that while global connectivity is improving, stronger inclusion of underrepresented regions, open-access data sharing, and equitable technology transfer are essential to build a cohesive international research ecosystem for medicinal plant cryobiology.

9.3.4 Identified Research Gaps and Future Priorities

Despite notable progress, global research remains taxonomically narrow and geographically uneven, focusing mainly on a limited number of high-value medicinal species while neglecting diverse regional taxa. Standardized cryopreservation and synthetic seed protocols for recalcitrant or aquatic species are still scarce, and long-term assessments of genetic stability and metabolite retention are rarely undertaken. Data integration also remains fragmented, with limited interoperability among cryogenic databases and biodiversity repositories. These gaps underscore the need for harmonized metadata frameworks, open-access repositories, and multi-year viability studies linking molecular and biochemical validation.

Future research must address both technical and governance challenges. Priorities include expanding taxonomic coverage, enhancing participation of underrepresented regions, and embedding cryobiological work within access-and-benefit-sharing frameworks to ensure compliance with the Nagoya Protocol. Emerging frontiers, such as nanomaterial-based encapsulation, omics-assisted cryobiology, and AI-supported digital cryobanks, offer promising directions. Achieving an inclusive and transparent global conservation strategy will depend on equitable collaboration, standardized methodologies, and integration of ethical and policy dimensions into cryogenic and synthetic seed research.

9.4 Cryopreservation Technologies for Medicinal Plant Germplasm

Cryopreservation is an advanced biotechnological approach that enables the long-term conservation of plant genetic resources at ultra-low temperatures, typically in liquid nitrogen (−196 °C). It effectively halts all cellular metabolic activities, thereby maintaining the viability and genetic integrity of biological materials for extended periods. In medicinal and aromatic plants, cryopreservation serves as a crucial tool for preserving elite genotypes, rare and endangered species, and valuable germplasm that are often threatened by habitat loss or overexploitation. Various explants, such as seeds, shoot tips, somatic embryos, pollen, and callus tissues, can be cryopreserved using methods like encapsulation-dehydration, vitrification, encapsulation-vitrification, and droplet freezing. These techniques not only ensure genetic stability and viability after recovery but also support biotechnological applications in breeding, phytochemical research, and sustainable utilization of medicinal biodiversity.

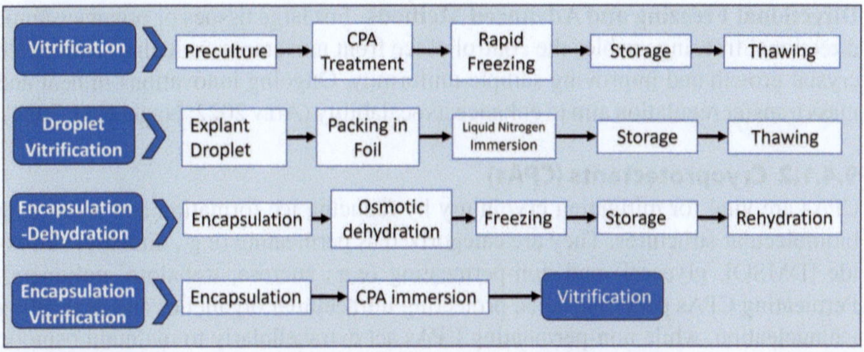

Fig. 9.2 Key steps in different cryopreservation methods

9.4.1 Fundamental Principles: Freezing Techniques and Cryoprotectants

Cryopreservation is a core technique for the long-term preservation of biological materials, including cells, tissues, and germplasm, at ultra-low temperatures, typically in liquid nitrogen (−196 °C) (Rajasekharan and Rohini 2023). At such temperatures, all metabolic and biochemical reactions are effectively arrested, maintaining structural and functional integrity over time. The major challenge lies in mitigating ice crystal formation, which can damage cellular membranes and organelles. Two fundamental components define the success of cryopreservation: the freezing technique employed and the choice and concentration of cryoprotective agents.

9.4.1.1 Freezing Techniques

Cryopreservation protocols primarily utilize slow freezing and vitrification, with emerging methods such as directional freezing developed to address limitations of traditional approaches (Fig. 9.2).

Slow Freezing This technique involves controlled, gradual cooling using low concentrations of CPAs. Controlled dehydration allows water to migrate out of cells, minimizing intracellular ice formation. "Seeding" techniques are often applied to promote extracellular ice formation, preventing lethal intracellular freezing. However, extracellular ice growth and solute concentration effects remain potential risks (Engelmann 2004; Bojic et al. 2021).

Vitrification This ultra-rapid cooling method employs high CPA concentrations to transform water into an amorphous glass-like state without ice formation. While this avoids ice-induced damage, it may introduce CPA toxicity and osmotic stress, particularly in large or complex samples (Roque-Borda et al. 2021; Zamecnik et al. 2021; Kim 2023).

Directional Freezing and Advanced Methods For large tissues or organ systems, directional freezing enables the control of ice front movement, reducing random ice crystal growth and improving sample uniformity. Ongoing innovations in heat and mass transfer regulation aim to enhance its scalability (Arav 2022; Nagel et al. 2024).

9.4.1.2 Cryoprotectants (CPAs)

CPAs are vital for mitigating cryoinjury by reducing ice formation and stabilizing biomolecular structures. They are categorized as permeating (e.g., dimethyl sulfoxide [DMSO], glycerol) and non-permeating (e.g., sucrose, trehalose, polymers). Permeating CPAs penetrate cells, protecting intracellular organelles and preventing ice nucleation, while non-permeating CPAs act extracellularly to maintain osmotic balance and inhibit ice propagation (Elliott et al. 2017; Marcantonini et al. 2022). Their protective mechanisms include lowering the freezing point, increasing viscosity, and elevating the glass transition temperature, which promotes vitrification. However, excessive CPA concentrations can cause cytotoxicity, necessitating optimization of exposure time and temperature (Klbik et al. 2022; Mahanta et al. 2024). Recent innovations explore antifreeze proteins, synthetic polymers, and nanomaterial-based CPAs, which enhance vitrification efficiency while reducing cytotoxic effects (Chang and Zhao 2021; Matsumura et al. 2021; Murray and Gibson 2022) (Table 9.2).

Effective cryopreservation depends on balancing freezing rates with optimal CPA use to prevent ice-induced injury and cytotoxicity. Ongoing advances in cryoprotectant chemistry, vitrification kinetics, and temperature control are enhancing survival rates and structural integrity across diverse biological systems, including medicinal plant germplasm. To translate these principles into practice, species-specific cryopreservation approaches have been developed for medicinally important taxa. The choice of technique often depends on tissue type, cellular water content, and species-specific tolerance to dehydration and CPA exposure.

9.4.2 Protocol Development for Specific Tissues (Shoot Tips, Embryogenic Calli, Pollen)

Cryopreservation plays a pivotal role in safeguarding the genetic resources of medicinal plants, particularly those with recalcitrant seeds, complex propagation systems, or high pharmaceutical and economic value. Tissue-specific protocols have been successfully developed for select species such as *Panax ginseng* and

Table 9.2 Key freezing techniques and roles of cryoprotectants in cryopreservation

Technique	CPA type(s) used	Main principle	Key risks/challenges
Slow freezing	Low–moderate, permeating	Gradual dehydration, ice control	Extracellular ice, solute effects
Vitrification	High, permeating/non-permeating	Glass formation, no ice formation	CPA toxicity, osmotic stress
Directional freezing	Variable	Controlled ice front movement	Heat/mass transfer, scale-up issues

Ginkgo biloba, while others, including *Withania somnifera*, *Curcuma longa*, *Phyllanthus emblica*, and *Centella asiatica*, remain underexplored in this context. The effectiveness of a cryopreservation protocol depends on the physiological state of the tissue, preconditioning treatments, such as sucrose preculture or cold acclimation, and the type and concentration of cryoprotectants used.

For *Panax ginseng*, extensive research has led to the establishment of robust cryopreservation methods for adventitious roots, embryogenic calli, and shoot tips. Vitrification and PVS2-based protocols, combined with sucrose preculture and cold acclimation, have achieved impressive regrowth rates, up to 90% in adventitious roots and 68% in embryogenic calli (Joshi and Teng 2000; Lu et al. 2009; Oh et al. 2009; Uchendu et al. 2011; Le et al. 2019; Lei et al. 2021). These findings highlight the importance of balancing cryoprotectant toxicity with dehydration efficiency to maintain post-thaw viability.

In *Ginkgo biloba*, successful cryopreservation has been demonstrated for callus cultures using sucrose and abscisic acid (ABA) preculture followed by controlled desiccation prior to liquid nitrogen immersion. Post-thaw assessments reveal stable genetic fidelity, suggesting that desiccation-based approaches can effectively preserve this species (Popova et al. 2009).

In *Withania somnifera*, recent work by Navya and Rajasekharan (2025) represents a pioneering effort in pollen cryopreservation. Pollen collected during peak anthesis (9–11 a.m., April–June) was dehydrated using zeolite granules to 10–12% moisture content before cryostorage in liquid nitrogen. Post-thaw viability was confirmed via in vitro germination on Brewbaker and Kwack's medium and validated using Alexander's stain. This protocol offers a foundation for maintaining pollen viability for breeding programs aimed at developing high-yield, stress-tolerant Ashwagandha cultivars.

For *Curcuma longa*, *Phyllanthus emblica*, and *Centella asiatica*, no direct cryopreservation protocols for shoot tips, embryogenic calli, or pollen have been reported. Existing studies primarily focus on in vitro propagation, phytochemical profiling, and metabolomic analyses (Shasmita et al. 2018; Tetali et al. 2021) (Table 9.3).

9.4.3 Post-cryopreservation Viability, Regeneration, and Field Recovery

Success of cryopreservation ultimately depends on post-thaw viability, regenerative capacity, and successful acclimatization under natural or controlled conditions (Rajasekharan and Ganeshan 2018). Existing studies reveal species-specific responses that highlight both the potential and limitations of cryogenic preservation. In *Panax ginseng*, vitrification-based cryopreservation has demonstrated notably high post-thaw survival rates. Adventitious roots exhibit up to 90% survival, while hairy roots show around 60% viability following recovery. These regenerated roots and plantlets maintain ginsenoside biosynthesis and genetic integrity comparable to non-cryopreserved controls, confirming the metabolic and genetic stability

Table 9.3 Overview of cryopreservation methods and their application

Technique	Description	Advantages	Limitations	Suitable plant material
Vitrification	Encapsulation of explants in alginate beads followed by vitrification using cryoprotectants like PVS2	High survival rate, avoids ice crystallization	Requires precise protocol optimization	Shoot tips, somatic embryos, callus
Encapsulation-dehydration	Small droplets of cryoprotectant with explants are rapidly frozen in liquid nitrogen	Easy handling, suitable for synthetic seeds	Sensitive to dehydration parameters	Pollen, shoot tips, meristems, embryonic axes
Droplet vitrification	Encapsulated explants are dehydrated before freezing in liquid nitrogen	Fast, suitable for shoot tips or embryos	Requires specialized handling	Somatic embryos, shoot tips
Encapsulation-vitrification	Combines encapsulation with vitrification, where encapsulated explants are treated with vitrification solutions to enhance protection and survival during freezing	Combines merits of both methods	Variable encapsulation efficiency	Somatic embryos, pollen

of cryopreserved material (Joshi and Teng 2000; Oh et al. 2009; Lee et al. 2023a, b, c; Xu et al. 2023). Such findings reinforce the suitability of vitrification protocols for preserving high-value medicinal species with complex root systems.

For *Ginkgo biloba*, callus cultures subjected to sucrose and abscisic acid (ABA) pretreatment followed by controlled desiccation exhibit moderate post-thaw regrowth, reaching up to 54%. Importantly, molecular analyses confirm the retention of genetic stability in regenerated callus tissues (Popova et al. 2009; Lu et al. 2009). This demonstrates that optimized desiccation-based cryopreservation can effectively maintain genetic fidelity in woody medicinal species with slow growth and limited seed viability.

In *Withania somnifera*, recent studies have established pollen cryopreservation as a viable approach for genetic conservation. Pollen collected during anthesis, desiccated to 10–12% moisture content, and cryostored in liquid nitrogen retains high viability upon rewarming (Navya and Rajasekharan 2025). This method ensures the availability of viable pollen for hybridization and breeding programs aimed at developing stress-tolerant or high-yielding Ashwagandha varieties.

Although laboratory recovery has been well documented for *Panax ginseng* and *Ginkgo biloba*, field-level data remain scarce. In *P. ginseng*, in vitro-derived plantlets can be successfully acclimatized to soil with survival rates exceeding 85%, though most reports do not explicitly confirm whether the regenerated material was cryopreserved (Lee et al. 2023c). For *G. biloba*, direct evidence of field

Table 9.4 Established and potential cryopreservation protocols for medicinal plant tissues

Species	Tissue type	Protocol highlights	Regrowth/ survival	Citations
Panax ginseng	Adventitious roots	Vitrification with sucrose preculture and optimized loading	Up to 90%	Joshi and Teng (2000), Oh et al. (2009), and Le et al. (2019)
	Embryogenic calli	Cold acclimation, PVS2 vitrification	~68%	Lei et al. (2021) and Lee et al. (2023a, b, c)
	Shoot tips	Sucrose/ABA pretreatment, PVS2 vitrification	~60%	Uchendu et al. (2011)
Ginkgo biloba	Callus	Sucrose + ABA preculture, desiccation before freezing	Stable post-thaw	Popova et al. (2009)
Withania somnifera	Pollen	Zeolite-based dehydration (10–12% MC), LN storage, viability via Brewbaker and Kwack's medium	Stable post-thaw	Navya and Rajasekharan (2025)

establishment following cryopreservation is lacking, though maintained genetic stability suggests its potential for successful acclimatization. In *W. somnifera*, pollen remains viable after cryostorage, supporting its continued use in reproductive applications, even though whole-plant recovery is not relevant for this tissue type.

Collectively, these findings demonstrate that while cryopreservation effectively safeguards cellular and genetic integrity in selected medicinal plants, comprehensive assessments linking post-thaw recovery, regeneration efficiency, and field performance are still needed. Establishing such integrated post-cryopreservation evaluations will strengthen the reliability of cryogenic storage as a core strategy for the long-term conservation of valuable medicinal germplasm (Table 9.4).

9.4.4 Challenges in Technology Transfer and Scaling

While laboratory-scale cryopreservation protocols have demonstrated success in preserving genetic resources of medicinal plants, transitioning these methods into large-scale, routine conservation systems presents multiple challenges. These obstacles span technical, operational, and institutional domains, restricting the widespread adoption of cryogenic storage in germplasm banks and commercial applications.

Protocol Standardization and Reproducibility A major limitation in scaling cryopreservation lies in the lack of standardized, universally applicable protocols. Most procedures are species-specific, and often genotype-dependent, requiring extensive optimization for explant type, pretreatment duration, cryoprotectant combinations, and cooling–warming regimes (Bettoni et al. 2021; Białoskórska et al.

2024; Malhotra et al. 2024). Even minor deviations in these parameters can drastically affect survival or regrowth outcomes, leading to inconsistency across laboratories. Furthermore, reproducibility remains a concern, as many reported protocols cannot be replicated outside the original research environment due to variations in materials, culture media, and operator expertise (El Merzougui et al. 2023).

Infrastructure, Cost, and Expertise Constraints The establishment of cryogenic facilities demands significant initial investment, including the acquisition of liquid nitrogen storage systems, temperature control devices, and specialized cryovials (Zhang et al. 2023). Although long-term maintenance costs are relatively low, the setup phase remains prohibitive for many institutions in developing regions. Moreover, successful operation requires trained personnel familiar with both in vitro culture and cryogenic techniques (Bettoni et al. 2021; El Merzougui et al. 2023). Reliable infrastructure, especially a consistent liquid nitrogen supply and robust backup systems, is essential for preventing sample loss or temperature fluctuations.

Biological and Technical Limitations At the biological level, cryoprotectant toxicity and incomplete dehydration are among the most persistent bottlenecks. Common cryoprotectants, such as DMSO and glycerol, can induce cytotoxic effects, while safer alternatives are still under investigation (Xiang et al. 2023; Białoskórska et al. 2024; Gowthami et al. 2024). Many medicinal plants possess recalcitrant or highly differentiated tissues that respond poorly to freezing, resulting in reduced post-thaw viability or regeneration potential (Kundu et al. 2018). Additionally, pre- and post-cryopreservation culture stages are susceptible to microbial contamination and somaclonal variation, complicating downstream regeneration and field validation (Kundu et al. 2018; Wang et al. 2021).

Documentation and Knowledge Sharing Another critical barrier to scaling is the lack of comprehensive, transparent documentation. Many studies report successful cryopreservation outcomes but omit detailed procedural information, including exact cryoprotectant formulations, exposure times, and recovery conditions (El Merzougui et al. 2023; Białoskórska et al. 2024). This lack of procedural clarity hinders reproducibility and technology transfer between research groups. Moreover, most published work emphasizes laboratory recovery, with very limited evidence of long-term stability or field performance of regenerated plants (Wang et al. 2021; Mosa et al. 2023).

The effective scaling and technology transfer of cryopreservation in medicinal plants are constrained by species-specific protocol dependencies, infrastructure and expertise limitations, biological challenges, and inadequate documentation. Overcoming these issues will require global collaboration, standardized reporting frameworks, and the integration of cryogenic preservation into broader conservation networks. Addressing these barriers is essential to ensure reliable, large-scale conservation and sustainable utilization of valuable medicinal plant germplasm.

9.5 Synthetic Seed Technology: Principles and Applications

Synthetic seed technology involves the encapsulation of somatic embryos, shoot buds, or other plant propagules in a protective gel matrix to form artificial seeds capable of germinating into complete plants under suitable conditions. This technique bridges the gap between in vitro culture and conventional propagation by enabling easy handling, storage, and transportation of valuable plant materials. The core principle lies in mimicking the natural seed's protective and nutritive functions using materials such as sodium alginate and calcium chloride for encapsulation. Synthetic seeds offer immense potential for the mass propagation, conservation, and exchange of elite genotypes and rare medicinal species, especially those that are difficult to propagate sexually or produce few viable seeds. Thus, the technology serves as an important tool in plant biotechnology, germplasm conservation, and sustainable utilization of medicinal plant resources.

9.5.1 Encapsulation Techniques Using Sodium Alginate and Other Matrices

Synthetic seed technology represents a pivotal innovation in plant biotechnology, enabling the encapsulation of somatic embryos, shoot tips, or other regenerable tissues in protective matrices for propagation, conservation, and germplasm exchange. This approach bridges the gap between in vitro micropropagation and practical field deployment, offering a scalable, contamination-free alternative to conventional tissue culture methods (Fig. 9.3). Among various encapsulating agents, sodium alginate, a natural polysaccharide derived from brown algae, remains the most widely employed due to its biocompatibility, ease of gel formation, and cost-effectiveness. When cross-linked with calcium chloride ($CaCl_2$), sodium alginate forms hydrogel beads that provide both mechanical support and a semi-permeable environment conducive to tissue survival and regrowth.

Recent advancements in synthetic seed research have focused on optimizing alginate–calcium formulations, refining encapsulation procedures, and integrating alternative or supplementary materials to enhance encapsulation efficiency,

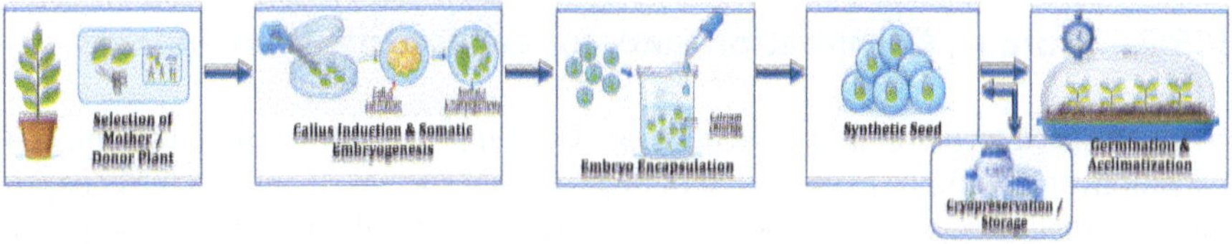

Fig. 9.3 Overview of synthetic seed technology

storability, and regeneration. The drip method is the most commonly used technique, wherein explants or embryos are suspended in sodium alginate and dropped into a calcium chloride solution to induce rapid cross-linking. Studies indicate that 2–4% sodium alginate combined with 75–100 mM $CaCl_2$ yields firm, uniform beads with high germination and regrowth efficiency across multiple species (Siddique and Bukhari 2018; Iqbal et al. 2019; Behera et al. 2020; Aisy et al. 2022; Tadda et al. 2021; Asadi et al. 2022; Ramírez-Mosqueda et al. 2023; Dutra et al. 2024; Wasi et al. 2024). Moreover, supplementing the encapsulation matrix with Murashige and Skoog (MS) salts, plant growth regulators, or carbohydrate sources enhances shoot emergence, rooting, and conversion efficiency during germination (Siddique and Bukhari 2018; Nazirwan et al. 2020; Badhepuri et al. 2025).

Beyond sodium alginate, researchers are exploring hybrid or alternative matrices to improve encapsulation performance and add functional properties. Incorporating starch, chia seed mucilage, pectin, or biodegradable polymers enhances water retention, nutrient exchange, and the gradual release of growth regulators or bioactive compounds (Szopa et al. 2022; Perea-Flores et al. 2023; Farousha et al. 2024; Gholivand et al. 2025; Kou et al. 2024; Nigar et al. 2024). Such innovations expand the potential of synthetic seeds beyond propagation, into bioinoculant delivery systems, germplasm shipping, and temporary ex situ conservation.

Experimental results demonstrate that alginate-based synthetic seeds achieve conversion rates of up to 99%, with high genetic fidelity in species such as *Satureja khuzistanica*, *Hedychium coronarium*, *Capparis decidua*, *Ipomoea batatas* (sweet potato), and *Coleus forskohlii* (Siddique and Bukhari 2018; Behera et al. 2020; Tadda et al. 2021; Asadi et al. 2022; Wasi et al. 2024; Badhepuri et al. 2025). Additionally, food-grade alginate performs comparably to analytical-grade reagents, providing an affordable option for large-scale and commercial applications (Aisy et al. 2022).

The following table summarizes key encapsulation parameters and recent innovations influencing synthetic seed performance (Table 9.5).

Synthetic seed technology, driven by advances in alginate encapsulation and hybrid matrix development, has matured into a versatile platform for medicinal and horticultural plant conservation. Its scalability, reproducibility, and adaptability make it an emerging complement to cryopreservation, especially for short-term germplasm storage and distribution.

9.5.2 Storage, Germination, and Conversion into Whole Plants

Synthetic seed technology represents a powerful approach for the propagation, conservation, and germplasm exchange of diverse plant species. It enables the encapsulation of plant tissues or embryos in protective matrices, allowing their storage and regeneration under controlled conditions. The efficiency of synthetic seed systems depends on key parameters including storage environment, germination protocols, and the conversion rate into genetically stable, fully developed plants.

Table 9.5 Comparison of encapsulation parameters and their effects across studies

Parameter/ innovation	Typical range/ approach	Impact/notes	References
Sodium alginate concentration	2–4%	Produces firm, uniform beads with high viability	Siddique and Bukhari (2018), Iqbal et al. (2019), Behera et al. (2020), Tadda et al. (2021), Asadi et al. (2022), Ramírez-Mosqueda et al. (2023), Dutra et al. (2024), Wasi et al. (2024), and Badhepuri et al. (2025)
$CaCl_2$ concentration	75–100 mM (up to 200 mM)	Ensures optimal cross-linking and bead stability	
Matrix additives (starch, mucilage)	1:1 ratio with alginate	Improves entrapment, mechanical strength, and controlled release	Szopa et al. (2022), Perea-Flores et al. (2023), Farousha et al. (2024), Gholivand et al. (2025), Kou et al. (2024), and Nigar et al. (2024)
Storage conditions	4 °C for 2–8 weeks	Maintains tissue viability and regeneration potential	Behera et al. (2020), Wasi et al. (2024), and Badhepuri et al. (2025)

Storage Conditions and Longevity The viability and regeneration potential of synthetic seeds are largely influenced by storage temperature and duration. Low-temperature storage, typically between 4 °C and 8 °C, is widely recognized as optimal for short- to medium-term preservation, maintaining high germination and conversion rates for up to 4–5 months across various species such as *Rubus fruticosus* (blackberry), *Cannabis sativa*, *Capparis decidua*, *Caladium bicolor*, and several orchids (Faisal et al. 2012; Gantait et al. 2015; Bhattacharyya et al. 2018; Priyadharshini et al. 2020; Zarei et al. 2022; Nagananda and Satishchandra 2023; Regni et al. 2024). Cryopreservation at −196 °C in liquid nitrogen extends long-term storage potential, with reports of high post-thaw viability in certain species Özbay et al. (2024). However, prolonged cold storage can gradually reduce conversion efficiency, emphasizing the need for species-specific optimization of encapsulation and storage parameters (Maqsood et al. 2015; Di Sario et al. 2025).

Germination and Conversion Efficiency Successful conversion of synthetic seeds into plantlets depends on the composition of the germination medium, the presence of growth regulators, and the physiological status of the encapsulated explant. High germination and conversion rates, ranging from 70% to 100%, are typically achieved on nutrient-enriched, sterile substrates such as agar-solidified Murashige and Skoog (MS) medium, cocopeat, or soilrite mixtures (Maqsood et al. 2015; Das et al. 2016; Siddique and Bukhari 2018; Bhattacharyya et al. 2018; Subrahmanyeswari et al. 2023). Supplementation with plant growth regulators such as BA, NAA, or GA_3 enhances shoot and root induction, while additives like acetylsalicylic acid, iron oxide nanoparticles, and elicitors promote seedling vigor and stress resilience (Gantait et al. 2021; Zarei et al. 2022; Kulus et al. 2025). Plants regenerated from optimized systems exhibit robust acclimatization and high survival rates under greenhouse or field conditions, highlighting the technology's potential for large-scale propagation (Asadi et al. 2022; Badhepuri et al. 2025).

Genetic Stability and Plant Quality Ensuring genetic fidelity is central to the success of synthetic seed-derived plants. Molecular marker analyses, including ISSR, RAPD, and SCoT profiling, consistently confirm the clonal stability of regenerated plants compared to their donor counterparts (Bhattacharyya et al. 2018; Badhepuri et al. 2025; Reshi et al. 2025). Moreover, regenerated plants typically exhibit normal morphology, physiological behavior, and in some cases, elevated levels of secondary metabolites due to the stress-associated metabolic modulation during encapsulation (Subrahmanyeswari et al. 2023; Kulus et al. 2025). Collectively, these findings reinforce synthetic seed technology as a dependable method for clonal propagation and conservation of elite germplasm.

Synthetic Seed Technology in Key Medicinal Plants: Current Status and Protocols Synthetic seed systems have been effectively adapted for several medicinally important species that exhibit poor natural propagation or seed set. These include *Withania somnifera*, *Curcuma longa*, *Phyllanthus emblica*, *Centella asiatica*, *Panax ginseng*, and *Ginkgo biloba*. Encapsulation typically employs sodium alginate matrices (3–4%) cross-linked with calcium chloride (75–100 mM), forming durable synthetic beads that encapsulate nodal, bud, or embryogenic explants.

9.5.2.1 Protocols and Success in Target Species

Withania somnifera Nodal segments encapsulated in 3% sodium alginate and 100 mM $CaCl_2$ yield synthetic seeds with conversion rates up to 86% when cultured on MS medium containing 2.5 μM BA and 0.5 μM NAA. The regenerated plants display stable morphology and confirmed genetic fidelity (Fatima et al. 2013; Saxena et al. 2025).

Curcuma longa Synthetic seeds derived from somatic embryos exhibit high viability and regeneration when stored for 30–60 days, producing disease-free, genetically uniform plantlets (Nirmal Babu et al. 2016; Jena et al. 2018; Shajahan et al. 2019).

Phyllanthus emblica Although direct synthetic seed protocols remain limited, established micropropagation systems through organogenesis and somatic embryogenesis provide a strong foundation for future encapsulation studies (Gantait et al. 2021).

Centella asiatica Axillary buds or somatic embryos encapsulated in 4% sodium alginate and 75 mM $CaCl_2$ can be stored up to 200 days with over 85% germination. The regenerated plants retain morphological, chemical, and genetic fidelity (Joshee et al. 2007; Prasad et al. 2014; Ganie et al. 2022).

Table 9.6 Summary of synthetic seed protocols and outcomes in key medicinal plants

Species	Explant/matrix	Storage (days)	Conversion rate	Genetic stability
Withania somnifera	Nodal/3% alginate	28–60	86%	Confirmed
Curcuma longa	Embryo/alginate	30–60	High	Confirmed
Centella asiatica	Bud/4% alginate	200	>85%	Confirmed
Panax ginseng	Embryo/alginate	30–60	High	Confirmed

Panax ginseng Somatic embryo-based synthetic seeds demonstrate efficient plant regeneration and acclimatization, overcoming the species' low seed viability and slow growth (Kim 2023; Lee et al. 2023a).

Ginkgo biloba While direct synthetic seed protocols are limited, encapsulation techniques developed for other gymnosperms indicate strong potential for adaptation in this medicinally important species (Maqsood et al. 2020) (Table 9.6).

9.5.3 Integrating Synthetic Seed Technology with Micropropagation Systems

Synthetic seed technology and micropropagation represent two highly complementary in vitro propagation approaches that, when integrated, provide a versatile platform for large-scale clonal propagation, germplasm conservation, and efficient plant delivery. This integration is particularly valuable for species with recalcitrant seeds, low seed viability, or limited natural propagation capacity.

In an integrated system, the source of propagules originates from micropropagation processes, either through organogenesis or somatic embryogenesis, that yield large numbers of uniform, disease-free explants such as somatic embryos, shoot tips, or nodal segments. These in vitro-derived micro propagules are ideal candidates for encapsulation as synthetic seeds, ensuring both genetic uniformity and phytosanitary safety (Rihan et al. 2017; Qahtan et al. 2019; Abbas et al. 2022; Micheli et al. 2022; Maheshwari and Garg 2023; Regni and Cesarini 2025).

The encapsulation step involves immersing these micro propagules in a hydrogel, most commonly sodium alginate cross-linked with calcium chloride, to form synthetic seeds that can be handled, stored, and transported much like true seeds. Following storage, the regeneration and acclimatization phase completes the micropropagation cycle: synthetic seeds are germinated either in vitro on nutrient-enriched media or ex vitro under controlled humidity, leading to the development of complete plantlets that are subsequently acclimatized and transferred to soil (Siddique and Bukhari 2018; Valdés et al. 2021; Abbas et al. 2022; Ahmad et al. 2023; Meucci et al. 2024).

Integrating synthetic seed technology with micropropagation systems offers multiple operational and biological advantages. The foremost benefit is mass propagation, as it enables the large-scale, uniform production of genetically identical plants from elite donor material (Rihan et al. 2017; Qahtan et al. 2019; Abbas et al. 2022; Regni and Cesarini 2025). Furthermore, it supports germplasm conservation and exchange, allowing encapsulated propagules to be safely stored or transported across regions without risk of contamination or genetic drift (Abbas et al. 2022; Meucci et al. 2024; Regni and Cesarini 2025).

From an economic standpoint, the integration enhances cost and labor efficiency by simplifying handling, reducing dependence on continuous culture maintenance, and enabling direct sowing of synthetic seeds (Rihan et al. 2017; Micheli et al. 2022). Biologically, it helps overcome seed dormancy and recalcitrance, a major limitation in several economically and medicinally important plants with poor natural seed set or desiccation-sensitive seeds (Qahtan et al. 2019). Importantly, this approach maintains disease-free, clonal propagation, as both the micropropagation stage and encapsulation process preserve genetic fidelity and eliminate microbial contaminants.

9.5.4 Advantages and Constraints Compared to Traditional Propagation

Synthetic seed technology represents a promising alternative to conventional propagation methods, particularly for plants that exhibit seed recalcitrance, low germination rates, or propagation constraints. By integrating synthetic seed systems with micropropagation, it enables the encapsulation of in vitro-derived propagules such as somatic embryos, shoot tips, or nodal segments in hydrogel to produce artificial seeds that can be stored, transported, and sown similarly to natural seeds (Rihan et al. 2017; Qahtan et al. 2019; Abbas et al. 2022).

One of the principal advantages of synthetic seed technology lies in its capacity for large-scale, clonal propagation of genetically uniform and disease-free plants. This approach overcomes several limitations associated with conventional seed propagation, such as dormancy, seasonal dependence, and variability in genetic composition (Singh et al. 2020; Khan et al. 2021; Sahu et al. 2025). The in vitro origin of micro propagules ensures phytosanitary quality and genetic fidelity, making the technology particularly valuable for high-value commercial and medicinal crops (Rihan et al. 2017; Zarei et al. 2022). Moreover, synthetic seeds support ex situ germplasm conservation and international germplasm exchange by allowing safe, efficient storage and transport of elite genotypes or endangered species, often in combination with cryopreservation strategies. The simplicity of handling and potential for direct sowing also reduce labor requirements and costs compared to traditional micropropagation, thereby improving overall operational efficiency (Gantait et al. 2015; Regni et al. 2024; Özbay et al. 2024).

Despite these advantages, several technical and biological constraints limit the widespread adoption of synthetic seed systems. Germination and conversion rates are often inconsistent, particularly when synthetic seeds are exposed to non-sterile field conditions (Regni et al. 2024; Sahu et al. 2025). Microbial contamination during storage and sowing remains a major obstacle, emphasizing the need for stringent aseptic handling and optimized matrix formulations. Additionally, the absence of natural dormancy mechanisms in synthetic seeds can reduce storage longevity and post-germination vigor (Khan et al. 2021). Developing standardized encapsulation parameters remains species-specific and technically demanding (Rihan et al. 2017; Abbas et al. 2022). These challenges, coupled with the high cost of micropropagule production, currently limit the scalability and commercial viability of synthetic seed systems (Maheshwari and Garg 2023).

Nevertheless, the integration of synthetic seed and micropropagation technologies has demonstrated considerable success in several medicinal species. For instance, encapsulated shoot tips or somatic embryos derived from *Withania somnifera*, *Curcuma longa*, *Centella asiatica*, and *Panax ginseng* exhibit high conversion and regeneration efficiency upon reintroduction into culture or soil (Rihan et al. 2017; Maqsood et al. 2020; Gantait et al. 2021). However, comparable progress remains limited for *Phyllanthus emblica* and *Ginkgo biloba*, indicating a need for further optimization and standardization in underexplored medicinal taxa. As demonstrated in horticulture, forestry, and conservation contexts, this integrated system offers scalable, sustainable solutions for clonal fidelity, rapid propagation, and the preservation of valuable or endangered genotypes (Meucci et al. 2024; Regni and Cesarini 2025; Ahmad et al. 2023).

Synthetic seed technology provides an efficient, biologically sound platform for clonal propagation and genetic resource management but still requires refinement in encapsulation, storage, and germination protocols for broader practical application. Continuous advances in formulation chemistry, aseptic techniques, and automation will likely enhance the reproducibility, scalability, and economic feasibility of this technology, positioning it as a critical complement to traditional propagation systems in the sustainable management of medicinal and economically important plants.

9.6 Integrated Case Studies: Germplasm Conservation and Exchange

Conservation and exchange of medicinal plant germplasm are critical for safeguarding biodiversity, ensuring sustainable use, and supporting pharmaceutical and horticultural research. Modern approaches, including synthetic seed technology and cryopreservation, complement traditional ex situ systems such as medicinal botanical gardens, specialized biobanks, and field gene banks, forming a robust framework for preserving high-value plant resources (Panis et al. 2020; Engels and Ebert 2021a, b; Priyanka et al. 2021).

Medicinal Botanical Gardens maintain living collections of rare, endemic, or pharmaceutically important species, providing material for research, education, and public engagement. These gardens curate genetically verified populations of species such as *Withania somnifera*, *Centella asiatica*, *Curcuma longa*, and *Panax ginseng*, supporting propagation, chemical profiling, and ex situ conservation (Engels and Ebert 2021a, b; Priyanka et al. 2021; Yadav et al. 2024a). The integration of seed banking, tissue culture, and cryopreservation within these gardens enables long-term preservation and safe distribution of elite medicinal genotypes. For instance, collections of *Withania somnifera* and *Centella asiatica* often include both in vitro propagated material and encapsulated synthetic seeds to enhance storage and dissemination (Asadi et al. 2022; Badhepuri et al. 2025).

Biobanks and seed banks provide secure, long-term storage of medicinal germplasm, especially for species producing orthodox seeds. Globally, gene banks store millions of accessions of medicinally valuable plants, safeguarding genetic diversity and enabling international germplasm exchange (Mascher et al. 2019; Priyanka et al. 2021; Engels and Ebert 2024). Advanced conservation strategies, including cryopreservation of seeds, pollen, or somatic embryos, extend storage to recalcitrant and vegetatively propagated species such as *Panax ginseng*, *Ginkgo biloba*, and *Curcuma longa* (Panis et al. 2020; Lee et al. 2023a, b, c; Özbay et al. 2024). These facilities allow global sharing of elite or endangered medicinal genotypes under strict phytosanitary and regulatory standards, supporting both research and commercial applications.

Field gene banks remain indispensable for clonal and recalcitrant medicinal species that cannot be stored as seeds. Species like *Withania somnifera*, *Phyllanthus emblica*, and tuberous *Curcuma* species are maintained in living collections, allowing direct evaluation of growth, metabolite content, and propagation potential (González-Arnao et al. 2017; Engels and Ebert 2021a, b; Yadav et al. 2024b). Field gene banks provide a platform for large-scale propagation, integration with in vitro conservation techniques, and distribution of planting material for restoration or cultivation programs. However, these collections require intensive management to mitigate risks from pests, diseases, and environmental stresses.

Germplasm exchange networks and policy frameworks facilitate global sharing of medicinal plant resources while ensuring legal and ethical compliance. International agreements, including the Nagoya Protocol and the International Treaty on Plant Genetic Resources for Food and Agriculture, govern access and benefit-sharing, while collaborative networks such as CGIAR and the Millennium Seed Bank Partnership coordinate distribution and documentation (Galluzzi et al. 2016; Liu et al. 2018; Halewood et al. 2020; Kumar et al. 2021; Smith et al. 2021; Engels et al. 2024). Digital platforms such as *Genesys* and *BGCI PlantSearch* enhance transparency, improve accession tracking, and support informed utilization of medicinal germplasm (Ebert and Engels 2020; Weise et al. 2020).

By integrating botanical gardens, specialized biobanks, and field gene banks, researchers and conservationists can secure both short- and long-term preservation of high-value medicinal species. When combined with modern techniques such as synthetic seed technology, micropropagation, and cryopreservation, this network allows safe distribution, clonal propagation, and maintenance of genetic and phytochemical fidelity, supporting both biodiversity conservation and sustainable utilization of medicinal plant resources (Wambugu et al. 2018; Engels and Ebert 2021; Volk et al. 2021).

9.6.1 Strategies for Technology Transfer to Low-Income and Biodiversity-Rich Regions

Effective technology transfer of cryopreservation and synthetic seed techniques is vital for conserving medicinal plant germplasm in biodiversity-rich, low-income regions. Despite their potential for long-term, secure, and cost-effective preservation, implementation is often limited by technical, infrastructural, and capacity-related constraints. Key strategies for successful transfer include the following:

Capacity Building and Training: Regional training programs and workshops are essential to develop local expertise and ensure species-specific protocol application. Simplified, validated, and locally adapted protocols enhance reproducibility and encourage broader adoption (Normah et al. 2019; Bettoni et al. 2021; Coelho et al. 2020; Malik et al. 2025).

Infrastructure and Resource Optimization: Adapting cryopreservation protocols to low-cost, locally available materials and using shared regional cryobanks can reduce costs and improve accessibility in resource-limited laboratories (Coelho et al. 2020; El Merzougui et al. 2023; Malik et al. 2025).

Collaborative Networks and Policy Support: Partnerships with established cryobanks, botanical gardens, and research institutions can facilitate knowledge exchange and germplasm sharing. Sustained funding and supportive policies are critical for long-term implementation and scaling (Bettoni et al. 2021; Zeng et al. 2024).

Community Engagement and Local Relevance: Involving local communities, traditional healers, and conservation groups ensures that transferred technologies align with cultural practices and regional conservation priorities, promoting acceptance and sustainable use (El Merzougui et al. 2023; Zeng et al. 2024).

9.6.2 Legal, Ethical, and Compliance Considerations Under the Nagoya Protocol

The Nagoya Protocol provides a binding international framework governing access to genetic resources and the fair and equitable sharing of benefits (ABS) arising from their use. Its implementation has profound legal, ethical, and practical implications for medicinal plant germplasm exchange, conservation, and research.

Legal and Compliance Requirements Users must obtain *Prior Informed Consent* (PIC) and establish *Mutually Agreed Terms* (MAT) with provider countries, outlining benefit-sharing obligations before accessing genetic resources (Smith et al. 2017; Brink and Van Hintum 2020; Ebert et al. 2023; Lachenmeier and Montagnon 2024). Compliance depends on adherence to national ABS laws, which vary widely in documentation, access procedures, and enforcement mechanisms (Friso et al. 2020; Knight et al. 2023). Transparent documentation and traceability of all exchanges are essential to ensure legal provenance, while monitoring systems and checkpoints help enforce compliance (Smith et al. 2017; Movilla Pateiro 2020).

Ethical Considerations The Protocol emphasizes respect for Indigenous and local knowledge associated with genetic resources, ensuring equitable benefit-sharing with the communities that safeguard them (Robinson et al. 2021; Schebesta 2021). Benefits may include both monetary compensation and non-monetary contributions such as technology transfer and capacity building (Ebert et al. 2023). Importantly, the Protocol seeks to prevent biopiracy and ensure that all resource use occurs with consent and fairness (Kariyawasam and Tsai 2018; Heinrich et al. 2020).

Practical Challenges Despite its intent, implementation remains complex. Navigating variable national ABS systems can be time-consuming and discourage international collaboration (Friso et al. 2020; Knight et al. 2023). Uneven enforcement and administrative burdens often create uncertainty for researchers and institutions (Kiehn et al. 2021; MacDonald 2025). In some cases, restrictive regulations may inadvertently hinder conservation and non-commercial research (Heinrich et al. 2020). Ensuring compliance with the Nagoya Protocol therefore requires balancing legal obligations with ethical commitments and practical feasibility like promoting fair, transparent, and sustainable germplasm exchange while safeguarding biodiversity and Indigenous rights.

9.7 Future Directions and Scaling Pathways

9.7.1 Innovations in Protocol Optimization: Role of Artificial Intelligence

Recent advances in artificial intelligence (AI) and machine learning (ML) are transforming the optimization of cryopreservation and synthetic seed protocols for medicinal plants. These tools address longstanding challenges related to species variability, protocol complexity, and resource efficiency by enabling data-driven decision-making and predictive modelling.

AI in Protocol Design and Optimization AI-based models such as artificial neural networks (ANNs), support vector machines, and decision tree algorithms have been used to predict optimal experimental parameters, including cryoprotectant concentration, cooling rates, and regrowth media across diverse plant tissues. Such predictive approaches minimize the need for extensive trial-and-error experimentation and enhance reproducibility (Sambu 2015; Hesami and Jones 2020; García-Pérez et al. 2020; Pepe et al. 2021). Similarly, advanced optimization algorithms like differential evolution and Bayesian learning enable high-throughput screening of parameter combinations, substantially reducing experimental time and cost (Murray and Gibson 2022; Hesami and Jones 2020).

Data Integration and Knowledge Extraction AI-driven natural language processing and large language models (LLMs) have emerged as valuable tools for mining existing literature to identify effective cryoprotectants, media formulations, and methodological innovations, thereby accelerating protocol development (Ashikhmina et al. 2024).

Real-Time Monitoring and Process Control ML-based monitoring systems are increasingly applied to control culture and cryopreservation conditions in real time, improving process stability and recovery success (Pepe et al. 2021).

AI-guided optimization enhances customization for recalcitrant or physiologically complex medicinal species, allowing the design of tailored protocols that improve survival and regeneration outcomes (Hesami and Jones 2020; García-Pérez et al. 2020). Moreover, standardized, AI-derived models enhance protocol transferability across laboratories and species, facilitating broader application in germplasm conservation and exchange (Kim and Popova 2023).

AI and ML innovations are enabling a paradigm shift in cryobiotechnology by supporting more efficient, reproducible, and species-specific conservation approaches. Their integration into medicinal plant cryopreservation pipelines promises to accelerate both conservation and applied biotechnological research.

9.7.2 Public–Private Partnerships and Capacity Building for Technology Adoption

Adoption of cryopreservation and synthetic seed technologies for medicinal plant conservation requires not only scientific innovation but also strong institutional collaboration and sustained capacity building. Public–private partnerships (PPPs) play a pivotal role in bridging gaps in infrastructure, expertise, and resource allocation, particularly in biodiversity-rich regions with limited research capacity.

Public–private partnerships enable resource pooling to establish and maintain cryobanks and tissue culture facilities that would otherwise be financially unfeasible for single institutions. Such collaborations combine public-sector research expertise with private-sector investment to create sustainable conservation infrastructure (El Merzougui et al. 2023; Zeng et al. 2024). Joint initiatives also promote the development, validation, and dissemination of standardized cryopreservation and synthetic seed protocols applicable across species and regions (Bettoni et al. 2021; Özbay et al. 2024). Moreover, PPPs provide stable, long-term funding mechanisms that support research continuity and the scaling of conservation programs (El Merzougui et al. 2023).

Building human and institutional capacity is essential for effective technology transfer. Training workshops, hands-on laboratory programs, and technical exchange initiatives strengthen local expertise needed to adapt protocols for native medicinal species (Bettoni et al. 2021). Simplified and accessible protocols, tailored to local infrastructure and material availability, further enhance adoption in resource-limited contexts (Özbay et al. 2024). Engaging local researchers, communities, and industry stakeholders ensures that technology transfer aligns with regional conservation priorities and socio-economic goals (El Merzougui et al. 2023).

Despite progress, ensuring the validation and reproducibility of protocols across laboratories remains a significant challenge, necessitating ongoing collaborative refinement and feedback (Bettoni et al. 2021). Long-term success will depend on sustained financial commitment, institutional coordination, and continuous training efforts (Özbay et al. 2024). PPPs and capacity-building initiatives represent critical enablers for the large-scale adoption of cryopreservation and synthetic seed technologies. By integrating shared resources, expertise, and local engagement, these approaches foster sustainable conservation and equitable utilization of medicinal plant genetic resources.

9.7.3 Commercialization Strategies

Commercialization of cryopreservation and synthetic seed technologies presents a transformative opportunity for ensuring the sustainable supply, genetic conservation, and biotechnological utilization of medicinal plants. However, successful market integration requires overcoming significant technical, infrastructural, and policy challenges while fostering innovative business models and institutional support mechanisms.

Technical Bottlenecks and Infrastructure Needs Current cryopreservation and synthetic seed protocols are highly species-specific, requiring precise optimization of explant type, encapsulation materials, and cryoprotectants, which limits scalability and standardization (Gantait et al. 2015; Bettoni et al. 2021; Kaviani and Kulus 2022; El Merzougui et al. 2023). High costs associated with specialized equipment, such as freeze dryers and ultra-low temperature storage systems, remain prohibitive for small enterprises and low-resource laboratories (El Merzougui et al. 2023; Xiang et al. 2023). Moreover, cryoprotectant-induced cellular damage and variable recovery rates, particularly in recalcitrant species, continue to impede efficiency, prompting research into safer cryoprotectants and refined protocols (Kaviani and Kulus 2022; Zeng et al. 2024). A persistent shortage of trained personnel and well-equipped facilities further constrains large-scale implementation (Bettoni et al. 2021).

Business Models and Value Chain Integration Several commercialization pathways have emerged for integrating cryopreservation and synthetic seed technologies into medicinal plant value chains. *Contract production* models allow specialized laboratories to supply cryopreserved or synthetic seed stock to growers and industries but face challenges related to high setup costs and technology transfer (Nausch and Buyel 2021). Licensing and franchising approaches enable the dissemination of optimized protocols to pharmaceutical or horticultural enterprises, though intellectual property (IP) management and quality control remain critical (Khan et al. 2021). Vertical integration, where companies oversee the entire pipeline from germplasm conservation to product development, offers control and traceability but demands substantial investment (Xiang et al. 2023). Public–private partnerships continue to provide effective frameworks for shared infrastructure, research, and market access, promoting equitable benefit-sharing (El Merzougui et al. 2023).

Policy and Institutional Support Mechanisms Robust regulatory, financial, and institutional frameworks are essential for commercial scalability. Standardized guidelines for genetic stability, biosafety, and quality assurance are required to facilitate trade and regulatory compliance (Nausch and Buyel 2021; Kang et al. 2025). Public funding, targeted grants, and fiscal incentives can mitigate high startup costs and promote infrastructure development (Bettoni et al. 2021). Capacity-building initiatives focusing on workforce training and protocol standardization will further support sustainable adoption (El Merzougui et al. 2023). Additionally, IP policies must ensure fair access to genetic resources while respecting benefit-sharing obligations under international agreements such as the Nagoya Protocol (Kang et al. 2025).

Commercializing cryopreservation and synthetic seed technologies thus requires coordinated efforts among researchers, policymakers, and industry stakeholders. Overcoming technical bottlenecks, fostering innovative business ecosystems, and implementing enabling policies are central to unlocking the full potential of these technologies for conservation and the sustainable utilization of medicinal plant biodiversity.

9.8 Conclusion and Future Research Guidelines

Integrating cryopreservation and synthetic seed technologies presents a promising and sustainable strategy for the global conservation, exchange, and utilization of anti-aging medicinal plants. Together, these approaches can ensure the long-term preservation of elite germplasm, rare genotypes, and bioactive-rich accessions, safeguarding them against genetic erosion and environmental threats. Future research should focus on standardizing cryogenic protocols for recalcitrant species, optimizing encapsulation matrices and nutrient formulations for synthetic seeds, and developing combined cryo-encapsulation systems for enhanced viability and regeneration efficiency. Additionally, the integration of molecular markers, metabolomic profiling, and AI-assisted viability assessment tools can accelerate progress in identifying and maintaining high-value phytochemical traits. Strengthening international germplasm exchange frameworks and cryobank networks will further support collaborative research, ensuring that these biotechnological innovations contribute effectively to global health, sustainable bioresource management, and future drug discovery from anti-aging medicinal plants.

Competing Interests The authors have no relevant financial or non-financial interests to disclose.

References

Abbas MK, Mahood HE, Alhasan AS (2022) Production of synthetic seeds in vegetable crops: a review. In: IOP conference series: earth and environmental science, vol 1060, no 1. IOP Publishing, p 012099

Ahmad A, Ansari FA, Anis M, Khan AS (2023) Micropropagation of *Pterocarpus marsupium* Roxb. through synthetic seeds and its novel antibiofilm activities against ESKAPE pathogens. Ind Crops Prod 198:116681

Aisy AR, Ratnasari E, Dewi SK (2022) The effect of the use of sodium alginate on the encapsulation of *Phalaenopsis* sp. synthetic seeds. Lentera Bio Berkala Ilmiah Biol 11:131–138

Alum EU (2025) Sustainable harvesting of medicinal plants: balancing therapeutic benefits with environmental conservation. Agroecol Sustain Food Syst 49:380–385

Applequist W, Brinckmann J, Cunningham A, Hart R, Heinrich M, Katerere D, Van Andel T (2019) Scientists' warning on climate change and medicinal plants. Planta Med 86:10–18

Arav A (2022) Cryopreservation by directional freezing and vitrification focusing on large tissues and organs. Cells 11:1072

Argyropoulou A, Aligiannis N, Trougakos IP, Skaltsounis AL (2013) Natural compounds with anti-ageing activity. Nat Prod Rep 30:1412–1437

Asadi R, Abdollahi MR, Moosavi SS, Mirzaie-Asl A (2022) Alginate encapsulation of micro-cuttings in endangered *Satureja khuzistanica* species: a promising method for obtaining genetically stable plants with high rosmarinic acid content. Plant Cell Tissue Organ Cult 151:307–320

Ashikhmina MS, Zenkin AM, Ivanova AO, Pavlishina IR, Orlova OY, Pantiukhin IS, Skorb EV (2024) Large language model for automating the analysis of cryoprotectants. J Chem Inform Model 65:162–172

Babich O, Larina V, Ivanova S, Tarasov A, Povydysh M, Orlova A et al (2022) Phytotherapeutic approaches to the prevention of age-related changes and the extension of active longevity. Molecules 27:2276

Badhepuri MK, Aileni M, Beeravelli PR, Durgam SK, Sandhya D, Shekhawat MS, Singisala NR (2025) Production of non-embryogenic synthetic seeds for short-term cold storage and ISSR marker-assisted genetic fidelity analysis of *coleus forskohlii* (Willd) Briq plants. Plant Cell Tissue Organ Cult 160:41

Balasubramaniam M, Sapuan S, Hashim IF, Ismail NI, Yaakop AS, Kamaruzaman NA, Mokhtar AMA (2024) The properties and mechanism of action of plant immunomodulators in regulation of immune response–a narrative review focusing on *Curcuma longa* L., *Panax ginseng* CA Meyer and *Moringa oleifera* Lam. Heliyon 10:e28261

Barbalho SM, Direito R, Laurindo LF, Marton LT, Guiguer EL, Goulart RDA et al (2022) *Ginkgo biloba* in the aging process: a narrative review. Antioxidants 11:525

Behera S, Rout KK, Panda PC, Naik SK (2020) Production of non-embryogenic synthetic seeds for propagation and germplasm transfer of *Hedychium coronarium* J Koenig. Appl Res Med Aromat Plants 19:100271

Bettoni JC, Bonnart R, Volk GM (2021) Challenges in implementing plant shoot tip cryopreservation technologies. Plant Cell Tissue Organ Cult 144:21–34

Bhattacharyya P, Kumar V, Van Staden J (2018) In vitro encapsulation based short term storage and assessment of genetic homogeneity in regenerated *Ansellia africana* (leopard orchid) using gene targeted molecular markers. Plant Cell Tissue Organ Cult 133:299–310

Białoskórska M, Rucińska A, Boczkowska M (2024) Molecular mechanisms underlying freezing tolerance in plants: implications for cryopreservation. Int J Mol Sci 25:10110

Bjørklund G, Shanaida M, Lysiuk R, Butnariu M, Peana M, Sarac I et al (2022) Natural compounds and products from an anti-aging perspective. Molecules 27:7084

Bojic S, Murray A, Bentley BL, Spindler R, Pawlik P, Cordeiro JL et al (2021) Winter is coming: the future of cryopreservation. BMC Biol 19:56

Brink M, Van Hintum T (2020) Genebank operation in the arena of access and benefit-sharing policies. Front Plant Sci 10:1712

Buranasudja V, Rani D, Malla A, Kobtrakul K, Vimolmangkang S (2021) Insights into antioxidant activities and anti-skin-aging potential of callus extract from *Centella asiatica* (L.). Sci Rep 11:13459

Cahyaningsih R, Phillips J, Brehm JM, Gaisberger H, Maxted N (2021) Climate change impact on medicinal plants in Indonesia. Glob Ecol Conserv 30:1752

Chang T, Zhao G (2021) Ice inhibition for cryopreservation: materials, strategies, and challenges. Adv Sci 8:2002425

Chen G, Sun W, Wang X, Kongkiatpaiboon S, Cai X (2019) Conserving threatened widespread species: a case study using a traditional medicinal plant in Asia. Biodivers Conserv 28:213–227

Coelho N, Gonçalves S, Romano A (2020) Endemic plant species conservation: biotechnological approaches. Plants 9:345

Cruz-Cruz CA, González-Arnao MT, Engelmann F (2013) Biotechnology and conservation of plant biodiversity. Resources 2:73–95

Dad JM, Rashid I (2022) Differential responses of kashmir himalayan threatened medicinal plants to anticipated climate change. Environ Conserv 49(1):33–41

Dar NJ, Hamid A, Ahmad M (2015) Pharmacologic overview of *Withania somnifera*, the Indian ginseng. Cell Mol Life Sci 72:4445–4460

Das DK, Rahman A, Kumari D, Kumari N (2016) Synthetic seed preparation, germination and plantlet regeneration of litchi (*Litchi chinensis* Sonn.). Am J Plant Sci 7:1395–1406

Das G, Kameswaran S, Ramesh B, Bangeppagari M, Nath R, Das Talukdar A et al (2024) Anti-aging effect of traditional plant-based food: an overview. Foods 13:3785

Di Sario L, Zubillaga MF, Moreno CFZ, Pizzio GA, Boeri PA (2025) Micropropagation of Mapuche hop and evaluation of synthetic seed storage conditions. Plant Cell Tissue Organ Cult 160:39

Durg S, Bavage S, Shivaram SB (2020) *Withania somnifera* (Indian ginseng) in diabetes mellitus: a systematic review and meta-analysis of scientific evidence from experimental research to clinical application. Phytother Res 34:1041–1059

Dutra FB, Dona DBJ, Shiguehara APP, Inocente MC, de Oliveira ART, da Silva JMS, Piña-Rodrigues FCM (2024) Seeds technologies: performance of sodium alginate and calcium chloride as seed coating. Acta Biol Catarin 11:34–43

Ebert AW, Engels JM (2020) Plant biodiversity and genetic resources matter. Plants 9:1706

Ebert AW, Engels JM, Schafleitner R, Hintum TV, Mwila G (2023) Critical review of the increasing complexity of access and benefit-sharing policies of genetic resources for genebank curators and plant breeders–a public and private sector perspective. Plants 12:2992

El Merzougui S, Benelli C, El Boullani R, Serghini MA (2023) The cryopreservation of medicinal and ornamental geophytes: application and challenges. Plants 12:2143

Elliott GD, Wang S, Fuller BJ (2017) Cryoprotectants: a review of the actions and applications of cryoprotective solutes that modulate cell recovery from ultra-low temperatures. Cryobiology 76:74–91

Engelmann F (2004) Plant cryopreservation: progress and prospects. In Vitro Cell Dev Biol Plant 40:427–433

Engels JM, Ebert AW (2021a) A critical review of the current global ex situ conservation system for plant agrobiodiversity. II. Strengths and weaknesses of the current system and recommendations for its improvement. Plants 10:1904

Engels JM, Ebert AW (2021b) A critical review of the current global ex situ conservation system for plant agrobiodiversity. I. History of the development of the global system in the context of the political/legal framework and its major conservation components. Plants 10:1557

Engels JM, Ebert AW (2024) How can we strengthen the global genetic resources' conservation and use system? Plants 13:702

Engels JM, Ebert AW, van Hintum T (2024) Collaboration between private and public genebanks in conserving and using plant genetic resources. Plants 13:247

Faisal M, Alatar AA, Ahmad N, Anis M, Hegazy AK (2012) Assessment of genetic fidelity in *Rauvolfia serpentina* plantlets grown from synthetic (encapsulated) seeds following in vitro storage at 4 C. Molecules 17:5050–5061

Farousha K, Rangaraj VM, Mazumder JA, Haija MA, Banat F (2024) Date seed extract encapsulated-MCM-41 incorporated sodium alginate/starch biocomposite films for food packaging application. Int J Biol Macromol 282:136785
Fatima N, Ahmad N, Anis M, Ahmad I (2013) An improved in vitro encapsulation protocol, biochemical analysis and genetic integrity using DNA based molecular markers in regenerated plants of *Withania somnifera* L. Ind Crop Prod 50:468–477
Friso F, Mendive F, Soffiato M, Bombardelli V, Hesketh A, Heinrich M et al (2020) Implementation of Nagoya protocol on access and benefit-sharing in Peru: implications for researchers. J Ethnopharmacol 259:112885
Galluzzi G, Halewood M, Noriega IL, Vernooy R (2016) Twenty-five years of international exchanges of plant genetic resources facilitated by the CGIAR genebanks: a case study on global interdependence. Biodivers Conserv 25:1421–1446
Ganie IB, Ahmad Z, Shahzad A, Zaushintsena A, Neverova O, Ivanova S et al (2022) Biotechnological intervention and secondary metabolite production in *Centella asiatica* L. Plants 11:2928
Gantait S, Kundu S, Ali N, Sahu NC (2015) Synthetic seed production of medicinal plants: a review on influence of explants, encapsulation agent and matrix. Acta Physiol Plant 37:98
Gantait S, Mahanta M, Bera S, Verma SK (2021) Advances in biotechnology of *Emblica officinalis* Gaertn syn *Phyllanthus emblica* L: a nutraceuticals-rich fruit tree with multifaceted ethnomedicinal uses. 3 Biotech 11:62
García-Pérez P, Lozano-Milo E, Landín M, Gallego PP (2020) Machine learning technology reveals the concealed interactions of phytohormones on medicinal plant in vitro organogenesis. Biomolecules 10:746
Gaurav H, Yadav D, Maurya A, Yadav H, Yadav R, Shukla AC et al (2023) Biodiversity, biochemical profiling, and pharmaco-commercial applications of *Withania somnifera*: a review. Molecules 28:1208
Gholivand S, Tan TB, Yusoff MM, Qoms MS, Wang Y, Liu Y et al (2025) Eco-friendly encapsulation: investigating plant-based protein-alginate shells for efficient delivery and digestion of hemp seed oil encapsulated via supercritical CO_2 dispersion. Food Chem 463:141515
González-Arnao MT, Dolce N, González-Benito ME, Castillo Martínez CR, Cruz-Cruz CA (2017) Approaches for in vitro conservation of woody plants germplasm. In: Biodiversity and conservation of woody plants, vol 1, no 1. Springer, Cham, pp 355–419
Gowthami R, Rajasekharan PE, Chander S, Shankar M, Srivastava V, Agrawal A (2024) Cryopreservation of two-celled pollen: a model system for studying the cellular mechanisms of cryoinjury and recovery. 3 Biotech 14:304
Groner VP, Nicholas O, Mabhaudhi T, Slotow R, Akçakaya HR, Mace GM, Pearson RG (2022) Climate change, land cover change, and overharvesting threaten a widely used medicinal plant in South Africa. Ecol Appl 32:2545
Guo J, Huang X, Dou L, Yan M, Shen T, Tang W, Li J (2022) Aging and aging-related diseases: from molecular mechanisms to interventions and treatments. Signal Transduct Target Ther 7:391
Halewood M, Jamora N, Noriega IL, Anglin NL, Wenzl P, Payne T et al (2020) Germplasm acquisition and distribution by CGIAR genebanks. Plants 9:1296
Hay FR, Sershen D (2021) New technologies to improve the ex situ conservation of plant genetic resources. In: Plant genetic resources. Burleigh Dodds Science Publishing, Cambridge
Heinrich M, Scotti F, Andrade-Cetto A, Berger-González M, Echeverría J, Friso F, García-Cardona F, Hesketh A, Hitziger M, Maake C, Politi M, Spadafora C, Spadafora R (2020) Access and benefit sharing under the Nagoya protocol—Quo Vadis? Six Latin American case studies assessing opportunities and risk. Front Pharmacol 11:765
Hesami M, Jones AMP (2020) Application of artificial intelligence models and optimization algorithms in plant cell and tissue culture. Appl Microbiol Biotechnol 104:9449–9485
Iqbal M, Ali A, Rashid H, Raja NI, Naveed NH, Mashwani ZUR et al (2019) Evaluation of sodium alginate and calcium chloride on development of synthetic seeds. Pakistan J Bot 51:1569–1574

Iweala EJ, Adurosakin OE, Innocent U, Omonhinmin CA, Dania OE, Ugbogu EA (2024) Anti-aging potential of bioactive phytoconstituents found in edible medicinal plants: a review. Sci 6:36

Jena S, Ray A, Sahoo A, Sahoo S, Kar B, Panda PC, Nayak S (2018) High-frequency clonal propagation of *Curcuma angustifolia* ensuring genetic fidelity of micropropagated plants. Plant Cell Tissue Organ Cult 135:473–486

Joshee N, Biswas BK, Yadav AK (2007) Somatic embryogenesis and plant development in *Centella asiatica* L., a highly prized medicinal plant of the tropics. HortScience 42:633–637

Joshi A, Teng WL (2000) Cryopreservation of Panax ginseng cells. Plant Cell Rep 19:971–977

Kang P, Kim SJ, Park HJ, Han SJ, Kim IC, Lee H, Yim JH (2025) Trends and challenges in plant cryopreservation research: a meta-analysis of cryoprotective agent development and research focus. Plants 14:447

Kariyawasam K, Tsai M (2018) Access to genetic resources and benefit sharing: implications of Nagoya protocol on providers and users. J World Intellect Pr 21:289–305

Karsono AH, Tandrasasmita OM, Berlian G, Tjandrawinata RR (2021) Potential antiaging effects of DLBS1649, a *Centella asiatica* bioactive extract. J Exp Pharmacol 18:781–795

Kaviani B, Kulus D (2022) Cryopreservation of endangered ornamental plants and fruit crops from tropical and subtropical regions. Biology 11:847

Khalid MU, Sultan MT, Baig I, Abbas A, Noman AM, Zinedine A et al (2025) A comprehensive review on the bioactivity and pharmacological attributes of *Withania somnifera*. Nat Prod Res 1:1–15

Khan SA, Verma P, Rahman LU, Parasharami V (2021) Synthetic seed-mediated synchronized multiplication under in vitro conditions: an efficient technique for conservation, multiplication and storage of medicinal plants. In: Medicinal and aromatic plants: healthcare and industrial applications, vol 1, no 1. Springer, Cham, pp 645–667

Kiehn M, Fischer F, Smith PP (2021) The Nagoya protocol and access and benefit sharing regulations of the convention of biological diversity (CBD) and its impacts on botanic gardens' collections and research. CABI Rev 16:2116034

Kim H (2023) Optimizing the droplet-vitrification procedure by balancing the cryoprotection and cytotoxicity of alternative plant vitrification solutions based on the nature of donor plant vigor. Plants 12:4040

Kim HH, Popova E (2023) Unifying principles of cryopreservation protocols for new plant materials based on alternative cryoprotective agents (CPAs) and a systematic approach. CryoLetters 44:1–12

Kibik I, Cechova K, Milovska S, Rusnak J, Vlasac J, Melichercik M et al (2022) Cryoprotective mechanism of DMSO induced by the inhibitory effect on eutectic NaCl crystallization. J Phys Chem Lett 13:11153–11159

Knight J, Flack-Davison E, Engelbrecht S, Visagie RG, Beukes W, Mwale M (2023) Compliance with the Nagoya protocol in the southern African development community. S Afr J Bot 159:302–318

Kou F, Wang W, You S, Wei X, Wu X (2024) Preparation and characterization of metal-polyphenol networks encapsulated in sodium alginate microbead hydrogels for catechin and vitamin C delivery. Int J Biol Macromol 276:133870

Kougioumoutzis K, Tsakiri M, Kokkoris IP, Trigas P, Iatrou G, Lamari FN et al (2024) Assessing the vulnerability of medicinal and aromatic plants to climate and land-use changes in a Mediterranean biodiversity hotspot. Land 13:133

Kulkarni SK, Dhir A (2008) *Withania somnifera*: an Indian ginseng. Prog Neuro-Psychopharmacol Biol Psychiatry 32:1093–1105

Kulus D, Tymoszuk A, Kulpińska A, Osial M (2025) Synergistic effects of iron oxide nanoparticles and indole-3-acetic acid on the germination and development of cold-stored chrysanthemum synthetic seeds. Plant Cell Tissue Organ Cult (PCTOC) 160:18

Kumar PL, Cuervo M, Kreuze JF, Muller G, Kulkarni G, Kumari SG et al (2021) Phytosanitary interventions for safe global germplasm exchange and the prevention of transboundary pest spread: the role of CGIAR germplasm health units. Plants 10:328

Kundu S, Salma U, Gantait S (2018) Cryopreservation of medicinal herbs: major breakthroughs, hurdles and future. In: Biotechnological approaches for medicinal and aromatic plants: conservation, genetic improvement and utilization, vol 1, no 1. Springer, Singapore, pp 353–381

Lachenmeier DW, Montagnon C (2024) Convention on biological diversity (CBD) and the Nagoya protocol: implications and compliance strategies for the global coffee community. Foods 13:254

Le KC, Kim HH, Park SY (2019) Modification of the droplet-vitrification method of cryopreservation to enhance survival rates of adventitious roots of *Panax ginseng*. Hortic Environ Biotechnol 60:501–510

Lee JW, Kim JU, Bang KH, Kwon N, Kim YC, Jo IH, Park YD (2023a) Efficient somatic embryogenesis, regeneration and acclimatization of *Panax ginseng* Meyer: true-to-type conformity of plantlets as confirmed by ISSR analysis. Plants 12:1270

Lee JW, Kwon N, Kim JU, Bang KH, Jung SM, Lee SW et al (2023b) In vitro micropropagation of commercial ginseng cultivars (*Panax ginseng* Meyer) via somatic embryogenesis compared to traditional seed production. Horticulturae 9:435

Lee JW, Kim JU, Bang KH, Kim DH, Jo IH, Park YD (2023c) Efficient micropropagation of genetically stable *Panax ginseng* meyer by somatic embryogenesis. Agronomy 13:1139

Lei X, Wang Q, Yang H, Qi Y, Hao X, Wang Y (2021) Vitrification and proteomic analysis of embryogenic callus of Panax ginseng CA Meyer. In Vitro Cell Dev Biol Plant 57:118–127

Lerose V, Ponticelli M, Benedetto N, Carlucci V, Lela L, Tzvetkov NT, Milella L (2024) *Withania somnifera* (L.) Dunal, a potential source of phytochemicals for treating neurodegenerative diseases: a systematic review. Plants 13:771

Li Y, Tian X, Luo J, Bao T, Wang S, Wu X (2024) Molecular mechanisms of aging and anti-aging strategies. Cell Commun Signal 22:285

Liu U, Breman E, Cossu TA, Kenney S (2018) The conservation value of germplasm stored at the millennium seed bank, royal botanic gardens, Kew, UK. Biodivers Conserv 27:1347–1386

Lorite J (2024) Overharvesting is the leading conservation issue of the endangered flagship species *Artemisia granatensis* Boiss. Diversity 16:744

Lu ZW, Popova EV, Wu CH, Lee EJ, Hahn EJ, Paek KY (2009) Cryopreservation of *Ginkgo biloba* cell culture: effect of pretreatment with sucrose and ABA. Cryo Lett 30:232–243

MacDonald H (2025) Conceptual uncertainties and practical challenges in voluntary Nagoya protocol compliance: the Australian situation. Biopreserv Biobank 23:23–30

Mahanta D, Brown D, Webber T, Pezzotti S, Schwaab G, Han S, Shell M, Havenith M (2024) Bridging the gap in cryopreservation mechanism: unraveling the interplay between structure, dynamics, and thermodynamics in cryoprotectant aqueous solutions. J Phys Chem 128:3720–3731

Maheshwari S, Garg R (2023) Synthetic seed technology, application and future trends. EPH-Int J Agric Environ Res 9:1–10

Maldonado E, Morales-Pison S, Urbina F, Solari A (2023) Aging hallmarks and the role of oxidative stress. Antioxidants 12:651

Malhotra EV, Bansal S, Gupta S (2024) Plant cryopreservation: a molecular perspective. Plant Cell Tissue Organ Cult 158:19

Malík M, Tlustoš P (2023) Nootropic herbs, shrubs, and trees as potential cognitive enhancers. Plants 12:1364

Malik SK, Bansal S, Malhotra EV, Singh AM, Singh GP (2025) Indian cryogenebank conserving diverse plant genetic resources for the last three decades: achievements and way forward. Cryobiology 118:105205

Manish K (2022) Medicinal plants in peril due to climate change in the Himalaya. Ecol Inform 68:101546

Maqsood M, Mujib A, Khusrau M (2015) Preparation and low temperature short-term storage for synthetic seeds of *Caladium bicolor*. Not Sci Biol 7:90–95

Maqsood M, Khusrau M, Mujib A, Kaloo ZA (2020) Synthetic seed technology in some ornamental and medicinal plants: an overview. Propag Genet Manip Plants 1:19–31

Marcantonini G, Bartolini D, Zatini L, Costa S, Passerini M, Rende M, Luca G, Basta G, Murdolo G, Calafiore R, Galli F (2022) Natural cryoprotective and cytoprotective agents in cryopreservation: a focus on melatonin. Molecules 27:3254

Mascher M, Schreiber M, Scholz U, Graner A, Reif JC, Stein N (2019) Genebank genomics bridges the gap between the conservation of crop diversity and plant breeding. Nat Genet 51:1076–1081

Matsumura K, Hayashi F, Nagashima T, Rajan R, Hyon SH (2021) Molecular mechanisms of cell cryopreservation with polyampholytes studied by solid-state NMR. Commun Mater 2:15

McGraw J, Lubbers A, Van Der Voort M, Mooney E, Furedi M, Souther S, Turner J, Chandler J (2013) Ecology and conservation of ginseng (*Panax quinquefolius*) in a changing world. Ann N Y Acad Sci 1286:62–91

Meccariello R, D'Angelo S (2021) Impact of polyphenolic-food on longevity: an elixir of life. An overview. Antioxidants 10:507

Meucci A, Ghelardi C, Chietera G, Mensuali A (2024) Synthetic seed production and slow growth storage of in vitro cultured plants of *Iris pallida* Lam. Horticulturae 10:272

Micheli M, Regni L, da Silva DF (2022) Encapsulation in calcium alginate of nodes from stolons of *Mentha spicata* L. Horticulturae 8(5):456

Moise G, Jîjie AR, Moacă AE, Predescu IA, Dehelean CA, Hegheş A et al (2024) Plants' impact on the human brain—exploring the neuroprotective and neurotoxic potential of plants. Pharmaceuticals 17:1339

Mondal SC, Singh P, Kumar B, Singh SK, Gupta SK, Verma A (2014) Ageing and potential anti-aging phytochemicals: an overview. World J Pharm Pharm Sci 4:426–454

Mosa K, Ahmed A, Hazem Y, Kanawati I, Abdullah A, Hernandez-Sori L, Ali M, Vendrame W (2023) Insights into cryopreservation, recovery and genetic stability of medicinal plant tissues. Fitoterapia 169:105555

Movilla Pateiro L (2020) Advances and uncertainties in compliance measures for users from the Nagoya protocol in the European Union. Rev Eur Comp Int Environ Law 29:282–290

Murray K, Gibson M (2022) Chemical approaches to cryopreservation. Nat Rev Chem 6:579–593

Mykhailenko O, Jalil B, McGaw LJ, Echeverría J, Takubessi M, Heinrich M (2025) Climate change and the sustainable use of medicinal plants: a call for "new" research strategies. Front Pharmacol 15:1496792

Nagananda GS, Satishchandra N (2023) In vitro direct plant regeneration from somatic embryos developed from the leaf explants of Flickingeria nodosa (Dalz.) Seidenf. through synthetic seed technology. S Afr J Bot 153:77–82

Nagel M, Pence V, Ballesteros D, Lambardi M, Popova E, Panis B (2024) Plant cryopreservation: principles, applications, and challenges of banking plant diversity at ultralow temperatures. Annu Rev Plant Biol 75:797

Natarajan K, Purushotham P, Rajendran A, Suresh S, Senthil K (2022) Comparative adaptogenic properties of *Withania somnifera* and Panax ginseng. Curr Bot 13:12–18

Nausch H, Buyel JF (2021) Cryopreservation of plant cell cultures–diverse practices and protocols. New Biotechnol 62:86–95

Navya BL, Rajasekharan PE (2025) Pollen cryopreservation in Ashwagandha (*Withania somnifera* L.). In: Pollen cryopreservation protocols, vol 2, no 1. Springer, New York, pp 727–735

Nazirwan, Wahyudi A, Wahyuni A, Yusanto (2020) Sprout results characterization of in vitro shoots encapsulation of broccoli using sodium alginate. In Journal of Physics: Conference Series (Vol. 1450, No. 1, p. 012061). IOP Publishing

Ngezahayo E, Komariah K, Masykuri M (2025) The climate change influence on Indonesia medicinal plants: a review. In: E3S web of conferences, vol 605. EDP Sciences, p 03062

Niazian M (2019) Application of genetics and biotechnology for improving medicinal plants. Planta 249:953–973

Nigar S, Subrahmanyam K, Choudhury DB, Gul K, Sehrawat R (2024) Alginate-millet starch composite matrices as novel carriers for controlled delivery of grape seed polyphenols: preparation, characterization, and in vitro release behaviour. Food Struct 42:100385

Nirmal Babu K, Divakaran M, Pillai GS, Sumathi V, Praveen K, Raj RP et al (2016) Protocols for in vitro propagation, conservation, synthetic seed production, microrhizome production, and molecular profiling in turmeric (*Curcuma longa* L.). In: Protocols for in vitro cultures and secondary metabolite analysis of aromatic and medicinal plants, vol 2. Springer, New York, pp 387–401

Normah MN, Sulong N, Reed BM (2019) Cryopreservation of shoot tips of recalcitrant and tropical species: advances and strategies. Cryobiology 87:1–14

Oh SY, Wu CH, Popova E, Hahn EJ, Paek KY (2009) Cryopreservation of *Panax ginseng* adventitious roots. J Plant Biol 52:348–354

Özbay S, Isak MA, Şimşek Ö (2024) Synthetic seed production and cryopreservation for Myrtle (*Myrtus communis* L.) genotypes. Çukurova Tarım ve Gıda Bilimleri Dergisi 39(1):126–136

Panis B, Nagel M, van Den Houwe I (2020) Challenges and prospects for the conservation of crop genetic resources in field genebanks, in in vitro collections and/or in liquid nitrogen. Plants 9:1634

Paul S, Chakraborty S, Anand U, Dey S, Nandy S, Ghorai M, Saha S, Patil M, Kandimalla R, Proćków J, Dey A (2021) *Withania somnifera* (L.) Dunal (Ashwagandha): a comprehensive review on ethnopharmacology, pharmacotherapeutics, biomedicinal and toxicological aspects. Biomed Pharmacother 143:112175

Pepe M, Hesami M, Jones AMP (2021) Machine learning-mediated development and optimization of disinfection protocol and scarification method for improved in vitro germination of cannabis seeds. Plants 10:2397

Perea-Flores MDJ, Aguilar-Morán HF, Calderón-Domínguez G, García-Hernández AB, Díaz-Ramírez M, Romero-Campos HE et al (2023) Entrapment efficiency (EE) and release mechanism of rhodamine b encapsulated in a mixture of chia seed mucilage and sodium alginate. Appl Sci 13:1213

Phu HT, Thuan DT, Nguyen TH, Posadino AM, Eid AH, Pintus G (2020) Herbal medicine for slowing aging and aging-associated conditions: efficacy, mechanisms and safety. Curr Vasc Pharmacol 18:369–393

Pientaweeratch S, Panapisal V, Tansirikongkol A (2016) Antioxidant, anti-collagenase and anti-elastase activities of *Phyllanthus emblica, Manilkara zapota* and silymarin: an in vitro comparative study for anti-aging applications. Pharm Biol 54:1865–1872

Popova EV, Lee EJ, Wu CH, Hahn EJ, Paek KY (2009) A simple method for cryopreservation of *Ginkgo biloba* callus. Plant Cell Tissue Organ Cult 97:337–343

Prasad A, Singh M, Yadav NP, Mathur AK, Mathur A (2014) Molecular, chemical and biological stability of plants derived from artificial seeds of *Centella asiatica* (L.) urban-an industrially important medicinal herb. Ind Crop Prod 60:205–211

Priyadharshini S, Manokari M, Shekhawat MS (2020) In vitro conservation strategies for the critically endangered Malabar river lily (*Crinum malabaricum* Lekhak Yadav) using somatic embryogenesis and synthetic seed production. S Afr J Bot 135:172–180

Priyanka V, Kumar R, Dhaliwal I, Kaushik P (2021) Germplasm conservation: instrumental in agricultural biodiversity—a review. Sustainability 13:6743

Qahtan AA, Abdel-Salam EM, Alatar AA, Wang QC, Faisal M (2019) An introduction to synthetic seeds: production, techniques, and applications. In: Synthetic seeds: germplasm regeneration, preservation and prospects, vol 1. Springer, Cham, pp 1–20

Rajasekharan PE, Ganeshan S (2018) Current perspectives on pollen cryopreservation in horticultural species. In: III international symposium on plant cryopreservation, vol 1, pp 47–56

Rajasekharan PE, Rohini MR (2023) Pollen cryobanking—implications in genetic conservation and plant breeding. In: Pollen cryopreservation protocols, vol 1. Springer, New York, pp 19–42

Ramírez-Mosqueda MA, de la Cruz D-SM, Gómez-Merino FC, Hernandez-Rosas F, Rodríguez-Lagunes DA, Bello-Bello JJ (2023) Nodal segment size and encapsulation matrix affect sugarcane bud emergence from synthetic seeds. Sugar Tech 25:827–833

Regni L, Cesarini A (2025) Over half a century of research on blackberry micropropagation: a comprehensive review. Horticulturae 11:556

Regni L, Micheli M, Del Pino A, Facchin S, Rabica E, Camilloni L, Cesarini A, Proietti P (2024) Blackberry synthetic seeds storage: effects of temperature, time, and sowing substrate. Plant Cell Tissue Organ Cult 158:17

Reshi ZA, Ahmad W, Husain FM, Khanam MN, Javed SB (2025) Optimizing storage and regeneration of clonal propagules of *Salix tetrasperma* through double-layered encapsulation. Horticulturae 11:486

Rihan HZ, Kareem F, El-Mahrouk ME, Fuller MP (2017) Artificial seeds (principle, aspects and applications). Agronomy 7:71

Robinson D, Raven M, Makin E, Kalfatak D, Hickey F, Tari T (2021) Legal geographies of kava, kastom and indigenous knowledge: next steps under the Nagoya protocol. Geoforum 118:169–179

Roque-Borda CA, Kulus D, Vacaro de Souza A, Kaviani B, Vicente EF (2021) Cryopreservation of agronomic plant germplasm using vitrification-based methods: an overview of selected case studies. Int J Mol Sci 22:6157

Ruta C, Lambardi M, Ozudogru EA (2020) Biobanking of vegetable genetic resources by in vitro conservation and cryopreservation. Biodivers Conserv 29:3495–3532

Sahu S, Monalisa SP, Jigile S, Kar S, Parida SS, Swain SK (2025) Exploring the potential of synthetic seeds: influence of explant, encapsulating agent and matrix, advantages and challenges. Plant Sci Today 12:2348-1900

Salehi B, Azzini E, Zucca P, Varoni E, Kumar N, Dini L, Panzarini E, Rajković J, Fokou P, Peluso I, Mishra AP, Nigam M, Rayess Y, Beyrouthy M, Setzer W, Polito L, Iriti M, Sureda A, Quetglas-Llabrés M, Martorell M, Martins N, Sharifi-Rad M, Estevinho L, Sharifi-Rad J (2020) Plant-derived bioactives and oxidative stress-related disorders: a key trend towards healthy aging and longevity promotion. Applied Sci 10:947

Salgotra RK, Chauhan BS (2023) Genetic diversity, conservation, and utilization of plant genetic resources. Genes 14:174

Sambu S (2015) A Bayesian approach to optimizing cryopreservation protocols. PeerJ 3:1039

Saxena H, Kumar A, Kumar R, Kumar P, Lingaraju M, Shukla R, Shankar K, Gupta M (2025) Exploring the remyelinating efficacy of nanoliposomes encapsulating *Withania somnifera* and *Ginkgo biloba* in a murine model of induced demyelination. J Liposome Res 35:1–23

Schebesta H (2021) The potential of private standards for valorizing compliance with access and benefit sharing obligations of genetic resources and traditional knowledge. Agronomy 11:1823

Seevaratnam V, Banumathi P, Premalatha MR, Sundaram SP, Arumugam T (2012) Functional properties of *Centella asiatica* (L.): a review. Int J Pharm Pharm Sci 5:8–14

Shajahan A, Raju CS, Mehaboob VM, Aslam A (2019) Somatic embryogenesis and synthetic seed Technology of *Curcuma* spp. In: Synthetic seeds: germplasm regeneration, preservation and prospects, vol 1. Springer, Cham, pp 363–375

Sharifi-Rad J, Quispe C, Ayatollahi S, Kobarfard F, Staniak M, Stępień A, Czopek K, Sen S, Acharya K, Matthews K, Şener B, Devkota H, Kirkin C, Özçelik B, Victoriano M, Martorell M, Suleria H, Alshehri M, Chandran D, Kumar M, Cruz-Martins N, Cho W (2021) Chemical composition, biological activity, and health-promoting effects of *Withania somnifera* for pharma-food industry applications. J Food Qual 2021:8985179

Shasmita N, Rai MK, Naik SK (2018) Exploring plant tissue culture in *Withania somnifera* (L.) Dunal: in vitro propagation and secondary metabolite production. Crit Rev Biotechnol 38:836–850

Shi C, Xiao S, Liu J, Guo K, Wu F, Yew DT, Xu J (2010) Ginkgo biloba extract EGb761 protects against aging-associated mitochondrial dysfunction in platelets and hippocampi of SAMP8 mice. Platelets 21:373–379

Siddique I, Bukhari NAW (2018) Synthetic seed production by encapsulating nodal segment of *Capparis decidua* (Forsk.), in vitro regrowth of plantlets and their physio biochemical studies. Agrofor Syst 92:1711–1719

Singh D, Pal S, Sinha A (2020) Artificial seed/synthetic seed production–brief procedure–advantages and limitations. Biotica Res Today 2:422–424

Singh N, Yadav SS, Rao AS, Nandal A, Kumar S, Ganaie SA, Narasihman B (2021) Review on anticancerous therapeutic potential of *Withania somnifera* (L.) Dunal. J Ethnopharmacol 270:113704

Smith D, da Silva M, Jackson J, Lyal C (2017) Explanation of the Nagoya protocol on access and benefit sharing and its implication for microbiology. Microbiology 163:289–296

Smith S, Nickson TE, Challender M (2021) Germplasm exchange is critical to conservation of biodiversity and global food security. Agron J 113:2969–2979

Subrahmanyeswari T, Laha S, Kamble SN, Singh S, Bhattacharyya S, Gantait S (2023) Alginate encapsulation of shoot tips and their regeneration for enhanced mass propagation and germplasm exchange of genetically stable *Stevia rebaudiana* Bert. Sugar Tech 25:542–551

Szopa D, Mielczarek M, Skrzypczak D, Izydorczyk G, Mikula K, Chojnacka K, Witek-Krowiak A (2022) Encapsulation efficiency and survival of plant growth-promoting microorganisms in an alginate-based matrix—a systematic review and protocol for a practical approach. Ind Crop Prod 181:114846

Tadda SA, Kui X, Yang H, Li M, Huang Z, Chen X, Qiu D (2021) The response of vegetable sweet potato (*Ipomoea batatas* Lam) nodes to different concentrations of encapsulation agent and MS salts. Agronomy 12:19

Tetali SD, Acharya S, Ankari AB, Nanakram V, Raghavendra AS (2021) Metabolomics of *Withania somnifera* (L.) Dunal: advances and applications. J Ethnopharmacol 267:113469

Thong-on W, Sakchaisri K, Prathanturarug S (2024) The effects of glycoside-rich green extract from *Centella asiatica* (L.) urban on wound healing and anti-aging activity. Phytomedicine Plus 4:100628

Tripathi P, Lodhi A, Rai S, Nandi N, Dumoga S, Yadav P, Tiwari A, Singh S, El-Shorbagi A, Chaudhary S (2024) Review of pharmacotherapeutic targets in Alzheimer's disease and its management using traditional medicinal plants. Degener Neurol Neuromuscul Dis 14:47–74

Tsaltaki C, Katsouli M, Kekes T, Chanioti S, Tzia C (2019) Comparison study for the recovery of bioactive compounds from *Tribulus terrestris, Panax ginseng, Gingko biloba, Lepidium meyenii, Turnera diffusa* and *Withania somnifera* by using microwav: e-assisted, ultrasound-assisted and conventional extraction methods. Ind Crop Prod 142:111875

Uchendu EE, Brown DCW, Saxena PK (2011) Cryopreservation of shoot tips and cotyledons of the north American ginseng (*Panax quinquefolius* L.). CryoLetters 32:463–472

Valdés YC, Shukla MR, González Vega ME, Saxena PK (2021) Improved conservation of coffee (*Coffea arabica* L.) germplasm via micropropagation and cryopreservation. Agronomy 11:1861

van Wyk AS, Prinsloo G (2018) Medicinal plant harvesting, sustainability and cultivation in South Africa. Biol Conserv 227:335–342

Verma RK, Kumari P, Maurya RK, Kumar V, Verma RB, Singh RK (2018) Medicinal properties of turmeric (*Curcuma longa* L.): a review. Int J Chem Stud 6:1354–1357

Volk GM, Byrne PF, Coyne CJ, Flint-Garcia S, Reeves PA, Richards C (2021) Integrating genomic and phenomic approaches to support plant genetic resources conservation and use. Plants 10:2260

Wambugu PW, Ndjiondjop MN, Henry RJ (2018) Role of genomics in promoting the utilization of plant genetic resources in genebanks. Brief Funct Genomics 17:198–206

Wang MR, Bi W, Shukla MR, Ren L, Hamborg Z, Blystad DR et al (2021) Epigenetic and genetic integrity, metabolic stability, and field performance of cryopreserved plants. Plants 10:1889

Wang Y, Cao X, Ma J, Liu S, Jin X, Liu B (2024) Unveiling the longevity potential of natural phytochemicals: a comprehensive review of active ingredients in dietary plants and herbs. J Agric Food Chem 72:24908–24927

Wani ZA, Pant S, Bhat JA, Shukla G (2024) Distribution and survival of medicinal and aromatic plants is threatened by the anticipated climate change. Trees Forests People 16:100549

Wasi A, Shahzad A, Tahseen S (2024) Synthesis of nonembryonic synseed, short term conservation, phytochemical evaluation and assessment of genetic stability through SCoT markers in *Decalepis salicifolia*. Plant Cell Tissue Organ Cult 156:101

Weise S, Lohwasser U, Oppermann M (2020) Document or lose it—on the importance of information management for genetic resources conservation in genebanks. Plants 9:1050

Wu M, Cai J, Fang Z, Li S, Huang Z, Tang Z et al (2022) The composition and anti-aging activities of polyphenol extract from *Phyllanthus emblica* L. fruit. Nutrients 14:857

Xiang J, Mlambo R, Shaw I, Seid Y, Shah H, He Y, Kpegah J, Tan S, Zhou W, He B (2023) Cryopreservation of bioflavonoid-rich plant sources and bioflavonoid-microcapsules: emerging technologies for preserving bioactivity and enhancing nutraceutical applications. Front Nut 10:1232129

Xu F, Valappil AK, Mathiyalagan R, Tran TNA, Ramadhania ZM, Awais M, Yang DC (2023) In vitro cultivation and ginsenosides accumulation in *Panax ginseng*: a review. Plants 12:3165

Yadav N, Tripathi S, Sangwan NS (2024a) Phyto-therapeutic potential of *Withania somnifera*: molecular mechanism and health implications. Phytother Res 38:1695–1714

Yadav B, Pandit D, Banjade D, Mehata D, Bhattarai S, Bhandari S, Ghimire N, Yadav P, Paudel P (2024b) Insights into the germplasm conservation and utilization: implications for sustainable agriculture and future crop improvement. Arch Agric Environ Sci 9:180–193

Yu M, Zhang H, Wang B, Zhang Y, Zheng X, Shao B, Zhuge Q, Jin K (2021) Key signaling pathways in aging and potential interventions for healthy aging. Cells 10:660

Zamecnik J, Faltus M, Bilavcik A (2021) Vitrification solutions for plant cryopreservation: modification and properties. Plants 10:2623

Zarei A, Feyissa BA, Davis B, Tavakouli Dinani E (2022) Cannabis synthetic seeds: an alternative approach for commercial scale of clonal propagation and germplasm conservation. Plants 11:3186

Zeng L, Sun Z, Fu L, Gu Y, Li R, He M, Wei J (2024) Cryopreservation of medicinal plant seeds: strategies for genetic diversity conservation and sustainability. Plants 13:2577

Zhang H, Abid S, Ahn JC, Mathiyalagan R, Kim YJ, Yang DC, Wang Y (2020) Characteristics of *Panax ginseng* cultivars in Korea and China. Molecules 25:2635

Zhang AL, Wang MR, Li Z, Panis B, Bettoni JC, Vollmer R et al (2023) Overcoming challenges for shoot tip cryopreservation of root and tuber crops. Agronomy 13:219

Zheng Y, Pan C, Zhang Z, Luo W, Liang X, Shi Y et al (2020) Antiaging effect of *Curcuma longa* L. essential oil on ultraviolet-irradiated skin. Microchem J 154:104608

Zuo W, Yan F, Zhang B, Li J, Mei D (2017) Advances in the studies of *Ginkgo biloba* leaves extract on aging-related diseases. Aging Dis 8:812

Green Synthesis of Nanoparticles from Medicinal Plants and Their Applications in Skin Aging and Regeneration

10

Özlem Ünlü, Ekrem Gürel, and Songül Gürel

Abstract

The combination of nanotechnology and phytotherapy provides new insights for anti-aging treatments. Medicinal plant extracts have long been used for alternative medicine. Now, they can be a promising resource for green synthesis of metallic nanoparticles such as silver, gold, and zinc oxide. Available methods for the green synthesis are eco-friendly since they use phytochemicals from plant extracts. These extracts hold metabolites that work as reducing and stabilizing agents for the synthesis mechanisms. There is no need for harmful chemicals or high-energy processes, which support current trends for reducing carbon trace. The nanoparticles produced through green synthesis can provide high levels of antioxidants, and they can hold anti-inflammatory and antimicrobial capacities. These are usually associated with preventing skin aging by boosting physiological responses. In this chapter, characterization and action mechanisms of these nanoparticles will be discussed. The important factors that affect nanoparticle synthesis, such as the type of plant and method of extraction, and/or reaction conditions, are also covered in the chapter. Some factors that may affect biocompatibility, possible toxicity, and some regulatory concerns will bring a better look at their potential in cosmetics and medicine, and how they may influence user preferences. By bringing together natural chemistry and nanotechnology, this

Ö. Ünlü (✉)
Faculty of Arts and Science, Department of Chemistry, Düzce University, Düzce, Türkiye

Faculty of Arts and Science, Department of Chemistry, Bolu Abant İzzet Baysal University, Bolu, Türkiye
e-mail: ozlemilkin@duzce.edu.tr

E. Gürel · S. Gürel
Faculty of Agriculture, Department of Agricultural Biotechnology, Iğdır University, Iğdır, Türkiye

S. Gantait et al. (eds.), *Medicinal Plants with Anti-Ageing Properties*,
https://doi.org/10.1007/978-981-95-8064-4_10

chapter presents a new and environment-friendly alternatives for developing future anti-aging treatments.

Keywords

Green synthesis · Medicinal plants · Nanocosmetics · Nanoparticles · Phytochemicals · Skin aging

10.1 Introduction

Safe, sustainable, and biocompatible materials in the biomedical and cosmetic industries have gained scientific attention due to public awareness about the ongoing concerns about health and the pollution crisis. This awareness has forced the cosmetic sector to shift to alternative methods and products to convince its customers. Within the last decade, industrial marketing strategies have shifted to a customer-centric center. Previously, product-centered approaches relied on selling their products by placing their product brand and/or companies at the center. Today's new trend enforces companies to rely on a customer-centered marketing strategy. Thus, companies can only exist by understanding individual customer demands and reshaping their strategies by creating value-based relationships. This kind of relationship is only possible if the products are satisfactory for a variety of needs in addition to their primary goal.

Indeed, the shift from a product-oriented to a more customer-centered mindset has changed the way industries proceed in terms of new research and development strategies. Today's consumers are evidently more conscious about health problems and environmental issues related to the products they use. Thus, their expectations have altered from mere functionality to a more comprehensive level. All over the globe, customers now demand cosmetic solutions that are effective and—at the same time—safe for their bodies, and harmless to the environment. In response to the demands, companies have been increasingly integrating eco-friendly alternatives, mostly relying on natural resources to their product strategies. These processes are now covered under green technologies and plant-based innovations, especially in the cosmetic and medical sectors.

The collaboration of green technology with traditional medicinal practices offers an alternative against skin aging. Main goal of green technology is to reduce or eliminate industrial processes and associated products to minimize the negative impacts on the environment. Scientists can improve innovative solutions that can be effective, safe, and sustainable for the treatment of aging skin and boost tissue regeneration. Collaborating the principles of green chemistry into the synthesis of various nanoparticle practices provides an essential point in creating a sustainable future for nanotechnology. Since there is no need for harmful chemicals or high-energy processes, these steps will also environmentally act to support reducing carbon trace during mass production. In other words, it is possible to enhance the safety and efficacy of nanoparticles while mitigating the environmental impact traditionally associated with nanomaterial production.

Herbal medicine and nanotechnology present a novel alternative for developing sustainable and innovative solutions in medical practices. Indeed, understanding of phytochemical dynamics, extraction methodologies, and the relative effectiveness of different medicinal plants as resources for nanoparticle synthesis will support the field of green nanotechnology. The benefits associated with these therapeutic applications will change our success rate in disease treatment. Of course, improving environmentally friendly synthesis protocols will significantly impact aging research, as well. Continued research and collaboration will provide alternative solutions for the development of nanocosmetics products that will help both consumers and the environment.

This chapter will provide an up-to-date review of crosstalk between nanoparticles and their applications for preventing skin aging.

Skin aging represents a complex phenomenon usually affected by several intrinsic and extrinsic factors. Intrinsic aging factors are associated with chronological aging. It is driven by genetic factors and associated biological processes. On the other hand, extrinsic aging factors are usually associated with environmental influences such as UV radiation, pollution, and/or lifestyle choices (Boo 2019; Galili et al. 2024). Over time, these factors can damage tissue degeneration processes, followed by decreased collagen production. It results in the loss of elasticity and with appearance of fine lines and wrinkles that are associated with the appearance of aging skin (Aljuffali et al. 2015). Table 10.1 presents a summarized list of factors driving skin aging, linking causes to biological effects and visible outcomes.

The key mechanisms of skin aging are related to the accumulation of free radicals, such as reactive oxygen species (ROS), along with the influence of inflammatory pathways (Zhang et al. 2022). Over the years, we learned that environmental influences such as exposure to UV and environmental pollutants are the major contributors to oxidative stress in the skin. Since skin is the first interface to these environmental effects, they lead to cellular damage and accelerate aging (Sree et al. 2021). It is significant to address oxidative stress through innovative treatments to provide viable solutions against skin aging.

Chronic inflammation is also an important factor in other skin conditions associated with aging by complicating the skin's structural integrity (Vallinayaki et al. 2024). It can be said that the biological response to oxidative stress is associated with an imbalance in the production of antioxidants and oxidative agents. The

Table 10.1 Key mechanisms and manifestations of skin aging

Factor type	Key drivers	Biological effects	Visible/clinical signs
Intrinsic aging	Genetics, metabolic processes, hormonal changes	Slower cell turnover, decreased collagen/elastin production, reduced sebum	Fine wrinkles, thin and dry skin, gradual tissue atrophy
Extrinsic aging	UV radiation (photoaging), pollution, lifestyle (e.g., smoking)	Oxidative stress (ROS), chronic inflammation, DNA damage, ECM degradation	Deep wrinkles, pigmentation issues, loss of elasticity, rough texture

inflammatory response leads to an alteration in the function of skin, causing long-term skin damage (Williams et al. 2022).

Since ancient times, people have always had the eagerness to protect their beauty. As a result, they rely on many products that help to preserve their skin health. Traditional anti-aging agents such as retinoids, peptides, and antioxidants are often relied upon to slow down skin aging. Even though they demonstrate a positive impact on skin vitality and, in turn, reduce visible signs of aging, these widely used chemicals for skincare often have limitations in terms of safety, efficacy, and sustainability (Chakraborty et al. 2022). For example, among the mostly used anti-aging agents, retinoids may have some harmful effects on skin, including irritation and photosensitivity (Wang et al. 2019). Aging is a complex process, and it is essential to highlight the need for advanced anti-aging solutions.

The cost of skin care products ranges from low to sky-high prices. For instance, it is known that peptide-based therapies are usually very expensive, yet they may not penetrate deep enough through the tissues where actual active skin regeneration occurs (Ghramh et al. 2020). On the other hand, there is a cheaper alternative to these high-cost skin care remedies, using Vitamin C and E because of their high antioxidant properties. Even though they are cheap or not, their bioavailability and skin penetration capacities still may not be efficient enough to get the expected anti-aging outcomes (Aljuffali et al. 2015). Yet, these old-fashioned artificial stabilizers and solvents are still used in conventional formulations, raising sustainability concerns and environmental pollution by hazardous waste (Kamli et al. 2021). That's why it is certainly an urgent subject to promote the investigation of better alternatives for anti-aging actions.

Biotechnological studies in the skin-care field are aimed at producing peptide-based smart materials, targeting regeneration at the cellular level. It is reported that promising nanoparticle-based systems for drug delivery have been available as a solution for increasing the effectiveness of these compounds by targeting improvement of their skin absorption and stability (Boo 2019). Green nanomaterial technology has a promising result in biotechnological studies as it contributes to obtaining biosafe, superior and cost-effective materials.

The collaboration of green nanotechnology practices with phytotherapy represents an exciting crossroad for developing effective and sustainable solutions in regenerative dermatology. This synergy-based approach applies plant-based compounds with advanced antioxidant properties and leverages the ability of nanoparticles to improve the delivery and efficacy of these compounds (Chakraborty et al. 2022).

In recent years, medicinal plant-based synthesis of nanoparticles has proven their potential for creating biocompatible and non-toxic materials with anti-inflammatory and antioxidant activities (Chandraker et al. 2021; Firmino et al. 2021). These green nanoparticles are also beneficial for human health since they may lower the side effects while aiming at the cellular level adsorption for increasing bioavailability.

Silver and gold nanoparticles have been known for centuries since they are the first synthesized metallic nanoparticles that have antibacterial activity. Unfortunately, chemical routes for the production of metal nanoparticles involve energy-consuming processes and hazardous chemical waste. With advanced technology in green chemistry, fabricating these valuable metal nanoparticles by using medicinal plants decreases the environmental burden on the planet. For instance, when we use extracts from medical plants to synthesize silver and gold nanoparticles, they gain antibacterial properties. As a result, by using biogenic synthesis for metal nanoparticles, we can keep that important antimicrobial activity, which is critical for supporting wound healing processes during applications and obtain biosafe nanomaterials at the same time (Gami et al. 2022). This advanced technology leads researchers to investigate new ecologically friendly and highly effective materials.

Encapsulation of phytochemicals within metal nanoparticles can be shown as an example of this new research. This encapsulation method increases their target-delivery efficiencies while improving skin penetration and antioxidant capacities (Chamizo et al. 2021). Recent advances in green nanotechnology have revealed that green nanomaterials improve skin hydration, restoring elasticity, and combating visible signs of aging. These revolutionary findings in green nanotechnology have a significant contribution to the cosmetic industry. The utilization of green nanoparticles in a number of products has increased in the market (Vallinayaki et al. 2024). Natural phytochemicals such as flavonoids and phenolic compounds present in the extracts obtained from plants provide therapeutic benefits thanks to their anti-inflammatory properties (Hama et al. 2014; Tokudome et al. 2018). We can conclude that the natural compounds in green nanotechnology increased the tendency for preferences for eco-friendly products in cosmetics.

10.2 Fundamentals of Green Nanomaterials

Green nanomaterials is a field that integrates green chemistry with advanced materials. The main principle of the synthesis of nanoparticles by green initiatives is using sustainable and environmentally friendly processes. The primary goal is the elimination of hazardous substances while limiting the energy requirements.

The scarcity of fossil resources leads the whole research world to work on greener solutions. The implementation of green chemistry principles extends energy efficiency by emphasizing the utilization of less energy-intensive processes. The conventional methods suggested for nanoparticle synthesis mainly require the usage of hazardous chemicals in addition to too much energy; however, green synthesis often occurs at ambient temperature and pressure (Thatyana et al. 2023). A direct comparison of synthesis approaches, highlighting the advantages of green synthesis across multiple criteria, is shown in Table 10.2.

It is our hope that green nanotechnology will build a more sustainable future. Thus, studies in green chemistry allow us to apply more reliable approaches to produce nanoparticles (Gulia et al. 2022). Natural resources such as plant extracts, microorganisms, and enzymes are important sources for developing green chemistry procedures. The abundance and low cost of raw materials, since these are highly available from nature, is one of the most important advantages of using green chemistry in green nanotechnology approaches. On top of that, a green approach will

Table 10.2 Comparison of conventional vs. green synthesis approaches for nanoparticles

Parameter	Conventional synthesis	Green synthesis
Reducing/stabilizing agents	Chemical agents (e.g., $NaBH_4$, citrate)	Plant phytochemicals (e.g., polyphenols, flavonoids)
Solvents and conditions	Often toxic solvents, high energy, high temp/pressure	Aqueous solvents, ambient temperature/pressure
Environmental impact	High energy use, hazardous waste generation	Low energy, biodegradable byproducts, eco-friendly
Biocompatibility and toxicity	Often higher cytotoxicity, requires purification	Generally higher biocompatibility, inherent bioactivity
Cost and scalability	Established but costly raw materials/processes	Low-cost, renewable raw materials; scalability is a current challenge
Key advantages	Good size/shape control, high yield, scalable	Sustainable, non-toxic, simple process, multifunctional nanoparticles

help us minimize health risks and environmental damage while allowing us to produce nanoparticles with advanced biocompatibility (Mao and Mooney 2015). The ingredients selected for the green synthesis techniques involve biosafe and biocompatible components (Thatyana et al. 2023). In this way, they reduce the use of toxic chemicals and minimize ecological pollution. Green nanoparticle technologies offer both environmentally friendly and economical resources. No one can argue about the benefits of using inexpensive, renewable biological resources that will minimize reliance on high-cost chemicals and processes (Gulia et al. 2022).

10.3 Medicinal Plants and Nanoparticle Synthesis

Medicinal plants have been used in healthcare and disease treatment not only by pharmaceutical industries but also by the public in traditional therapy. They have a rich phytochemical composition. Their rich phytochemical composition makes them valuable for manufacturing new materials. So, they are considered strong candidates for green nanoparticle synthesis. Many studies have revealed that medicinal plants have proven themselves as efficient biofactories.

Medicinal plants are taken into account for their rich phytochemical composition. Polyphenols, flavonoids, terpenoids, alkaloids, and proteins are the primary classes of phytochemicals. They act as significant dual-function materials. They serve as reducing and capping agents in nanoparticle synthesis. We know that nanoparticles considered for green synthesis may be directly or indirectly affected by the composition of the plant in question. Genotypes, growth conditions, and processing strategies can significantly change the content of plant bioactive compounds. Studies in some wheat varieties indicate that environmental and genetic interactions directly influence phenolic profiles and antioxidant capacity (Ünüvar and Ünlü 2022). A study on *Digitalis ferruginea*, a plant that contains active ingredients used in cardiovascular disease, reported that ZnO NPs exhibited biostimulatory properties for plants at optimal concentrations. This study demonstrates the indirect effects of NPs on a medicinal plant, rather than their direct use in humans (Biswas et al. 2025).

Polyphenols show particularly high antioxidant properties. In the nanoparticle production path, polyphenols facilitate the reduction of metal ions. In the reduction process, polyphenolic compounds inhibit agglomeration and stabilize the nanoparticle as well (Kumari et al. 2024). Flavonoids are also an essential type of phytochemical. They increase the antimicrobial activity of nanoparticles while acting as a reducing agent. Terpenoids exhibit strong reducing activity, which is important for the formation of nanoparticles (Javed et al. 2020). Alkaloids are bioactive and have the capacity to form metal complexes. They also increase the stability and functionality of nanoparticles (García-Pérez et al. 2022). Proteins are natural stabilizers, and they help to preserve the shape and size of the nanoparticles since both are critical for effectiveness in biomedical applications (Kumari et al. 2024). Phytochemicals are important for both the efficiency and functionality of nanoparticles. Applications across medicine, catalysis, and environmental remediation simply rely on the interaction between phytochemicals during green synthesis (Zuhrotun et al. 2023).

Optimization of the extraction procedures is essential to get the best from medicinal plants if we want to consider them as biofactories. Over the years, no one argues about the fact that the yield and bioactivity of phytochemicals are highly dependent on how the extraction methods proceed. Choosing the right solvents, such as water, ethanol, methanol, and hexane, significantly affects phytochemical extraction efficiency. Polar solvents like ethanol are often preferred for extracting polyphenols and flavonoids over non-polar solvents. Similarly, the selection of solvent is the key to achieving maximum extraction yield for nanoparticle synthesis. Experimental parameters significantly affect the extraction process. If faster extraction is needed, other parameters such as high temperatures can be considered to enhance the diffusion of solutes. We must consider that excessive heat may degrade sensitive phytochemicals (El-Baghdady et al. 2018). Similarly, pH affects solubility and stability. For example, acidic media can increase the solubility of certain flavonoids, thereby promoting extraction (Ovais et al. 2018). Moreover, ultrasound-assisted and microwave-assisted extraction methods have shown promising results while reducing the risk of degradation (Khanahmad et al. 2024). In summary, it should be well noted that what would be expected from the protocol should be finely detailed before choosing the extraction protocol and changing the parameters accordingly.

Several medicinal plants have been examined for their potential in green nanoparticle synthesis. Knowing the importance of these medicinal plants is indeed important, but understanding their potential in terms of their potential for producing reducing agents is a totally distinct subject that we need to investigate regarding the synthesis of nanoparticles.

Aloe vera has been considered a conventional medical plant for centuries, thanks to its healing properties. *Aloe vera* plant has been popular among medical plants for the green synthesis of metal nanoparticles (Añibarro-Ortega et al. 2019). These nanoparticles derived from *Aloe vera* extracts show high biocompatibility, meaning that they provide antiviral, antibacterial, and anti-inflammatory behaviors as well. As a rich alternative in terms of phenolic compounds, *Curcuma longa* serves as a

Table 10.3 Examples of medicinal plants used in green nanoparticle synthesis

Plant species	Key phytochemicals	Nanoparticles synthesized	Applications in skin aging and regeneration
Aloe vera	Polysaccharides, anthraquinones, enzymes	AgNPs, AuNPs	Wound healing, antimicrobial, anti-inflammatory
Curcuma longa (turmeric)	Curcuminoids (curcumin)	AuNPs, ZnONPs	Antioxidant, anti-inflammatory, wound healing
Camellia sinensis (green tea)	Catechins, polyphenols	AgNPs, AuNPs	Antioxidant, antimicrobial, anti-aging creams
Centella asiatica	Triterpenoids (asiaticoside)	AgNPs, ZnONPs	Collagen synthesis, wound healing, scar reduction

reducing and stabilizing agent in the synthesis of gold nanoparticles. These nanoparticles have been reported to exhibit excellent antioxidants and anti-inflammatory properties, as well (Kheiripour et al. 2021). *Centella asiatica* is recognized for its wound-healing properties, as *Aloe vera,* and the extracts from this plant have shown promising results in yielding stable nanoparticles suitable especially for cosmetic applications (Witkowska et al. 2024). *Camellia sinensis* is usually preferred for more robust nanoparticle synthesis due to the presence of high catechin levels (Umadevi and Krishna 2024).

In summary, if we want to use these nanoparticles in therapeutic and cosmetic formulations, we should always consider increasing the yield and stability by choosing not only the right conditions but also the appropriate plant depending on the type of nanoparticle. In Table 10.3, specific plants, their key phytochemicals, the nanoparticles produced, and their primary applications regarding medicinal plants used in green nanoparticle synthesis are summarized.

10.4 Mechanisms of Green Synthesis of Metal Nanoparticles

Silver, gold, and zinc oxide nanoparticles are the most popular nanomaterials. They are, of course, considered for green nanotechnology procedures since we know that they all show tremendous properties in skin aging and regeneration fields.

Silver nanoparticles have been synthesized via green methods by using plant extracts. Flavonoids and phenols for the reduction of silver ions are the main contents that are used as reducing agents during the green synthesis of silver nanoparticles (AgNPs). The action mechanisms involve the electrostatic interactions between silver ions and the negatively charged functional groups of bioactive compounds. The reducing agents in the biomaterial stabilized the silver nanoparticles while preventing agglomeration and ensuring a uniform dispersion. If an industrial application is considered, we have to have control over the size and shape of AgNP since the functionality is highly dependent on these two parameters. Many factors can affect the Particle size of AgNPs (e.g., concentration of silver ions, the type and concentration of reducing agents, and reaction conditions such as temperature and pH) (Ali et al. 2019; Guleria et al. 2022). Additionally, the variations in the reaction environment (acidic or basic medium) can have an impact on particle morphology that determines if the product is going to have as spherical, rod, or triangular shape. Regarding particle size, it should be considered that different shapes mean distinct properties and functionalities for these nanoparticles (Zhang and Liu 2023).

Gold nanoparticles have been demonstrating a variety of important properties, such as antifungal, antibacterial, and anticancer. Gold nanoparticles (AuNPs) via green methods employ principles that are parallel to those used for AgNPs. In the green synthesis of AuNPs, phytochemicals from plant extracts act as both reducing and stabilizing agents that involve the reduction of gold ions (Au^{3+}) to AuNPs. The capping agents present in the extracts are highly responsible for the mode of action. They are the key factors for stabilization of nanoparticle formation and preventing aggregation during the process (Guleria et al. 2022; Mallah et al. 2025). Thus, we can say that the stability of AuNP is dependent on the nature of the capping agents. Other factors, such as temperature, can indeed have an influence on the nucleation process and growth kinetics. It is important to note that green synthesis of AuNPs usually exhibits high biocompatibility and lower toxicity, which enhances their usage potential for biomedical applications. Another important property of AuNPs is that they show a unique optical activity, which is particularly localized surface plasmon resonance. This phenomenon is greatly influenced by the size and morphology of the AuNPs. As long as we know where to look, such factors can be handy to understand the dynamics of nanoparticle properties and control these parameters to create new opportunities in nanotechnology (Sood and Agarwal 2018).

Zinc oxide nanoparticles (ZnONPs) have been known for their excellent semiconductor properties. These properties make them excellent choices for applications, especially in photocatalysis, UV protection, and water treatment. In addition, they have been widely used in the degradation of organic pollutants during wastewater treatment. It is proven that they have strong potential for environmental remediation processes. The green synthesis of ZnONPs involves the use of natural extracts rich in reactive phytochemicals (Bukhari et al. 2021). The photocatalytic behavior of ZnONPs is associated with the size and morphology of the nanoparticles. While the band gap energy determines the photocatalytic activity in a ZnONP, the content of the extract in terms of the abundance of reducing and stabilizing agents would change the photocatalytic behavior of ZnONPs. Studies show that the production of smaller particles creates a better photocatalytic efficiency (Tanwar et al. 2024).

We can control the crystalline structure of nanoparticles by simply controlling the reaction parameters such as temperature, the concentration of the ingredients, the nature of the stabilizing agent, and the duration of incubation (Tokumoto et al. 2002; Singh et al. 2021). The biomolecules in plant compositions are directly related to the reaction mechanism of nanomaterial synthesis. The concentration of reducing agents and metal precursors can directly affect nucleation and the size of the nanoparticles. Eventually, nanoparticle size mostly depends on the concentration of reagents. A restricted growth of a particle is usually associated with a lower concentration of the precursors (Guleria et al. 2022). Time required for incubation during extraction is also an important factor in maintaining the duration of reaction time. It surely affects the growth of NPs. There must be a fine adjustment for incubation time since shorter times may lead to smaller nanoparticles, and over-incubation could result in further growth and agglomeration. pH is usually associated with the state of ionization of desolvated phytochemicals. This basically changes the nucleation and growth rates of nanoparticles. For instance, acidic conditions may favor the reduction of metal ions, whereas alkaline conditions may lead to particle agglomeration due to an increase in hydroxide formation (Khan et al. 2022). As in most chemical processes, temperature is directly related to reaction kinetics. Increasing the temperature indeed can lead to a faster formation of nanoparticles. However, it should always be considered, especially with natural products, that excessive heat can degrade sensitive phytochemicals or alter nanoparticle morphology (Reddy et al. 2022).

Synthesis of an adequate amount and high quality of desired nanoparticles with desired size and morphology depends on fine adjustment of these parameters, along with some others not listed here. It depends on the expertise to establish a balance between environmental sustainability and functional efficiency (Saxena et al. 2021). While considering the laboratory set-up conditions, we also must consider the environmental parameters affecting the natural contents of plant extracts used for synthesis. For instance, differing phytochemical compositions in a plant extract may be affected by solvents as well as being controlled by geographic and seasonal factors. Neglecting these parameters will cause inconsistencies in the production of nanoparticles (Mustapha et al. 2022; Salve et al. 2022). Thus, we must set up standardized protocols for extract preparation and related parameter adjustments, such as metal ion concentration, reaction conditions, and characterization methods. This will provide a better solution for everyone in terms of reproducibility and reliability in experimental outcomes.

10.5 Characterization of Green-Synthesized Nanoparticles

Characterization of green-synthesized nanoparticles is indeed important to define their structural and functional features. As mentioned above, challenges remain regarding standardization and reproducibility, but if we consider the potential advantages offered by green synthesis methods for nanoparticles, we must come up with solutions accordingly. Unfortunately, there are significant variations in particle morphology and other related functional characteristics, such as their effectiveness in different applications, due to inconsistent synthesis protocols (Melkamu and Bitew 2021). Nonetheless, we are aware of which techniques to be adapted into the standardization processes through characterization. Today, there are advanced analytical techniques such as UV-VIS spectroscopy, Fourier transform infrared spectroscopy (FTIR), transmission electron microscopy (TEM), scanning electron microscopy (SEM), X-ray diffraction (XRD), dynamic light scattering (DLS), and zeta potential measurements that are used to examine the physical and chemical properties.

UV-VIS and FTIR spectroscopies are the most common first characterization tools, most likely due to their availability. While UV–VIS spectroscopy can provide some insight into the surface plasmon resonance (SPR) characteristic of the nanoparticles to estimate variations related to variations in size and shape, FTIR spectroscopy technology is more related to detecting the presence of functional groups of the synthesized nanoparticles (Khane et al. 2022; Mustapha et al. 2022).

TEM and SEM analyses provide a deeper look at the morphology, size, and distribution of nanoparticles. Especially, the information gathered about the surface morphology would provide valuable insights into the biological activity of nanoparticles.

If we are interested in determining the crystalline structure of synthesized nanoparticles, XRD analysis becomes very handy in terms of the quantification of crystalline phases in metal nanoparticles associated with different metal phases. As a result, XRD analysis gives essential data about their physicochemical properties (Seçkin and Meydan 2021).

Hydrodynamic size and surface charge of nanoparticles can also be detected with current technologies such as DLS and zeta potential measurements. While information obtained from DLS is on the particle size distribution, stability of colloidal systems via analyzing the electrostatic repulsion between particles through zeta potential measurement (Rajurkar et al. 2023). These measurements become necessary when the target is to detect the presence of functional groups on the surface for biological interactions.

In addition to these analytical methods, some biochemical and microbiological activities should also be included in the analysis protocols in order to create a correlation between morphological properties and their impact on living systems. For example, the gold nanoparticles (AuNPs) synthesized using plant extracts are proven to exhibit varying antioxidant activities related to their specific shapes (Singh et al. 2018). Therefore, understanding the relationship between the structural properties of nanoparticles and their functional outcomes is crucial for improving their applications.

10.6 Biological Activities and Mechanisms of Action

Green-synthesized metal nanoparticles have important potential in medicinal usage due to their biological activities. Their potential in regenerative therapeutics and cosmeceuticals has been widely studied in green nanotechnology. The mechanisms in green nanomaterials can be listed as antioxidant mechanisms, anti-inflammatory mechanisms, antimicrobial mechanisms, and tissue regeneration mechanisms. Antioxidant mechanisms play an important role in the green synthesis of metal nanoparticles. They are known to help reduce oxidative stress by inhibiting lipid peroxidation and neutralizing reactive oxygen species (ROS). In the previous studies, they found that Ag NPs have strong antioxidant activities due to their ability to donate electrons and form stable complexes with ROS (Vankar and Shukla 2011; Chopade and Kamble 2021). Interaction with free radicals during the reaction helps neutralize the nanoparticles, preventing subsequent cellular damage. Furthermore, the phytochemicals in reducing agents are reported to contribute to the overall antioxidant potential by improving electron transfer pathways (Mohan et al. 2007; Shah and Malik 2019).

The inhibition mechanism is a protective mechanism that limits tissue damage. In this process, lipid peroxidation prevents free radical interactions with fatty acids within cellular membranes. Through its interaction with membrane phospholipids, it reduces membrane flow and sensitivity to oxidative damage. Most of the metal nanoparticles have been reported to stabilize lipid bilayers. Studies on these mechanisms show that metal nanoparticles can certainly influence membrane structures. Although it may vary according to the specific cellular conditions and nanoparticle characteristics (Park et al. 2011). From another perspective, it has been observed that the production of some antioxidants produced by plants (such as betaine and betalain) increases because of the application of ZnO and/or TiO_2 NP to seeds during the plant germination stage (Alawad et al. 2024).

Anti-inflammatory mechanisms typically suppress inflammation through the production of cytokines. The anti-inflammatory effects of the nanoparticles are primarily obtained by the cytokine signaling pathways. Silver nanoparticles can regulate the expression of pro-inflammatory cytokines, including TNF-α and IL-6. They reduce inflammation in tissues, which is especially helpful in dermatological applications (Marimuthu et al. 2010).

Antimicrobial mechanisms are important for many therapeutic applications. Studies demonstrate the efficacy of metal nanoparticles against a variety of bacterial and fungal pathogens. Green-synthesized metal nanoparticles exhibit significant anti-inflammatory and antimicrobial properties in wound healing and infection control, and multiple pathways, including those regulating oxidative stress responses and inflammatory processes during the formation of metal nanoparticles, yield visible results in skin health and regeneration (Durán et al. 2016).

Antibacterial activity is associated with the breakdown of bacterial cell membranes that result in the loss of cellular integrity of selected bacteria. It is proven that the formation of ROS during exposure to various nanoparticles promotes bacterial cell death (Munusamy et al. 2023). In addition, production of metal nanoparticles through the green processes increases the impact of the intrinsic antimicrobial agents that come with the plant extracts (Gabi et al. 2024).

Controlling the mechanisms of tissue regeneration is the ultimate goal in regenerative medicine applications. There have been reports associated with molecular interactions and pathways that can be regulated by green-synthesized nanoparticles. For instance, AgNPs are competent to increase growth associated with tissue repair and remodeling via impacting some signaling pathways (e.g., activation of transforming growth factor-beta (TGF-β)) (Bose and Chatterjee 2015). Fibroblast cells are considered the principal cells in tissue repair processes. It is evident that nanoparticles may accelerate fibroblast proliferation and migration, which unarguably contributes to the regeneration process of epithelial tissue (Lestari et al. 2019). In addition, we know that AgNPs can promote angiogenesis, which is a mechanism associated with the formation of new blood vessels. These vessels play a crucial role in supplying nutrients and oxygen to regenerated tissues (Song and Kim 2008). As seen, understanding these molecular mechanisms in green-synthesized nanoparticles will help us to develop more sophisticated medicinal strategies for various skin-healing treatments.

10.7 Applications in Skin Aging and Regenerative Medicine

Since the beginning of life, aging cannot be stopped. I believe that well-applied skin care helps slow aging in the skin. With the advanced innovations in the anti-aging area, there are numerous new applications in skin aging. Topical nano-scale formulations utilizing green-synthesized nanoparticles have been accepted as a sustainable approach in the field of anti-aging treatments. High penetration capacity into the cell membrane makes these nano-scale formulations more valuable than conventional ones. Various formulations, including creams, hydrogels, nanoemulsions, and liposomal delivery systems, harness the properties of nanoparticles to promote the delivery and effectiveness of active ingredients such as antioxidants and peptides directly to the skin (Kumar et al. 2023).

In every skin care routine, there are creams and gels. Creams and gels enriched with silver and gold nanoparticles have been investigated for their potential to improve skin texture and elasticity, as well as to provide hydration. The formulations with active ingredients to enhance diffusion capacity will provide deeper penetration into the skin (Kumar et al. 2020). It is necessary to improve the formulations and associated technologies in the cream and gel industries, such as developing hydrogel-containing products. Green-synthesized zinc oxide nanoparticles embedded in hydrogels have a better ability to create a moist environment on the skin. Advancement of the formulations via encapsulation technologies enables us to apply both hydrophilic and lipophilic compounds in skin regeneration therapies. Advancements in encapsulation techniques provide an effective solution for the controlled release of active compounds to ensure better healing capacities. Studies show that green-synthesized nanomaterials play an important role in stabilizing active agents just like coenzyme Q10 (Masoumi et al. 2021; Hashemi et al. 2022). These advantages are promising in delaying aging by helping skin regeneration processes. It is important to note that the green synthesized nanoparticles are also effective in wound healing and tissue regeneration in a similar manner. Silver nanoparticles (AgNPs) can enhance epithelial regeneration and wound closure (Thomas et al. 2022). When it comes to wounded injuries, infection is another issue to be controlled for effective healing. Studies show that silver nanoparticles significantly contribute to minimizing the complications from secondary infections during the healing process thanks to their antimicrobial properties (Naderifar et al. 2025). The effect of herbal compounds on skin healing is not only at the in vitro level; it has been clinically reported that *Morus alba* extract supports epidermal healing in topical treatment, especially in atopic dermatitis, and reduces the risk of secondary infection caused by *S. aureus* (Ezairjawi et al. 2025). As mentioned earlier, green-synthesized nanoparticles that are used in wound healing improve angiogenesis which increase blood flow which is a key factor affecting the healing process (Swetha et al. 2021). These nanoparticles are capable of releasing bioactive molecules that boost vascular endothelial growth factor (VEGF) signaling pathway, which is crucial for new blood vessel formation (Thomas et al. 2022). All these processes, particularly important in managing inflammation and oxidative stress during the healing process, can be associated with delaying aging and/or boosting healing processes during older ages (Chatterjee et al. 2023).

It is obvious that green-synthesized nanoparticles have been gaining popularity in the cosmetics industry. Their integration into cosmeceutical and dermatological products is attracting significant interest. Trends in market indicate a growing support among consumers for the sustainable and environmentally friendly practices of these green nanoparticles. There is an increased number of patents for the use of green-synthesized nanoparticles, particularly silver and zinc oxide nanoparticles. These—so-called superior—nanoparticles are also appealing to consumers with environmental concerns. These nanoparticles are now essential components of different cosmetic products from sunscreens and moisturizers to anti-aging creams.

There is obvious concern about sustainability all around the world which enforces companies to take adequate actions. Indeed, several commercial products have been affected by this integration. Skincare products offering formulations that utilize plant-derived nanoparticles known for their antioxidant and anti-inflammatory effects now offer a gateway for sustainability and biocompatibility. Skincare is a daily routine for people with sensitive skin, but the content and effectiveness of skincare products can be a concern for them. With the integration of green nanoparticles, the biocompatibility and low toxicity concerns have been mostly eliminated making the products particularly attractive for formulations targeted at sensitive skin types (Virkutytė and Varma 2011).

Today, there are several delivery systems (e.g., micelles, solid lipid nanoparticles, and nanofibers) that can be adapted for nanoparticles in skincare products (Kumar et al. 2023). It is obvious that green-synthesized nanoparticles hold an irreplaceable place in cosmetic and dermatological products. It seems that the demand for innovative skin therapies and regenerative solutions will continue to grow in the following years, and these advanced materials will shape the future of cosmetic and dermatological industries.

In summary, the application of green-synthesized nanoparticles in skin aging and regenerative medicine presents a promising area. Nanoparticles are gaining attention with their integration into topical formulations and their role in tissue regeneration.

10.8 Biocompatibility, Toxicity, and Safety Considerations

Nanoparticles are materials that are less than 100 nm in size. Because of their tiny structures they can penetrate many systems easily. Because of their very small size and more advanced properties, there is a concern about the safety and toxicity of nanomaterials. It is essential to evaluate the safety and cytotoxicity of green-synthesized nanoparticles before they can be applied in medical and cosmetic products. They have been in the market for only two decades; however, there are many publications that reveal their toxicity and hazardous effects.

Common in vitro models, such as keratinocyte and fibroblast cell line assays, are instrumental in examining the biocompatibility of nanoparticles. Various studies indicate that green-synthesized AgNPs exhibit relatively low cytotoxicity when tested on keratinocyte and fibroblast cultures. Keratinocytes are cell types that exist predominantly in the epidermis. They are particularly relevant for testing topical formulations aimed at skin applications (Moulton et al. 2010). These studies also show that the antioxidant properties of the phytochemicals used in synthesis play a role in protecting cells from nanoparticle-induced toxicity (Soltys et al. 2021).

Cell viability and metabolic activity can be easily monitored by some conventional laboratory techniques. There are several standardized approaches such as the MTT assay or lactate dehydrogenase (LDH) leakage tests for this purpose (Dhiman et al. 2021). These tests provide essential data on the concentration-dependent effects of nanoparticles, and we can get information about potential dose limits for safe human use. In vivo studies are essential for validating green-synthesized nanoparticles in complex biological systems in terms of their safe usage. Animal models, such as mice and rats, are usually used by researchers—indeed controlled by strict ethical regulations—to evaluate the toxicity the distribution of nanoparticles within biological systems following systemic or topical administration (Idris and Roy 2023).

In vivo studies have aligned with the three 'R's—Replacement, Reduction, and Refinement principles. There is significant attention directed toward ethical considerations in these studies. It is aimed at minimizing animal suffering while maximizing the scientific output of research (Ali et al. 2020). With the increasing popularity of green nanotechnology, there is a rise of public concern for animal welfare. Increasing emphasis on discussions about the methods of synthesis and testing. They are not considered as ethical and environmentally sustainable (Tyavambiza et al. 2023).

Furthermore, studies have shown that green-synthesized nanoparticles may lead to lower systemic toxicity levels compared to their counterparts, which are the products of chemical synthesis. Their biocompatibility is derived from the natural compounds used during fabrication (Ashikbayeva et al. 2022). The usage of plant extracts for the synthesis is a more reliable path in terms of sustainability. Additionally, it is patient safety and acceptance in potential medicinal applications.

An essential aspect of the evaluation of green-synthesized nanoparticles is their long-term exposure implications for both human health and environmental impact. The fate of nanoparticles in ecosystems can vary significantly based on their chemical properties, especially on the surface. Along with size and shape, they can affect their bioavailability and potential toxicity to aquatic and terrestrial organisms (Veeragoni et al. 2022). Studies have indicated that longer exposure to nanoparticles can lead to bioaccumulation in aquatic systems, raising concerns about trophic transfer and potential impacts on nutrients (Wypij et al. 2021).

The consequences of nanoparticle interaction with environmental factors like soil and water are under consideration. Understanding how these materials interact with biological systems in nature can help us control experimental environments. They are critical for addressing potential ecological risks regarding the exposure to nanoparticles released during manufacturing processes and/or consumer use (Mukherjee et al. 2012). The degradation mechanisms of green nanoparticles in environmental conditions are another issue to be considered. Nanoparticles that exhibit enhanced stability might persist in ecosystems longer. It is necessary to make ecological risk assessments to ensure their long-term safety.

Overall, understanding the biocompatibility, toxicity, and environmental impact of green-synthesized nanoparticles is crucial. Therefore, they can show their promising characteristics in applicable solutions in medicine and beyond. Research focusing on safety, efficiency, and ethical considerations will increase public trust and acceptance of nanotechnology applications.

10.9 Regulatory and Socioeconomic Perspectives

Control of current regulations in cosmetics and pharmaceuticals is usually carried out by agencies such as the U.S. Food and Drug Administration (FDA) and the European Medicines Agency (EMA). Organizations like ISO are responsible for setting up the standards. The role of these regulators is essential for maintaining safe products for the consumers. When the issue is about the nanoproducts, setting standard regulations for all nanoparticles is not applicable since each nanoparticle can have a unique property associated with its chemical structure and application. As suggested by the FDA, the data on the physicochemical properties of nanoparticles, including size, shape, surface charge, and reactivity, are crucial to making these kinds of decisions in the regulatory process. A similar example can be given for the EMA that they have protocols to evaluate specific parameters of nanomaterials, in terms of pharmacokinetics and toxicity analysis (Samuel and Verma 2016).

Although there are standards, there is a need for harmonizing the regulatory frameworks across different regions of the globe since there are significant differences exist between the FDA and EMA in terms of timelines and criteria for approval. For example, while the FDA is more concerned about the in vivo toxicological evaluations, the EMA is focused more on environmental risk assessments. Thus, an application to the FDA typically takes less time to consider the applications compared to the EMA, since long-term stability data is required before market authorization is completed.

As in any other product consumption, public awareness of environmental issues regarding nanoparticles becomes more obvious, and consumers demand to reach transparent fact sheets for the products they choose in cosmetics and pharmaceuticals. Thus, industrial authorities put that request at the center of their marketing strategies by providing eco-labeling and sustainability claims (Shweta et al. 2025). In other words, companies tend to share the details about the products as labels and provide extra support regarding the eco-compatibility of their products.

In order to increase the effectiveness of these strategies, educating consumers about the benefits and safety of green-synthesized nanoparticles is essential in terms of ethical marketing strategies. It would not be effective to provide all these scientific claims made about product efficacy and environmental impact and expect to be understood by the consumers as stated (Shweta et al. 2025).

Sustainable solutions in cosmetic formulation increase brand image and create competitive advantages in the nanomaterial market. Nowadays, consumers value their purchased products with their ecological concerns (Junkes and Sousa 2025). The increase in digital markets' partnerships with influencers can amplify these messages. Build a deeper connection between consumers and brands committed to sustainability (Brancu and Turcu 2025).

There are several problems with commercialization green-synthesized nanoparticles that remain unsolved. The issues are about scale-up challenges, intellectual property issues, and associated costs. Scaling production processes from laboratory to industrial scale can be complex due to variability in green synthesis processes. Achieving consistent quality and stability of nanoparticles at larger volumes presents significant logistical problems. Furthermore, the optimized methodology for green synthesis requires a balance between efficiency and the economic viability of large-scale production.

Intellectual property rights also pose challenges since companies embark on developing novel formulations and methods of synthesis. Patent laws can differ significantly across jurisdictions. It complicates the ability to safeguard private technology and may inhibit innovation (Natsir et al. 2025). Additionally, the initial costs for research and development of green synthesis processes may involve higher investments. It can discourage businesses from pursuing these sustainable alternatives (Yang and Hamid 2024).

Green nanotechnology products in cosmetics and pharmaceuticals are highly demanding innovations and a market entry. Therefore, it is essential to forge collaborations between academia, industry, and regulatory agencies to streamline processes, enhance consumer trust, and facilitate sustainable practices (Yang and Hamid 2024). By addressing these barriers effectively, the market for environmentally friendly products can expand rapidly. That gives a promotion for the broader adoption of nanotechnology across diverse applications.

In conclusion, while the regulatory landscape for nanomaterials continues to evolve, there remains a crucial need for informed policies that balance innovation with safety. Consumer awareness aligns with ethical marketing practices that should be developed. In the end, removing the barriers to commercialization will ultimately enhance the prospects for green-synthesized nanoparticles in the market.

10.10 Conclusion and Prospect

System biology is an approach to understanding the interactions and functions of biological components that act as a whole system. Comprehensive modeling of cell networks interacting with nanomaterials can be possible by system biology.

Moreover, there are also omics technologies that help to analyze bulky molecules such as proteins, DNA, and metabolites in a very short time. Of course, we have nano-bioinformatics to examine these bulky data on computers. The convergence of biological systems with nanotechnology, omics, and nano-bioinformatics is leading new paths for predicting nanoparticle-plant interactions. These advanced approaches can enhance the understanding of biological responses to nanoparticles. By this, skincare applications and tailor-made therapies might be easily designed. At present, this technology is costly, but it is expected to become more developed and accessible for wider application in mass-market cosmetics in the future.

AI-driven designs in creating personalized nanocosmetics reflect the trajectory toward sustainable practices that match consumer preferences. Biomimetic strategies can lead to many innovations in skin care. Precise phytotherapy can lead to innovative formulations for individual skin types and conditions, establishing a new revolution in cosmetics.

The synergies between green nanotechnology and biomedicine hold a huge potential for advancing aging research in the cosmetic industry. As the landscape of nanotechnology evolves, eco-friendly solutions and sustainable practices will remain popular. Continued exploration reshapes the future of skincare and regenerative medicine.

Competing Interests The authors have no relevant financial or non-financial interests to disclose.

References

Alawad N, Ünüvar ÖC, Gürel S et al (2024) Effect of nanoparticles on sugar beet cultured in vitro and its betalain and betaine production. Turk J Agric For 48:129–138

Ali J, Ali N, Wang L et al (2019) Revisiting the mechanistic pathways for bacterial mediated synthesis of noble metal nanoparticles. J Microbiol Methods 159:18–25

Ali MA, Ahmed T, Wu W et al (2020) Advancements in plant and microbe-based synthesis of metallic nanoparticles and their antimicrobial activity against plant pathogens. Nanomaterials 10:1146

Aljuffali IA, Hsu C, Lin Y, Fang J (2015) Cutaneous delivery of natural antioxidants: the enhancement approaches. Curr Pharm Des 21:2745–2757

Añibarro-Ortega M, Pinela J, Barros L et al (2019) Compositional features and bioactive properties of *Aloe vera* leaf (fillet, mucilage, and rind) and flower. Antioxidants (Basel) 8:444

Ashikbayeva Z, Aitkulov A, Atabaev TS et al (2022) Green-synthesized silver nanoparticle–assisted radiofrequency ablation for improved thermal treatment distribution. Nanomaterials 12:426

Biswas DK, Kaya Ö, Ünüvar ÖC et al (2025) Application of nanoparticles as biostimulator for growth of *Digitalis ferruginea* subsp. *ferruginea* L under in vitro conditions. 3 Biotech 15:408

Boo YC (2019) Can plant phenolic compounds protect the skin from airborne particulate matter? Antioxidants 8:379

Bose D, Chatterjee S (2015) Antibacterial activity of green synthesized silver nanoparticles using vasaka (*Justicia adhatoda* L.) leaf extract. Indian J Microbiol 55:163–167

Brancu C, Turcu O (2025) Exploring the economic performance and market trends in the cosmetics industry: insights and analysis. BMC 69:23–32

Bukhari A, Ijaz I, Gilani E et al (2021) Green synthesis of metal and metal oxide nanoparticles using different plants' parts for antimicrobial activity and anticancer activity: a review article. Coatings 11:1374

Chakraborty A, Haque SM, Dey D et al (2022) Phytogenic silver nanoparticles from tissue-cultured *Kaempferia Angustifolia*—an underutilized medicinal herb: a comparative antibacterial study on urinary pathogens. J Genet Eng Biotechnol 20:131

Chamizo LL, Morato YL, Paredes KO et al (2021) Ionotropic gelation-based synthesis of chitosan-metal hybrid nanoparticles showing combined antimicrobial and tissue regenerative activities. Polymers 13:3910

Chandraker SK, Ghosh MK, Lal M, Shukla R (2021) A review on plant-mediated synthesis of silver nanoparticles, their characterization and applications. Nano Express 2:022008

Chatterjee S, Jaiganesh R, Rajeshkumar S (2023) Green synthesis of zinc oxide nanoparticles using chamomile and green tea extracts and evaluation of their anti-inflammatory and antioxidant activity: an in vitro study. Cureus 15(9):e46088

Chopade V, Kamble D (2021) Application of green silver nanoparticles synthesized using leaf extract of *Tridax procumbens* for preparation of clinical antimicrobial bandages. Int J Pharm Investig 11:10–13

Dhiman S, Singh S, Varma A, Goel A (2021) Phytofabricated zinc oxide nanoparticles as a nanofungicide for management of *Alternaria* blight of *Brassica*. Biometals 34:1275–1293

Durán N, Nakazato G, Seabra AB (2016) Antimicrobial activity of biogenic silver nanoparticles, and silver chloride nanoparticles: an overview and comments. Appl Microbiol Biotechnol 100:6555–6570

El-Baghdady K, El-Shatoury EH, Abdullah OM, Khalil MM (2018) Biogenic production of silver nanoparticles by enterobacter cloacae ism26. Turk J Biol 42:319–321

Ezairjawi MSM, Ünüvar ÖC, Akben C et al (2025) Mulberry's healing touch: exploring ethnobotanical roots and medicinal potentials in the treatment of atopic dermatitis. J Ethnopharmacol 337:118981

Firmino JP, Fernández-Alacid L, Vallejos-Vidal E et al (2021) Carvacrol, thymol, and garlic essential oil promote skin innate immunity in gilthead seabream (*Sparus aurata*) through the multifactorial modulation of the secretory pathway and enhancement of mucus protective capacity. Front Immunol 12:633621

Gabi B, Aliyu AM, Orasugh JT, Abdullahi Z (2024) Green synthesis, characterization, and antibacterial activity of silver nano particles against some bacterial isolates. Sci World J 18:653–661

Galili U, Li J, Schaer GL (2024) Regeneration in mice of injured skin, heart, and spinal cord by α-gal nanoparticles recapitulates regeneration in amphibians. Nano 14:730

Gami B, Bloch K, Mohammed S et al (2022) *Leucophyllum frutescens* mediated synthesis of silver and gold nanoparticles for catalytic dye degradation. Front Chem 10:932416

García-Pérez P, Lozano-Milo E, Zhang L et al (2022) Neurofuzzy logic predicts a fine-tuning metabolic reprogramming on elicited *Bryophyllum* PCSCs guided by salicylic acid. Front Plant Sci 13:991557

Ghramh HA, Ibrahim EH, Kilnay M et al (2020) Silver nanoparticle production by *Ruta graveolens* and testing its safety, bioactivity, immune modulation, anticancer, and insecticidal potentials. Bioinorg Chem Appl 2020:1–11

Guleria A, Sachdeva H, Saini K et al (2022) Recent trends and advancements in synthesis and applications of plant-based green metal nanoparticles: a critical review. Appl Organomet Chem 36:e6778

Gulia K, James A, Pandey S et al (2022) Bio-inspired smart nanoparticles in enhanced cancer theranostics and targeted drug delivery. J Funct Biomater 13:207

Hama S, Kimura Y, Mikami A et al (2014) Electric stimulus opens intercellular spaces in skin. J Biol Chem 289:2450–2456

Hashemi M, Paskeh MDA, Babaei N et al (2022) Protective effects of coenzyme Q10 along with Fe2O3 nanoparticles on sperm parameters in rats with scrotal hyperthermia. Galen Med J 11:1–7

Idris DS, Roy A (2023) Synthesis of bimetallic nanoparticles and applications—an updated review. Crystals 13:637

Javed R, Zia M, Naz S et al (2020) Role of capping agents in the application of nanoparticles in biomedicine and environmental remediation: recent trends and future prospects. J Nanobiotechnol 18(1):172

Junkes P, Sousa B (2025) Green marketing and the competitive edge in the cosmetics and wellness products sector. In: Design, development, and deployment of cutting-edge medical devices. IGI Global Scientific Publishing, Hershey, pp 49–66

Kamli MR, Srivastava V, Hajrah NH et al (2021) Phytogenic fabrication of Ag–Fe bimetallic nanoparticles for cell cycle arrest and apoptosis signaling pathways in *Candida auris* by generating oxidative stress. Antioxidants 10:182

Khan F, Shariq M, Asif M et al (2022) Green nanotechnology: plant-mediated nanoparticle synthesis and application. Nanomaterials 12:673

Khanahmad A, Ebrahimi Z, Abreshteh A et al (2024) Biological applications of nanoparticles synthesized via *Olea europaea* plant: a comprehensive review. Int J Polym Sci 2024:418466

Khane Y, Benouis K, Albukhaty S et al (2022) Green synthesis of silver nanoparticles using aqueous *Citrus limon* zest extract: characterization and evaluation of their antioxidant and antimicrobial properties. Nanomaterials 12:2013

Kheiripour N, Plarak A, Heshmati A et al (2021) Evaluation of the hepatoprotective effects of curcumin and nanocurcumin against paraquat-induced liver injury in rats: modulation of oxidative stress and Nrf2 pathway. J Biochem Mol Toxicol 35:e22739

Kumar H, Bhardwaj K, Kuča K et al (2020) Flower-based green synthesis of metallic nanoparticles: applications beyond fragrance. Nanomaterials 10:766

Kumar P, Ashawat MS, Pandit V et al (2023) Recent trends in nanocarriers for the management of atopic dermatitis. Pharm Nanotechnol 11:397–409

Kumari M, Sadhu P, Talele C, Shah N (2024) Green nanotechnology: how plants can help synthesize nanoparticles for biomedical and environmental purposes. J Nat Med 24:1021–1034

Lestari TP, Tahlib FA, Sukweenadhi J et al (2019) Physical characteristic and antibacterial activity of silver nanoparticles from green synthesis using ethanol extracts of *Phaleria Macrocarpa* (Scheff.) Boerl leaves. Majalah Obat Tradisional 24:22

Mallah S, Mchaouri ME, Meziani S et al (2025) Green photocatalysis: a comprehensive review of plant-based materials for sustainable water purification. Reactions 6:55

Mao AS, Mooney D (2015) Regenerative medicine: current therapies and future directions. Proc Natl Acad Sci 112:14452–14459

Marimuthu S, Rahuman AA, Rajakumar G et al (2010) Evaluation of green synthesized silver nanoparticles against parasites. Parasitol Res 108:1541–1549

Masoumi M, Salehi M, Angaji SA, Hashemi M (2021) Effects of coenzyme Q10 and diamond nanoparticles on ischemia-reperfusion-induced testicular damages in rats. Galen Med J 10:e2029

Melkamu WW, Bitew LT (2021) Green synthesis of silver nanoparticles using *Hagenia abyssinica* (Bruce) J.F. Gmel plant leaf extract and their antibacterial and anti-oxidant activities. Heliyon 7:e08459

Mohan YM, Raju KM, Sambasivudu K et al (2007) Preparation of *Acacia*-stabilized silver nanoparticles: a green approach. J Appl Polym Sci 106:3375–3381

Moulton MC, Braydich-Stolle LK, Nadagouda MN et al (2010) Synthesis, characterization and biocompatibility of "green" synthesized silver nanoparticles using tea polyphenols. Nanoscale 2:763

Mukherjee S, Sushma V, Patra S et al (2012) Green chemistry approach for the synthesis and stabilization of biocompatible gold nanoparticles and their potential applications in cancer therapy. Nanotechnology 23:455103

Munusamy T, Rajeshkumar S, Al-Ghanim KA et al (2023) *Terminalia chebula*-assisted silver nanoparticles: biological potential, synthesis, characterization, and ecotoxicity. Biomedicine 11:1472

Mustapha T, Misni N, Ithnin NR et al (2022) A review on plants and microorganisms mediated synthesis of silver nanoparticles, role of plants metabolites and applications. Int J Environ Res Public Health 19:674

Naderifar M, Ghahramani A, Nik MM et al (2025) *Ephedra* herb as a sustainable source: insights into its role in metal nanoparticle synthesis—a review. Environ Prog Sustain Energy 44:e14624

Natsir SA, Takai A, Seo E et al (2025) How awareness of organic JAS and RSPO labels influences Japanese consumers' willingness to pay more for organic cosmetics. Sustainability 17:7466

Ovais M, Khalil AT, Islam NU et al (2018) Role of plant phytochemicals and microbial enzymes in biosynthesis of metallic nanoparticles. Appl Microbiol Biotechnol 102:6799–6814

Park Y, Hong Y, Weyers A et al (2011) Polysaccharides and phytochemicals: a natural reservoir for the green synthesis of gold and silver nanoparticles. IET Nanobiotechnol 5:69–78

Rajurkar A, Gogri D, Jamdade N, Pathak AP (2023) Green synthesis of silver nanoparticles: their characterization, antimicrobial, antioxidant activity and nanogel formulation. Nano Biomed Eng 15:42–50

Reddy GB, Ramakrishna D, Kondaiah S, Mangatayaru KG (2022) Fungal mediated synthesis of nanoparticles: characterization and bioapplications. In: Sustainable utilization of fungi in agriculture and industry. Bentham Science Publishers, Québec, pp 328–358

Salve P, Vinchurkar A, Raut RW et al (2022) An evaluation of antimicrobial, anticancer, anti-inflammatory and antioxidant activities of silver nanoparticles synthesized from leaf extract of *Madhuca longifolia* utilizing quantitative and qualitative methods. Molecules 27:6404

Samuel N, Verma S (2016) Cross-comparison of cancer drug approvals at three international regulatory agencies. Curr Oncol 23:454–460

Saxena M, Jain K, Saxena R (2021) Green synthesized nanomaterial-based colorimetric sensors for detection of environmental toxicants. ChemNanoMat 7:392–414

Seçkin H, Meydan İ (2021) Synthesis and characterization of *Veronica beccabunga* green synthesized silver nanoparticles for the antioxidant and antimicrobial activity. Türkiye tarım araşt derg 8:49–55

Shah PJ, Malik R (2019) Study of antibacterial activity of *Phyllanthus emblica* and its role in green synthesis of silver nanoparticles. J Drug Deliv Ther 9:76–81

Shweta YK, Ojha V, Mathur A (2025) Cosmetic choices in the green era: evaluating sustainable and green practices' influence on consumers. Comm Res Rev 2:45–58

Singh J, Dutta T, Kim K et al (2018) Green' synthesis of metals and their oxide nanoparticles: applications for environmental remediation. J Nanobiotechnol 16:84

Singh A, Madhavi BLR, Nithin Sagar MN (2021) An overview of green synthesis mediated metal nanoparticles preparation and its scale up opportunities. J Drug Deliv Ther 11:304–314

Soltys LM, Olkhovyy O, Татарчук T, Naushad M (2021) Green synthesis of metal and metal oxide nanoparticles: principles of green chemistry and raw materials. Magnetochemistry 7:145

Song JY, Kim BS (2008) Rapid biological synthesis of silver nanoparticles using plant leaf extracts. Bioprocess Biosyst Eng 32:79–84

Sood A, Agarwal S (2018) Metallic nanoparticle synthesis by green chemistry. Int J Pharm Sci Nanotechnol 11:4287–4294

Sree K, Vedhanayagam M, Nair BU, Anandasadagopan SK (2021) Dendrimer functionalized metal oxide nanoparticle mediated self-assembled collagen scaffold for skin regenerative application: function of metal in metal oxides. Appl Biochem Biotechnol 194(1):266–290

Swetha TV, Jeevitha M, Rajeshkumar S, Jayaraman S (2021) Antioxidant activity of *Abies webbiana* mediated zinc oxide nanoparticles. J Pharm Res Int 33:135–143

Tanwar SN, Parauha YR, There Y et al (2024) Plant-based biosynthesis of metal and metal oxide nanoparticles: an update on antimicrobial and anticancer activity. ChemBioEng Rev 11:e202400012

Thatyana M, Dube NP, Kemboi D et al (2023) Advances in phytonanotechnology: a plant-mediated green synthesis of metal nanoparticles using *Phyllanthus* plant extracts and their antimicrobial and anticancer applications. Nanomaterials 13:2616

Thomas AA, Babu H, Rajeshkumar S (2022) Evaluation of cytotoxic and antioxidant properties of copper nanoparticles using green tea and neem formulation. Int J Health Sci 6:773–780

Tokudome Y, Komi T, Omata A, Sekita M (2018) A new strategy for the passive skin delivery of nanoparticulate, high molecular weight hyaluronic acid prepared by a polyion complex method. Sci Rep 8:2336

Tokumoto MS, Pulcinelli SH, Santilli CV, Briois V (2002) Catalysis and temperature dependence on the formation of ZnO nanoparticles and of zinc acetate derivatives prepared by the sol–gel route. J Phys Chem B 107:568–574

Tyavambiza C, Meyer M, Wusu AD et al (2023) The cytotoxicity of cotyledon *Orbiculata aqueous* extract and the biogenic silver nanoparticles derived from the extract. Curr Issues Mol Biol 45:10109–10120

Umadevi G, Krishna KG (2024) *Camellia sinensis*-mediated green synthesis of ZnO nanoparticles: an eco-friendly approach for high-performance gas sensing. Sens Actuators A Phys 374:115479

Ünüvar ÖC, Ünlü ES (2022) From hologenomes to biofertilizers in wheat production. In: Zencirci N, Ulukan H, Baloch FS et al (eds) Ancient wheats. Springer, Cham, pp 181–196

Vallinayaki K, Shanmugam R, Tharani M (2024) Biosynthesis of strontium nanoparticles using *Mimosa pudica* and its antioxidant effect. J Pharm Bioallied Sci 16:S1330–S1334

Vankar PS, Shukla D (2011) Biosynthesis of silver nanoparticles using lemon leaves extract and its application for antimicrobial finish on fabric. Appl Nanosci 2:163–168

Veeragoni D, Deshpande S, Rachamalla HK et al (2022) In vitro and in vivo anticancer and genotoxicity profiles of green synthesized and chemically synthesized silver nanoparticles. ACS Appl Bio Mater 5:2324–2339

Virkutytė J, Varma RS (2011) Green synthesis of metal nanoparticles: biodegradable polymers and enzymes in stabilization and surface functionalization. Chem Sci 2:837–846

Wang W, Lu K, Yu C et al (2019) Nano-drug delivery systems in wound treatment and skin regeneration. J Nanobiotechnology 17:82

Williams L, Arshad MS, JeejaRani AT, Joseph A (2022) Biogenic MnO_2 nanoparticles derived from *Acedrus Deodarapine* needle extract and their composites with polyaniline/activated charcoal as an electrode material for supercapacitor applications. New J Chem 46:4325–4333

Witkowska K, Paczkowska-Walendowska M, Garbiec E, Cielecka-Piontek J (2024) Topical application of *Centella asiatica* in wound healing: recent insights into mechanisms and clinical efficacy. Pharmaceutics 16:1252

Wypij M, Jędrzejewski T, Trzcińska-Wencel J et al (2021) Green synthesized silver nanoparticles: antibacterial and anticancer activities, biocompatibility, and analyses of surface-attached proteins. Front Microbiol 12:632505

Yang J, Hamid MBB (2024) Sustainable beauty: a conceptual paper of how sustainable marketing impact consumer behaviour in the cosmetic industry. Adv Econ Manag Politi Sci 93:54–59

Zhang D, Liu P (2023) Biosynthesis of metal nanoparticles: bioreduction and biomineralization. Nanotechnol Rev 12:20230170

Zhang H, Ma W, Ma H et al (2022) Spindle-like zinc silicate nanoparticles accelerating innervated and vascularized skin burn wound healing. Adv Healthc Mater 11:e2102359

Zuhrotun A, Oktaviani DJ, Hasanah AN (2023) Biosynthesis of gold and silver nanoparticles using phytochemical compounds. Molecules 28:3240

Extraction Techniques and Chromatography for Anti-aging Metabolites

11

Mostafa A. Asmaey and Masaki Kita

Abstract

The isolation and characterization of anti-aging secondary metabolites from medicinal plants are pivotal for the development of natural therapeutics targeting oxidative stress, inflammation, and cellular senescence. This chapter presents a comprehensive overview of chemical extraction techniques and chromatographic methods employed to obtain and analyze bioactive compounds—particularly flavonoids, polyphenols, alkaloids, and terpenoids—recognized for their antioxidant, anti-inflammatory, and cytoprotective properties. Conventional extraction approaches, such as maceration and Soxhlet extraction, are discussed alongside advanced methodologies, including ultrasound-assisted extraction (UAE), microwave-assisted extraction (MAE), and supercritical fluid extraction (SFE), each evaluated for their efficiency in enhancing metabolite yield and preserving structural integrity. The integration of chromatographic platforms—thin-layer chromatography (TLC), high-performance liquid chromatography (HPLC), gas chromatography-mass spectrometry (GC-MS), and liquid chromatography-tandem mass spectrometry (LC-MS/MS)—is explored in relation to their resolution, sensitivity, and applicability to complex phytochemical matrices. Emphasis is placed on method optimization to improve extraction selectivity and chromatographic performance, thereby facilitating the reliable identification and purification of pharmacologically active metabolites. This chapter aims to guide researchers in phytochemistry, pharmacognosy, and natural product drug discovery by outlining strategic methodologies for the efficient isolation and characterization of anti-aging plant-derived compounds.

M. A. Asmaey (✉)
Department of Chemistry, Faculty of Science, Al-Azhar University, Assiut, Egypt
e-mail: asmaey.mostafa.2142@azhar.edu.eg

M. Kita
Department of Applied Biosciences, Graduate School of Bioagricultural Sciences, Nagoya University, Nagoya, Japan

S. Gantait et al. (eds.), *Medicinal Plants with Anti-Ageing Properties*,
https://doi.org/10.1007/978-981-95-8064-4_11

Keywords

Anti-aging metabolites · Chromatographic analysis · Extraction techniques · Medicinal plants · Natural product isolation · Phytochemicals

11.1 Introduction

Aging is a multifaceted biological process marked by a gradual decline in physiological functions, increased susceptibility to chronic diseases, and ultimately, mortality. Far from being a simple accumulation of wear and tear, aging involves a constellation of interconnected cellular and molecular events—including genomic instability, telomere attrition, epigenetic drift, mitochondrial dysfunction, and chronic low-grade inflammation. These hallmarks contribute to the breakdown of homeostasis and set the stage for age-related disorders such as neurodegeneration, metabolic syndromes, and cardiovascular disease (López-Otín et al. 2013; Maldonado et al. 2023; Sanada et al. 2025).

In response to the growing societal demand for strategies that not only extend lifespan but also preserve healthspan—the years lived in good health—scientific interest has turned toward natural products. Among these, secondary metabolites derived from medicinal plants have emerged as particularly promising (Asmaey et al. 2024a, b, c). Long valued in traditional medicine systems such as Ayurveda and Traditional Chinese Medicine, these botanicals are now under the scrutiny of modern science. Compounds like flavonoids, polyphenols, alkaloids, and terpenoids exhibit potent antioxidant, anti-inflammatory, and cytoprotective effects that directly target the molecular drivers of aging (Bjørklund et al. 2022; Das et al. 2024).

However, the therapeutic utility of these bioactive metabolites hinges not only on their inherent biological properties but also on the methods used to extract and purify them. Crude plant extracts are chemically diverse, often containing hundreds of constituents, many of which may be inactive or antagonistic. To harness the pharmacological potential of these metabolites, selective extraction and precise chromatographic purification are essential. Efficient extraction allows for the targeted recovery of desired compounds while minimizing degradation and contamination (Kumar et al. 2023; Montaño-Pérez and Yáñez-Ibarra 2025). Sophisticated chromatographic techniques—ranging from thin-layer chromatography (TLC) to high-resolution platforms like HPLC, LC-MS, and GC-MS—enable the isolation, identification, and quantification of these active compounds with high precision (Zhang et al. 2018; Asmaey 2024e).

This chapter provides a comprehensive overview of the chemical extraction and chromatographic strategies employed in the isolation of anti-aging metabolites from medicinal plants. Section 11.2 introduces the major classes of phytochemicals implicated in anti-aging effects. Sections 11.3, 11.4, and 11.5 review both conventional and advanced extraction techniques, such as maceration, Soxhlet, ultrasound-assisted extraction (UAE), microwave-assisted extraction (MAE), and supercritical fluid extraction (SFE). Following sections explore chromatographic methods, from rapid screening techniques like TLC to advanced hyphenated systems including HPLC, GC-MS, and LC-MS/MS. The subsequent section discusses strategies for optimizing yield, purity, and bioactivity throughout the extraction and purification process. Next, real-world case studies illustrate how these tools are applied to high-value plants such as *Withania somnifera* (ashwagandha), *Curcuma longa* (turmeric), and *Phyllanthus emblica* (amla). The chapter concludes in Sects. 11.6 and 11.7 with a discussion of current challenges and future directions.

In integrating phytochemistry, pharmacognosy, and analytical science, this chapter aims to serve as a practical and critical resource for researchers and professionals working in the expanding field of natural anti-aging therapeutics.

11.2 Anti-aging Secondary Metabolites from Medicinal Plants

Medicinal plants produce a vast array of secondary metabolites, which are organic compounds not directly involved in the normal growth, development, or reproduction of the organism, but often play crucial roles in defense, signaling, and adaptation to environmental stress (Asmaey et al. 2024a, b, c, d). Many of these compounds possess remarkable pharmacological activities, including potent antioxidant, anticancer, and anti-aging properties. Their diverse chemical structures and mechanisms of action make them prime targets for isolation and characterization in the development of natural anti-aging agents (Asmaey et al. 2023; Xie et al. 2024; Eldeen et al. 2025). This section details the major classes of these bioactive compounds and their relevance to anti-aging.

11.2.1 Polyphenols: Diversity, Properties, and Sources

Polyphenols, a diverse and vital class of plant secondary metabolites characterized by the presence of multiple phenolic hydroxyl groups, include flavonoids, phenolic acids, lignans, and stilbenes. Their structural variety imparts a broad spectrum of biological activities, making them essential targets in anti-aging research.

Flavonoids represent one of the most ubiquitous and diverse groups of plant phenolic compounds, characterized by a C6-C3-C6 carbon skeleton, consisting of two benzene rings (A and B) linked by a three-carbon chain that usually forms a heterocyclic oxygen-containing ring (C) (Asmaey et al., 2024b, c, d). Based on the oxidation level of the C-ring and the position of the B-ring attachment, flavonoids are classified into several subclasses, including flavones, flavonols, flavanones, flavanonols, isoflavones, anthocyanidins, and chalcones (Asmaey et al., 2024b, c).

Flavonoids are renowned for their potent antioxidant and anti-inflammatory activities, which are central to their anti-aging effects. Their chemical structure, particularly the presence of multiple hydroxyl groups on the aromatic rings, enables them to directly scavenge reactive oxygen species (ROS) and reactive nitrogen species (RNS), chelate metal ions that catalyze free radical formation, and upregulate endogenous antioxidant enzymes (e.g., superoxide dismutase, catalase, glutathione peroxidase) (Hossain et al. 2025). By mitigating oxidative stress, flavonoids protect cellular components like DNA, proteins, and lipids from damage, thereby preventing premature cellular senescence and maintaining cellular function (Liu et al. 2021).

Moreover, flavonoids can modulate various inflammatory pathways by inhibiting pro-inflammatory enzymes such as cyclooxygenase (COX) and lipoxygenase (LOX), and by suppressing the production of pro-inflammatory cytokines (e.g., TNF-α, IL-6, IL-1β) and chemokines (Al-Khayri et al. 2022; Singh et al. 2022). Chronic low-grade inflammation, often termed inflammaging, is a significant contributor to age-related diseases and tissue degeneration. By reducing inflammation, flavonoids (like quercetin, luteolin, etc.) inhibit pro-inflammatory cytokines, oxidative stress, and downregulate NF-κB, leading to reduced chronic inflammation and help preserve tissue integrity and function (Singh et al. 2022; Jomova et al. 2025).

Certain flavonoids, such as those found in *Centella asiatica* (asiaticoside, madecassoside) and *Hibiscus*, have been shown to inhibit collagenase and elastase enzymes, thus preventing the degradation of collagen and elastin fibers in the skin (Kumar et al. 2025). Furthermore, some flavonoids can stimulate collagen synthesis, contributing to skin elasticity and reducing the appearance of wrinkles (Izumi et al. 2007; Moraes et al. 2009). For instance, a 12-week oral regimen of 40 mg soy isoflavone aglycones led to noticeable improvements in fine lines and skin firmness among middle-aged Japanese women (Izumi et al. 2007). In contrast, applying genistein topically for 24 weeks did not outperform estradiol—and actually proved less effective—in enhancing epidermal thickness as well as increasing dermal papillae, fibroblast count, and blood vessels in postmenopausal women (Moraes et al. 2009).

Additionally, flavonoids can absorb ultraviolet (UV) radiation, acting as natural sunscreens, and reduce UV-induced oxidative damage to skin cells (Li et al. 2020; Eldeen et al. 2025). In a recent study, a crude extract of *Salsola imbricata*—found to be especially rich in phenolic acids and naringenin (40.42 μg/mL) as the primary flavonoid—has shown remarkable power in reducing histological and ultrastructural damage caused by UV-C radiation in earthworm models. These findings suggest that the high flavonoid content, particularly naringenin, plays a key role in counteracting the harmful effects of UV radiation and protecting cellular structure and function. Some flavonoids have demonstrated the ability to inhibit cellular senescence or promote the clearance of senescent cells, which accumulate with age and contribute to tissue dysfunction (Mbara et al. 2022).

Flavonoids with anti-aging properties are abundantly found across a variety of medicinal plants, each contributing unique bioactive compounds. For instance, *amla* (*P. emblica*), widely recognized in traditional medicine, is rich in potent antioxidants such as gallic acid, ellagic acid, and quercetin derivatives. These compounds confer strong antioxidative and anti-inflammatory effects that help delay cellular aging and skin degeneration (Chaikul et al. 2021). *Ginkgo biloba*, another well-studied botanical, contains ginkgo flavone glycosides, including quercetin, kaempferol, and isorhamnetin, which have demonstrated protective effects against oxidative stress and support cognitive health during aging (Smith and Luo 2004).

C. asiatica is notable for its triterpenoid saponins like asiaticoside and madecassoside, which often act synergistically with co-occurring flavonoids to enhance skin regeneration and reduce signs of aging (Brinkhaus et al. 2000). Hibiscus species are rich sources of anthocyanins, a flavonoid subclass with powerful antioxidant activity, as well as other phenolic compounds that protect against environmental stressors (Widowati et al. 2017; Li et al. 2020). Lastly, *turmeric* (*C. Longa* Linn), although predominantly valued for its curcuminoids, also contains various flavonoids that contribute to its overall anti-inflammatory and skin-protective effects (Anggraeni et al. 2025). Together, these medicinal plants and several others represent promising natural sources of flavonoids with the potential to mitigate aging processes at the cellular and tissue levels.

Phenolic acids, such as hydroxybenzoic acids (gallic acid, protocatechuic acid) and hydroxycinnamic acids (caffeic acid, ferulic acid, *p*-coumaric acid), are potent antioxidants that inhibit enzymes involved in inflammation and carcinogenesis (Asmaey et al. 2023; Hossain et al. 2025). Lignans, dimers of phenylpropanoid units, possess antioxidant, anti-inflammatory, and phytoestrogenic effects that may alleviate hormone-related aging imbalances (Domínguez-López et al. 2020). Stilbenes, notably resveratrol found in grapes and *Polygonum cuspidatum*, activate sirtuin pathways that mimic caloric restriction, promoting longevity and lifespan extension in various models (Baur and Sinclair 2006; Liu et al. 2021).

Collectively, polyphenols contribute to anti-aging by enhancing mitochondrial biogenesis and efficiency, modulating longevity-associated gene expression, stimulating autophagy to clear cellular debris, and fostering gut microbiota health as emerging key players in systemic aging (Baur and Sinclair 2006; Davinelli et al. 2013).

11.2.2 Alkaloids: Characteristics and Anti-aging Roles

Alkaloids are nitrogen-containing heterocyclic compounds with diverse structures classified by amino acid precursors or ring systems, such as indole, quinoline, isoquinoline, and tropane alkaloids. This chemical diversity underpins their broad spectrum of biological activities, many of which have relevance to anti-aging. Numerous alkaloids demonstrate potent antioxidant and anti-inflammatory properties, essential for mitigating oxidative stress and chronic inflammation—key contributors to aging. For example, bacosides from *Bacopa monnieri* scavenge free radicals and regulate antioxidant enzymes, thereby protecting cells from oxidative damage (Banerjee et al. 2021; Das et al. 2024). Neuroprotective alkaloids facilitate cognitive health by enhancing neurotransmission, shielding neurons from oxidative insults, and promoting neural plasticity, which preserves brain function with advancing age (Utpal et al. 2025).

Ashwagandha (*W. somnifera*) contains withanolides and alkaloids such as somniferine that serve as adaptogens, helping the body resist stress—a significant contributor to accelerated aging. The adaptogenic properties of these alkaloids promote physiological resilience and support homeostasis (Kuśmierska et al. 2024). Similarly, ginseng's alkaloid content complements its ginsenosides, imparting combined adaptogenic, anti-fatigue, and immunomodulatory effects, enhancing resilience against age-related stressors (Lu et al. 2021).

Alkaloids additionally inhibit advanced glycation end products (AGEs), which accumulate with aging and promote tissue damage and inflammation. Some alkaloids from Piper species have been identified as senomodulators, which regulate cellular senescence and alleviate age-associated diseases. Many alkaloids impact inflammatory signaling pathways, including NF-κB and MAPK, effectively dampening chronic inflammation and oxidative damage linked to aging (Freitas et al. 2020).

11.2.3 Other Key Metabolites (e.g., Terpenoids, Saponins, Glycosides, Vitamins)

Beyond the major classes, several other types of secondary metabolites and essential nutrients contribute significantly to the anti-aging properties of medicinal plants.

Terpenoids, derived from isoprene units, range widely from monoterpenes and sesquiterpenes (in essential oils like limonene and menthol, with antioxidant, anti-inflammatory effects) to triterpenes (saponins like ginsenosides and asiaticosides that promote collagen synthesis and wound healing) (Kim et al. 2011; Liu et al. 2022). Tetraterpenes such as carotenoids (β-carotene, lycopene) provide potent antioxidant protection for skin and eyes (Tufail et al. 2024). Furthermore, terpenoids as a class reduce oxidative stress and inflammation, and promote cellular repair, thereby supporting youthful tissue maintenance (Raza et al. 2024).

Additional metabolites vital for anti-aging include saponins with adaptogenic and immune-boosting effects found in ginseng and *B. monnieri* (Das et al. 2024); glycosides, which enhance solubility and bioavailability of flavonoids and terpenoids (Chen et al. 2022); and vitamins such as vitamin C from amla, essential for collagen synthesis and antioxidation (Talreja et al. 2021), and vitamin E, which protects cellular membranes and skin integrity from oxidative damage (Packer and Valacchi 2002). Plant polysaccharides, like those in *Aloe vera*, enhance immunity, gut health, and skin hydration (Liang et al. 2021).

11.3 Extraction Techniques for Anti-aging Metabolites

11.3.1 Principles of Chemical Extraction Techniques

The extraction of bioactive metabolites from plant matrices is the critical first step in natural product research. It involves separating the desired metabolites from the insoluble plant material using a suitable solvent system. The efficiency of this process significantly impacts the yield, purity, and ultimately, the bioactivity of the isolated compounds. A thorough understanding of the underlying principles and influencing factors is essential for developing effective extraction protocols.

11.3.1.1 Factors Influencing Extraction Efficiency

Several interconnected factors dictate the efficiency of chemical extraction from plant materials. Optimizing these parameters is crucial for maximizing the recovery of target anti-aging metabolites while minimizing the co-extraction of undesirable compounds.

Nature of the Plant Matrix The extraction of anti-aging compounds from plants is significantly influenced by the inherent nature of the plant matrix. The specific physical and chemical properties of the plant material, including the part used—whether leaves, roots, bark, or fruits—affect the efficiency of the extraction process (Abubakar and Haque 2020; Bhadange et al. 2024). Key factors such as the cellular structure, the composition of the cell wall (comprising cellulose, hemicellulose, and lignin), and the presence of protective layers like waxes or lipids can either facilitate or impede solvent penetration into the plant tissues (Abubakar and Haque 2020). For instance, woody stems with robust, lignified structures often necessitate finer grinding and may require more rigorous extraction conditions to release targeted compounds effectively. In comparison, soft leaves, with less dense cellular architecture, typically allow solvents to penetrate more readily, enabling easier recovery of bioactive molecules. Understanding these matrix characteristics is crucial for optimizing extraction protocols to maximize yield and preserve the bioactivity of anti-aging phytochemicals.

Particle Size of Plant Material The particle size of plant material is a critical factor influencing the efficiency of extracting anti-aging compounds. By reducing particle size through grinding or milling, the surface area exposed to the solvent is increased, which facilitates better solvent penetration and enhances mass transfer of the bioactive compounds into the extraction medium (Azwanida 2015). This results in improved extraction yields and more efficient recovery of target phytochemicals. However, excessively fine grinding may cause particle compaction, which restricts solvent flow and limits effective percolation. Additionally, the mechanical action during fine grinding can generate heat, potentially leading to the degradation of heat-sensitive (thermolabile) compounds. Therefore, determining an optimal particle size tailored to the specific plant matrix and desired compounds is essential to balance enhanced extraction efficiency with preservation of phytochemical integrity (Azwanida 2015; Abubakar and Haque 2020). This balance is vital to maximize both yield and quality in natural product extraction processes.

Solvent Type and Polarity The choice of solvent type and its polarity is crucial for effectively extracting anti-aging compounds from plants. Following the principle of like dissolves like, the polarity of the solvent must align with that of the target compounds to ensure efficient dissolution and extraction (Abubakar and Haque 2020). Polar solvents such as water, methanol, and ethanol are ideal for extracting highly polar compounds like polyphenols and glycosides, whereas non-polar solvents like hexane and chloroform are better suited for less polar compounds, including some terpenoids and fatty acids Fig. 11.1 (Abubakar and Haque 2020). Solvent mixtures are commonly used to extend the spectrum of extractable phytochemicals or to selectively isolate specific metabolite classes (Cao et al. 2025).

Extraction Temperature Extraction temperature significantly influences the solubility, diffusivity, and viscosity of the solvent, as well as the permeability of the plant cell walls. Generally, higher temperatures increase metabolite solubility and solvent diffusion, thereby enhancing extraction yield (Cao et al. 2025). However, excessive heat can degrade thermolabile compounds such as vitamin C and certain anthocyanins and may lead to the co-extraction of undesirable impurities (Abubakar and Haque 2020). Hence, optimizing temperature is essential to maximize yield without compromising the stability and purity of the anti-aging compounds.

Extraction Time Extraction time governs the extent to which bioactive compounds diffuse from the plant into the solvent. Initially, extraction proceeds rapidly due to the sharp concentration gradient but slows as equilibrium is approached (Bhadange et al. 2024). Prolonged extraction times, especially under high temperatures, may cause degradation of sensitive compounds, increased energy consumption, and extraction of impurities. Optimal time balances yield efficiency with compound integrity (Zhang et al. 2018).

Fig. 11.1 Sequential extraction workflow for anti-aging metabolites. (Source: Unpublished diagram of Mostafa Asmaey)

Solvent-to-Solid Ratio The solvent-to-solid ratio, which defines the volume of solvent per unit weight of plant material, is vital to ensure full immersion and maintain a favorable concentration gradient for mass transfer. Increasing the ratio typically improves yield up to a saturation point, after which gains diminish, and operational costs rise due to solvent handling and evaporation (Abubakar and Haque 2020).

Agitation or Mixing Agitation methods like stirring, shaking, or sonication enhance extraction by improving contact between solvent and plant surfaces and disrupting boundary layers around particles (Azwanida 2015; Elsoghiar et al. 2024). This leads to shorter extraction times and higher yields, especially in conventional extraction techniques such as maceration.

Optimizing these interconnected parameters is fundamental to maximizing the recovery and quality of anti-aging phytochemicals from botanical sources.

11.3.1.2 Solvent Selection and Optimization

The choice of extraction solvent is paramount and depends heavily on the chemical nature of the target anti-aging metabolites and the specific plant matrix. Solvents are typically classified by their polarity (Table 11.1), which dictates their ability to dissolve different types of compounds.

11.3.1.3 Optimization Strategies for Solvent Selection

- *Single Solvent Extraction:* Using a single solvent (e.g., 80% ethanol) is common for initial screening or when targeting a broad range of moderately polar compounds (Maes et al. 2012).
- *Sequential Extraction:* Involves using solvents of increasing polarity to fractionate the extract. For example, a plant material might first be extracted with hexane (to remove lipids), followed by ethyl acetate (for semi-polar compounds like flavonoid aglycones), and finally with ethanol or water (for polar compounds like glycosides and polyphenols) (Zhang et al. 2018). This approach helps in preliminary purification and enrichment of specific compound classes.
- *Binary or Ternary Solvent Mixtures:* Blending solvents with different polarities (e.g., ethanol-water mixtures) can often achieve higher extraction yields for a wider range of compounds or selectively extract specific metabolites by fine-tuning the overall polarity of the solvent system (Boo 2022). For instance, a mixture of ethanol and water is commonly used for extracting polyphenols and flavonoids due to its ability to dissolve both moderately polar and highly polar compounds (Darwin et al. 2025).
- *Green Solvents:* With increasing environmental concerns, there is a growing trend towards using green solvents, such as water, ethanol, supercritical CO_2, and deep eutectic solvents (DESs), which are safer, less toxic, and more environmentally friendly (Boonpisuttinant et al. 2023; Jovanović et al. 2025).

Table 11.1 Correlation between solvent polarity index and the selective extraction of anti-aging phytochemicals

Solvent type	polarity index	Examples	Typical metabolites extracted	Advantages	Disadvantages
Non-polar	0.0–2.0	Hexane, petroleum ether, chloroform	Lipids, waxes, chlorophyll, non-polar terpenoids (e.g., carotenoids)	Good for defatting, selective for non-polar compounds	Poor for polar anti-aging compounds, often flammable
Medium polar	2.1–5.0	Ethyl acetate, dichloromethane	Flavonoids (aglycones), alkaloids, some terpenoids	Good selectivity for semi-polar compounds	Can be toxic, volatile
polar	5.1–9.0	Ethanol, methanol, acetone, water	Polyphenols, flavonoid glycosides, saponins, alkaloid salts, glycosides, vitamin C	High extraction efficiency for many bioactive compounds, ethanol/water is food-grade	Can extract many unwanted polar compounds (sugars, proteins), water can lead to microbial growth

Note: Polarity index is an approximate scale and can vary slightly depending on the source. This table is a general guide

The careful selection and optimization of extraction parameters and solvent systems are paramount to obtaining high-quality extracts enriched in anti-aging metabolites, which are then ready for further purification and analysis by chromatographic techniques.

11.3.2 Conventional Extraction Techniques

Conventional extraction methods, often rooted in historical practices, form the bedrock of natural product isolation. While some may appear rudimentary compared to modern counterparts, their simplicity, cost-effectiveness, and ease of scalability make them indispensable, particularly for initial large-scale extractions or when dealing with robust plant materials. These techniques primarily rely on the principle of solid-liquid extraction, where the desired compounds diffuse from the solid plant matrix into a liquid solvent.

11.3.2.1 Maceration

Principle and Procedure Maceration is a simple, yet effective extraction technique commonly used to obtain bioactive compounds from plants. The solvent penetrates plant cells, dissolves soluble constituents, and these diffuse into the bulk solvent driven by the concentration gradient (Hidayat and Wulandari 2021). Occasional agitation enhances solvent contact and extraction efficiency. Typically, dried and coarsely powdered plant material is soaked in the chosen solvent and left to incubate at room temperature for hours to weeks, depending on the plant and target compounds. After incubation with intermittent stirring, the liquid extract is separated from the solid residue and concentrated by evaporation (Asmaey et al. 2024e).

Advantages and Limitations This method's advantages include operational simplicity, low equipment needs, suitability for thermolabile compounds, and easy scalability. However, maceration's longer extraction times, relatively low efficiency, high solvent usage, potential microbial contamination, and lack of selectivity limit its industrial application compared to advanced techniques (Cao et al. 2025).

Applications for Anti-aging Metabolites Despite its drawbacks, maceration remains valuable for extracting delicate anti-aging phytochemicals, such as withanolides from *Ashwagandha* (Kuśmierska et al. 2024) and vitamin C and hydrolyzable tannins from amla (Sharafzadeh 2013), preserving their bioactivity under gentle conditions. Recent studies comparing maceration with ultrasonic extraction highlight maceration's efficiency in yielding antioxidant compounds, albeit at longer durations. For example, a comparative study on green coffee beans showed maceration-derived serums retained active antioxidants and acceptable physicochemical properties, though ultrasonic extraction provided superior microbial safety and extraction yield (Firdaus et al. 2025). Another investigation optimizing maceration parameters in *Catharanthus roseus* demonstrated that temperature, solvent composition, and extraction time significantly influence total phenol content and antioxidant activity (Caeique et al. 2020).

11.3.2.2 Soxhlet Extraction

Principle and Procedure Soxhlet extraction, developed by Franz von Soxhlet in 1879, is a continuous solid-liquid extraction technique that has stood the test of time due to its efficiency, especially for extracting lipids and other natural products from solid plant materials (Asmaey et al. 2024a, b). The method relies on the continuous recirculation of solvent through the sample, ensuring thorough and exhaustive extraction. The setup involves a round-bottom flask containing the solvent, a Soxhlet extractor holding the plant material in a porous thimble, and a condenser at the top. When heated, the solvent vapor rises, condenses, and drips onto the plant material, dissolving the target compounds. Once the solvent in the extraction chamber reaches the siphon, it automatically flows back into the boiling flask, carrying the extracted compounds with it. This cycle repeats continuously until extraction is complete, typically indicated by colorless solvent dripping from the siphon (López-Bascón and De Castro 2020). In practice, the finely powdered, dried plant sample is placed in the thimble, assembled into the apparatus, and the solvent is heated to its boiling point. The continuous cycle effectively transfers bioactive compounds into the solvent without manual intervention, and the resultant extract is concentrated by evaporating the solvent, usually under reduced pressure.

Advantages and Limitations Soxhlet extraction offers several advantages; it provides high extraction efficiency by repeatedly exposing the material to fresh solvent, uses less solvent per extraction cycle than maceration, operates largely automatically, and is particularly effective for non-volatile or semi-volatile compounds (Alara et al. 2021; Bhadange et al. 2024). However, it does have limitations such as long extraction times, exposure of extracts to elevated temperatures—which may degrade thermolabile compounds like certain vitamins and polyphenols—and lack of selectivity, often extracting unwanted impurities like chlorophyll and waxes. Additionally, the method consumes relatively high energy due to continuous heating and involves using large volumes of organic solvents, raising environmental and safety concerns (López-Bascón and De Castro 2020; Alara et al. 2021).

Applications for Anti-aging Metabolites This technique finds applicability in extracting various stable anti-aging phytochemicals. For instance, Soxhlet extraction efficiently recovers curcuminoids from turmeric (*C. longa*) using solvents such as ethanol, methanol, or acetone. These polyphenolic compounds are relatively heat-stable, making Soxhlet extraction suitable for high yields (Waseem et al. 2023). Similarly, comprehensive extraction of ginsenosides from ginseng (*Panax ginseng*) can be achieved using ethanol or methanol Soxhlet extraction, recovering a broad spectrum of polar and semi-polar adaptogenic compounds; however, attention must be paid to the temperature sensitivity of specific ginsenosides (Jegal et al. 2019). Rosemary (*Rosmarinus officinalis*), rich in phenolic diterpenes like carnosic acid and rosmarinic acid, is also effectively processed by Soxhlet extraction, leveraging its continuous solvent exposure to achieve substantial yields of potent antioxidant compounds (Waseem et al. 2023).

Together with maceration, Soxhlet extraction represents a foundational approach widely used for natural product isolation. While effective and reliable, their drawbacks in time, energy, and solvent use have motivated the advancement of more selective, rapid, and environmentally friendly extraction methods, the discussion of which follows.

11.3.3 Advanced Extraction Techniques

While conventional extraction methods have served natural product chemistry well for centuries, their limitations—such as long extraction times, high solvent consumption, and potential degradation of thermolabile compounds—have spurred the development of more efficient and environmentally friendly advanced techniques. These modern methods often leverage physical forces (like ultrasound or microwaves) or unique solvent properties (like supercritical fluids) to enhance mass transfer, reduce extraction time, and improve selectivity.

11.3.3.1 Ultrasound-Assisted Extraction (UAE)

Principle and Mechanism Ultrasound-assisted extraction (UAE) is a modern approach that uses high-frequency sound waves, generally between 20 and 100 kHz, to accelerate the recovery of bioactive constituents from plant tissues. The technique relies on acoustic cavitation, where sound waves passing through a solvent generate microscopic bubbles that grow and then collapse violently. This collapse produces localized zones of high temperature and pressure and generates strong mechanical forces. These conditions rupture cell walls, increase solvent penetration into the matrix, and promote faster mass transfer of intracellular compounds into the extraction medium (Mostafa et al. 2026). The efficiency of UAE depends on several parameters, including ultrasound frequency and intensity, solvent properties, extraction time, temperature, and the particle size of the plant material (Chemat and Khan 2011; Wen et al. 2018).

Advantages and Limitations Compared with conventional techniques, UAE reduces solvent usage and extraction time while improving energy efficiency. It is particularly suitable for heat-sensitive compounds, as it often operates at relatively mild temperatures, thereby protecting delicate metabolites such as polyphenols, vitamins, and terpenoids relevant to anti-aging applications. The method is relatively simple, cost-effective, and adaptable to both laboratory and industrial scales. Nevertheless, prolonged sonication or excessive power can damage sensitive molecules, and cavitation may generate free radicals that compromise extract stability. Moreover, the uniformity of extraction may be influenced by the sample's geometry and solvent type (Vilkhu et al. 2008; Chemat and Khan 2011).

Applications for Anti-aging Metabolites UAE has been applied widely to extract antioxidant and anti-aging phytochemicals. For instance, enhanced yields of carnosic and rosmarinic acids from *R. officinalis* and catechins from *Camellia sinensis* have been reported, with extracts showing stronger antioxidant activity compared to conventional methods (Pizani et al. 2024; Prawira-Atmaja and Puangpraphant 2025). The method is also effective for obtaining triterpenoids from *C. asiatica*, which support collagen synthesis, and flavonoids from *G. biloba*, known for neuroprotective effects. Given its eco-friendly nature, compatibility with food-grade solvents, and ability to preserve metabolite activity, UAE is increasingly adopted in pharmaceutical, nutraceutical, and cosmetic industries focusing on anti-aging formulations (Chavan et al. 2022).

11.3.3.2 Microwave-Assisted Extraction (MAE)

Principle and Mechanism Microwave-assisted extraction (MAE) is a modern, efficient technique that employs microwave radiation to enhance the extraction of bioactive compounds from plant materials. The process is based on the interaction of microwave energy with polar molecules in the solvent and plant matrix. This energy causes dipolar rotation and ionic conduction, resulting in rapid, internal heating of the plant cells and solvent. The increase in temperature and pressure inside the cells leads to rupture of cell walls and membranes, facilitating the release of intracellular compounds into the surrounding solvent (Zhang et al. 2011). MAE often uses closed or open vessel systems, with the closed system allowing for controlled high temperatures and pressures, which can further improve extraction efficiency. Key parameters influencing the process include solvent polarity, microwave power, extraction time, temperature, and particle size of the plant material (Sadeghi et al. 2017).

Advantages and Limitations MAE offers several significant advantages over conventional extraction methods. It drastically reduces extraction time from hours or days to minutes, while consuming less solvent and energy, making it more sustainable and cost-effective. The rapid heating improves solvent penetration and mass transfer, often leading to higher yields and better preservation of thermolabile anti-aging compounds compared to prolonged high-temperature methods. In addition, MAE can be automated and scaled from laboratory to industrial use (Akhtar et al. 2019).

However, MAE has limitations such as the potential degradation of sensitive compounds if temperature and power are not carefully controlled. The equipment cost is higher than that of simpler methods like maceration. The penetration depth of microwaves is limited, which may affect extraction uniformity in larger samples. Also, certain solvents and plant matrices might not efficiently absorb microwave energy, reducing effectiveness. Safety concerns arise in closed-vessel systems due to pressure buildup, necessitating precise monitoring (Azwanida 2015).

Applications for Anti-aging Metabolites MAE has been successfully applied to extract a variety of anti-aging metabolites, including polyphenols, flavonoids, essential oils, and glycosides. For instance, MAE efficiently recovers phenolic antioxidants from medicinal plants such as turmeric, green tea, and *C. asiatica* by enhancing yield and shortening extraction times (Yingngam et al. 2020; Saleem et al. 2024). It is particularly valuable for preserving heat-sensitive vitamins and maximizing the extraction of collagen-promoting triterpenoids. Studies demonstrate MAE's effectiveness in extracting flavonoids and saponins from *G. biloba* and *C. asiatica*, key plants known for their anti-aging properties (Sati et al. 2019; Sabaragamuwa and Perera 2023). The technique's ability to optimize solvent use and extraction parameters makes it an excellent choice for developing high-quality, bioactive-rich anti-aging extracts in the pharmaceutical and cosmetic industries.

11.3.3.3 Supercritical Fluid Extraction

Principle of Supercritical Fluids (SFE) Supercritical fluid extraction (SFE) is a sophisticated technique that uses a fluid—most often carbon dioxide (CO_2)—in its supercritical state to draw out valuable compounds from plants. When CO_2 is kept above its critical temperature (31.1 °C) and pressure (73.8 bar), it becomes a supercritical fluid. This unique state allows CO_2 to behave both like a liquid (with the ability to dissolve substances effectively due to its density) and like a gas (with the diffusivity and low viscosity needed to easily penetrate even compact, porous plant materials) (McHugh et al. 2013). A major advantage here is the tunable solvent power—researchers can make the fluid more or less effective as an extractor simply by adjusting the temperature and pressure. This flexibility enables targeted, selective extractions for different classes of anti-aging metabolites, all within the same system (Montalbán and Víllora 2022).

Advantages and Limitations SFE stands out for its ability to produce exceptionally pure extracts: once the process is done, lowering the pressure releases the CO_2 as a gas, leaving no trace of solvent in the final product. This makes SFE especially attractive for pharmaceuticals, nutraceuticals, and cosmetics, which demand solvent-free purity (Geow et al. 2021). The method also minimizes the risk of thermal degradation, as the process runs at moderate temperatures—ideal for heat-sensitive anti-aging compounds. Additionally, CO_2 is safe, non-flammable, and can often be recycled within the system, making SFE a green, sustainable option that generates little environmental waste (McHugh et al. 2013).

However, there are a few caveats. Setting up an SFE system requires specialized, high-pressure equipment, which can be costly. For more polar compounds, pure CO_2 might not be sufficient, so small amounts of polar co-solvents (modifiers) like ethanol are often added to boost efficiency (Kadri et al. 2021).

11.3.3.4 Key Optimization Parameters

The effectiveness of SFE depends heavily on carefully adjusting multiple parameters. Pressure is a major factor—raising the pressure increases the density of the supercritical CO_2, which enhances its solvation power and typically improves extraction yield. Temperature influences both the density and diffusion rate of the fluid, creating a complex balance; for example, at constant pressure, increasing temperature may reduce solvent density but increase diffusivity, so optimal conditions require careful tuning (Ahangari et al. 2021). The flow rate of the supercritical fluid must be set to maximize contact with plant material, efficiently driving mass transfer without wasting time. Incorporating modifiers or co-solvents—usually small amounts of polar solvents like ethanol—can extend SFE's capability to dissolve more polar compounds that pure CO_2 alone struggles to extract. Finally, the particle size of the plant matrix should be optimized so that particles are fine enough to expose more surface area but not so fine that they impede solvent flow or cause channeling issues, which can reduce extraction efficiency (Abou Elmaaty et al. 2022).

Applications for Anti-aging Metabolites SFE is highly effective for extracting gingerols, shogaols, and essential oil components from ginger roots, producing clean, potent extracts without solvent residues (Samota et al. 2024). Because SFE can be carefully tuned, it's possible to focus on extracting just the most active or desirable components, important in maximizing the bioactivity of anti-aging extracts. Additionally, *B. monnieri*, a plant that contains bacosides, triterpenoid saponins linked to cognitive support and antioxidant effects. Though somewhat polar, bacosides can be successfully isolated with SFE when CO_2 is combined with a small percentage of ethanol as a cosolvent, helping not just with purity but with the preservation of delicate structures that might be lost in harsh chemical extractions (Das et al. 2024).

Beyond ginger and *B. monnieri*, SFE has been utilized with various other plants known for their ability to extract high-quality anti-aging compounds. For example, *Matricaria chamomilla* (chamomile) has been successfully processed using SFE to obtain extracts rich in phenols, flavonoids, and essential oils with notable antioxidant and antimicrobial properties, making it valuable for cosmetic and therapeutic applications that target skin aging (Pastare et al. 2023). Similarly, *Bidens pilosa*, a medicinal plant known for its anti-inflammatory and antioxidant compounds, has been extracted via supercritical CO_2 under optimized pressure and temperature conditions, yielding bioactive oils with strong potential to combat oxidative stress linked to aging (Paschoal et al. 2025). Additional applications include *Rheum ribes* flower, which has been subjected to ethanol-modified SFE to extract phenolic compounds with skin-protective and antioxidant effects that support anti-aging benefits (Hoseini et al. 2025).

The versatility of SFE to selectively isolate various classes of phytochemicals using co-solvents like ethanol allows efficient recovery of both non-polar and moderately polar anti-aging agents, including terpenoids, flavonoids, and saponins that are crucial in mitigating cellular aging phenomena. SFE's ability to selectively extract a wide array of plant metabolites inside a single, environmentally friendly system makes it one of the most advanced tools for producing high-quality, anti-aging botanical extracts.

11.3.4 Other Emerging Extraction Techniques

The field of natural product extraction continues to advance rapidly, with emerging techniques addressing key challenges such as improving extraction efficiency, selectivity, and sustainability. Notable among these newer methods are pressurized liquid extraction (PLE), deep eutectic solvents (DESs), and enzyme-assisted extraction (EAE), each offering distinct advantages for obtaining anti-aging compounds from plants.

Pressurized Liquid Extraction (PLE) Also called accelerated solvent extraction (ASE), it uses elevated temperatures and pressures to keep solvents in a liquid state above their normal boiling points. This allows for faster and more exhaustive extraction in shorter times and with less solvent compared to conventional methods (Bhadange et al. 2024). PLE is particularly effective for stable compounds ranging from semi-polar to polar, such as many polyphenols and flavonoids found in anti-aging botanicals. For example, PLE has been used to efficiently extract polyphenols from green tea leaves, improving yield while maintaining bioactivity (Dawidowicz and Wianowska 2005). Its ability to optimize solvent penetration under pressure reduces extraction times significantly, which is valuable for industrial-scale processes focused on sustainability.

Other examples include pressurized liquid extraction of *Moringa oleifera* leaves, yielding high concentrations of polyphenols, flavonoids, and ascorbic acid—all antioxidants with anti-aging potential. Optimization of parameters such as temperature (~150 °C), pressure (~1700 psi), and extraction time (~15 min) enabled the extraction of up to 24.6 mg gallic acid equivalents per gram of dry weight and considerable antioxidant activity, supporting the use of PLE for producing potent extracts for pharmaceutical and nutraceutical uses (Chatzimitakos et al. 2024). Similarly, coffee bean husk, a waste biomass with antioxidant polyphenols like gallic and chlorogenic acids, was efficiently processed using hydroethanolic PLE, showing superior mass yield and antioxidant activity, indicating potential for anti-aging nutraceutical ingredient development (Xu et al. 2019).

Moreover, PLE has been used to extract bioactive compounds from seaweeds, such as polysaccharides, phenolics, vitamins, and pigments, which possess antioxidant and anti-inflammatory activities beneficial in anti-aging formulations (Perez-Vazquez et al. 2023). In combination with supercritical CO_2 extraction, PLE enhances yield and economic feasibility for scaling up production of marine-derived anti-aging ingredients.

Deep Eutectic Solvents (DESs) DESs are a relatively novel and environmentally friendly class of extraction solvents that have gained considerable attention in recent years, especially for isolating anti-aging compounds from plants (Abouheif et al. 2022). DESs are typically formed by mixing two or more natural components, such as organic acids, sugars, amino acids, or choline derivatives, which interact through hydrogen bonding to produce a eutectic mixture with a melting point far lower than that of the individual components. This unique property allows DESs to remain liquid at room temperature, providing a tunable solvent system that can be tailored for different extraction targets (Ferreira and Sarraguça 2024).

DESs excel in extracting polyphenols, flavonoids, and other valuable antioxidants—key phytochemicals linked to anti-aging effects through their radical scavenging and anti-inflammatory activities. For example, DESs formulated with choline chloride and organic acids have demonstrated superior extraction of phenolics from medicinal plants like *Sideritis scardica* and *Plantago major*, leading to extracts with enhanced bioactivities, including antioxidant and antimicrobial effects, which support skin health and longevity (Guo et al. 2022; Zissi et al. 2025). Moreover, varying the hydrogen bond donor (HBD) or acceptor (HBA) components allows customization of polarity, enabling effective extraction of both hydrophilic and hydrophobic compounds. Applying DESs in combination with techniques such as ultrasound-assisted extraction further improves yield and selectivity by promoting cell wall disruption and enhancing mass transfer (Ferreira and Sarraguça 2024).

Another important benefit of DESs is the improved stability they often confer on the extracted compounds, increasing shelf life compared to conventional solvent extracts (Nahar and Thickett 2021). This stabilization is attributed to strong intermolecular interactions within the DES matrix that protect sensitive phytochemicals from oxidation or degradation. Consequently, DES-based extracts offer promising formulations for long-lasting, potent anti-aging products.

Enzyme-Assisted Extraction (EAE) EAE is an eco-friendly extraction method that utilizes specific enzymes to break down the structural components of plant cell walls, such as cellulose, hemicellulose, and pectin (Puri et al. 2012). This process enhances the release of bioactive metabolites from within the plant cells while maintaining mild conditions. The mild operating conditions of EAE (moderate temperature, near-neutral pH) are advantageous for preserving thermolabile and sensitive anti-aging compounds such as polyphenols, flavonoids, saponins, and vitamins. By doing so, EAE helps preserve sensitive anti-aging compounds, making it a valuable complement to other extraction techniques (Puri et al. 2012).

EAE has been notably successful in increasing the extraction yields of triterpenoid saponins from medicinal plants like *C. asiatica*, which are key for collagen synthesis and skin regeneration (Abu Bakar et al. 2022). Enzyme pre-treatment softens or partially hydrolyzes the matrix, allowing solvents better penetration and enhancing subsequent extraction efficiency, whether performed with water, ethanol, or other solvents.

Additionally, this approach has been applied to *P. ginseng* roots to improve the recovery of ginsenosides, bioactive saponins with prominent anti-aging and adaptogenic effects (Song et al. 2018). Optimization of enzyme type, concentration, incubation time, and temperature helps maximize yield while minimizing degradation. Furthermore, combining EAE with other emerging methods like ultrasound or microwave-assisted extraction often produces synergistic effects, reducing extraction time and solvent use, while increasing purity and potency of the extracts (Kalyniukova et al. 2021).

Together, these emerging extraction technologies represent significant progress toward more sustainable, efficient, and selective recovery of anti-aging phytochemicals. They offer complementary or alternative approaches to conventional and other advanced methods like ultrasound-assisted or supercritical fluid extraction, further expanding the toolkit available for natural product extraction. These innovations not only improve the sustainability and economics of extraction but also enable the formulation of more potent, stable, and pure extracts for health, cosmetic, and pharmaceutical applications.

11.4 Chromatographic Separation of Anti-aging Metabolites

Once anti-aging metabolites are extracted from plant material, the next crucial step involves their separation and purification from the complex mixture of co-extracted compounds. This is typically achieved through chromatography, a highly versatile and widely used analytical and preparative technique. Chromatography operates on the principle of differential partitioning of compounds between two distinct phases: a stationary phase and a mobile phase. The separation efficiency depends on variations in each analyte's affinity for these phases, leading to differences in migration rates and effective resolution of the mixture's components (Snyder et al. 2011).

11.4.1 Key Concepts: Stationary Phase, Mobile Phase, and Retention Parameters

To fully appreciate how chromatography achieves high-resolution separations, it is essential to understand the foundational concepts:

Stationary Phase This is the immobile phase within the chromatographic system, usually consisting of a solid sorbent (e.g., silica gel, alumina, or chemically modified resins) or a liquid immobilized on a solid support. In column chromatography, it is packed into a column, whereas in planar methods such as thin-layer chromatography (TLC), it forms a thin coating on a plate. The physicochemical properties of the stationary phase, including polarity, surface chemistry, and pore size, critically determine the retention behavior of analytes (Wu et al. 2022).

Mobile Phase The mobile phase is the fluid (liquid or gas) that moves through or across the stationary phase, transporting analytes. In liquid chromatography, this could be an organic solvent, an aqueous buffer, or a combination thereof; in gas chromatography, an inert carrier gas is used. The solvent composition, polarity, flow rate, and temperature strongly influence elution profiles (Coskun 2016).

Separation Mechanism The movement of analytes is governed by their comparative affinity for the stationary and mobile phases. Compounds with stronger interactions with the stationary phase migrate more slowly, whereas those with higher solubility in the mobile phase travel faster. This differential migration results in physical separation along the chromatographic medium.

Retention Parameters In TLC, the retention factor (R_f) is a characteristic value for a compound under specific chromatographic conditions. It is defined as the ratio of the distance traveled by the solute (spot) to the distance traveled by the solvent front; both measured from the origin (spotting line) (Coskun 2016). Values range between 0 and 1, where a higher R_f indicates lower retention by the stationary phase and stronger interaction with the mobile phase.

Retention Time (t_R) In high-performance liquid chromatography (HPLC) or gas chromatography (GC), retention time is the elapsed time between sample injection and the appearance of a peak in the detector. It is characteristic of a given compound under set operating conditions and can be used for compound identification when compared with standards (Asmaey 2024e).

11.4.2 Modes of Chromatography: Adsorption, Partition, Ion Exchange, and Size Exclusion

Chromatographic techniques are classified into distinct modes based on the dominant interaction mechanism between analytes and the stationary phase. Understanding these modes is essential for selecting the most effective approach for the purification of bioactive metabolites, including anti-aging compounds (Fig. 11.2).

Adsorption Chromatography One of the earliest and most widely applied chromatographic modes, adsorption chromatography operates through the differential adsorption of analytes onto a solid stationary phase such as silica gel or alumina. In normal-phase chromatography, polar analytes interact more strongly with the polar stationary phase and thus elute more slowly, whereas non-polar analytes migrate faster. The mobile phase competes with analytes for adsorption sites, and the choice of solvent polarity strongly influences resolution (Ismail 2024).

Partition Chromatography In this mode, separation depends on the partitioning of analytes between two immiscible liquid phases—one immobilized on a solid support (stationary phase) and the other flowing through the column (mobile phase). This approach underpins both normal-phase and reversed-phase liquid chromatography, with the latter (non-polar stationary phase, polar mobile phase) being the most prevalent in natural product isolation due to its compatibility with aqueous-organic solvent systems (Ismail 2024).

Ion-Exchange Chromatography Here, separation is driven by electrostatic interactions between charged analytes and oppositely charged groups on the stationary phase. Cation exchangers carry negatively charged functional groups (e.g., sulfonic

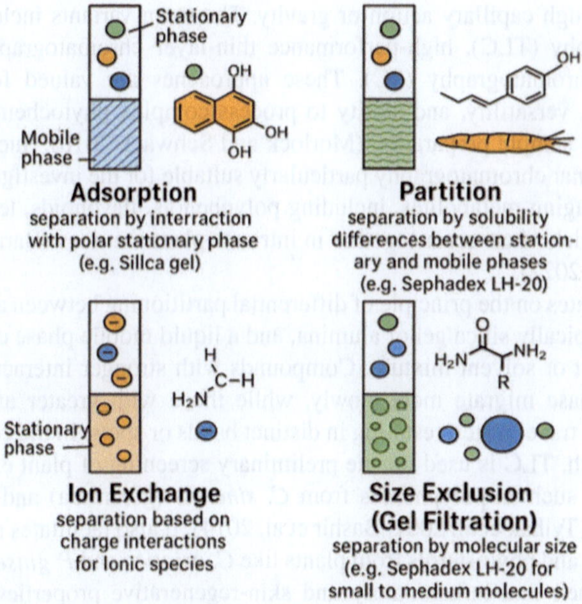

Fig. 11.2 Main chromatographic modes used in natural product separation and their mechanisms. (Source: Unpublished diagram of Mostafa Asmaey)

acids), whereas anion exchangers carry positively charged groups (e.g., quaternary ammonium). Elution is achieved by adjusting the ionic strength or pH of the mobile phase to alter analyte binding. This mode is particularly valuable for charged metabolites such as alkaloid salts and certain phenolic acids (Cummins et al. 2016).

Size-Exclusion Chromatography (SEC) Also known as gel filtration (for aqueous systems) or gel permeation chromatography (for organic systems), SEC separates molecules based on molecular size and shape. Porous stationary phases allow smaller molecules to diffuse into the pores, extending their elution time, whereas larger molecules are excluded and elute more quickly. SEC is particularly suited for macromolecular separations or desalting steps rather than for fine separation of small-molecule metabolites (Sun et al. 2004).

A clear understanding of these principles allows researchers to tailor chromatographic strategies for optimal recovery and purity of anti-aging metabolites.

11.4.3 Planar Chromatography Techniques

Planar chromatography refers to a group of analytical methods in which the stationary phase is spread as a thin layer over a flat surface, while the mobile phase migrates across it through capillary action or gravity. The main variants include thin-layer chromatography (TLC), high-performance thin-layer chromatography (HPTLC), and paper chromatography (PC). These approaches are valued for their cost-effectiveness, versatility, and ability to process complex phytochemical mixtures with minimal sample preparation (Morlock and Schwack 2010). Such characteristics make planar chromatography particularly suitable for the investigation of plant-derived anti-aging metabolites, including polyphenols, flavonoids, terpenoids, and alkaloids, which often occur together in intricate plant matrices (Barak et al. 2022; Dubois et al. 2022).

TLC operates on the principle of differential partitioning between a solid stationary phase, typically silica gel or alumina, and a liquid mobile phase composed of a single solvent or solvent mixture. Compounds with stronger interactions with the stationary phase migrate more slowly, while those with greater affinity for the mobile phase travel faster, resulting in distinct bands or spots. In the context of anti-aging research, TLC is used for the preliminary screening of plant extracts rich in antioxidants, such as polyphenols from *C. sinensis* (green tea) and *Vitis vinifera* (grape seed) (Tyihák et al. 2001; Bashir et al. 2014). It also facilitates rapid profiling of terpenoids and phytosterols from plants like *C. asiatica* and *P. ginseng*, which are valued for their anti-inflammatory and skin-regenerative properties (Song et al. 2018; Sabaragamuwa and Perera 2023). Furthermore, TLC enables monitoring of extraction efficiency and comparative chemotype analysis across plant cultivars to identify those with enhanced concentrations of bioactive metabolites.

Besides, HPTLC represents an advanced form of TLC that employs layers of smaller, uniformly sized particles, yielding higher resolution and reproducibility. The use of automated sample application, controlled development chambers, and densitometric scanning allows for semi-quantitative and quantitative assessment (Cebolla et al. 2021). In the study of anti-aging botanicals, HPTLC is employed for the quantification of compounds such as resveratrol in grape skins and curcuminoids in *C. longa* (Gandhi et al. 2024). It is also integral to the standardization of herbal formulations targeting skin health, stability testing of botanical cosmetic products, and the creation of chemical fingerprints to authenticate raw plant materials within the anti-aging sector (Lalhriatpuii 2019).

Furthermore, PC, one of the earliest planar techniques, uses cellulose fibers as the stationary phase, which retain a thin film of water. The mobile phase, typically an organic solvent or a solvent mixture, moves across the paper via capillary action, affecting separation through partitioning between these phases (Block et al. 2013). This method remains relevant for the detection of hydrophilic antioxidants such as vitamin C and phenolic acids from *Citrus* species and *Rosa canina* (Bitis et al. 2017). PC is also advantageous for field-based phytochemical screening in resource-limited settings and for the preliminary analysis of anthocyanins from berries and pigmented grains, which are known for their photoprotective and skin-brightening properties (Kowalska and Sajewicz 2022).

11.4.4 Column Chromatographic Techniques

While planar chromatographic methods such as TLC provide rapid qualitative information, column chromatography techniques are indispensable for achieving high-resolution separation, quantitative analysis, and isolation of individual compounds from complex mixtures. These methods utilize columns filled with stationary phases through which the mobile phase continuously flows, enabling precise and reproducible separations. The development of high-pressure systems has significantly advanced these techniques, revolutionizing natural product research.

High-Performance Liquid Chromatography (HPLC) HPLC is a highly sophisticated and broadly employed technique for separating, identifying, and quantifying components, especially non-volatile and thermally labile metabolites. HPLC operates on differential partitioning between a stationary phase packed inside a column and a mobile liquid phase propelled by a high-pressure pump. Unlike gravity-driven traditional liquid chromatography, HPLC uses high pressure to push the mobile phase through columns containing very fine stationary particles (typically 1.7–5 μm), resulting in enhanced separation efficiency and faster analysis times. The sample is injected into the mobile phase stream, and components separate based on their differing affinities to the stationary and mobile phases, eluting at characteristic retention times. Detectors then monitor eluents to provide both qualitative and quantitative data (Hanai 2007; Asmaey et al. 2021, 2024a, b).

The choice of HPLC mode is tailored to the chemical nature of target anti-aging metabolites. Reversed-phase HPLC (RP-HPLC) is the most commonly employed mode for natural products, utilizing a non-polar stationary phase such as C18 or C8 bonded silica, with a polar mobile phase composed of aqueous buffers and organic solvents like acetonitrile or methanol. This mode efficiently separates moderately to non-polar flavonoids, polyphenols, and terpenoids known for their antioxidant and anti-inflammatory properties (Moldoveanu and David 2022). Conversely, normal-phase HPLC (NP-HPLC) employs a polar stationary phase with a non-polar mobile phase, which is occasionally preferred for very non-polar compounds or preparative scale purifications. Additionally, hydrophilic interaction liquid chromatography (HILIC), a variant of normal-phase, is increasingly applied to highly polar anti-aging metabolites such as glycosides and highly hydroxylated polyphenols that exhibit poor retention in reversed-phase systems (Zhang et al. 2017).

Detection methods integrated with HPLC are crucial for accurate identification and quantification of anti-aging compounds. The ubiquitous UV-Visible detector is suitable for compounds possessing chromophores, including many flavonoids and phenolic acids. Photodiode array (PDA) detectors extend this capability by recording full spectra, enabling peak purity assessment and compound confirmation. Fluorescence detectors provide enhanced sensitivity for naturally fluorescent compounds or those derivatized accordingly. Other detectors, such as refractive index (RI) and evaporative light scattering detectors (ELSD), broaden detection capabilities to analytes lacking UV activity, such as certain saponins and polysaccharides frequently implicated in skin health and longevity.

Optimizing HPLC methods for anti-aging metabolite analysis requires careful consideration of stationary phase chemistry, mobile phase composition (including pH and organic solvent proportion), flow rate, temperature, and elution profile (isocratic or gradient). For instance, accurate quantification of curcuminoids in *C. longa* or anthocyanins in *H. sabdariffa* necessitates tailored mobile phases and column selections to achieve baseline separation (Ali et al. 2019; Bolat et al. 2024). Methodical optimization enhances resolution, reproducibility, and sensitivity, facilitating reliable quality control and bioactivity correlation studies.

Gas Chromatography-Mass Spectrometry (GC-MS) Beyond HPLC, GC-MS is a pivotal hyphenated technique used for volatile and semi-volatile anti-aging metabolites. GC-MS combines chromatographic separation based on volatility and interaction with a coated capillary column with mass spectrometric detection, which provides structural identification through fragmentation patterns (Asmaey 2024e). Though many polar plant metabolites require derivatization prior to GC-MS analysis, this method excels in profiling essential oils and volatile terpenoids from species like olive, *Ocimum basilicum* (basil) and *Cinnamomum* spp. (cinnamon), which are valued for their antioxidative and skin-protective properties (Stanojevic et al. 2017; Farag et al. 2022; Ahmad et al. 2023).

Liquid Chromatography–Mass Spectrometry/Mass Spectrometry (LC-MS/MS) Liquid chromatography coupled with tandem mass spectrometry (LC-MS/MS) represents one of the most advanced and versatile hyphenated analytical techniques available for the in-depth study of complex natural product mixtures (Nazeh et al. 2024). This method synergistically combines the high-resolution separation capabilities of HPLC with the exceptional sensitivity, selectivity, and structural elucidation power provided by tandem mass spectrometry (Elsoghiar et al. 2024; Nazeh et al. 2024). Such attributes render LC-MS/MS uniquely suited for the comprehensive profiling, identification, and quantification of a broad spectrum of anti-aging metabolites, especially those occurring at trace concentrations in botanical matrices.

The fundamental principle underlying LC-MS/MS involves the direct interfacing of an HPLC system with a mass spectrometer. The HPLC first achieves separation of the complex mixture's constituents, which are then ionized and introduced into the mass spectrometer for detection. Tandem mass spectrometry (MS/MS) further enhances this process by isolating specific parent ions and subjecting them to collision-induced dissociation (CID), generating characteristic fragment ions that facilitate detailed structural analysis (Asmaey 2024).

Instrumentation consists of several key components: an HPLC system provides chromatographic separation; the interface—typically electrospray ionization (ESI) or atmospheric pressure chemical ionization (APCI)—transfers analytes from liquid to gas phase under atmospheric pressure; the mass spectrometer houses the ion source, mass analyzers (e.g., triple quadrupole, quadrupole time-of-flight (Q-TOF),

Orbitrap), a collision cell for MS/MS fragmentation, and detectors to measure ion abundance. Data systems process and interpret complex chromatographic and spectral outputs, enabling robust qualitative and quantitative analyses (Asmaey et al. 2023, 2024a, b, c, d). Among ionization techniques, ESI is predominant due to its soft ionization process that preserves molecular ions intact, making it particularly effective for polar, thermolabile compounds such as flavonoids, polyphenols, glycosides, and saponins. APCI, on the other hand, is suited for less polar analytes and can accommodate higher solvent flow rates, extending the technique's applicability (Ayala-Cabrera et al. 2023).

Mass analyzers within LC-MS/MS systems vary in resolution, accuracy, and scan speed. Quadrupole analyzers excel in targeted quantification through multiple reaction monitoring (MRM). Ion traps offer multiple stages of fragmentation (MSn), useful for detailed structural investigations. TOF analyzers provide high mass accuracy and resolution beneficial in untargeted metabolomics, while Orbitrap systems deliver exceptional sensitivity and dynamic range, critical for complex mixture analyses (Li et al. 2021). The MS/MS capability significantly improves structural elucidation by fragmenting selected parent ions into product ions, whose fragmentation patterns constitute unique fingerprints that can confirm known compounds or help deduce novel structures.

LC-MS/MS finds widespread application in profiling anti-aging metabolites from diverse plant sources. For instance, broccoli (*Brassica oleracea* var. italica) contains polyphenols such as kaempferol and quercetin glycosides, along with glucosinolate precursors of sulforaphane, compounds extensively characterized and quantified by LC-MS/MS despite their low abundance in complex matrices (Salas-Millán and Aguayo 2024). Similarly, the adaptogenic properties of *W. somnifera* (ashwagandha) are attributed to a suite of withanolides (e.g., withaferin A, withanolide A), whose sensitive and selective analysis relies on LC-MS/MS methodologies (Bashir et al. 2023). Beyond these, the technique is instrumental in analyzing flavonoid glycosides and phenolic acids in hibiscus, amla, and *G. biloba*, as well as complex alkaloids and saponins such as ginsenosides from *P. ginseng* and bacosides from *B. monnieri*, where isomeric compounds and low-abundance metabolites necessitate the method's superior capabilities (Zhou et al. 2018; Ali et al. 2019; Banerjee et al. 2021; Lu et al. 2021; Hegde et al. 2024).

To provide a clearer perspective on the strengths and limitations of various chromatographic techniques commonly employed in the analysis of anti-aging metabolites, Table 11.2 summarizes key features of TLC, HPLC, GC-MS, and LC-MS/MS. The table compares these methods in terms of their separation principles, types of analytes suitable for analysis, resolution and sensitivity, identification and quantification capabilities, cost implications, sample throughput, and complexity of sample preparation. Such a comparative overview aids researchers in selecting the most appropriate analytical approach based on the specific characteristics of the anti-aging compounds under investigation and the goals of the study, whether they pertain to rapid screening, detailed profiling, or structural elucidation.

Table 11.2 Comparison of key chromatographic techniques for anti-aging metabolites

Feature	TLC	HPLC	GC-MS	LC-MS/MS
Separation principle	Adsorption/ partition on planar surface	Partition/ adsorption in column	Partition/ volatility in column	Partition in column + mass-to-charge ratio
Analyte type	Non-volatile, semi-volatile, stable	Non-volatile, thermally labile/stable	Volatile, thermally stable (or derivatized)	Non-volatile, thermally labile/ stable, wide polarity range
Resolution	Low to medium	High	Very high	Very high
Sensitivity	Low to medium	Medium to high (depending on detector)	High	Very high (especially MRM)
Identification	Rf value, color reaction, comparison with standards	Retention time, UV spectrum, comparison with standards	Retention time, mass spectrum (NIST library)	Retention time, molecular ion, fragmentation pattern (MS/MS)
Quantification	Semi-quantitative (densitometry)	Quantitative (peak area/ height)	Quantitative (peak area/ height)	Highly quantitative (especially MRM)
Cost	Low	Medium to high	High	Very high
Sample throughput	High (multiple samples per plate)	Medium to high	Medium to high	Medium to high
Sample prep	Simple	Moderate	Moderate (derivatization often needed)	Moderate to complex
Main application	Rapid screening, monitoring, preliminary method development	Routine quantitative/ qualitative analysis, purification	Volatile profiling, essential oil analysis	Comprehensive profiling, structural elucidation, trace analysis, metabolomics

Note: This table provides a general comparison. Specific instrument configurations and applications can vary

11.5 Optimization Strategies for Yield, Purity, and Bioactivity of Plant-Derived Anti-aging Metabolites

The optimization of extraction and purification processes for plant-based anti-aging metabolites requires more than the selection of isolated techniques; it involves the strategic integration of complementary approaches to simultaneously enhance yield, purity, and preservation of bioactivity. Given the structural diversity and chemical sensitivity of such metabolites, careful methodological design is essential to prevent degradation and ensure reproducibility. This section outlines key strategies that have proven effective in maximizing both analytical and therapeutic potential.

11.5.1 Sequential Extraction and Fractionation

As highlighted earlier (see Sect. 11.2), plant extracts are intrinsically complex, often containing bioactive molecules alongside unwanted constituents. Employing a sequential extraction approach—where solvents of increasing or decreasing polarity are used in a stepwise fashion—provides an effective preliminary fractionation method. By selectively solubilizing groups of compounds based on their polarity, researchers can enrich specific classes of metabolites and reduce impurities.

Example Workflow

Defatting: Initiate extraction with a non-polar solvent such as hexane or petroleum ether to remove fats, waxes, and chlorophyll—substances that may interfere with subsequent purification or analytical steps and generally lack anti-aging effects (Sasidharan et al. 2011).

Targeted Extraction: Proceed with solvents of increasing polarity (e.g., ethyl acetate, dichloromethane, methanol, water) to selectively isolate semi-polar and polar metabolites, such as flavonoids, phenolics, glycosides, and saponins (Sarker and Nahar 2012). This sequential strategy reduces sample complexity and streamlines subsequent chromatographic purification, improving selectivity and efficiency in isolating the compounds of interest.

11.5.2 Coupling Extraction and Chromatography

Integrating extraction directly with chromatographic separation—often in automated or online formats—can improve efficiency, minimize sample loss, and preserve thermolabile anti-aging metabolites. One example is the direct coupling of SFE with HPLC or LC–MS, enabling immediate analysis without intermediate drying steps (Herrero et al. 2006). Similarly, automated Solid Phase Extraction (SPE) linked with HPLC allows for simultaneous matrix cleanup and analyte concentration, improving sensitivity and extending column lifespan (Badawy et al. 2022). This approach is particularly advantageous for phenolic antioxidants and terpenoid-based anti-aging agents that are prone to oxidative degradation during manual handling.

11.5.3 Bioactivity-Guided Fractionation (BGF)

BGF combines chemical separation with iterative bioassays to focus isolation efforts on compounds that directly contribute to anti-aging effects, such as antioxidant capacity, anti-inflammatory activity, or inhibition of senescence-associated enzymes. The workflow begins with crude extract preparation and preliminary bioactivity screening (Fig. 11.3). Active extracts are fractionated, and each fraction is re-evaluated for biological activity. Iterative purification—via preparative HPLC, medium-pressure chromatography, or TLC—continues until highly enriched or

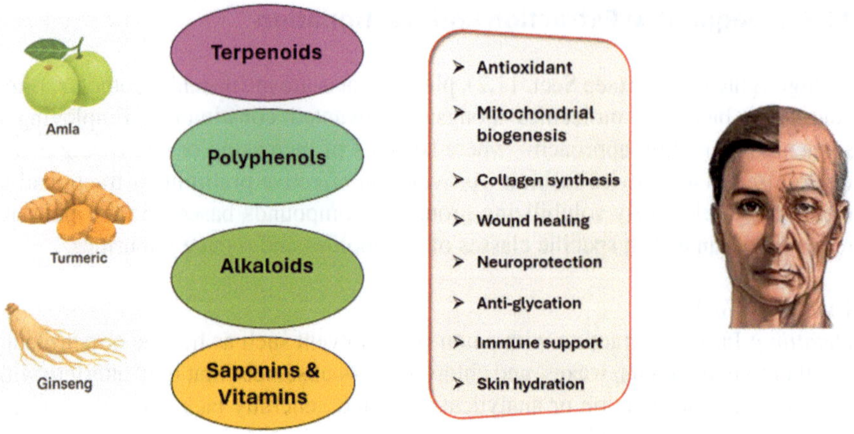

Fig. 11.3 Plant metabolites and their role in anti-aging effects. (Source: Unpublished diagram of Mostafa Asmaey)

pure bioactive compounds are obtained. Structural characterization using techniques such as NMR and MS follows (Cos et al. 2006). This approach not only increases discovery efficiency but also prevents unnecessary work on inactive constituents.

11.5.4 Scale-Up Considerations

While laboratory-scale extractions are invaluable for discovery, industrial-scale production of anti-aging ingredients—whether for nutraceuticals, cosmeceuticals, or pharmaceuticals—requires adjustments in process design. Conventional techniques can be scaled up by increasing vessel size while maintaining optimal solvent-to-solid ratios, whereas advanced methods like UAE or MAE demand proportional adjustments in energy delivery systems (Chemat et al. 2012).

Preparative chromatography, including flash and large-scale HPLC, prioritizes sample throughput over maximum resolution, employing larger columns and particles to manage back pressure. Scale-up decisions must balance yield, purity, and cost-effectiveness while ensuring compliance with good manufacturing practices (GMP). Importantly, bioactivity retention should remain a central performance metric during scale-up, as some anti-aging metabolites—particularly polyphenols—are susceptible to degradation under prolonged processing conditions (Snyder et al. 2011).

By integrating sequential extraction, coupled analytical workflows, bioactivity-guided purification, and robust scale-up strategies, researchers can efficiently progress from raw plant material to high-purity anti-aging metabolites ready for application in both research and commercial development.

11.6 Challenges and Future Perspectives

The study of anti-aging metabolites derived from medicinal plants remains a vibrant and rapidly evolving field, holding considerable therapeutic promise. However, progress is tempered by persistent scientific and technical challenges that require innovative solutions. Overcoming these obstacles while adopting emerging methodologies will be central to fully realizing the pharmacological potential of these natural products.

11.6.1 Challenges in Metabolite Extraction and Purification

Despite notable advancements, isolating bioactive metabolites from complex botanical matrices continues to present multiple difficulties. Plant extracts typically comprise thousands of chemical constituents—including primary metabolites, pigments, lipids, sugars, proteins, and diverse secondary metabolites—making the selective recovery of target compounds inherently complex (Azmir et al. 2013). The challenge is compounded when the desired molecules occur in trace concentrations, necessitating the use of large amounts of raw material and high-sensitivity analytical systems for reliable detection and quantification.

The thermolabile nature of many anti-aging agents, such as ascorbic acid, anthocyanins, and certain phenolic derivatives, demands careful control over extraction conditions to prevent degradation. Additionally, conventional extraction strategies often lack specificity, resulting in the co-extraction of unwanted compounds that complicate downstream purification and may interfere with biological assays (Chemat et al. 2019). Variability in plant material due to environmental, seasonal, and genetic factors further hampers reproducibility and standardization, creating significant quality control challenges for therapeutic applications.

Environmental and economic considerations also play a crucial role. Large-scale extraction processes typically rely on considerable volumes of organic solvents, many of which are costly, flammable, and environmentally harmful. High energy demands for heating and solvent removal further contribute to operational expenses and ecological impact. Finally, while hyphenated techniques such as LC–MS/MS provide powerful tools for compound identification, complete structural elucidation—especially of novel metabolites—remains resource-intensive, often requiring advanced NMR spectroscopy and other complementary analytical methods.

11.6.2 Emerging Technologies and Trends

Technological progress in extraction and chromatography is increasingly geared towards addressing these limitations through greener, more precise, and more integrated approaches. Green extraction technologies—such as subcritical water extraction (SWE), pressurized hot water extraction (PHWE), and SFE—are gaining prominence for their reduced solvent usage and minimal environmental footprint.

Novel solvent systems, including DESs and ionic liquids, offer tunable physicochemical properties and improved selectivity for polar and semi-polar anti-aging compounds (Ražić et al. 2023).

Biologically assisted methods, such as enzyme-assisted extraction (EAE) and targeted microbial fermentation, are emerging as powerful strategies to liberate or biotransform plant-derived precursors into more bioactive forms. On the analytical side, the integration of high-resolution mass spectrometry (e.g., Q-TOF, Orbitrap) with untargeted metabolomics provides deeper insight into complex phytochemical profiles, facilitating the discovery of novel anti-aging agents.

Advancements in hyphenated and multidimensional chromatography—including two-dimensional liquid chromatography (2D-LC)—are enabling improved separation of structurally similar or co-eluting metabolites. Furthermore, computational and AI-driven tools are beginning to influence the field by predicting optimal extraction parameters, automating data interpretation, and linking structural information to potential biological activity. Process analytical technology (PAT) is also expected to enhance reproducibility by enabling real-time monitoring of extraction and purification processes.

11.6.3 Sustainable Practices in Extraction and Chromatography

The shift toward sustainability is driven by both environmental responsibility and economic necessity. Future extraction platforms will likely focus on minimizing solvent use, incorporating renewable solvent sources such as bioethanol, and maximizing energy efficiency through optimized process design. Recycling solvent streams, reducing heating and cooling requirements, and implementing circular economy principles—such as valorizing residual plant biomass for secondary products like biofuels, animal feed, or secondary metabolite recovery—will be essential in reducing the environmental footprint.

By strategically adopting these advancements, the field can transition toward more efficient, reproducible, and eco-friendly production pipelines for plant-derived anti-aging agents. This convergence of sustainable engineering, advanced analytical science, and bioinformatics will not only accelerate discovery but also enhance the clinical and commercial viability of these natural compounds.

11.7 Conclusion

The journey from a raw medicinal plant to a purified anti-aging metabolite is a testament to the sophistication of modern natural product chemistry. This chapter has provided a comprehensive overview of the essential chemical extraction techniques and chromatographic separation methods that are indispensable for unlocking the therapeutic potential of botanical sources in the fight against aging.

We began by highlighting the immense promise of medicinal plants as rich reservoirs of diverse secondary metabolites, emphasizing their crucial roles as antioxidants, anti-inflammatories, and cellular protectors. The fundamental principles governing extraction, including the critical factors influencing efficiency and the strategic selection of solvents, were then thoroughly discussed. From the time-honored simplicity of maceration and the exhaustive nature of Soxhlet extraction, we transitioned to the cutting-edge efficiency of advanced techniques such as ultrasound-assisted extraction (UAE), microwave-assisted extraction (MAE), and supercritical fluid extraction (SFE). Each method, with its unique mechanism and advantages, offers tailored solutions for recovering a wide array of anti-aging compounds.

The narrative then delved into the intricate world of chromatographic separation, elucidating the basic concepts that underpin the differential movement of molecules. Planar techniques like thin-layer chromatography (TLC) were presented as invaluable tools for rapid screening and preliminary analysis, while the power of column chromatography, particularly high-performance liquid chromatography (HPLC), gas chromatography-mass spectrometry (GC-MS), and liquid chromatography-mass spectrometry/mass spectrometry (LC-MS/MS), was explored in detail. These advanced hyphenated techniques, with their unparalleled resolution, sensitivity, and structural elucidation capabilities, are the workhorses for precise qualitative identification and accurate quantitative analysis of complex anti-aging metabolite mixtures.

Furthermore, we emphasized the importance of integrated optimization strategies, including sequential extraction, coupling of techniques, and bioactivity-guided fractionation, to maximize yield, enhance purity, and crucially, preserve the inherent bioactivity of the target compounds. Practical case studies on ashwagandha, turmeric, and amla illustrated how these methodologies are applied in real-world research to isolate and characterize key anti-aging constituents.

Finally, we addressed the persistent challenges in this field, such as matrix complexity, compound instability, and reproducibility issues, while casting a forward glance at exciting future perspectives. The increasing adoption of green extraction technologies, high-resolution mass spectrometry, metabolomics, and AI-driven approaches, coupled with a strong emphasis on sustainable practices, promises to revolutionize the discovery and development of natural anti-aging agents. By continually refining these techniques, researchers can more effectively harness the vast chemical diversity of medicinal plants, paving the way for novel and effective interventions to promote healthier and longer lives.

Competing Interests The authors have no relevant financial or non-financial interests to disclose.

References

Abou Elmaaty T, Sayed-Ahmed K, Elsisi H, Magdi M (2022) Optimization of extraction of natural antimicrobial pigments using supercritical fluids: a review. Processes 10(10):2111

Abouheif SA, Sallam SM, El Sohafy SM, Kassem FF, Shawky E (2022) A green extraction approach using natural deep eutectic solvents enhances the in-vivo bioavailability of curcuminoids from turmeric extracts. Ind Crops Prod 189:115790

Abu Bakar IN, Ibrahim MF, Hakiman M, Abd-Aziz S, Prasongsuk S, Yen Tin LC, Jenol MA (2022) Characterization of asiaticoside concentration, total phenolic compounds, and antioxidant activity of different varieties of *Centella asiatica* (L.) and essential oil extraction using hydro-distillation with enzyme assisted. Biocat Agric Biotechnol 44:102474

Abubakar AR, Haque M (2020) Preparation of medicinal plants: basic extraction and fractionation procedures for experimental purposes. J Pharm Bioallied Sci 12(1):1–10

Ahangari H, King JW, Ehsani A, Yousefi M (2021) Supercritical fluid extraction of seed oils—a short review of current trends. Trends Food Sci Technol 111:249–260

Ahmad R, Alqathama A, Alam MM, Riaz M, Abdalla AN, Aldholmi M, Al-Said HM, Aljishi FS, Althomali EH, Alabdullah MM (2023) Biological quality and phytochemical profiling of olive fruits using gas chromatography–mass spectrometry (GCMS) analysis. Chem Biol Technol Agric 10(1):45

Akhtar I, Javad S, Yousaf Z, Iqbal S, Jabeen K (2019) Microwave assisted extraction of phytochemicals an efficient and modern approach for botanicals and pharmaceuticals. Pak J Pharm Sci 32(1):223–230

Al-Khayri JM, Sahana GR, Nagella P, Joseph BV, Alessa FM, Al-Mssallem MQ (2022) Flavonoids as potential anti-inflammatory molecules: a review. Molecules 27(9):2901

Alara OR, Abdurahman NH, Ukaegbu CI (2021) Extraction of phenolic compounds: a review. Curr Res Food Sci 4:200–214

Ali SM, Zainalabidin S, Latip J (2019) Quantitative analysis of phenolics content in two roselle varieties (*Hibiscus sabdariffa*) by high performance liquid chromatography. Malays J Anal Sci 23(4):715–724

Anggraeni GP, Sumarawati T, Wibowo JW, Zulaikah ST, Trisnadi S (2025) Effect of turmeric ethanol extract (*Curcuma longa* Linn) on IL-6, IL-10 levels and skin moisture. Gema Lingkungan Kesehatan 23(3):421–426

Asmaey MA, Abatis D, Abdel-Razek AS, Lambrinidis G, Chinou I, Fokialakis N, Tsafantakis N, Shaaban M, Aligiannis N (2021) Ochraceopyronide, a rare α-pyrone-C-lyxofuranoside from a soil-derived fungus Aspergillus ochraceopetaliformis. Molecules 26(13):3976

Asmaey MA, Salem MM, Emam M, El Shabrawy MO, Hussein SR, El Garf IA, Marzouk MM, Aligiannis N, Farid MM (2023) Chemical constituents from *Colchicum palaestinum* (Baker) C. Archer with the assessment of its antioxidant, wound scratch, and tyrosinase repressive potential. S Afr J Bot 157:209–218

Asmaey MA, Aati HY, Emam M, Tsafantakis N, Fokialakis N, Hasaballah AI (2024a) Larvicidal potency of four Egyptian herbs on *Culex pipiens* larvae: phytochemical composition and molecular networking for most potent extracts. Arab J Chem 17(10):105974

Asmaey MA, Elsoghiar AA, Shaaban M, Moharram AM, El-Gaby M (2024b) Phenolics and other structural compounds from leaves of *Olea europae*a L.: extraction techniques and pharmacological activities. Chem Afr 7(10):5123–5148

Asmaey MA, Hamed A, Shaaban M (2024c) Natural peroxides from plants: historical discovery, biosynthesis, and biological activities. Chem Biodivers 21(10):e202400644

Asmaey MA, Kalofolias DA, Charavgi M-D, Abdel-Rahim IR, Chrysina ED, Abatis D (2024d) Crystal structure of the cytotoxic macrocyclic trichothecene Isororidin A. Acta Crystallogr Sect C Struct Chem 80(8):407–411

Asmaey MA (2024e) Utilizing GC–MS and UPLC–MS for rapid metabolomic analysis in medicinal plants. In: Propagation to pharmacopeia. CRC Press, Boca Raton, pp 361–378

Ayala-Cabrera JF, Montero L, Meckelmann SW, Uteschil F, Schmitz OJ (2023) Review on atmospheric pressure ionization sources for gas chromatography-mass spectrometry. Part I: current ion source developments and improvements in ionization strategies. Anal Chim Acta 1238:340353

Azmir J, Zaidul ISM, Rahman MM, Sharif KM, Mohamed A, Sahena F, Jahurul MHA, Ghafoor K, Norulaini NA, Omar AK (2013) Techniques for extraction of bioactive compounds from plant materials: a review. J Food Eng 117(4):426–436

Azwanida N (2015) A review on the extraction methods use in medicinal plants, principle, strength and limitation. Med Aromat Plants 4:196. 2167-0412

Badawy ME, El-Nouby MA, Kimani PK, Lim LW, Rabea EI (2022) A review of the modern principles and applications of solid-phase extraction techniques in chromatographic analysis. Anal Sci 38(12):1457–1487

Banerjee S, Anand U, Ghosh S, Ray D, Ray P, Nandy S, Deshmukh GD, Tripathi V, Dey A (2021) Bacosides from *Bacopa monnieri* extract: an overview of the effects on neurological disorders. Phytother Res 35(10):5668–5679

Barak TH, Celep İK, Şentürk TB, Bardakci H, Celep E (2022) In vitro anti-aging potential evaluation of *Maclura pomifera* (Rafin.) Schneider 80% methanol extract with quantitative HPTLC analysis. Turk J Pharm Sci 19(4):400

Bashir S, Khan BM, Babar M, Andleeb S, Hafeez M, Ali S, Khan MF (2014) Assessment of bioautography and spot screening of TLC of green tea (*Camellia*) plant extracts as antibacterial and antioxidant agents. Indian J Pharm Sci 76(4):364

Bashir A, Nabi M, Tabassum N, Afzal S, Ayoub M (2023) An updated review on phytochemistry and molecular targets of *Withania somnifera* (L.) Dunal (Ashwagandha). Front Pharmacol 14:1049334

Baur JA, Sinclair DA (2006) Therapeutic potential of resveratrol: the in vivo evidence. Nat Rev Drug Discov 5(6):493–506

Bhadange YA, Carpenter J, Saharan VK (2024) A comprehensive review on advanced extraction techniques for retrieving bioactive components from natural sources. ACS Omega 9(29):31274–31297

Bitis L, Sen A, Ozsoy N, Birteksoz-Tan S, Kultur S, Melikoglu G (2017) Flavonoids and biological activities of various extracts from *Rosa sempervirens* leaves. Biotechnol Biotechnol Equip 31(2):299–303

Bjørklund G, Shanaida M, Lysiuk R, Butnariu M, Peana M, Sarac I, Strus O, Smetanina K, Chirumbolo S (2022) Natural compounds and products from an anti-aging perspective. Molecules 27(20):7084

Block RJ, Le Strange R, Zweig G (2013) Paper chromatography: a laboratory manual. Elsevier, Burlington

Bolat E, Sarıtaş S, Duman H, Eker F, Akdaşçi E, Karav S, Witkowska AM (2024) Polyphenols: secondary metabolites with a biological impression. Nutrients 16(15):2550

Boo YC (2022) Ascorbic acid (vitamin C) as a cosmeceutical to increase dermal collagen for skin antiaging purposes: emerging combination therapies. Antioxidants 11(9):1663

Boonpisuttinant K, Srisuttee R, Khong HY, Chutoprapat R, Ruksiriwanich W, Udompong S, Chompoo W, Boonbai R, Rakkaew R, Sangsee J (2023) In vitro anti-ageing activities of ethanolic extracts from Pink rambutan (*Nephelium lappaceum* Linn.) for skin applications. Saudi Pharm J 31(4):535–546

Brinkhaus B, Lindner M, Schuppan D, Hahn E (2000) Chemical, pharmacological and clinical profile of the East Asian medical plant *Centella aslatica*. Phytomedicine 7(5):427–448

Cacique AP, Barbosa ÉS, Pinho GP, Silvério FO (2020) Maceration extraction conditions for determining the phenolic compounds and the antioxidant activity of *Catharanthus roseus* (L.) G. Don. Cienc Agrotecnol 44:e017420

Cao S, Liang J, Chen M, Xu C, Wang X, Qiu L, Zhao X, Hu W (2025) Comparative analysis of extraction technologies for plant extracts and absolutes. Front Chem 13:1536590

Cebolla VL, Jarne C, Vela J, Garriga R, Membrado L, Galbán J (2021) Scanning densitometry and mass spectrometry for HPTLC analysis of lipids: the last 10 years. J Liq Chromatogr Relat Technol 44(3–4):148–170

Chaikul P, Kanlayavattanakul M, Somkumnerd J, Lourith N (2021) *Phyllanthus emblica* L.(amla) branch: a safe and effective ingredient against skin aging. J Trad Complement Med 11(5):390–399

Chatzimitakos T, Athanasiadis V, Kotsou K, Mantiniotou M, Kalompatsios D, Makrygiannis I, Bozinou E, Lalas SI (2024) Optimization of pressurized liquid extraction (PLE) parameters for extraction of bioactive compounds from *Moringa oleifera* leaves and bioactivity assessment. Int J Mol Sci 25(9):4628

Chavan P, Sharma P, Sharma SR, Mittal TC, Jaiswal AK (2022) Application of high-intensity ultrasound to improve food processing efficiency: a review. Foods 11(1):122
Chemat F, Khan MK (2011) Applications of ultrasound in food technology: processing, preservation and extraction. Ultrason Sonochem 18(4):813–835
Chemat F, Vian MA, Cravotto G (2012) Green extraction of natural products: concept and principles. Int J Mol Sci 13(7):8615–8627
Chemat F, Abert-Vian M, Fabiano-Tixier AS, Strube J, Uhlenbrock L, Gunjevic V, Cravotto G (2019) Green extraction of natural products. Origins, current status, and future challenges. TrAC Trends Anal Chem 118:248–263
Chen L, Cao H, Huang Q, Xiao J, Teng H (2022) Absorption, metabolism and bioavailability of flavonoids: a review. Crit Rev Food Sci Nutr 62(28):7730–7742
Cos P, Vlietinck AJ, Vanden Berghe D, Maes L (2006) Anti-infective potential of natural products: how to develop a stronger in vitro 'proof-of-concept'. J Ethnopharmacol 106(3):290–302
Coskun O (2016) Separation techniques: chromatography. North Clin Istanb 3(2):156
Cummins PM, Rochfort KD, O'Connor BF (2016) Ion-exchange chromatography: basic principles and application. In: Protein chromatography: methods and protocols. Springer, New York, pp 209–223
Darwin R, Valmon R, Chithanna S, Galla S, Syed S, Mohathasim Billah A, Kumar Reddy K, Naga Venkata Arjun U (2025) Sustainable extraction and purification of phytochemicals: a review of green solvents and techniques. Chem Methodol 9(5):356–385
Das G, Kameswaran S, Ramesh B, Bangeppagari M, Nath R, Das Talukdar A, Shin H-S, Patra JK (2024) Anti-aging effect of traditional plant-based food: an overview. Foods 13(23):3785
Davinelli S, Sapere N, Visentin M, Zella D, Scapagnini G (2013) Enhancement of mitochondrial biogenesis with polyphenols: combined effects of resveratrol and equol in human endothelial cells. Immun Ageing 10(1):28
Dawidowicz AL, Wianowska D (2005) PLE in the analysis of plant compounds: part I. The application of PLE for HPLC analysis of caffeine in green tea leaves. J Pharm Biomed Anal 37(5):1155–1159
Domínguez-López I, Yago-Aragón M, Salas-Huetos A, Tresserra-Rimbau A, Hurtado-Barroso S (2020) Effects of dietary phytoestrogens on hormones throughout a human lifespan: a review. Nutrients 12(8):2456
Dubois C, Plainfossé H, Delcroix M, Trinel M, Verger-Dubois G, Azoulay S, Burger P, Fernandez X (2022) Anti-aging potential of a Rosa centifolia stem extract with focus on phytochemical composition by bioguided fractionation. Chem Biodivers 19(7):e202200158
Eldeen AZE, El-Aziz FEZAA, Sayed AM, Mousa SA, Asmaey MA (2025) Exploring the phytochemical composition of *Salsola imbricata*: investigating its protective potential against UV-C radiation in earthworms and isopods models. 3 Biotech 15(4):1–11
Elsoghiar AA, El-Gaby MS, Shabaan MA, Moharram AM, Asmaey MA (2024) Eco-friendly dyeing of cotton fabrics using a blend of *Olea europaea* L. and *Ziziphus spina-christi* L. leaf extracts: chemical profiling and color fastness properties. Egypt J Chem 67(4):207–213
Farag MA, Kabbash EM, Mediani A, Döll S, Esatbeyoglu T, Afifi SM (2022) Comparative metabolite fingerprinting of four different cinnamon species analyzed via UPLC–MS and GC–MS and chemometric tools. Molecules 27(9):2935
Ferreira C, Sarraguça M (2024) A comprehensive review on deep eutectic solvents and its use to extract bioactive compounds of pharmaceutical interest. Pharmaceuticals (Basel) 17(1):124
Firdaus S, Permadi A, Sulistiawati E (2025) Comparative study of ultrasonic and maceration extraction in enhancing antioxidant and SPF properties of green coffee beans serum. Curr Res Biosci Biotechnol 6(2):36–34
Freitas L d, Valli M, Dametto AC, Pennacchi PC, Andricopulo AD, Maria-Engler SS, Bolzani VS (2020) Advanced glycation end product inhibition by alkaloids from Ocotea paranapiacabensis for the prevention of skin aging. J Nat Prod 83(3):649–656
Gandhi SP, Mate GD, Nagore DH, Chitlange SS (2024) Quantifying curcumin, gallic acid, and resveratrol in a polyherbal mixture: a robust HPTLC method. Eurasian J Chem 29(1 (113)):56–65

Geow CH, Tan MC, Yeap SP, Chin NL (2021) A review on extraction techniques and its future applications in industry. Eur J Lipid Sci Technol 123(4):2000302

Guo H, Feng C, Hu L, Zhao X, Tang X, Huang Y, Luo J, Xu M, Xie W (2022) Exploration of a ternary deep eutectic solvent for the efficient extraction of plantamajoside, acteoside, quercetin and kaempferol from *Plantago asiatica* L. Phytochem Anal 33(1):94–104

Hanai TT (2007) HPLC: a practical guide. Royal Society of Chemistry, Cambridge

Hegde SN, Lavanya Devi K, Choudhary M, Menon N, Singh G (2024) A comprehensive metabolome profiling of *Terminalia chebula*, *Terminalia bellerica*, and *Phyllanthus emblica* to explore the medicinal potential of Triphala. Sci Rep 14(1):31635

Herrero M, Cifuentes A, Ibañez E (2006) Sub-and supercritical fluid extraction of functional ingredients from different natural sources: plants, food-by-products, algae and microalgae: a review. Food Chem 98(1):136–148

Hidayat R, Wulandari P (2021) Methods of extraction: maceration, percolation and decoction. Eureka Herba Indones 2(1):68–74

Hoseini SA, Vazifedoost M, Hajirostamloo B, Didar Z, Nematshahi MM (2025) Supercritical fluid extraction and encapsulation of Rivas (*Rheum ribes*) flower: principal component analysis (PCA). Heliyon 11(2):e41746

Hossain MS, Wazed MA, Asha S, Amin MR, Shimul IM (2025) Dietary phytochemicals in health and disease: mechanisms, clinical evidence, and applications—a comprehensive review. Food Sci Nutr 13(3):e70101

Ismail BP (2024) Basic principles of chromatography. In: Nielsen's food analysis. Springer, Cham, pp 167–192

Izumi T, Saito M, Obata A, Arii M, Yamaguchi H, Matsuyama A (2007) Oral intake of soy isoflavone aglycone improves the aged skin of adult women. J Nutr Sci Vitaminol 53(1):57–62

Jegal J, Jeong EJ, Yang MH (2019) A review of the different methods applied in ginsenoside extraction from *Panax ginseng* and *Panax quinquefolius* roots. Nat Prod Commun 14(9):1934578X19868393

Jomova K, Alomar SY, Valko R, Liska J, Nepovimova E, Kuca K, Valko M (2025) Flavonoids and their role in oxidative stress, inflammation, and human diseases. Chem Biol Interact 413:111489

Jovanović J, Jović M, Trifković J, Smiljanić K, Gašić U, Krstić Ristivojević M, Ristivojević P (2025) Green extraction of bioactives from curcuma longa using natural deep eutectic solvents: unlocking antioxidative, antimicrobial, antidiabetic, and skin depigmentation potentials. Plants 14(2):163

Kadri F, Xavier G, Patel U, Prajapati P (2021) Super-critical fluid: an overview on analytical applications. Curr Green Chem 8(3):203–221

Kalyniukova A, Holuša J, Musiolek D, Sedlakova-Kadukova J, Płotka-Wasylka J, Andruch V (2021) Application of deep eutectic solvents for separation and determination of bioactive compounds in medicinal plants. Ind Crops Prod 172:114047

Kim YS, Cho IG, Jeong MJ, Jeong SJ, Nah SY, Cho YS, Kim SH, Go S, Kim SE, Kang SS (2011) Therapeutic effect of total ginseng saponin on skin wound healing. J Ginseng Res 35(3):360

Kowalska T, Sajewicz M (2022) Thin-layer chromatography (TLC) in the screening of botanicals–its versatile potential and selected applications. Molecules 27(19):6607

Kumar A, Nirmal P, Kumar M, Jose A, Tomer V, Oz E, Proestos C, Zeng M, Elobeid T, Sneha K (2023) Major phytochemicals: recent advances in health benefits and extraction method. Molecules 28(2):887

Kumar S, Yadav R, Singh VK, Yaduvanshi PS, Jain G, Loksh KR, Sharma S (2025) Fruits and herbs as potential human anti-aging compounds-a review. Cuestiones de Fisioterapia 54(4):438–445

Kuśmierska M, Kuśmierski J, Kwaśniewska O (2024) Exploring the therapeutic potential of Ashwagandha (*Withania somnifera*) supplementation in alleviating stress and stress-related disorders. Qual Sport 15:51854–51854

Lalhriatpuii T (2019) HPTLC fingerprint in herbal drug formulations. In: Herbal medicine in India: indigenous knowledge, practice, innovation and its value. Springer, Singapore, pp 337–362

Li J, Lu YR, Lin IF, Kang W, Chen H b, Lu HF, Wang HMD (2020) Reversing UVB-induced photoaging with *Hibiscus sabdariffa* calyx aqueous extract. J Sci Food Agric 100(2):672–681

Li C, Chu S, Tan S, Yin X, Jiang Y, Dai X, Gong X, Fang X, Tian D (2021) Towards higher sensitivity of mass spectrometry: a perspective from the mass analyzers. Front Chem 9:813359

Liang J, Cui L, Li J, Guan S, Zhang K, Li J (2021) *Aloe vera*: a medicinal plant used in skin wound healing. Tissue Eng Part B Rev 27(5):455–474

Liu L, Guo P, Wang P, Zheng S, Qu Z, Liu N (2021) The review of anti-aging mechanism of polyphenols on Caenorhabditis elegans. Front Bioeng Biotechnol 9:635768

Liu E, Gao H, Zhao Y, Pang Y, Yao Y, Yang Z, Zhang X, Wang Y, Yang S, Ma X (2022) The potential application of natural products in cutaneous wound healing: a review of preclinical evidence. Front Pharmacol 13:900439

López-Bascón M, De Castro ML (2020) Soxhlet extraction. In: Liquid-phase extraction. Elsevier, Amsterdam, pp 327–354

López-Otín C, Blasco MA, Partridge L, Serrano M, Kroemer G (2013) The hallmarks of aging. Cell 153(6):1194–1217

Lu G, Liu Z, Wang X, Wang C (2021) Recent advances in *Panax ginseng* CA Meyer as a herb for anti-fatigue: an effects and mechanisms review. Foods 10(5):1030

Maes J, Verlooy L, Buenafe OE, De Witte PA, Esguerra CV, Crawford AD (2012) Evaluation of 14 organic solvents and carriers for screening applications in zebrafish embryos and larvae. PLoS One 7:e43850

Maldonado E, Morales-Pison S, Urbina F, Solari A (2023) Aging hallmarks and the role of oxidative stress. Antioxidants 12(3):651

Mbara KC, Devnarain N, Owira PM (2022) Potential role of polyphenolic flavonoids as senotherapeutic agents in degenerative diseases and geroprotection. Pharmaceut Med 36(6):331–352

McHugh M, Krukonis V, Brenner H (2013) Supercritical fluid extraction: principles and practice. Butterworth-Heinemann, Boston

Moldoveanu SC, David V (2022) Essentials in modern HPLC separations. Elsevier, Amsterdam

Montalbán MG, Víllora G (2022) Supercritical fluids: properties and applications. In: Phase equilibria with supercritical carbon dioxide-application to the components of a biocatalytic process. IntechOpen, London

Montaño-Pérez CA, Yáñez-Ibarra G (2025) Phytochemical compounds of plants: extraction, purification, and identification methods. In: Improving health and nutrition through bioactive compounds. Elsevier, London, pp 323–335

Moraes AB, Haidar MA, Júnior JMS, Simões MJ, Baracat EC, Patriarca MT (2009) The effects of topical isoflavones on postmenopausal skin: double-blind and randomized clinical trial of efficacy. Eur J Obstet Gynecol Reprod Biol 146(2):188–192

Morlock G, Schwack W (2010) Hyphenations in planar chromatography. J Chromatogr A 1217(43):6600–6609

Mostafa A, Asmaey Abouelabbas AM, Elsoghiar Masaki, Kita Ahmad M, Moharram Ahmed I, Hasaballah Mohamed SA, El-Gaby (2026) Phytochemical profiling and larvicidal mechanisms of olive (*Olea europaea* L.) leaf extract against *Culex pipiens*: antioxidant enzymatic inhibition and computational docking insights. Chemistry Select 11(5). https://doi.org/10.1002/slct.202505097

Nahar Y, Thickett SC (2021) Greener, faster, stronger: the benefits of deep eutectic solvents in polymer and materials science. Polymers 13(3):447

Nazeh M, Asmaey MA, Eladly A, Afifi MM (2024) Metabolomic profiling of streptomyces griseorubens with the evaluation of their antioxidant and anticancer potentialities. Egypt J Chem 67(8):251–268

Packer L, Valacchi G (2002) Antioxidants and the response of skin to oxidative stress: vitamin E as a key indicator. Skin Pharmacol Physiol 15(5):282–290

Paschoal RG, Higino ML, Alves APG, Sampaio ALF, Teixeira DF, Lucchetti L, Mendes MF (2025) Supercritical fluid extraction of *Bidens pilosa* L. oil: kinetics evaluation and antitumor activity. J Appl Res Med Aromat Plants 47:100641

Pastare L, Berga M, Kienkas L, Boroduskis M, Ramata-Stunda A, Reihmane D, Senkovs M, Skudrins G, Nakurte I (2023) Exploring the potential of supercritical fluid extraction of matricaria chamomilla white ray florets as a source of bioactive (cosmetic) ingredients. Antioxidants 12(5):1092

Perez-Vazquez A, Carpena M, Barciela P, Cassani L, Simal-Gandara J, Prieto MA (2023) Pressurized liquid extraction for the recovery of bioactive compounds from seaweeds for food industry application: a review. Antioxidants 12(3):612

Pizani RS, Viganó J, Contieri LS, Strieder MM, Kamikawashi RK, Vilegas W, de Souza Mesquita LM, Rostagno MA (2024) New selective and sustainable ultrasound-assisted extraction procedure to recover carnosic and rosmarinic acids from Rosmarinus officinalis by sequential use of bio-based solvents. Food Chem 435:137540

Prawira-Atmaja MI, Puangpraphant S (2025) Effects of ultrasound-assisted bath, probe, and extraction time on bioactive compounds and antioxidant activity in white tea (*Camellia sinensis*) extracts from purple-and green-leaf cultivars. J Saudi Soc Agric Sci 24(3):10

Puri M, Sharma D, Barrow CJ (2012) Enzyme-assisted extraction of bioactives from plants. Trends Biotechnol 30(1):37–44

Raza A, Chohan TA, Zaidi SHH, Hai A, Alzahrani AR, Abida, Imran M, Saleem H (2024) A systematic review on biochemical perspectives on natural products in wound healing: exploring phytochemicals in tissue repair and scar prevention. Chem Biodivers 21(10):e202400615

Ražić S, Arsenijević J, Mračević SĐ, Mušović J, Trtić-Petrović T (2023) Greener chemistry in analytical sciences: from green solvents to applications in complex matrices. Current challenges and future perspectives: a critical review. Analyst 148(14):3130–3152

Sabaragamuwa R, Perera CO (2023) Total triterpenes, polyphenols, flavonoids, and antioxidant activity of bioactive phytochemicals of *Centella asiatica* by different extraction techniques. Foods 12(21):3972

Sadeghi A, Hakimzadeh V, Karimifar B (2017) Microwave assisted extraction of bioactive compounds from food: a review. Int J Food Sci Nutr Eng 7(1):19–27

Salas-Millán JÁ, Aguayo E (2024) Bioaccessibility and unravelling of polyphenols, sulforaphane, and indoles biotransformation after in vitro gastrointestinal digestion of a novel lactofermented broccoli beverage. Food Funct 15(24):11949–11960

Saleem M, Ullah M, Kamreen H, Hajri AK, Alanazi AN, Alraih AM, Alsuhaibani AM, Albedair LA, ur Rehman K, Khan D (2024) Microwave-assisted extraction of green tea catechins and antioxidant activity of tea extracts: the role of solvents, microwave power, and contact time. Microchem J 203:110906

Samota MK, Rawat M, Kaur M, Garg D (2024) Gingerol: extraction methods, health implications, bioavailability and signaling pathways. Sustain Food Technol 2(6):1652–1669

Sanada F, Hayashi S, Morishita R (2025) Targeting the hallmarks of aging: mechanisms and therapeutic opportunities. Front Cardiovasc Med 12:1631578

Sarker SD, Nahar L (2012) An introduction to natural products isolation. In: Natural products isolation, pp 1–25. https://doi.org/10.1007/978-1-61779-624-1_1

Sasidharan S, Chen Y, Saravanan D, Sundram KM, Latha LY (2011) Extraction, isolation and characterization of bioactive compounds from plants' extracts. Afr J Tradit Complement Altern Med 8(1):1–10

Sati P, Dhyani P, Bhatt ID, Pandey A (2019) *Ginkgo biloba* flavonoid glycosides in antimicrobial perspective with reference to extraction method. J Tradit Complement Med 9(1):15–23

Sharafzadeh S (2013) Medicinal plants as anti-ageing materials: a review. Glob J Med Plant Res 1(2):234–236

Singh S, Bansal A, Singh V, Chopra T, Poddar J (2022) Flavonoids, alkaloids and terpenoids: a new hope for the treatment of diabetes mellitus. J Diabetes Metab Disord 21(1):941–950

Smith J, Luo Y (2004) Studies on molecular mechanisms of *Ginkgo biloba* extract. Appl Microbiol Biotechnol 64(4):465–472

Snyder LR, Kirkland JJ, Dolan JW (2011) Introduction to modern liquid chromatography. Wiley, Hoboken

Song YR, Sung SK, Jang M, Lim TG, Cho CW, Han CJ, Hong HD (2018) Enzyme-assisted extraction, chemical characteristics, and immunostimulatory activity of polysaccharides from Korean ginseng (*Panax ginseng* Meyer). Int J Biol Macromol 116:1089–1097

Stanojevic LP, Marjanovic-Balaban ZR, Kalaba VD, Stanojevic JS, Cvetkovic DJ, Cakic MD (2017) Chemical composition, antioxidant and antimicrobial activity of basil (*Ocimum basilicum* L.) essential oil. J Essent Oil Bear Plants 20(6):1557–1569

Sun T, Chance RR, Graessley WW, Lohse DJ (2004) A study of the separation principle in size exclusion chromatography. Macromolecules 37(11):4304–4312

Talreja S, Kumari S, Srivastava P, Pandey S (2021) A complete pharmacognostic review on amla. World J Pharm Pharm Sci 8:622–637

Tufail T, Bader Ul Ain H, Noreen S, Ikram A, Arshad MT, Abdullahi MA (2024) Nutritional benefits of lycopene and beta-carotene: a comprehensive overview. Food Sci Nutr 12(11):8715–8741

Tyihák E, Káy G, Király-Véghely Z, Németh Z, Albert L, Szende B (2001) *Vitis vinifera* as a medicinal plant. International conference on medicinal and aromatic plants (part II) 597

Utpal BK, Al Amin M, Zehravi M, Sweilam SH, Arjun UV, Madhuri YB, Gupta JK, Yaidikar L, Tummala T, Suseela R, Durairaj A (2025) Alkaloids as neuroprotectors: targeting signaling pathways in neurodegenerative diseases. Mol Cell Biochem 480:1–26

Vilkhu K, Mawson R, Simons L, Bates D (2008) Applications and opportunities for ultrasound assisted extraction in the food industry—a review. Innov Food Sci Emerg Technol 9(2):161–169

Waseem M, Majeed Y, Nadeem T, Naqvi LH, Khalid MA, Sajjad MM, Sultan M, Khan MU, Khayrullin M, Shariati MA, Lorenzo JM (2023) Conventional and advanced extraction methods of some bioactive compounds with health benefits of food and plant waste: a comprehensive review. Food Front 4(4):1681–1701

Wen C, Zhang J, Zhang H, Dzah CS, Zandile M, Duan Y, Ma H, Luo X (2018) Advances in ultrasound assisted extraction of bioactive compounds from cash crops – a review. Ultrason Sonochem 48:538–549

Widowati W, Rani AP, Hamzah RA, Arumwardana S, Afifah E, Kusuma HS, Rihibiha DD, Nufus H, Amalia A (2017) Antioxidant and antiaging assays of *Hibiscus sabdariffa* extract and its compounds. Nat Prod Sci 23(3):192–200

Wu Y, Zhang N, Luo K, Liu Y, Bai Z, Tang S (2022) Recent advances of innovative and high-efficiency stationary phases for chromatographic separations. TrAC Trends Anal Chem 153:116647

Xie M, Jiang Z, Lin X, Wei X (2024) Application of plant extracts cosmetics in the field of anti-aging. J Dermatol Sci Cosmet Technol 1(2):100014

Xu JL, Kim TJ, Kim J-K, Choi Y (2019) Simultaneous roasting and extraction of green coffee beans by pressurized liquid extraction. Food Chem 281:261–268

Yingngam B, Chiangsom A, Brantner A (2020) Modeling and optimization of microwave-assisted extraction of pentacyclic triterpenes from *Centella asiatica* leaves using response surface methodology. Ind Crop Prod 147:112231

Zhang H-F, Yang X-H, Wang Y (2011) Microwave assisted extraction of secondary metabolites from plants: current status and future directions. Trends Food Sci Technol 22(12):672–688

Zhang Q, Yang FQ, Ge L, Hu YJ, Xia ZN (2017) Recent applications of hydrophilic interaction liquid chromatography in pharmaceutical analysis. J Sep Sci 40(1):49–80

Zhang Q-W, Lin L-G, Ye W-C (2018) Techniques for extraction and isolation of natural products: a comprehensive review. Chin Med 13(1):20

Zhou G, Ma J, Tang Y, Wang X, Zhang J, Duan JA (2018) Multi-response optimization of ultrasonic assisted enzymatic extraction followed by macroporous resin purification for maximal recovery of flavonoids and ginkgolides from waste *Ginkgo biloba* fallen leaves. Molecules 23(5):1029

Zissi L, Dimaki VD, Birba VS, Galani VC, Magafa V, Hatziantoniou S, Lamari FN (2025) Natural deep eutectic solvents as green alternatives for extracting bioactive compounds from sideritis taxa with potential cosmetic applications. Antioxidants 14(1):68

In Vitro Mass Propagation of *Cannabis sativa* L. for Enhanced Production of Anti-aging Metabolites: The Strategic Role of Plant Growth Regulators

12

Agustina Fernández, Alexander Aguila Wharton, Luciano Protti Cosenza, Luisa Franchi, Luciana Mentasti, Ignacio Scandura, and Gastón Barreto

Abstract

Cannabis sativa L. is a valuable source of anti-aging metabolites, particularly cannabinoids and terpenes, which exert potent antioxidant and anti-inflammatory effects by mitigating oxidative stress, inflammation, and cellular damage. However, conventional cultivation fails to deliver standardized, high-quality bio-

---

The authors "Agustina Fernández" and "Alexander Aguila Wharton" contributed equally to this chapter.

A. Fernández (✉)
Centro de Investigaciones en Física e Ingeniería del Centro de la Provincia de Buenos Aires CIFICEN (UNCPBA-CICPBA-CONICET), Buenos Aires, Argentina

Comisión de Investigaciones Científicas de la Provincia de Buenos Aires (CICPBA), Buenos Aires, Argentina

Universidad Nacional del Centro de la Provincia de Buenos Aires (UNCPBA), Facultad de Ingeniería, Buenos Aires, Argentina
e-mail: afernandez@fio.unicen.edu.ar

A. A. Wharton · L. P. Cosenza
Centro de Investigaciones en Física e Ingeniería del Centro de la Provincia de Buenos Aires CIFICEN (UNCPBA-CICPBA-CONICET), Buenos Aires, Argentina

Consejo Nacional de Investigaciones Científicas y Técnicas (CONICET), Buenos Aires, Argentina

L. Franchi
Centro de Investigaciones en Física e Ingeniería del Centro de la Provincia de Buenos Aires CIFICEN (UNCPBA-CICPBA-CONICET), Buenos Aires, Argentina
Universidad Nacional del Centro de la Provincia de Buenos Aires (UNCPBA), Facultad de Ingeniería, Buenos Aires, Argentina

S. Gantait et al. (eds.), *Medicinal Plants with Anti-Ageing Properties*,
https://doi.org/10.1007/978-981-95-8064-4_12

mass due to genetic heterogeneity, phenotypic plasticity, and susceptibility to pathogens and environmental fluctuations. This chapter outlines how advanced biotechnological interventions, most notably in vitro mass propagation, offer a robust and scalable solution for the consistent and enhanced production of these bioactive metabolites. We emphasize the central role of plant growth regulators (PGRs) as precise modulators of morphogenesis, detailing how specific cytokinins and auxins regulate high-frequency shoot multiplication, in vitro rooting, and overall developmental stability. Particular attention is given to the challenges posed by genotype-dependent responses and physiological disorders such as hyperhydricity, both of which require carefully optimized PGR regimes and culture conditions. In addition, we describe improved acclimatization strategies using engineered substrates to ensure ex vitro survival and vigorous growth, and we highlight the importance of verifying genetic and metabolic fidelity through molecular markers, cytological screening, and phytochemical profiling. Together, these components define an integrated pipeline that ensures the production of true-to-type, metabolically consistent cannabis plantlets. By framing micropropagation not merely as a cloning tool but as a controlled and scalable bioproduction platform, this chapter provides a comprehensive roadmap for deploying in vitro systems to generate reliable, standardized, and pathogen-free cannabis biomass for emerging anti-aging applications in the cosmetic and pharmaceutical industries.

Keywords

Antioxidant activity · Biomass production · Genetic fidelity · Metabolite consistency · Micropropagation · Secondary metabolites

12.1 Introduction

Cannabis sativa—**A Reservoir of Bioactive Compounds** *Cannabis sativa* L. (1753), known as cannabis or marijuana/hemp, has been cultivated in the pharmaceutical field for the production of bioactive compounds, such as cannabinoids

L. Mentasti · G. Barreto
Centro de Investigaciones en Física e Ingeniería del Centro de la Provincia de Buenos Aires CIFICEN (UNCPBA-CICPBA-CONICET), Buenos Aires, Argentina
Universidad Nacional del Centro de la Provincia de Buenos Aires (UNCPBA), Facultad de Ingeniería, Buenos Aires, Argentina

Consejo Nacional de Investigaciones Científicas y Técnicas (CONICET), Buenos Aires, Argentina

I. Scandura
Centro de Investigaciones en Física e Ingeniería del Centro de la Provincia de Buenos Aires CIFICEN (UNCPBA-CICPBA-CONICET), Buenos Aires, Argentina

Comisión de Investigaciones Científicas de la Provincia de Buenos Aires (CICPBA), Buenos Aires, Argentina and terpenes (Andre et al. 2016), that have demonstrated significant therapeutic potential in treating various health conditions, including chronic pain, epilepsy, and inflammatory disorders. In the specific case of phytocannabinoids, they represent a family of compounds that has garnered significant interest, owing to their biochemical effects (Pisanti et al. 2017; Pascual et al. 2018). Pharmacological modulation mediated by phytocannabinoids constitutes a promising therapeutic avenue, with substantial evidence supporting the analgesic, neuroprotective, anti-inflammatory, and antibacterial properties of these compounds (Appendino and Taglialatela-Scafati 2013; Pisanti et al. 2017; Fernández-Ruiz et al. 2020). These properties have led to the incorporation of cannabis-derived compounds in cosmetic formulations, particularly in anti-aging creams and hair care products (Schettino et al. 2021).

Regarding its botany, cannabis is a C3 herbaceous species from the Cannabaceae family that follows a short-day photoperiodic cycle. It is generally divided into three primary types: *C. sativa, C. indica*, and *C. ruderalis*, all of which can be considered subspecies of the same species, *C. sativa*. The latest genome-based evidence suggests that the genus consists of a highly diverse monotypic species (Lapierre et al. 2023). Although its exact origin is still unclear, it is thought to have emerged from temperate zones in Asia, particularly areas such as the southern Caspian region, Siberia, China, or the Himalayan mountains (Fordjour et al. 2023). This plant has been cultivated for centuries for its fibers, edible seeds, medicinal uses, and psychoactive effects (Small 2017; McPartland 2018).

The plant stem is erect and fibrous, allowing it to reach heights of up to 5 meters under optimal growing conditions (Russo 2011). The leaves are arranged oppositely, shifting to an alternate arrangement during development (Hesami et al. 2024). They are palmate with serrated edges, and their size and shape vary depending on the plant's genetics and the growing environment (Hanuš et al. 2016). It is a wind-pollinated, naturally dioecious species, with male and female flowers on separate plants, but it can also be monoecious. Male flowers feature five yellowish petals and five anthers that produce pollen. In contrast, female inflorescences develop greenish-yellow at the tips of the stems and in the axils, which can turn red to violet through maturing and senescence (depending on the cultivar). They consist of a single ovule and a tightly fused perianth. Female flowers contain the highest concentrations of cannabinoids, which are produced in highly specialized structures called trichomes.

The plant life cycle can be divided into four phases: (i) germination, (ii) vegetative stage, (iii) flowering stage, and finally (iv) senescence. During the vegetative stage, which lasts for several weeks, the plant experiences rapid growth, developing its aerial structure and deep roots. The flowering stage begins when female plants receive signals from changes in the photoperiod, specifically when the number of light hours decreases, requiring a dark period of around 12 hours to induce flowering (Caplan et al. 2017). Phytocannabinoid production significantly rises during this stage, as these compounds act as secondary metabolites that help protect the plant from pathogens and regulate pollination (Zuardi 2008). Therefore, a key objective in cannabis cultivation is to maximize biomass production, with a particular focus on mature, unfertilized female flowers, which concentrate the metabolites of interest in glandular trichomes.

Early reports have described three varieties of glandular trichomes on female flowers, distinguished by their surface morphology: bulbous, sessile, and pedunculate or stalked (Hammond and Mahlberg 1973). The bulbous trichomes are the smallest in size and synthesize only a limited quantity of specialized metabolites (Potter 2009). Sessile trichomes are located within the epidermis on vegetative leaves and flowers, and produce more sesquiterpenes, feature a short stalk, and are characterized by a globular head. This head consists of a multicellular disc composed of secretory cells and a subcuticular cavity that serves as a reservoir for metabolites (Livingston et al. 2020). In contrast, stalked trichomes possess a similarly shaped globose head that is somewhat larger, and they produce the highest amounts of cannabinoids and monoterpenes (Fig. 12.1a–d). This structure is elevated several hundred microns above the epidermal surface by a multicellular stalk (Mahlberg and Kim 2004; Potter 2009). The shift from a translucent initial state to an opaque milky white appearance, and, eventually, to an amber shade, indicates the synthesis and accumulation of cannabinoids and other resinous compounds within the glandular trichomes. This alteration in resin color implies the maturation and chemical transformation of secondary metabolites, including cannabinoids and terpenes (Mahlberg and Kim 2004; Livingston et al. 2020; Punja et al. 2023).

This structural heterogeneity reflects the high degree of metabolic specialization occurring within glandular trichomes. Their organization and maturation strongly correlate with the chemical diversity produced in these secretory tissues. Consequently, cannabis exhibits an exceptionally complex secondary metabolite profile, with cannabinoids and terpenoids being the most biologically and commercially significant groups (Andre et al. 2016). Comprehensive phytochemical analyses have revealed more than 500 metabolites in cannabis, with approximately 125 identified as cannabinoids. Table 12.1 summarizes the main groups of secondary metabolites reported in *C. sativa*, highlighting representative compounds, their concentration ranges, and the specific plant organs where they predominantly accumulate. The biosynthesis of the major cannabinoids begins with cannabigerolic acid (CBGA), which serves as the central precursor (Gülck and Møller 2020). From CBGA, enzymes such as THCA synthase and CBDA synthase catalyze the formation of tetrahydrocannabinolic acid (THCA) and cannabidiolic acid (CBDA), respectively (Sirikantaramas et al. 2005). These acidic compounds naturally decarboxylate into their neutral, more bioactive forms—Δ^9-tetrahydrocannabinol (THC), cannabidiol (CBD), and cannabigerol (CBG)—upon heating, drying, or aging. Other significant cannabinoids include cannabichromene (CBC) and cannabinol (CBN), the latter being primarily an oxidation product of THC (Andre et al. 2016). Quantitative analyses indicate that cannabinoid concentrations are highest in glandular trichomes, especially in stalked ones (Livingston et al. 2020; Jin et al. 2020).

Terpenoid profiles are equally complex, comprising up to 47 compounds, including monoterpenes and sesquiterpenes, which are volatile and contribute to the fragrance of cannabis flowers and cannabis products (Hillig 2004; Fischedick et al. 2010; Booth et al. 2017; Booth and Bohlmann 2019; Mudge et al. 2019; Zager et al. 2019). Although present at much lower concentrations than major cannabinoids, terpenoids constitute a chemically rich and highly diverse fraction of the

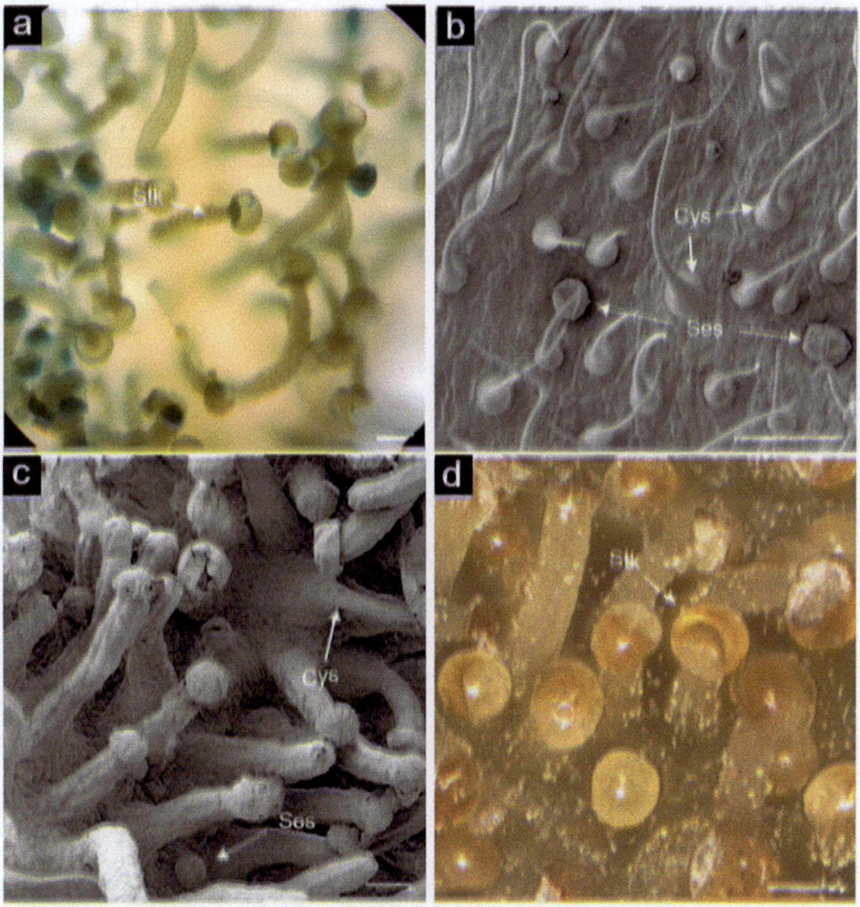

Fig. 12.1 *Cannabis sativa* trichomes. (**a**) Optical microscopy, (**b, c**) Scanning electron microscopy, (**d**) Laser scanning confocal microscopy. Scale bar: 100 μm. *Stk* stalked glandular trichome, *Cys* cystolith trichome, *Ses* sessile glandular trichome. (*Source:* unpublished figures of Aguila Wharton)

metabolome, with dozens of mono- and sesquiterpenes contributing to the sensory and therapeutic attributes of cannabis through entourage effect (Russo 2011).

Additionally, flavonoids, sterols, triterpenoids, lignamides, phenolic compounds, alkaloids, and stilbenoids have also been detected, particularly in leaves, stem bark, seeds, and roots (Radwan et al. 2021; Boucher et al. 2025). Although they occur in much lower abundance than cannabinoids and terpenoids, they remain biologically relevant and contribute to distinct functional properties, including antioxidant, anti-inflammatory, and structural roles. Notably, chemical composition varies significantly across plant tissues, and it is precisely due to this extensive phytochemical diversity and complexity that modern taxonomy increasingly classifies cannabis varieties not as traditional cultivars but as distinct chemovars, defined by their unique and reproducible chemical profiles (Hazekamp and Fischedick 2012;

Table 12.1 Major classes of secondary metabolites identified in *Cannabis sativa* L., their concentration ranges, and predominant plant locations

Metabolite family	Compound identified	Concentration range	Plant location	References
Cannabinoids	Tetrahydrocannabinolic acid (THCA), cannabidiolic acid (CBDA), cannabigerolic acid (CBGA), tetrahydrocannabinol (THC), Cannabidiol (CBD), cannabigerol (CBG), cannabichromene (CBC), cannabicyclolic acid (CBLA), cannabinol (CBN), cannabichromenic acid (CBCA), tetrahydrocannabivarin (THCV), cannabidivarin (CBDV).	Inflorescences: 15.77–20.37%; leaves: 1.10–2.10%	Glandular trichomes in flowers, young leaves, roots	Andre et al. (2016), Jin et al. (2020), Gülck and Møller (2020), Gul et al. (2023)
Terpenes	Monoterpenes (myrcene, limonene, α-pinene, β-pinene), sesquiterpenes (β-caryophyllene, humulene), oxygenated derivatives (linalool, terpineol), triterpenes (β-amyrin, epifriedelanol, friedelin).	Inflorescences: 1.28–3.5%; leaves: 0.13–0.28%	Flowers and leaves (glandular trichomes), roots	Booth et al. (2017), Booth and Bohlmann (2019), Jin et al. (2020)
Flavonoids	Cannflavins A, B, and C; apigenin, luteolin, kaempferol, quercetin, caflanone.	Trace-0.48% dry weight (variable among tissues)	Leaves, inflorescences, pollen, stem bark	Ross et al. (2005), Flores-Sanchez and Verpoorte (2008), Zagórska-Dziok et al. (2021), Lowe et al. (2021)
Sterols	Campesterol, stigmasterol, β-sitosterol, fucosterol	Trace-1.2 mg/g dry weight (approx.)	Leaves, stems, seeds, roots	Fischedick et al. (2010), Andre et al. (2016), Kornpointner et al. (2021)
Lignanamides and phenolic compounds	Cannabisin A–G, N-trans-coumaroyloctopamine, N-trans-coumaroyltyramine N-trans-feruloyltyramine N-trans-caffeoyltyramine	0.02–0.3 mg/g dry weight	Seeds, roots, leaves	Yan et al. (2015), Gülck and Møller (2020), Isidore et al. (2021)

(continued)

Table 12.1 (continued)

Metabolite family	Compound identified	Concentration range	Plant location	References
Alkaloids	Cannabisativine, anhydrocannabisativine	Trace–not quantified	Leaves, roots	Lowe et al. (2021)
Stilbenoids	Denbinobin, canniprene, cannabispirone, cannabispirenone	Trace–not quantified	Stem, leaves and flower	Radwan et al. (2021), Isidore et al. (2021)

Hazekamp et al. 2016). In parallel with the chemovar concept, which reflects the full metabolomic profile, the notion of chemotype remains valuable, particularly for distinguishing plants based on their predominant cannabinoid ratios. Cannabis is traditionally described as comprising three principal chemotypes defined by their THC:CBD ratio: chemotype I (THC/CBD > 1), chemotype II (THC/CBD ≈ 1), and chemotype III (THC/CBD < 1). These categories provide a practical framework for linking chemical composition to therapeutic applications, as different THC/CBD proportions modulate pharmacological and psychoactive outcomes (Russo 2011). Nevertheless, genotype × environment interactions—including light, nutrition, and developmental stage—can modulate cannabinoid content, introducing variability within chemotypic classes (Pacifico et al. 2008). Beyond the classical I–III system, more recent classifications recognize two additional groups: chemotype IV, characterized by a predominance of CBG (cannabigerol) and very low THC/CBD levels, and chemotype V, which encompasses plants with negligible or undetectable amounts of major cannabinoids (Salamone et al. 2022; Chirambo et al. 2024). These extended categories broaden the spectrum of chemical phenotypes and are especially relevant for fiber-type hemp, industrial applications, or research of minor cannabinoids and non-psychotropic chemovars (Aizpurua-Olaizola et al. 2016).

An overview of the main biosynthetic routes involved in secondary metabolism is shown in Fig. 12.2. Panel **a** summarizes the general metabolic framework in plants, highlighting the connections between primary carbohydrate metabolism and secondary metabolite biosynthesis through the shikimic acid, malonic acid, mevalonate (MVA), and 2-C-methyl-D-erythritol 4-phosphate (MEP) pathways (Khare et al. 2020). Panel **b** focuses on *C. sativa*, illustrating the convergence of these routes in the formation of its main classes of bioactive compounds. The malonic acid pathway yields olivetolic acid, which condenses with the prenyl intermediate geranyl pyrophosphate (GPP) —originated from the MEP/MVA pathways—to form cannabigerolic acid (CBGA), the central precursor of the major cannabinoids. Meanwhile, terpenoids derive from the condensation of isopentenyl diphosphate (IPP) and dimethylallyl diphosphate (DMAPP), sterols arise from the cyclisation of squalene, and flavonoids originate from phenylpropanoid intermediates through the shikimic acid pathway (Jin et al. 2020). This integrated metabolic network underlies the complex chemical diversity and tissue-specific distribution of secondary metabolites in cannabis, with terpenes and cannabinoid production within glandular trichomes, whereas flavonoids and sterols are mainly produced in other tissues, such as leaves, stem bark, and root.

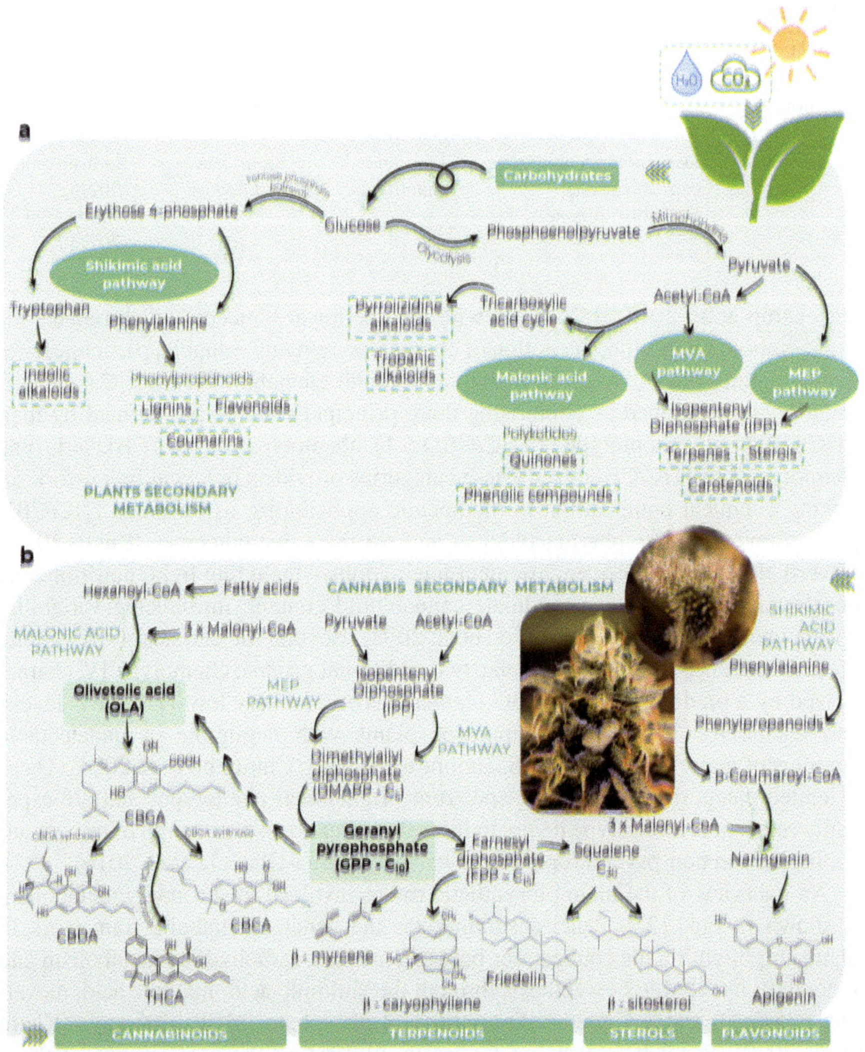

Fig. 12.2 Overview of secondary metabolism in plants and specific biosynthetic routes leading to the major bioactive compounds in *Cannabis sativa*. (**a**) General secondary metabolism pathways in plants derived from carbohydrate metabolism. (**b**) Detailed scheme of the main biosynthetic pathways in cannabis, showing the convergence of the malonic acid, mevalonate (MVA), and 2-C-methyl-D-erythritol 4-phosphate (MEP) pathways leading to the biosynthesis of cannabinoids, terpenoids, and sterols. Flavonoid biosynthesis involves the shikimic acid pathway. (*Source:* unpublished figure of Aguila-Wharton & Fernández)

12.2 Anti-aging Properties of *Cannabis sativa*

The aging process is intrinsically linked to the progressive accumulation of molecular damage, primarily driven by oxidative stress and a state of chronic, low-grade inflammation referred to as "inflammaging". These processes are central to the pathogenesis of a spectrum of age-related conditions affecting the nervous system, cardiovascular health, metabolism, and skin integrity. Within this context, secondary metabolites from cannabis have emerged as molecules of significant scientific interest. Their potent antioxidant, anti-inflammatory, and cytoprotective properties position them as promising candidates for mitigating age-associated molecular dysfunctions and potentially promoting health span. This premise is supported by a growing body of preclinical evidence, including studies showing that cannabinoids can extend lifespan and delay age-related decline in model organisms (Wang and Arnold 2024).

The antioxidant capacity of cannabinoids like CBD, THC, and CBN is well-documented, often exceeding that of classic antioxidant vitamins such as C and E (Kopustinskiene et al. 2022; Pagano et al. 2023; Ni et al. 2023). This activity is not a singular phenomenon but operates through two primary and complementary mechanisms. Direct free radical scavenging, driven by the phenolic chemistry of cannabinoids, which allows them to act as electron donors, directly neutralizing reactive oxygen species (ROS). For instance, Zagórska-Dziok et al. (2021) demonstrated that cannabis extracts exhibited a strong, dose-dependent ability to scavenge DPPH radicals, attributed to the high content of phenolic compounds and flavonoids. Also, extracts from various cultivars, such as those analyzed by Beleggia et al. (2023) and Izzo et al. (2020), showed substantial DPPH and ABTS radical scavenging activities, confirming that different plant chemovars retain significant direct antioxidant power.

A more potent and sustained mechanism involves the induction of endogenous antioxidant enzymes. Rather than merely acting as suppressible antioxidants, compounds like CBD activate the Nrf2 signaling pathway, a master regulator of cellular defense. This leads to the upregulation of cytoprotective enzymes, like heme oxygenase-1 (HO-1) and superoxide dismutase (SOD) (Casares et al. 2020; Khaksar et al. 2022; Ni et al. 2023). This mechanism is evolutionarily conserved, as demonstrated in *Caenorhabditis elegans* models where CBD promotes longevity by enhancing the expression of antioxidant enzymes like SOD-3 (Wang and Arnold 2024).

Both direct and indirect scavenging potential have been consistently reviewed across numerous studies (Atalay et al. 2019; Kopustinskiene et al. 2022; Pagano et al. 2023; Ni et al. 2023; Wang and Arnold 2024; Bukowska 2024), some of which are summarized in Table 12.2.

Chronic inflammation is a hallmark of aging, and phytocannabinoids modulate a wide network of pro-inflammatory signals. CBD, for example, inhibits the central NF-κB pathway, leading to the downregulation of key pro-inflammatory cytokines like IL-6, TNF-α, and IL-1β (Ni et al. 2023). This has direct implications for skin aging, where inflammation driven by environmental stressors like UVA radiation accelerates degradation of the extracellular matrix (Luangpraditkun et al. 2024).

The confluence of antioxidant and anti-inflammatory properties translates into significant neuroprotective effects, with evidence pointing to benefits in both health span and life span (Wang and Arnold 2024). In models of Alzheimer's disease, CBD reduces β-amyloid plaques and protects against Aβ-induced oxidative stress and

Table 12.2 Antioxidant activity and anti-aging potential reported from cannabis secondary metabolites

Starting material	Main compounds	Key finding	Reference
In vitro *model*			
scCO_2 extractions of inflorescences and isolated cannabinoids. *cv. Not reported.*	CBD and THC	ABTS (IC_{50} µg mL^{-1}): 3.4 ± 0.8–7.1 ± 0.5; DPPH (IC_{50} µg mL^{-1}): 141.9 ± 10.2–640.1 ± 16.9; HOCl (IC_{50} µg mL^{-1}): 2.9 ± 0.2–6.5 ± 0.5.	Hacke et al. (2019)
Ethanol, hexane, and scCO_2 root extracts. *cv. Futura 75, Felina 32, and Uso 31.*	Triterpenes and phytosterols	FRAP (mg VCE 100 g^{-1} DW): 20.6–97.1; ABTS (mg VCE 100 g^{-1} DW): 22.9–89.	Kornpointner et al. (2021)
Water-ethanol inflorescence extracts, ultrasound (UAE), and magnetic stirrer-assisted (MAE). *cv. not reported.*	Phenols, flavonoids, chlorophylls, and cannabinoids	AA: UAE > MAE. DPPH: UAE >40%; MAE ~30%.	Zagórska-Dziok et al. (2021)
Seed extracts. *cv. Futura 75.*	Tocopherols, carotenoids, polyprenols	ABTS (mg TE 100 g^{-1}) 734.2; FRAP (mg TE 100 g^{-1}): 574.5.	Irakli et al. (2019)
Cold-pressed seed oil. Seed oil (FHSO), lipophilic fraction (LF), and hydrophilic fraction (HF). *cv. Finola.*	Phenolic acids, flavanones, flavonols, isoflavones, and flavones	FHSO ↑ AA. DPPH (mmol TE 100 g^{-1} oil): 146.76; BCB: BHT > LF > FHSO > HF; CC (µmol EDTA 100 g^{-1} FW): 158.2 ± 2.11; ABTS (µmol TE 100 g^{-1} oil): 695.2; FRAP (µmol TE 100 g^{-1} oil): 3690.6	Smeriglio et al. (2016)
Cold-pressed seed oil, methanol, and ethanol extractions. *cv. not reported.*	Phenolic compounds	DPPH (IC_{50} vol extract/vol mixture): 3.433 ± 0.017; ABTS (%): 93.301 ± 1.099; FRAP (µmol Fe^{2+} L^{-1} extract): 1063.883 ± 39.225; CUPRAC (µmol TE L^{-1} extract): 420.471 ± 1.765.	Kalinowska et al. (2022)
Methanol inflorescence extraction. *cv. Kompoti, Tiborszallasi, Antal, and Carmagnola Cs.*	Flavonoids	DPPH (mmol TE kg^{-1}): 27.5–77.6.	Izzo et al. (2020)

(continued)

Table 12.2 (continued)

Starting material	Main compounds	Key finding	Reference
Ethanol extraction of the aerial part. *cv. KC Dora.*	CBDA and CBD	FRAP (%): 55.0 ± 1.3; CC (%): 27.26 ± 0.2; SRSA (%): 69.1 ± 3; HOSA (%): 221.5 ± 2.56; LPA (%):59.77 ± 2.0.	Petrovici et al. (2021)
Methanol and acetone leave extractions. *cv. not reported.*	Polyphenols	FRAP (%): 0.202–0.866. MeOH > acetone >50% MeOH >50% acetone.	Mkpenie et al. (2012)
Acetic acid/ hexane inflorescence extractions *cv. G-309.*	CBDVA, THCVA, THCV, CBCV	ABTS (IC_{50} μg mL^{-1}): 4.42–5.66; ORAC (IC_{50} μg mL^{-1}): 4.94–18.66; BCB (IC_{50} μg mL^{-1}): 21.22–9.94; CC (IC_{50} μg mL^{-1}): 32.95–34.32; DPPH (IC_{50} μg mL^{-1}): 373.66–291.25; FRAP (IC_{50} μg mL^{-1}): 108.31–86.87.	Muscarà et al. (2021)
Hydroalcoholic leaves and inflorescences extracts. *cv. strawberry.*	CBDA and CBD	FRAP (EC_{50} μg mL^{-1}): 95 ± 50; ABTS (EC_{50} μg mL^{-1}): 2 ± 0.1; DPPH (EC_{50} μg mL^{-1}): 67 ± 5.	Serventi et al. (2023)
Ethanol inflorescence extraction. *cv. Carmagnola.*	Terpenes, Cannabidiol, polyphenols	DPPH (EC_{50} mg mL^{-1}): 2.47 ± 0.10; CC (EC_{50} mg mL^{-1}): 0.39 ± 0.01. Peroxide value kinetics in linseed oil: With extract < with α-tocopherol < stripped oil.	Cantele et al. (2020)
Methanol inflorescence extract. *cv. Codimono, Carmaleonte, Futura 75, Santhica 27, Fibrante, and Carmagnola Selezionata.*	Phenols, terpenes, cannabinoids, tocopherols, and phytosterols	ABTS (mg TE g^{-1} DW): 139.2–211.2. DPPH (mg TE g^{-1} DW): 100.5–122.6.	Beleggia et al. (2023)
Methanol female (F) and male (M) stem, leaves, and inflorescence extracts. Subsequently partitioned in buthanol. *cv. not reported.*	Polyphenols	ABTS (mM TE): 144.46–1.47; DPPH: M-stem-Buthanol > F-stem-Buthanol > F-leaves-Buthanol > F-inflorescence-Buthanol> M-leaves-Buthanol > M-inflorescence-Buthanol; CC (%): 9.46–84.94. Peroxidation inhibition for all fractions.	Nadeem et al. (2012)
Seeds, flour, and cold-pressed seed oil. *cv. Fedora.*	Phytosterols, polyphenols	DPPH (%): Seeds: 51.5%; flour: 46.8%; oil: 8.2%.	Siano et al. (2018)

(continued)

Table 12.2 (continued)

Starting material	Main compounds	Key finding	Reference
Heptane inflorescence extraction. *cv. not reported.*	THCA, THC, CBD	DPPH (μg TE mg^{-1} substance): THC: 669 ± 11.1; THCA: 16 ± 3.2; CBD: 356 ± 29.5.	Moldzio et al. (2012)
Methanol extract of aerial parts during flowering. *cv. not reported.*	Cannabinoids, phenolics, flavonoids	FRAP (μmol $Fe^{2+}g^{-1}$ DW): 12.80 to 21.58. DPPH (%): 52.80–76.58.	Mirzamohammad et al. (2021)
Methanol seeds and sprouts extracts. *cv. Futura 75.*	Cannabisin A/B/C, caffeoyl-tyramine	DPPH (%): Seeds 40, sprouts 48–52; ORAC (μmol TE g^{-1} DW): Seeds 134 ± 3; sprouts 239 ± 2–243 ± 3.	Frassinetti et al. (2018)
Isolated compounds.	Cannabinoids	ABTS: All cannabinoids > Trolox (except THCA).DPPH: Only CBG and THC > Trolox; ORAC: Neutrals > Trolox; BCB: Neutrals < Trolox; CUPRAC and FRAP: Neutrals > Trolox.	Dawidowicz et al. (2021)
Methanol extracts of cell suspensions. *cv. Futura.*	Phenolics	*Elicitation with methyl jasmonate and tyrosine.* DPPH scavenging ↑ 51% with elicitation.	Gabotti et al. (2019)
Essential oils (EOs) of inflorescences and leaves. *cv. not reported.*	Terpenes	DPPH (IC_{50} mg mL^{-1}): 0.981 ± 0.059 mg mL^{-1}; FRAP (EC_{50} mg ml^{-1}): 1.74 ± 0.05; TAC (mg AAE g^{-1}): 0.101 ± 0.001.	El-Mernissi et al. (2024)
Ethanol, chloroform, and hexane extracts of cannabis flowers. *cv. Beldeya.*	Phenolics	DPPH (IC_{50} μg mL^{-1}): Ethanol: 231.39; hexane: 376.40; chloroform: 769.60.	Ahidar et al. (2024)
Methanol, pentane, and EOs from leaves, inflorescences, stems, seeds, and roots. *cv. Futura 75 and Felina.*	Terpenes, cannabinoids, and sterols	Leaf EOs had ↑ AA; DPPH (mmol L^{-1}): 15.03–35.04; root extract scavenged H_2O_2 as effectively as hemp stem extract.	Judžentienė et al. (2023)
EOs of inflorescences. *cv. Futura 75.*	Terpenes	FRAP (mg TE g^{-1}): 706.25; DPPH (mg TE g^{-1}): 11.70; ABTS (mg TE g^{-1}): 137.2.	Palmieri et al. (2021)
EOs of inflorescences. *cv. Futura 75.*	Terpenes	DPPH (mg TE g^{-1}): 63.38 ± 0.08; FRAP (mg TE g^{-1}): 438.52 ± 6.92.	Pellegrini et al. (2021)
Ethanol extraction of inflorescences. *cv. CAT1.*	CBD and THC	FRAP (μmol mg^{-1} protein): RD: 465.7 ± 5.61, SRD: 395.9 ± 8.53; SRD + Ca: 467.3 ± 16.63.	Degrave et al. (2023)

(continued)

Table 12.2 (continued)

Starting material	Main compounds	Key finding	Reference
EOs and the aromatic water (AW) extract of aerial parts. *cv. Futura 75.*	Cannabinoids, terpenes, phenolic, and flavonoid patterns	DPPH (mg TE g^{-1} sample): AW: 39.71 ± 1.54, EO: 5.56 ± 0.42; ABTS (mg TE g^{-1} sample): AW: Not active, EO: 103.38 ± 0.07; CUPRAC (mg TE g^{-1} sample): AW: 109.15 ± 1.78, EO: 141.15 ± 2.74; TAC (mmol TE g^{-1} extract or oil): AW: 1.42 ± 0.05, EO: 35.12 ± 1.63; FRAP (mg TE g^{-1} sample): AW: 82.93 ± 1.70, EO: 57.02 ± 0.69.	Zengin et al. (2018)
Hexane extract from hulled ground seeds. *cv. not reported.*	Phenolics	TAC (μM TE): 2720.95 ± 85.58; ABTS (%): 60.00 ± 2.73 at a dilution of 1:2.	Frazzini et al. (2024)
Microwave-assisted extraction of inflorescences. Aqueous (AE) and hexane (HE) extracts. *cv. Futura 75.*	Terpenes and CBD	DPPH (μg mL^{-1}): AE: 60, HE: 97; SRSA (μg mL^{-1}): AE: 9, HE: 127; ORAC (μmol TE g^{-1}): AE: 0.91, HE: 7.37; FRAP (μmol Fe^{2+} g^{-1}): AE: 17.75, HE: 21.94.	Cásedas et al. (2022)
EOs of aerial parts. *cv. not reported.*	Terpenes	DPPH (mg mL^{-1}): 1.6 ± 0.1; BCB (mg mL^{-1}): 1.8 ± 0.2; FRAP (mg mL^{-1}): 0.9 ± 0.1.	Nafis et al. (2019)
EtOH extract of leaves into an aqueous solution. *cv. RPF3.*	THC, CBD	DPPH (EC_{50} μg mL^{-1}): 277.9 ± 2.41.	Luangpraditkun et al. (2024)
Cell-based model			
Methanol extracts of cell suspensions. *cv. Futura 75.*	Phenolics	*Elicitation with methyl jasmonate and tyrosine.* CAA-RBC (%): 80 ↑ AA after elicitation. Anti-hemolytic effect in RBC.	Gabotti et al. (2019)
Ethanol, hexane, and $scCO_2$ root extracts. *cv. Futura 75.*	Triterpenes and phytosterols	*Saccharomyces cerevisiae.* Ethanol-Futura 75 ↑ intracellular oxidation by 20%.	Kornpointner et al. (2021)
Methanol seeds and sprouts extracts. *cv. Futura 75.*	Cannabisin A/B/C, caffeoyl-tyramine	*Red blood cells (RBC) and Saccharomyces cerevisiae.* CAA-RBC: Seeds 82 ± 4; sprouts 34 ± 2 to 42 ± 3.5; hemolysis inhibition (%): 52 seeds, 30–39 sprouts; both expressed antimutagenic effects.	Frassinetti et al. (2018)

(continued)

Table 12.2 (continued)

Starting material	Main compounds	Key finding	Reference
Isolated compounds.	CBD, THC	*Rat cortical neuron cultures.* CBD and THC ↑ glutamate-induced neuronal death (EC_{50} 2–4 mM); protective effect: CBD ≈ BHT > α-tocopherol and ascorbate.	Hampson et al. (1998)
Heptane inflorescence extraction. *cv. not reported.*	THCA, THC, CBD	*Mice mesencephalic cultures.* THC and THCA protect dopamine neurons from MPP+ toxicity.	Moldzio et al. (2012)
Ethanol extraction of the aerial part. *cv. KC Dora.*	CBDA and CBD	*Fibroblasts (NHDF) and cancer lines.* CBD ↑ H_2O_2 in all cell lines. CBD ↑ GSR in NHDF. CBD ↑ apoptosis in cancer lines.	Petrovici et al. (2021)
Isolated compound.	CBD	*Pheochromocytoma PC12 rat cells.* Pre-treatment with CBD (10^{-7}–10^{-4} M) before β-amyloid peptide exposure: ↑ cell survival; ↓ ROS production; ↓ lipid peroxidation; ↓ caspase 3 levels; ↓ DNA fragmentation; ↓ intracellular calcium.	Iuvone et al. (2004)
Ethanol extract of leaves into an aqueous solution. *cv. RPF3.*	THC, CBD	*Fibroblasts (NHDF).* Pre-treatment with extract (20 μg mL^{-1}) ↓ UVA damage; ↑ I procollagen; ↓ MMP-1 ↓ IL-6; ↓ cellular senescence. No cytotoxicity at 20 μg mL^{-1} or lower.	Luangpraditkun et al. (2024)
Isolated compounds.	THC, CBN, and CBD	*Rat cerebellar granule cells.* Cannabinoids ↑ AA in the brain lipid oxidation assay; ↓ cellular senescence.	Marsicano et al. (2002)
EOs and the aromatic water (AW) extract of aerial parts. *cv. Futura 75.*	Cannabinoids, terpenes, phenolic, and flavonoid patterns	*Cancer lines: MCF-7, MDA-MB-468, Caco-2, Mz-ChA-1.* Cytotoxicity after 24 h exposure (IC_{50} μg mL^{-1}): MCF-7*:* 83.2, MDA-MB-468: 53.0, Caco-2: 28.7, Mz-ChA-1: 22.3.	Zengin et al. (2018)
Microwave-assisted extraction of inflorescences. Aqueous (AE) and hexane (HE) extracts. *cv. Futura 75.*	Terpenes and CBD	*Mouse neuroblastoma cell line.* AE: No cytotoxicity; AE (25–50 μg ml^{-1}): ↑ mitochondrial activity vs. control cells; HE showed cytotoxicity; AE (25–200 μg ml^{-1}) ↑ cell viability in 20–30% against H_2O_2 exposure; ↓ 200% ROS production.	Cásedas et al. (2022)

(continued)

Table 12.2 (continued)

Starting material	Main compounds	Key finding	Reference
Water-ethanol inflorescence extracts, ultrasound-assisted (UAE) and magnetic stirrer-assisted (MAE). *cv. not reported.*	Phenols, flavonoids, chlorophylls, and cannabinoids	*Keratinocytes and fibroblasts.* SOD activity (%): UAE 178, MAE 165 (dose-dependent). ↓ ROS: Fibroblasts: At 1 and 100 μg mL^{-1}; keratinocytes: 1–250 μg mL^{-1}.	Zagórska-Dziok et al. (2021)
In vivo *model*			
Isolated compound.	CBD	*Spinal cord injury in rats.* Acute treatment with CBD (2.5–20 mg kg^{-1}): ↓ nociceptive behaviors, ↓ LP, ↑ GSH in injured tissue 15 days after injury.	Baron-Flores et al. (2022)
Isolated compound.	THC	*Type 2 diabetic rats (DB).* Erythrocyte GSH (nmol g^{-1} Hb): DB:3.93 ± 0.39, DB + THC: 5.53 ± 0.22; plasma CAT (U ml^{-1}): DB: 112.5 ± 15.98, DB + THC: 84.78 ± 15.68; plasma SOD (U ml^{-1}): DB: 3.70 ± 1.39, DB + THC: 8.64 ± 0.16.	Coskun and Bolkent (2016)
EOs of inflorescences and leaves. *cv. not reported.*	*Terpenes*	*Acute oral toxicity study on mice.* No abnormalities or toxicity at 2 mg kg^{-1}.	El-Mernissi et al. (2024)
Isolated compound.	CBD.	*Ischemic stroke in rats.* ↓ infarct volume and lipid peroxidation, ↑ SOD and CAT; anti-apoptosis role: ↓ caspase-3 and P53, and ↓ pyknotic neurons.	Khaksar et al. (2022)
Ethanol extraction of inflorescences. *cv. CAT1.*	Full-spectrum oil. CBD and THC.	*Metabolic syndrome (MS) model induced by a sucrose-rich diet.* ↓ serum triglyceride, total cholesterol, LDL-c, ALT, uric acid, and TBARS levels; ↑ serum HDL-c.	Degrave et al. (2023)

$scCO_2$ supercritical carbon dioxide, *cv.* cultivar, *ABTS* 2,2′-azino-bis(3-ethylbenzothiazoline-6-sulfonic acid), *DPPH* 2,2-diphenyl-1-picrylhydrazyl, *HOCl* hypochlorous acid, *FRAP* ferric reducing antioxidant power, *VCE* vitamin C equivalent, DW dry weight, *TE* trolox equivalents, *AA* antioxidant activity, *BCB* β-carotene bleaching assay, *BHT* Butylated hydroxytoluene, *CC* chelating capacity, *FW* fresh weight, *CUPRAC* cupric reducing antioxidant capacity, *SRSA* superoxide radical scavenging activity, *HOSA* hydroxyl radical scavenging activity, *LPA* lipid peroxidation inhibitory assay, *ORAC* oxygen radical absorbance capacity, *EOs* essential oils, *TAC* total antioxidant capacity, *AAE* ascorbic acid equivalents, *RD* reference diet, *SRD* sucrose-rich diet, *SRD + Ca* sucrose rich diet + cannabis, *CAA-RBC* cellular antioxidant activity in red blood cells, *RBC* red blood cells, *MPP+* 1-methyl-4-phenylpyridinium, *GSR* glutathione reductase, *ROS* reactive oxygen species, *UVA* ultraviolet a radiation, *MMP-1* matrix metalloproteinase-1, *IL-6* interleukin-6, *SOD* superoxide dismutase, *LP* lipid peroxidation, *GSH* glutathione, *Hb* hemoglobin, *CAT* catalase, *P53* cellular tumor antigen p53, *LDL-c* low-density lipoprotein cholesterol, *ALT* alanine aminotransferase, *TBARS* thiobarbituric acid reactive substances, *HDL-c* high-density lipoprotein cholesterol

neurotoxicity (Wang et al. 2021; Zhang et al. 2022). Similarly, in Parkinson's disease models, CBD and CBDV protect dopaminergic neurons, reduce α-synuclein accumulation, and attenuate oxidative stress, resulting in improved motor function and extended lifespan (Muhammad et al. 2022; Wang et al. 2023). These profound effects in invertebrates are supported by cell-based studies.

Beyond their neuroprotective and anti-aging properties, phytocannabinoids are increasingly recognized for their significant analgesic potential, contributing to improved quality of life and health span. Chronic pain, particularly neuropathic pain, is often driven by underlying inflammation and oxidative stress. Pagano et al. (2023) highlight that cannabinoids modulate pain pathways through multiple mechanisms, including the suppression of pro-inflammatory cytokines and the activation of CB1 receptors in the central and peripheral nervous systems, which inhibits the release of neurotransmitters involved in pain signaling. This mechanistic understanding is supported by in vivo studies compiled in Table 12.2. The study of Baron-Flores et al. (2022) demonstrated that CBD administration significantly reduced nociceptive behaviors in a rat model of spinal cord injury, a finding directly linked to the reduction of oxidative stress markers like lipid peroxidation. Similarly, Khaksar et al. (2022) reported the anti-nociceptive and anti-apoptotic effects of CBD in a rat model of ischemic stroke. Bukowska (2024) reinforces this by detailing the clinical translation of these effects, noting the approval of Sativex® (nabiximols, an oromucosal spray that contains a 1:1 ratio of THC and CBD) for neuropathic pain and spasticity treatment in multiple sclerosis patients, and discussing the palliative use of cannabinoids in cancer patients to alleviate pain (Boivin 2021).

Taken together, this compiled evidence demonstrates that the therapeutic potential of cannabis derivatives is driven by a multimodal mechanism of action. They concurrently target oxidative stress (via direct scavenging and Nrf2 pathway activation), inflammation (via NF-κB suppression and MMP inhibition), and fundamental anti-aging processes like the enhancement of autophagy and the reduction of cellular senescence. The consistency of these findings across different model organisms, from nematodes to mammalian cells and vertebrates, strongly supports the premise that cannabinoids can promote health span.

The research summarized not only confirms that antioxidant activity is a widespread property across different plant parts, extraction methods, and cultivars, but also provides the foundational in vitro and in vivo data that help explain their broad therapeutic potential. While the preclinical research is compelling, the next crucial steps involve standardizing formulations, exploring the synergistic effects of the full cannabinoid and terpene spectrum, and rigorously translating these promising findings into clinical applications for age-related and pain-related human conditions.

12.3 A Biotechnological Approach for Anti-aging Metabolites Production

The scientific and industrial interest in cannabis has expanded remarkably in recent decades, driven by its diverse applications in medicine, biotechnology, and industry. Although cultivated since antiquity, the species is currently experiencing renewed global attention due to the growing recognition of its therapeutic potential, advances in plant biotechnology, and progressive legal reforms. However, this surge in interest has also exposed persistent challenges associated with conventional cultivation, including strict regulatory constraints, pronounced genetic heterogeneity, variability in metabolite profiles, susceptibility to pathogens, and the gradual loss of vigor observed in aging mother plants (Gao et al. 2014; Punja 2021; Crowley et al. 2024).

In this context, in vitro culture systems have emerged as powerful biotechnological tools capable of overcoming many of these limitations. Techniques such as micropropagation, somatic embryogenesis, and cell suspension cultures provide a controlled, scalable, and pathogen-free environment for the uniform production of plant biomass and its associated secondary metabolites. Critically, these systems allow for the clonal propagation of elite genotypes with stable and enhanced contents of key anti-aging compounds—such as cannabinoids, flavonoids, and other phenolics—addressing one of the major limitations of field-grown plants: metabolite inconsistency.

Despite the increasing number of studies on cannabis biotechnology, comprehensive reviews that integrate techniques, challenges, applications, and future perspectives remain scarce. Therefore, this chapter aims to deliver an in-depth and critical synthesis of the current state of cannabis in vitro propagation systems, with a specific focus on their role in supporting the standardized production of anti-aging metabolites. Beyond a general biotechnological overview, it places particular emphasis on the strategic role of plant growth regulators (PGRs) as central determinants of direct organogenesis and high-efficiency shoot multiplication.

More specifically, we (i) outline the core steps for establishing aseptic cultures; (ii) examine the mechanisms of action and key effects of auxins, cytokinins, and other PGR derivatives in shoot and root development; (iii) discuss substrates and protocols for the acclimatization of micropropagated plantlets; and (iv) underscore the importance of ensuring genetic fidelity through morphological, biochemical, and molecular markers to guarantee the true-to-type nature and metabolic consistency of the regenerated plants. Together, these components aim to guide the development of robust micropropagation platforms capable of reliably supplying high-quality, genetically stable, and metabolically consistent plant material for anti-aging research and applications.

12.3.1 The Foundation: Establishment of Aseptic Cultures

Given the biotechnological relevance of cannabis for the production of high-value metabolites, the first and most decisive step is the successful establishment of viable in vitro cultures. Success at this stage determines whether subsequent multiplication and rooting can proceed efficiently, as it relies on the careful selection of explants, effective surface-disinfection, and the optimization of basal media components. The key steps and critical factors involved in the initiation phase of cannabis are summarized in Fig. 12.3, outlining the workflow from explant selection through propagation, highlighting the main challenges encountered during cannabis culture establishment. In vitro establishment provides the foundation upon which all subsequent processes—shoot multiplication, rooting, and acclimatization—depend. Cannabis is notably challenging during initiation due to its high sensitivity to contamination, genotypic variability, and physiological disorders such as hyperhydricity. As a result, establishing standardized, genotype-tailored protocols is indispensable for enabling reliable large-scale micropropagation systems.

12.3.2 Selection and Disinfection of Explants

The successful establishment of in vitro cultures of cannabis is a critical first step for mass propagation, and it is highly dependent on two key factors: explant selection and effective disinfection protocols. The anatomical location of the stem explant significantly impacts morphogenetic performance. Studies by Hesami et al. (2023a) demonstrated that basal and near-basal stem segments from mature mother plants (~2.5 years old) yielded the highest multiplication rates, tallest shoots, and wider leaves. Furthermore, the nodal position (e.g., positions 1–5 along the stem) directly affects shoot regeneration success, with lower nodal segments often showing superior performance (Mestinsek-Mubi et al. 2020). Other promising explants include

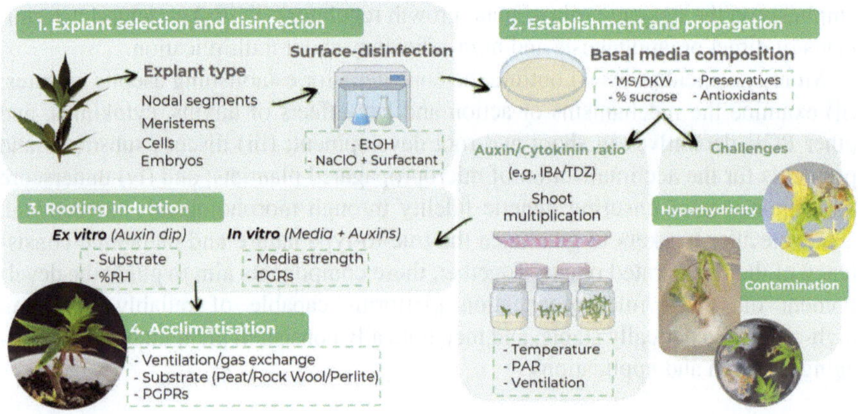

Fig. 12.3 Schematic overview of the main stages involved in the establishment and propagation of *C. sativa* in vitro. The workflow highlights explant selection and surface-disinfection (1), culture initiation and shoot multiplication (2), rooting induction (3), and acclimatization (4), along with key influencing factors and common challenges such as hyperhydricity and contamination. (*Source*: unpublished figure of Aguila-Wharton & Fernández)

hypocotyls from seedlings, which showed high spontaneous shoot regeneration rates, and floral tissues via floral reversion (Galán-Ávila et al. 2020; Monthony et al. 2021a).

Effective surface disinfection is paramount due to high susceptibility to fungal and bacterial contamination. Multiple studies consistently report multi-step protocols. A typical and effective method involves a brief rinse in ethanol (70–75% for 1–2.5 min) followed by a longer immersion in a sodium hypochlorite solution (0.625%–3% active chlorine for 10–25 min). While the incorporation of systemic fungicides like carbendazim has been explored to enhance disinfection, Das et al. (2024) demonstrated that its use, even at low concentrations (0.5–1 $g.L^{-1}$), significantly increased tissue mortality and reduced explant viability in vegetative tissues, ultimately diminishing the number of suitable explants for subculture. Their findings support the reliance on sodium hypochlorite as the primary disinfecting agent for sensitive shoot tips and nodal segments. It is noteworthy, however, that for more robust explants like seeds, a pretreatment with a fungicide like Captan® (2 $g.L^{-1}$) proved effective, highlighting that disinfection protocols must be optimized for the specific explant type and its initial microbial load (Pedrosa et al. 2024). Importantly, the growing environment of the mother plant was a more critical factor than bleach concentration alone; explants from plants maintained in a clean growth chamber showed near-total absence of contamination (1.7%), whereas those from a greenhouse environment reached 88.3%, highlighting the paramount importance of the initial phytosanitary status of the stock plant (Stephen et al. 2023). The addition of a surfactant like Tween 20 (0.05%–0.1%) is crucial to improve surface contact and efficacy. Multiple rinses (at least three) with sterile water are essential to remove toxic residues. It is critical to note that disinfection success is highly genotype-dependent and influenced by the explant type itself, necessitating protocol optimization for specific varieties (Ioannidis et al. 2020; Holmes et al. 2021).

This body of evidence suggests that the initial phytosanitary status of the donor plant is paramount, as the internal tissues of cannabis plants host a diverse community of endophytic fungi and bacteria. These endophytes, including genera like *Fusarium, Penicillium,* and *Aspergillus,* are protected within the plant's vascular system and pith tissues, making them difficult to eradicate with surface sterilization alone (Buirs and Punja 2025). Studies have shown that cuttings and seeds inherently carry these microbial communities from the mother plant, which can lead to contamination during in vitro establishment.

Notably, while tissue culture from meristem tips can eliminate most background endophytes, the resulting sterile plantlets can be rapidly recolonized upon exposure to a non-sterile environment, sometimes by detrimental pathogens (Buirs and Punja 2025). This underscores that a holistic strategy—combining clean stock plants, effective explant disinfection, and sometimes the use of systemic fungicides in the initial media—is essential to manage the endophytic load and ensure successful culture initiation.

For a comprehensive discussion, it is valuable to note that meristematic tips are valuable explants for establishing virus-free cultures. In other words, pre-acclimatizing donor plants in controlled conditions can reduce endogenous microbial load. Alternative disinfecting agents like chlorine dioxide (ClO_2) and the incorporation of systemic antibiotics or fungicides (e.g., PPM®) into the initial culture media are often necessary to combat endophytic contaminants.

In conclusion, the meticulous selection of explants from optimal positions and the implementation of a robust, genotype-tailored disinfection protocol are fundamental pillars for successfully establishing sterile cannabis tissue cultures, forming the essential foundation for subsequent PGR-mediated multiplication and rooting stages. The evidence underscores that an integral approach, which includes maintaining clean stock plants as a primary strategy, is equally as important as the surface sterilization protocol itself for improving initiation rates.

12.3.3 Basal Media Composition

The success of micropropagation depends critically on the composition of the basal media (Das et al. 2024), which must provide an adequate balance of nutrients, a carbon source, and growth regulators to direct in vitro morphogenesis.

Published studies report that sucrose concentrations between 2% and 3% (w/v) most often support shoot, callus, and root development in cannabis tissue culture. In solid or semisolid systems, Pepe et al. (2021) demonstrated that 3% sucrose yields superior growth metrics compared to 1%, significantly enhancing shoot length, root length, node number, and canopy surface area. These findings are partially supported by Stephen et al. (2023), who reported optimal growth at 1.5–3.0% sucrose, with 1.5% being preferable due to reduced hyperhydricity (a physiological disorder characterized by excessive tissue hydration, discussed in detail in the following subsection) and lower cost. Conversely, in liquid temporary immersion systems, Rico et al. (2022) found that a lower sucrose concentration of 0.5% supported shoot proliferation effectively while reducing hyperhydricity, whereas the omission of sucrose led to explant death.

Regarding basal media composition, several screening protocols have compared Murashige and Skoog (MS) with Driver and Kuniyuki Walnut (DKW) formulations. Some studies reported DKW's superiority, noting that—especially when supplemented with 0.5 μM thidiazuron (TDZ)—it produced up to 1.5 times higher multiplication rates and twice the canopy area compared to MS (Page et al. 2021). However, contrasting results were reported by Stephen et al. (2023) in a hemp cultivar, who found that MS and Linsmaier and Skoog (LS) media produced superior fresh weights, shoot lengths, and numbers of shoots compared to DKW and Woody Plant Medium (WPM). A comprehensive mineral optimization study by Zarei et al. (2023) identified specific limitations in MS formulation and demonstrated that elevated levels of ammonium nitrate (1.65 times the MS concentration) and mesonutrients ($CaCl_2$, KH_2PO_4, $MgSO_4$ at 1.5 times the MS concentration) significantly improved canopy area (up to 3.34-fold), fresh weight, and multiplication number (52% increase) across three diverse cultivars compared to standard MS.

These previous findings contrast with results from Das et al. (2024), which identified a modified MS-based medium as more effective for sustained multiplication, the optimized formulation—consisting of MS basal salts supplemented with calcium nitrate (0.71 g/L) and calcium gluconate (1.35 g/L), and solidified with 0.95% agar without plant growth regulators—promoted healthier explants, improved growth, and higher multiplication rates while minimizing hyperhydricity.

Environmental conditions also significantly influence media performance. Stephen et al. (2023) identified temperature as a critical factor, with 28 °C producing substantially better growth metrics compared to lower temperatures. This highlights that media optimization cannot be isolated from environmental parameters, as temperature conditions may modulate nutritional requirements and metabolic responses to different media formulations. Based on the evidence, cannabis micropropagation tends to perform better with mineral formulations richer in nitrogen and calcium than standard MS. However, optimizing media remains complex due to multifactorial interactions. Machine learning has been proposed as a powerful tool to navigate this complexity efficiently (Hesami and Jones 2020; Monthony et al. 2021b).

12.3.4 Technical Challenges and Limitations in Cannabis In Vitro Establishment

In vitro culture performance can be technically constrained by several factors, including medium composition, genotype-specific responses, contamination risks, and physiological disorders such as hyperhydricity and somaclonal variation.

One of the most frequent challenges is microbial contamination, where fungi, bacteria, and endophytes compromise explant viability (Das et al. 2024). Studies demonstrate that tailored sterilization protocols using sodium hypochlorite or mercuric chloride can significantly reduce contamination, although optimization is required for each genotype and explant type (Pepe et al. 2022; Wongchaya et al. 2024). Broad-spectrum antimicrobial agents, such as PPM®, may also be incorporated during early culture stages.

Hyperhydricity is another frequent disorder in cannabis, characterized by water retention within the apoplast, leading to translucent and brittle tissues (Isah 2019; Monthony et al. 2021b). This physiological alteration, commonly associated with suboptimal culture conditions, can be mitigated by regulating vessel ventilation (through gas exchange), temperature, and the concentration of cytokinins, gelling agents, and mineral nutrition (Zarei et al. 2023; Ahsan et al. 2025). Recent evidence suggests that even small modifications—such as lowering the temperature by 2 °C during the initiation phase—can significantly improve plantlet quality (Abiri et al. 2025). The Ahsan et al. (2025) review extensively covers the benefits of photoautotrophic micropropagation (PAM)—using sugar-free media and increased ventilation. The practical benefits of PAM and enhanced ventilation extend beyond mitigating hyperhydricity. Research by Murphy and Adelberg (2021) demonstrated that increased light intensity and ventilation in vitro directly translated to higher-quality microcuttings, impacting subsequent stages.

In our systems, hyperhydricity was strongly exacerbated under 3% sucrose in full-strength MS medium, producing severely water-soaked and malformed tissues in the MK × PP genotype (Fig. 12.4g, h). A 2% sucrose concentration also resulted in hyperhydric explants (Fig. 12.4e, f), suggesting that sucrose levels ≥2% likely acted as osmotic stressors that suppressed photosynthetic activity (Pepe et al. 2022).

Fig. 12.4 Morphology of *Cannabis sativa* explants (genotype MK×PP) after 30 days of culture under different treatments. (**a**, **c**, **e**, **g**) Top and (**b**, **d**, **f**, **h**) lateral views of cannabis explants grown on: (**a**, **b**) half-strength MS medium supplemented with 1% sucrose, (**c**, **d**) full-strength MS medium with 1% sucrose, (**e**, **f**) full-strength MS medium with 2% sucrose, and (**g**, **h**) full-strength MS medium with 3% sucrose. Scale bar = 1 cm. (Source: unpublished photos of Aguila-Wharton & Fernández)

This contrasts with reports in species such as orchids, figs, and even in other cannabis genotypes (Pinto et al. 2010; Elazab and Shaaban 2015; Lata et al. 2016), where comparable sucrose concentrations promote growth, underscoring the narrow and genotype-dependent response window that is typical of cannabis. At 1% sucrose, MK × PP plantlets exhibited markedly improved structure and overall development; however, mild hyperhydricity symptoms were still present, indicating that reducing sucrose alone was insufficient to fully restore normal morphology (Fig. 12.4c, d). By contrast, lowering the ionic strength of the medium (MS/2) emerged as the most effective strategy for controlling hyperhydricity in MK × PP, generating vigorous shoots with minimal or no symptoms (Fig. 12.4a, b). Similar improvements under reduced-strength media have been reported in clove and cucumber, likely due to enhanced water balance and improved development of the cuticle and functional stomata (Polivanova and Bedarev 2022).

Beyond physicochemical factors, recent studies indicate that hyperhydricity also has a hormonal basis. Stress-induced alterations in endogenous auxin biosynthesis and signaling can disrupt cell wall development and water homeostasis, while excessive ethylene accumulation under limited gas exchange conditions further exacerbates tissue hyperhydricity (Johnson et al. 2025).

Another strategy for addressing hyperhidricity is through mineral supplementation. Zarei et al. (2023) found that the addition of 4.86 mM potassium silicate to their optimized medium significantly reduced apoplastic water volume (up to 62%) and improved leaf morphology without negatively impacting growth parameters. Altogether, these results reinforce the idea that hyperhydricity management in cannabis requires a fine adjustment of both the carbon source and the mineral strength of the medium, in combination with environmental parameters.

Tissue recalcitrance is another commonly reported challenge in cannabis, impairing shoot proliferation or regeneration and resulting in limited responsiveness to culture media (Mestinsek-Mubi et al. 2020; Monthony et al. 2021b). Strategies such as the use of juvenile explants or fine-tuned PGR combinations can improve outcomes, though responses remain strongly genotype-dependent.

Browning caused by phenolic oxidation is also a limiting factor that can inhibit cell division and regeneration (Skrzypek et al. 2007); adding antioxidants such as ascorbic acid or activated charcoal has proven effective in reducing this effect (Poniewozik et al. 2022). Several of these issues can be further mitigated by incorporating specific medium additives—antioxidants, adsorbents, osmotic regulators, or signaling modulators.

Finally, somaclonal variation may compromise genetic uniformity, particularly cannabinoid profiles. To minimize this risk, the number of subcultures should be restricted, and axillary bud proliferation should be favored over callus-based systems. Ultimately, ensuring that micropropagated plants remain true-to-type requires formal genetic fidelity assessment.

Together, these limitations emphasize the importance of developing robust, genotype-specific protocols to ensure reliable and scalable cannabis micropropagation. Beyond the technical challenges, regulatory contexts can also act as barriers. While cannabis extracts exhibit therapeutic effects, their use remains restricted due to the psychotropic activity of THC. In contrast, CBD is a non-psychotropic compound, classified by the WHO Expert Committee on Drug Dependence (2017) as non-harmful and without abuse potential.

12.3.5 The Role of Plant Growth Regulators in Shoot Multiplication

The successful regulation of morphogenesis in cannabis in vitro relies fundamentally on the precise manipulation of Plant Growth Regulators (PGRs), which act as the primary biochemical switches controlling cell division, differentiation, and developmental fate. Since the classical work of Skoog and Miller (1957), the auxin–cytokinin balance has been recognized as the cornerstone of in vitro regeneration, determining whether cultured tissues undergo callogenesis, root induction, or de novo shoot formation. In micropropagation systems focused on shoot multiplication, exogenous cytokinins promote axillary bud activation by suppressing apical dominance, while low auxin levels act synergistically to sustain meristematic activity without redirecting development toward callus formation (George et al. 2008).

The optimization of PGR remains particularly challenging in cannabis due to strong genotype-dependent variability, a lack of protocol standardization, and frequent physiological disorders associated with cytokinin overexposure, such as hyperhydricity (Lata et al. 2010a; Hesami et al. 2021). Nonetheless, advances have demonstrated significant improvements in multiplication efficiency through the strategic use of aromatic cytokinins like *meta*-topolin (*m*T) or low concentrations of phenylurea-type cytokinins such as thidiazuron (TDZ), and refined auxin formulations tailored to specific chemotypes (Lata et al. 2009; Lata et al. 2016). Against this backdrop, understanding the mechanistic roles and interactive effects of individual PGRs is essential for establishing robust, scalable systems capable of producing true-to-type plant material with stable metabolic profiles. It is noteworthy, however, that this pursuit of optimization does not always necessitate their inclusion; as evidenced in protocols for cultivars like "Cherry Soda" and "Purple" (Johnson et al. 2025), a cytokinin-free medium can itself be the optimal formula for maximizing shoot proliferation, underscoring that the best PGR regimen is profoundly genotype-specific.

The following subsections examine the functional principles governing the auxin–cytokinin balance, the specific effects of commonly used cytokinins, the role of auxins in supporting organogenesis and rooting, and the contribution of additional regulators and additives that enhance shoot quality and overall culture performance.

The Cytokinin-Auxin Balance: The Cornerstone of In Vitro Regeneration The orchestration of plant growth and development in vitro is predominantly governed by the interplay between two central classes of PGRs: auxins and cytokinins. This dynamic balance is not merely additive but synergistic and/or antagonistic, dictating the fate of undifferentiated plant cells. The ratio of exogenous auxins to cytokinins in the culture medium serves as the primary switch determining the pathway of regeneration, either promoting unorganized cell proliferation (callogenesis) or directing organized organ development (organogenesis).

In the context of cannabis micropropagation for shoot multiplication, a high cytokinin-to-auxin ratio is universally employed to induce and promote axillary bud breaking and adventitious shoot formation. Cytokinins dominate this regime by stimulating cell division and suppressing the apical dominance of the main shoot apex. This releases the dormancy of lateral meristems, leading to the proliferation of multiple shoots. Their mechanism of action involves activating cell cycle genes and promoting the biosynthesis of endogenous cytokinins, creating a conducive environment for shoot initiation.

Conversely, auxins play a secondary but essential supportive role during the multiplication phase. At low concentrations, they act synergistically with cytokinins to stimulate meristematic activity and promote robust shoot development. However, their principal function becomes critical in the subsequent rhizogenesis stage, where a high auxin-to-cytokinin ratio is required to induce root formation. The precise equilibrium is critical. An imbalance, such as excessively high auxin levels during the multiplication phase, can inadvertently promote callus formation at the base of the explant, increasing the risk of somaclonal variation and hindering the development of healthy shoots. Therefore, optimizing the specific type, concentration, and ratio of these PGRs for each specific cannabis chemotype is the basis of developing an efficient and reliable micropropagation protocol aimed at producing true-to-type plants with high fidelity (George et al. 2008; Lata et al. 2010a; Hesami et al. 2021).

The following subsections will delve into the specific mechanisms and effects of individual cytokinins and auxins commonly used in *C. sativa* micropropagation protocols.

Cytokinins and Their Derivatives: Mechanisms and Morphogenic Effects Cytokinins promote cell division primarily by stimulating the transition in the cell cycle from the G1 to the S phase, activating key cell cycle genes and enabling the re-entry of differentiated cells into a proliferative state (Velappan et al. 2017). In in vitro systems, this results in enhanced meristematic activity, the suppression of apical dominance, and the formation of multiple axillary shoots. These mechanisms are central to direct regeneration in cannabis.

Cytokinin perception occurs through membrane-bound histidine-kinase receptors (e.g., AHK2, AHK3, AHK4/CRE1), which activate a phosphorelay cascade leading to the induction of type-B response regulators (Argueso and Kieber 2024). These transcription factors promote the expression of genes associated with cell division and shoot meristem identity, while type-A regulators provide negative feedback control (Hutchison et al. 2006).

In cannabis, cytokinins are particularly effective in overriding strong apical dominance and activating dormant axillary meristems, a species-specific constraint that often limits shoot proliferation. Low doses of TDZ or *m*T are consistently reported as the most responsive cytokinin treatments for direct shoot organogenesis (Lata et al. 2009, 2016). Cytokinin type and concentration profoundly influence cannabis micropropagation efficiency, with optimal treatments varying by outcome measure, basal medium composition, and explant type.

Among adenine-type cytokinins, BAP remains the most widely used due to its low cost, commercial availability, and long-standing effectiveness across a broad range of plant species. In shoot-multiplication systems, BAP promotes axillary-bud outgrowth, making it a reliable first-line cytokinin during protocol establishment. For cannabis, concentrations between 0.5–5.0 μM have been reported to induce high shooting frequencies (Ioannidis et al. 2020; Stephen et al. 2023), although its responsiveness is generally lower than that achieved with TDZ or with *m*T. Despite this general trend, BAP can outperform *m*T or TDZ in certain genotypes, underscoring the strong genotype dependency of cytokinin responses in cannabis. In a comparative study of breeding lines, Mestinsek-Mubi et al. (2020) found that TDZ- and *m*T-based media produced weak or inconsistent regeneration, whereas BAP—particularly when supplemented with low-dose indole-3-acetic acid (IAA), an auxin—yielded higher shoot numbers and markedly longer shoots. Similarly, Ioannidis et al. (2020) reported that 4.0 μM BAP produced the highest shoot number and length in both full-strength and half-strength MS media, emphasizing that cytokinin performance is influenced not only by concentration but also by the ionic strength and overall composition of the basal medium.

At higher doses, BAP may lead to compact shoots, occasional hyperhydricity, or callus formation when combined with auxins, reinforcing the need for genotype-specific optimization. Nevertheless, BAP remains a practical and robust cytokinin, particularly for routine micropropagation and in laboratories prioritizing cost-effectiveness and protocol reproducibility.

Unlike BAP, kinetin (KIN) frequently results in comparatively weaker shoot multiplication in *C. sativa*, producing fewer axillary shoots and limited elongation. However, its milder physiological effect may be advantageous in genotypes prone to hyperhydricity or excessive compaction, making KIN a secondary but occasionally valuable alternative in genotype-specific optimization. In some species, KIN can outperform classical cytokinins, such as 1 mg.L^{-1} yielding the highest regeneration percentage in cucumber (Abu-Romman et al. 2015) or 0.8–1.0 mg.L^{-1} supporting optimal shoot multiplication in citrus (Nugroho et al. 2024). Although in other species this phytoregulator has demonstrated strong shoot-regeneration potential, its use in cannabis remains somewhat limited to callus induction, in combination with auxins, rather than direct organogenesis (Lata et al. 2010a; Ioannidis et al. 2022). More recent comparisons further confirm that, at equivalent molarities, KIN produces fewer and shorter shoots than TDZ or BAP (Lata et al. 2009). As highlighted by Ndlovu et al. (2024), KIN remains underexplored in cannabis micropropagation, with limited evidence supporting its use for robust shoot regeneration. Nevertheless, its moderate cytokinin strength and low incidence of physiological disorders indicate that KIN may still hold untapped potential in genotype-specific frameworks, particularly where excessive cytokinin activity (e.g., from TDZ) is detrimental.

Although not a classical adenine-type cytokinin, TDZ exhibits exceptionally strong cytokinin-like activity, and it is widely recognized as one of the most potent inducers of shoot organogenesis in *C. sativa*. Across multiple studies, low to moderate TDZ concentrations (0.1–2.5 μM) consistently promote high shoot induction but often at the expense of shoot elongation and overall morphological quality. Kastelec et al. (2025) reported that supplementation with TDZ (0.1 $mg.L^{-1}$) markedly increased shoot multiplication, yet concurrently reduced shoot elongation and promoted callus formation. These effects are consistent with TDZ's dual cytokinin- and auxin-like activities, as well as its capacity to inhibit cytokinin oxidase and activate genes associated with heightened cell proliferation (Ali et al. 2022).

TDZ has also proven effective for regeneration from non-traditional explants. Sorokin and Kovalchuk (2025) reported that the combination of TDZ (0.4 $mg.L^{-1}$) and NAA (0.2 $mg.L^{-1}$) yielded the most effective callus-mediated shoot regeneration from hypocotyl explants, particularly when the cotyledonary node and shoot apical meristem were present, resulting in viable adult plants.

At higher concentrations, TDZ often intensifies physiological disorders such as hyperhydricity and shoot compaction. In this context, Stephen et al. (2023) found that 5 μM TDZ triggered the highest axillary-bud activation in nodal explants of hemp, yet this response came at the expense of shoot quality. Shoots produced under this treatment were markedly shorter and displayed severe hyperhydricity symptoms.

Collectively, the evidence so far indicates that TDZ is unmatched in its ability to trigger rapid and prolific shoot induction in cannabis, but its strong morphogenic activity also increases the risk of hyperhydricity, fasciation, stunted shoots, and unwanted callogenesis. Thus, low-dose TDZ (0.1–0.5 μM) or limited-exposure strategies generally offer the best balance between multiplication rate and shoot quality, though final optimization remains strongly genotype-dependent.

Importantly, *m*T has emerged as one of the most promising cytokinins for cannabis micropropagation, owing to its ability to promote vigorous shoot proliferation while minimizing physiological disorders commonly observed with other cytokinins (Gantait and Mitra 2021). Structurally, it derives from BAP, but it is hydroxylated at the meta position. This phytoregulator is associated with improved shoot elongation, enhanced chlorophyll development, and lower incidence of hyperhydricity. These features are particularly useful in cannabis, where tissue response is highly genotype-dependent and prone to this physiological disorder (Koç 2021; Krishna Vrundha et al. 2021). In nodal explants of cannabis, *m*T has been reported to induce high shoot formation frequency and superior shoot numbers per explant compared with BAP, while maintaining better morphological quality. Lata et al. (2016) documented the best response in terms of the number of shoots on MS medium supplemented with 2.0 μM *m*T, demonstrating its strong capacity for direct organogenesis.

Recent studies have expanded the use of *m*T to more intensive propagation systems. In temporary-immersion bioreactors, 2 μM *m*T enabled efficient large-scale multiplication (Rico et al. 2022), and sequential subculturing on DKW-based media supplemented with 0.5–5 μM *m*T supported sustained proliferation across multiple cycles. Plants regenerated under *m*T-based protocols have also shown strong ecophysiological performance, high acclimatization success, and both genetic and biochemical fidelity, underscoring its suitability for pharmaceutical-grade clonal propagation (Lata et al. 2016). An additional advantage of *m*T is its reduced formation of toxic cytokinin degradation products, which have been associated with lower oxidative stress, improved physiological stability, and reduced risk of somaclonal variation (Aremu et al. 2012).

Although *m*T generally promotes high-quality shoots, high concentrations can also be detrimental. Recent studies reported that 5 µM *m*T induced marked hyperhydricity in cannabis, emphasizing the need for careful dose optimization (Das et al. 2024; Johnson et al. 2025). Increased cytokinin levels may boost proliferation but can also delay rooting or induce localized callusing, reinforcing the importance of genotype-specific fine-tuning. Compared with TDZ, *m*T typically produces fewer shoots per explant, yet with substantially better physiological quality, reduced hyperhydricity tendency, and a smoother transition to rooting, making it especially advantageous for commercial micropropagation pipelines.

Evidence positions *m*T as one of the most balanced cytokinins for cannabis shoot multiplication, particularly for genotypes sensitive to TDZ-induced hyperhydricity or where downstream rooting and acclimatization efficiency are priorities. Nonetheless, even though *m*T consistently delivers favorable results across multiple reports, no single cytokinin can be regarded as universally optimal for cannabis. Strong genotype-dependence means that empirical comparison remains essential, as different cultivars may respond more efficiently to BAP, *m*T, KIN, TDZ, or combined regimes. Therefore, cytokinin selection should be determined through targeted optimization that considers genotype behavior, production goals, and practical constraints such as laboratory resources and regulatory requirements.

Consistent with the strong genotype dependency reported throughout cannabis tissue-culture research, Codesido et al. (2020) demonstrated that in vitro introduction is dictated by the interaction between basal medium composition, PGR supplementation, and cultivar identity. Working with axillary buds from six genotypes, the authors observed pronounced differences in explant survival and "reactiveness," with no single combination of salts, vitamins, or cytokinin treatments performing universally well. Their results reinforce that cannabis cannot be reliably established on a one-size-fits-all medium; instead, successful introduction requires fine-tuning of basal components and PGR regimes according to each genotype's physiological profile. Taken together, current evidence indicates that cytokinin choice in cannabis cannot rely on extrapolation from other species and requires systematic and empirical genotype-specific optimization.

Auxins and Their Role in Organogenesis and In Vitro Rooting Auxins constitute the second central pillar of morphogenic regulation in plant tissue culture, acting in close interplay with cytokinin-directed pathways. Their effects are intensely dose-dependent and highly sensitive to developmental context. At low concentrations, they act synergistically with cytokinins to sustain meristematic activity and stabilize shoot development, whereas at higher levels they redirect cellular fate toward callogenesis, somatic embryogenesis, or rhizogenesis (George et al. 2008).

At the molecular level, auxins trigger a broad reprogramming of somatic plant cells by modulating hormone-responsive signaling networks, reactivating the cell cycle, and reshaping epigenetic and transcriptional landscapes to enable organogenesis or callus formation. Upon perception by the receptor TIR1/AFB (Transport Inhibitor Response 1 and Auxin Signaling F-Box proteins), auxins promote the degradation of Aux/IAA repressors, thereby releasing auxin response factors (ARFs) that activate gene networks associated with cell proliferation, meristem identity, and developmental reprogramming (Weijers and Wagner 2016; Leyser 2018). Auxin accumulation also interacts with cytokinin signaling to establish the hormonal balance that dictates cell fate: high auxin-to-cytokinin ratios favor root initiation or unorganized cell proliferation, whereas lower ratios enable shoot organogenesis (Skoog and Miller 1957; Ikeuchi et al. 2019).

In cannabis, auxins play essential roles in two key processes: (i) the induction of adventitious roots in micropropagated shoots and (ii) callus initiation for indirect regeneration or transformation pipelines. Both natural auxins (such as indole-3-acetic acid, IAA) and synthetic ones (including indole-3-butyric acid, IBA; naphthaleneacetic acid, NAA; and 2,4-dichlorophenoxyacetic acid, 2,4-D) exhibit distinct stability, uptake dynamics, and morphogenic signatures, making their selection and dosage critical determinants of protocol success.

The precise dosage of auxins is critical for predictable morphogenesis. Excessive auxin may cause callus overproduction, root inhibition, or hyperhydricity, while insufficient auxin can fail to trigger rhizogenesis or callogenesis (George et al. 2008; Weijers and Wagner 2016).

Root induction is strongly modulated by auxin type and concentration, basal medium composition, and culture conditions. IBA has consistently emerged as the most reliable auxin for rhizogenesis in cannabis, owing to its chemical stability, low phytotoxicity, and strong ability to induce root primordia with minimal callus proliferation (Lata et al. 2009; Monthony et al. 2021a). Optimal concentrations range from 0.5 to 4.0 μM, typically in half-strength MS or DKW media, achieving rooting frequencies between 80% and 96% with physiologically normal roots and high acclimatization success.

By contrast, NAA can induce rooting but shows more variable outcomes across genotypes and media. Low concentrations (0.25–0.5 μM) may enhance root thickness but frequently induce callusing at the shoot base, whereas higher concentrations often entirely suppress root initiation (Mestinsek-Mubi et al. 2020).

IAA alone generally induces weak or inconsistent rooting and is seldom used due to its instability in culture media (Ioannidis et al. 2022). In some cases, combinations of IAA and NAA yield additive effects on root number or length (Ślusarkiewicz-Jarzina et al. 2005).

It is worth mentioning that medium strength also affects root responses: half-strength MS or DKW typically outperform full-strength formulations, likely due to reduced osmotic stress and improved hormone uptake (Stephen et al. 2023). Among basal salts, DKW consistently supports higher rooting percentages at equivalent IBA concentrations. Photoperiod also plays a role: a 10-day dark incubation increases root induction, likely by reducing light-driven auxin degradation and altering polar auxin transport (Monteuuis and Bon 2000; Liang et al. 2025).

2,4-Dichlorophenoxyacetic acid (2,4-D) is another major auxin used in cannabis tissue culture, though almost exclusively for callogenesis rather than for organogenic or rooting purposes. As a potent synthetic auxin, effective callus induction typically requires concentrations of 1–3 mg.L^{-1}, particularly when combined with cytokinins such as BAP, TDZ, or KIN (Ioannidis et al. 2022), which synergistically enhance dedifferentiation and tissue proliferation. Nevertheless, 2,4-D in certain formulations can stimulate rooting (Holmes et al. 2021), highlighting strong genotype- and protocol-dependency. Overall, 2,4-D is foundational for callus induction in cannabis, especially in indirect regeneration and transformation systems, but is generally unsuitable for direct organogenesis or rooting.

Collectively, current evidence highlights a clear trend across studies: IBA remains the gold-standard auxin for consistent, high-quality rooting; NAA is potent but might be callogenic; IAA is weak unless combined with other auxins; and 2,4-D is indispensable for callogenesis but practically incompatible with direct shoot regeneration. As with cytokinins, auxin performance is profoundly genotype-dependent, with basal medium strength, photoperiod, and auxin–cytokinin interactions strongly modulating morphogenic outcomes.

Given the substantial genotype-dependent variability observed in cannabis tissue culture responses, Table 12.3 summarizes the key properties, concentration ranges, and morphogenic effects of the major auxins and cytokinins. This comparative synthesis underscores the lack of a universal PGR formulation in cannabis and serves as a practical guide for tailoring protocols to specific cultivars.

Auxin–Cytokinin Interactions and Combined Treatments in Cannabis The morphogenic outcome in plant tissue culture, as aforementioned, is largely determined not by auxins or cytokinins individually but by their relative ratio and molecular interaction. In cannabis, combined treatments play a central role in modulating shoot induction, callogenesis, and rooting, and these responses remain strongly genotype-, dose-, and medium-dependent.

Low-to-moderate cytokinin levels supplemented with trace amounts of auxin often enhance axillary shoot proliferation, particularly in chemotypes with strong apical dominance. For instance, the addition of low-dose IAA to BAP increased both shoot number and elongation in breeding lines that responded weakly to TDZ or *m*T (Mestinsek-Mubi et al. 2020). Comparable interactions have been reported for TDZ combined with very low concentrations of NAA (0.1–0.2 mg.L^{-1}), which can stimulate shoot-bud formation or callus-mediated organogenesis; however, slightly higher NAA levels rapidly redirect tissues toward unorganized callus proliferation (Wang et al. 2021; Ioannidis et al. 2022). These observations align with classical auxin–cytokinin models, where minor shifts in auxin level can increase cell division without overriding shoot-organogenic programs.

In rooting systems, interactions among auxins themselves are equally relevant. Although IBA remains the most reliable auxin for cannabis rhizogenesis, mixtures such as IAA + IBA have been reported to increase root number and elongation in particular cultivars (Ioannidis et al. 2022). Conversely, combinations of IBA + NAA tend to reduce rooting efficiency in agar-based systems and frequently intensify

Table 12.3 Profiles of plant growth regulators (PGRs) commonly used in *Cannabis sativa* micropropagation: concentration ranges, morphogenic responses, and practical considerations for shoot multiplication

Plant growth regulator (PGR)	Concentration range (μM)	Morphological response	Advantages	Disadvantages	Genotype dependency	References
Cytokinins						
Benzylaminopurine (BAP)	0.5–5.0	Moderate axillary shoot proliferation; shoots can be compact.	Low cost; widely available; reproducible performance.	Moderate response; may cause hyperhydricity or excessive compaction.	Medium	Ioannidis et al. (2020), Stephen et al. (2023)
Thidiazuron (TDZ)	0.1–2.5	Very high shoot induction; multiple but short shoots; callusing.	Maximum potency; activates recalcitrant explants; strong bud break.	High hyperhydricity, fasciation, stunted shoots, unwanted callogenesis.	High	Lata et al. (2009), Kastelec et al. (2025)
Meta-Topolin (mT)	0.5–5.0	Vigorous proliferation of high-quality shoots; good elongation.	Excellent shoot quality; improves rooting; reduces oxidative stress.	Higher cost; fewer shoots than TDZ in some genotypes.	Medium-high	Lata et al. (2016), Rico et al. (2022), Das et al. (2024)
Kinetin (KIN)	1.0–10.0	Low to moderate proliferation; useful for maintaining explant health.	Mild cytokinin activity; low stress; useful in hyperhydricity-prone genotypes.	Generally low multiplication response in cannabis.	High	Ioannidis et al. (2020), Ndlovu et al. (2024)
Auxins						
Indole-3-butyric acid (IBA)	0.5–4.0	Efficient adventitious rhizogenesis with minimal basal callus. Performs best in ½-strength MS or DKW	"Gold standard" for rooting; high stability; well-structured rooting; high rooting frequencies	Less effective for callus induction.	Low-medium	Lata et al. (2009), Monthony et al. (2021b), Stephen et al. (2023)

(continued)

Table 12.3 (continued)

Plant growth regulator (PGR)	Concentration range (μM)	Morphological response	Advantages	Disadvantages	Genotype dependency	References
Indole-3-acetic acid (IAA)	0.1–1.0	Promotes shoot elongation; synergistic with cytokinins (especially BAP); occasionally used in callus–shoot transitions.	Natural auxin; gentle effect; enhances shoot elongation with minimal basal callus.	Highly unstable; rapid photodegradation and oxidation; weak for rooting.	Medium-high	Lata et al. (2010b), Mestinsek-Mubi et al. (2020), Ioannidis et al. (2020)
Naphthaleneacetic acid (NAA)	0.5–5.0	Rhizogenesis or callogenesis (dose-dependent); induces more callus than IBA.	Versatile (root/callus); stable; potent.	Significant basal callusing during rooting.	Medium	Lata et al. (2010a); Mestinsek-Mubi et al. (2020)
2,4-Dichlorophenoxyacetic acid (2,4-D)	4.5–15	Potent callus induction; suppresses direct organogenesis.	Superior auxin for initiating embryogenic callus.	Strong inhibitor of shoots; high somaclonal variation risk.	Medium	Ioannidis et al. (2020); Holmes et al. (2021)

Note 1: Morphological responses and concentration ranges are synthesized from the reviewed literature and are subject to genotype dependency. Empirical optimization for each cannabis variety is strongly recommended

Note 2: Genotype dependency is qualitatively classified as low, medium, or high, based on the consistency of the reported response across different studies and genotypes basal callusing at the shoot base, performing worse than IBA alone (Stephen et al. 2023). These contrasting outcomes underscore that auxin synergy is context-dependent and that many auxin mixtures tend to bias tissues toward callogenesis rather than rhizogenesis.

Additionally, Johnson et al. (2025) found the most successful root formation was achieved by exposing shoots to an initial 10-day dark period on a medium supplemented with either 4.0 μM NAA or a combination of 6.0 μM IBA and 1.0 μM NAA. The practice, widely applied in other species, relies on the physiological basis that darkness can decrease the light-induced degradation of auxins within the culture medium. It also alters polar auxin transport within the plant cells, ultimately establishing a more favorable hormonal balance for root initiation. Furthermore, the stress imposed by darkness can induce changes in carbohydrate and hormone metabolism, redirecting the plant's energy towards rhizogenesis. The results are promising, and the cost implications are beneficial, providing fertile ground for further evaluations to elucidate the applicability to other genotypes and conditions.

Cytokinin–auxin combinations are indispensable for callogenesis, particularly in protocols aimed at indirect regeneration or transformation. Highly potent auxins such as 2,4-D require the presence of a cytokinin (most commonly BAP, KIN, or TDZ) for robust dedifferentiation. Effective callus proliferation is typically achieved at high auxin-to-cytokinin ratios, such as 1–3 $mg.L^{-1}$ 2,4-D combined with 0.1–1.0 $mg.L^{-1}$ BAP (Lata et al. 2010a; Ioannidis et al. 2022). The strong activity of 2,4-D, combined with the morphogenic pressure of TDZ or BAP, yields rapidly proliferating callus that can be competent for indirect organogenesis, though subsequent regeneration success remains highly genotype-specific.

Across all combined treatments, three consistent principles emerge. First, ratios matter more than absolute concentrations: minute differences in auxin can shift the system from shoot organogenesis to callogenesis, especially when potent cytokinins such as TDZ are present. Second, basal medium strength modulates auxin–cytokinin balance, with half-strength MS or DKW often improving IBA-mediated rooting and reducing NAA-induced callusing compared with full-strength formulations (Stephen et al. 2023). Third, genotype dependency is unavoidable, as identical PGR combinations can yield vigorous shoots in one cultivar and hyperhydric callus in another (Codesido et al. 2020).

Collectively, current evidence shows that in *Cannabis*, auxin–cytokinin interactions function as a tunable regulatory system rather than a binary switch. Effective optimization must consider not only the absolute concentrations of each PGR but also their ratio, the duration of exposure, basal medium composition, photoperiod, and the intrinsic physiological sensitivity of each genotype.

Other Growth Regulators and Additives While the auxin–cytokinin axis forms the core regulatory module governing in vitro morphogenesis, advanced cannabis micropropagation systems increasingly rely on additional growth regulators and biochemical additives to refine developmental outcomes. These compounds are typically not morphogenic on their own, but they act as modulators of the cellular responses initiated by primary PGRs, enhancing shoot quality, mitigating physiological disorders such as hyperhydricity or oxidative browning, improving elongation when required, and stabilizing regeneration in genotype-sensitive systems.

Gibberellins, particularly gibberellic acid (GA_3), serve as strategic tools to counteract the compact growth habit often imposed by high-cytokinin regimes. Its primary role lies in promoting internode elongation, thereby converting tightly clustered shoots into longer, more manageable plantlets ideal for subsequent rooting and acclimatization (Geng et al. 2016). For instance, in cannabis, the combination of TDZ (0.5 μM) with GA_3 (7.0 μM) proved highly effective in enhancing shoot multiplication (Lata et al. 2009). More recently, a formulation of 2.5 μM TDZ and GA_3 (5 μM) was shown to improve shoot induction and proliferation in MS (Dasauni and Nailwal 2025). However, its application requires precise timing and genotype-specific optimization, as its premature or excessive use can promote apical dominance over axillary bud development and may induce hyperhydricity, yielding fragile shoots unsuitable for commercial propagation.

Polyamines (PAs)—primarily putrescine, spermidine, and spermine—are ubiquitous polycationic compounds that modulate key physiological processes and have shown consistent benefits in plant tissue culture. At low micromolar levels (typically 0–5 μM), they enhance shoot proliferation, improve explant vigor, and can support rooting by interacting synergistically with auxin and cytokinin signaling (Sreelekshmi and Siril 2020; Subrahmanyeswari et al. 2024). Their protective effects are strongly linked to their capacity to stabilize membranes, scavenge reactive oxygen species (ROS), and modulate ethylene biosynthesis. Recent work has shown that PAs mitigate hyperhydricity by reducing ethylene production and restoring stomatal functionality, while simultaneously decreasing H_2O_2 accumulation (Boucher et al. 2025). Higher concentrations have also been effective in improving regeneration efficiency in other high-phenolic crops such as stevia (Subrahmanyeswari et al. 2024). Although not yet widely explored in cannabis, evidence from other species indicates that putrescine, at both low and higher concentrations, could enhance regeneration, suggesting potential for future optimization.

Antioxidants complement these mechanisms and are widely used to prevent oxidative browning caused by phenolic accumulation—a recurrent challenge in cannabis in vitro culture. Ascorbic acid is among the most effective additives, reducing ROS and phenolic oxidation while promoting shoot elongation and increased node number in hemp at 5–30 $mg.L^{-1}$, with optimal performance near 30 $mg.L^{-1}$ (Mishchenko et al. 2022). Low-millimolar doses (≈5 mM) further suppress ROS production (Deguchi et al. 2020), and in other species, 100 $mg.L^{-1}$ combined with activated charcoal significantly improves shoot vigor and acclimatization success (Das and Srivastav 2015). Nevertheless, excessive concentrations can exhibit pro-oxidant behavior, as seen in *Rubus idaeus*, where high ascorbic acid triggered hydroxyl radical formation leading to browning and necrosis in sensitive cultivars (Fridia et al. 2014). Citric acid also acts as a chelating anti-browning agent; 0.1 $g.L^{-1}$ effectively reduced browning and enhanced regeneration in *Cymbidium floribundum*, whereas higher concentrations impaired growth (Ferdous Ona and Shimasaki 2023).

Polyvinylpyrrolidone (PVP) is another widely used antioxidant additive that acts primarily as a phenolic adsorbent, preventing the polymerization of oxidized phenols into toxic quinones that lead to browning and tissue necrosis. In sugarcane, PVP added at 0.2–0.3 $g.L^{-1}$ maximized explant survival, whereas higher concentrations (≥0.4–0.5 $g.L^{-1}$) reduced growth and increased mortality, likely due to excessive adsorption of essential metabolites from the medium (Kassahun Bantte and Feyissa 2015; Pérez-Pérez et al. 2025). Similar behavior has been reported in *Pyrus glabra*, where 200 $mg.L^{-1}$ PVP significantly increased meristem viability and reduced browning without interfering with organogenesis (Pardaz et al. 2015). Collectively, these findings indicate that PVP is an effective anti-browning agent, particularly during the establishment of highly phenolic genotypes, but requires careful optimization since its protective effects are strongly concentration- and species-dependent.

Activated charcoal (AC) is also used in cannabis micropropagation—particularly during initiation, elongation, and rooting—due to its ability to adsorb phenolics and oxidized quinones at 0.5–2.0 $g.L^{-1}$ (Thomas 2008). Although it is occasionally included to mitigate hyperhydricity, cannabis-specific studies indicate that this practice is not consistently effective: in the "Cheungsam" genotype, 0.5 $g.L^{-1}$ AC did not reduce hyperhydricity and even produced poorer morphological outcomes (Baek et al. 2024), a trend also observed in our own genotypes.

The judicious use of these supplementary compounds and methods allows for targeted fine-tuning of cannabis micropropagation protocols, addressing challenges such as oxidative stress, shoot hyperhydricity, and poor elongation, ultimately improving plantlet quality and ensuring reliable, genotype-responsive performance.

12.4 The Bridge to the Soil: Rooting and Acclimatization

The successful transition of a micropropagated plantlet from the controlled, heterotrophic environment of the culture vessel to the challenging photoautotrophic conditions of the greenhouse represents the most critical bottleneck in commercial micropropagation. This phase is where significant losses can occur if the process is not meticulously managed. The plant must undergo profound physiological and morphological changes, including the development of a functional root system for water and nutrient uptake, the formation of a protective cuticle to minimize water loss, and the activation of efficient photosynthetic machinery.

The following subsection covers the bridge to soil, ex vitro rooting methods, their comparative advantages over in vitro approaches, acclimatization protocols, and optimal substrate strategies for successful establishment.

12.4.1 Ex Vitro Rooting: Methods and Advantages

A significant advancement in cannabis micropropagation has been the development of robust ex vitro rooting protocols, which involve rooting microcuttings directly in a non-sterile substrate, bypassing the traditional in vitro rooting stage. This approach addresses several bottlenecks in the production pipeline and is highly effective.

A compelling demonstration of the efficiency and robustness of ex vitro rooting was provided by Murphy and Adelberg (2021), who employed a "hedging" technique in vitro to produce unrooted microcuttings. In this system, shoot tips were repeatedly harvested from the same rooted base over multiple cycles without subculturing, significantly reducing labor costs. Crucially, these microcuttings were then rooted ex vitro in phenolic foam plugs with remarkable success. Under optimal in vitro pre-conditioning—specifically, higher light intensities (up to 167 $\mu mol.m^{-2}.s^{-1}$) and vessel ventilation—rooting success reached nearly 100%. Furthermore, the quality of the root systems and the number of new leaves developed ex vitro were significantly enhanced by these in vitro physical factors, illustrating that the conditions during the micropropagation stage directly prime the cuttings for successful ex vitro establishment. This approach elegantly combines the labor-saving principle of hedging with the streamlined efficiency of ex vitro rooting.

Research has demonstrated that ex vitro rooting is not only feasible but can achieve high success rates. The protocol typically involves harvesting microcuttings from stabilized in vitro cultures and transferring them directly to a humidified, non-sterile environment to induce roots. Kurtz et al. (2022) provided a detailed evaluation, showing that rooting success is highly dependent on the cultivar and the age of the in vitro culture. For the cultivar "Wife", microcuttings harvested from young cultures (6–12 weeks old) rooted at rates of 80% or higher. However, this rate declined significantly to 50% and 30% for 15- and 18-week-old cultures, respectively, highlighting the impact of culture decline. In contrast, "Abacus" maintained a more constant rooting rate (47–70%) across the same timeframe, underscoring a strong genotypic influence. This profound cultivar-dependent response is a well-documented phenomenon in conventional propagation. Campbell et al. (2021), in a study on traditional stem cuttings, observed significant variation across eight cultivars, with "Cherry Wine" outperforming "Wife" by approximately 20%, confirming that genetic predisposition is a primary factor in rooting competence, regardless of the propagation method.

A powerful method to enhance the output of a micropropagation system is "retipping," which involves harvesting shoots from recently acclimated micropropagated plants and using them as cuttings. Kurtz et al. (2022) found that these retip cuttings of "Abacus" and "Wife" rooted at high rates of 76% to 81% even without auxin application. The application of 1000 ppm IBA could boost this rate to 85–88%. This demonstrates that retips combine the phytosanitary benefits of tissue culture with the practical handling of conventional cuttings, offering a scalable method for liner production.

The rooting environment is critical. Stephen et al. (2023) directly compared a humidity dome system with an intermittent mist system for ex vitro acclimatization. Their results were clear: a humidity dome was vastly superior, producing plants with greater shoot height, fresh weight, and overall quality ratings. The dome provides a stable, high-humidity microenvironment that prevents desiccation without the leaching and waterlogging problems associated with overhead mist. While a high-humidity environment is paramount, auxin can still play a beneficial role. Stephen et al. (2023) reported that treating the base of microcuttings with a 0.1% IBA rooting powder mildly increased fresh root weight, though it was less critical than the environment itself. Research on stem cuttings suggests that the mere application of auxin is a key driver. On traditional stem cuttings, Campbell et al. (2021) found that using any rooting hormone (including IBA and NAA formulations) resulted in a significant two-fold increase in rooting success compared to untreated controls, though differences between specific auxin types and concentrations were often non-significant.

12.4.2 Comparative Advantages of Ex Vitro vs. In Vitro Rooting

The choice between in vitro and ex vitro rooting has significant implications for efficiency, cost, and plant quality. Ex vitro rooting streamlines the process by eliminating an entire stage. This reduces the consumption of culture vessels, sterile media, and laboratory labor, significantly lowering production costs and shortening the production cycle. This labor reduction is a critical economic factor. It can constitute 60–70% of in vitro plantlet costs in developed countries, strategies that minimize subculturing, such as the hedging technique for producing ex vitro cuttings, offer substantial savings and scalability potential (Murphy and Adelberg 2021; Ahsan et al. 2025).

Plants rooted ex vitro often develop more robust and functional root systems that are better adapted to non-sterile greenhouse conditions. Kurtz et al. (2022) further demonstrated that micropropagated plants (from ex vitro rooted micro cuttings or retips), when given adequate time to establish, perform agronomically as well as, or even better than, those from traditional stem cuttings, with no differences in cannabinoid profile.

Ex vitro conditions naturally promote the development of a functional cuticle and normalized stomata, reducing the incidence of hyperhydricity and facilitating a smoother transition to the greenhouse. The primary advantage of in vitro rooting remains its high level of environmental control, which can be crucial for recalcitrant genotypes that fail to root ex vitro reliably. It ensures a sterile environment until the plantlet is fully formed.

12.4.3 Acclimatization: Tailoring the Transition to the Rooting Method

Successful acclimatization is the final critical step. The strategy must be tailored to the physiological state of the plant, which is fundamentally different for in vitro-rooted plantlets and ex vitro-rooted microcuttings.

Plantlets from in vitro conditions are highly vulnerable, possessing a poor cuticle, dysfunctional stomata, and roots adapted to a saturated, heterotrophic environment. Their acclimation is a careful "weaning" process. Plantlets are removed from agar, transplanted to a sterile, well-draining substrate, and immediately placed under a humidity dome to maintain near-100% RH. Then, over 7–10 days, humidity is slowly reduced by progressively ventilating the dome (Johnson et al. 2025). This forces the development of a functional cuticle and stomatal control. Light intensity is kept moderate initially, and irrigation is carefully managed to avoid waterlogging the adapting roots. A direct transfer would cause rapid desiccation.

For ex vitro-rooted plants, acclimation is more of a "hardening" phase, as the rooting and initial adaptation occur simultaneously. The protocol is integrated and more efficient. Microcuttings root directly under the protective environment of a humidity dome, as used by Stephen et al. (2023) and Lubell-Brand et al. (2021), which maintains turgor during root initiation. Once a root system is established (typically after 2–4 weeks), the dome is gradually removed over 5–10 days. This process is faster and has higher success rates than the in vitro weaning process because the plants have already begun to develop adaptive traits in a more natural environment. The research consistently shows that the use of a humidity dome, rather than misting, is the key environmental factor for high-quality ex vitro rooting and subsequent acclimation success.

12.4.4 Substrate Strategies for Successful Acclimatization

Acclimatization is the physiological and morphological adaptation of in vitro-derived plantlets to ex vitro conditions. This process is profoundly stressful, as plantlets must rapidly develop a fully functional root system, transition from heterotrophic to photoautotrophic growth, and develop a competent cuticle to reduce water loss, all while facing new microbial pressures (Debnath 2009). The choice of substrate is not merely a growing medium; it is a critical engineering tool designed to manage the water, air, and biological relationships around the vulnerable root system, ultimately determining the commercial viability of the entire micropropagation process (Klerk 2002).

An ideal acclimatization substrate is a sterile, well-aerated, and well-drained mixture that provides a careful balance between water retention and oxygenation. In vitro plantlets are accustomed to nearly 100% humidity and have roots adapted to a water-saturated environment. A substrate that retains too much water (e.g., pure fine peat) will lead to hypoxia, encourage root rot by pathogens like *Pythium* spp., and prevent the necessary morphological changes (Zobayed 2000). Conversely, a medium that drains too quickly (e.g., pure perlite) will subject the plantlet to immediate water stress before its roots and cuticle are functional. Therefore, the universal strategy is to use blended substrates that create optimal pore space. Some common components include: peat moss, which provides excellent water retention and a high cation exchange capacity; perlite, a lightweight, sterile, volcanic glass that increases aeration and drainage, preventing compaction and ensuring oxygen reaches the roots; vermiculite, which adds further water retention and aeration, and its plate-like structure also provides some mechanical support; coir (coconut fiber), a sustainable alternative to peat, offering good structure and moisture management with high porosity; phenolic foam plugs, which are stable, pathogen-free environment, and with excellent water retention and aeration; and rockwool, which minimizes root asphyxiation for fast, even growth. Its fibrous structure provides excellent drainage, preventing root rot and simplifying watering.

Specific substrates have been successfully employed. For instance, Kurtz et al. (2022) used 1-inch rockwool cubes for the ex vitro rooting of hemp microcuttings, leveraging its inert and well-aerated properties. The superiority of rockwool for rooting has been demonstrated in conventional propagation as well; on traditional stem cuttings, Campbell et al. (2021) found it outperformed several organic mixes, producing 7- to 13-fold greater root mass, attributed to its optimal air-to-water ratio and uniform structure. For organic blends, common successful ratios for various species include 1:1:1 peat:perlite:vermiculite or 3:1 peat:perlite, with the specific formula often optimized for the local environment and specific genotype.

Beyond physical properties, advanced strategies incorporate biological additives to mitigate transplant shock and boost establishment. Inoculating the substrate with arbuscular mycorrhizal fungi is a powerful strategy. These symbiotic fungi colonize the root system, dramatically increasing the effective root surface area for water and nutrient uptake (particularly phosphorus), thereby reducing water stress and improving plantlet vigor and survival rates (Nowak et al. 2007). Also, certain beneficial bacteria (e.g., *Bacillus* spp., *Pseudomonas* spp.) can be introduced to the substrate to suppress root pathogens through competition or antibiotic production, produce plant growth-stimulating hormones like auxins, and aid in nutrient solubilization, creating a protective "rhizosphere microbiome" for the young plant. Not only does the evaluation of beneficial bacteria constitute an area yet to be explored, but also the study of the balance and synergy of endophytic microorganisms with the plant deserves attention (Taghinasab and Jabaji 2020; Buirs and Punja 2025).

Another strategy includes incorporating controlled-release fertilizers into the substrate, which provides a gentle, steady supply of nutrients as the plantlet begins its autotrophic growth, avoiding the burn potential of immediate full-strength fertilization. This approach was used effectively by Kurtz et al. (2022), who top-dressed acclimated hemp plants with a controlled-release fertilizer after potting, supporting their subsequent vigorous growth.

The acclimatization protocol must complement the substrate choice. As demonstrated by several studies (Lubell-Brand et al. 2021; Stephen et al. 2023; Johnson et al. 2025), this involves maintaining high relative humidity using propagation domes initially and gradually reducing it over 1–2 weeks to harden the plantlets and promote cuticle development before their final transfer to standard greenhouse conditions. The synergy between a well-engineered substrate and a controlled acclimatization environment is the cornerstone of achieving high survival rates and producing robust, commercially viable plants. This protocol forms the basis for efficient ex vitro rooting and retipping, while reviews of the field, such as that by Monthony et al. (2021b), recognize ex vitro rooting as a standard and advantageous practice for commercial micropropagation. However, despite these promising foundational studies, the optimization of ex vitro protocols—particularly concerning the interaction of genotype-specific responses, substrate composition, and microbial amendments—remains a significant and underexplored frontier in cannabis biotechnology.

12.5 Ensuring Genetic Fidelity: From Morphology to Molecules

Somaclonal variation represents one of the central risks associated with plant tissue culture and constitutes a major challenge for maintaining genetic fidelity in clonally propagated cannabis. During in vitro culture, cells are exposed to non-physiological conditions, including high concentrations of PGRs, altered osmotic environments, continuous light, and repeated subculturing. These factors can induce both genetic and epigenetic alterations. These changes may manifest as chromosomal abnormalities, point mutations, or epigenetic reprogramming such as DNA methylation shifts, histone modifications, and RNA-interference–mediated regulation (Kumar et al. 2018; Hesami et al. 2021). Collectively, these alterations are referred to as somaclonal variation, defined as any heritable or non-heritable phenotypic divergence among regenerated plants arising from the tissue-culture process. While such variation can be exploited as a source of novel diversity in breeding programs, it is detrimental when the goal is the mass production of true-to-type elite genotypes. An integrated overview of the main phenotypic, cytological, biochemical, and molecular approaches used to assess genetic and metabolic stability in micropropagated cannabis is summarized in Table 12.4.

Ensuring the genetic fidelity of micropropagated material represents a fundamental requirement for any regeneration or multiplication protocol, particularly in cannabis, where agronomic and pharmaceutical value relies directly on the consistency of cannabinoid and terpene profiles. Regenerated plants must be true-to-type, maintaining the same genotype and, consequently, the same chemotype/chemovar as the elite mother plant. While morphological and biochemical analyses offer rapid and cost-effective screening tools, only DNA-based molecular methods definitively confirm genetic stability and are therefore considered the current gold standard (Wróbel et al. 2018; Hesami et al. 2021).

Table 12.4 Integrated validation methods for assessing genetic and metabolic stability in micropropagated *Cannabis sativa* plants

Level of analysis	Specific method	Detects	Resolution	Advantages	Main limitations	References
Phenotypic	Morphological assessment	Gross somaclonal variants (leaf shape, architecture, vigor).	Very low	Rapid and inexpensive; requires no specialized equipment.	Strongly influenced by environment and developmental stage; unable to detect subtle variation.	Hesami et al. (2023b)
Cytological	Ploidy analysis (chromosome counts, flow cytometry)	Polyploidy, aneuploidy or mitotic irregularities induced by PGRs (e.g., 2,4-D, TDZ)	Low	Confirms fundamental chromosomal stability (2n = 20 in *Cannabis*).	Cannot detect sequence-level mutations; may require technical expertise.	Adhikary et al. (2021)
Biochemical	Secondary metabolite profiling (HPLC, UHPLC, GC-MS)	Changes in chemotype/chemovar (cannabinoids, terpenes)	Medium	Directly relevant for pharmaceutical and industrial quality; provides functional validation.	Metabolite levels are affected by physiology and environment, not solely genetics.	Lata et al. (2010b); Monthony et al. (2021b)
Molecular	DNA-based markers (ISSR, SSR, AFLP, SNPs)	DNA sequence polymorphisms; high-resolution detection of somaclonal variation	Very high	Objective, sensitive, and environment-independent; detects subtle genomic changes.	Higher cost; requires specialized equipment and bioinformatic capacity.	Kojoma et al. (2002); Lata et al. (2010b); Ioannidis et al. (2020)

ISSR Inter Simple Sequence Repeat, *SSR* Simple Sequence Repeat, *AFLP* Amplified Fragment Length Polymorphism, *SNPs* Single Nucleotide Polymorphisms

12.5.1 Morphological and Cytological Assessment

Morphological and cytological evaluations constitute the first level of fidelity assessment in cannabis micropropagation. These methods are simple, relatively inexpensive, and useful for detecting major deviations associated with somaclonal variation, though they lack the resolution to detect subtle genetic changes.

Morphological markers involve the comparison of phenotypic traits between regenerated plants and the mother plant, including plant height, leaf shape and size, internode length, leaf coloration, trichome density, and general architecture. Such markers have been traditionally used in cannabis micropropagation (Hesami et al. 2023b) and remain relevant for early screening. However, morphological traits are strongly influenced by environmental and developmental conditions, meaning that stressed but genetically identical plants can present altered morphologies. For this reason, morphology can detect only gross variants and cannot reliably confirm true-to-type status (Hesami et al. 2021).

Cytological assessment provides a more structured evaluation of genomic stability by examining chromosome number and organization. Cannabis is a diploid species with $2n = 20$ chromosomes; deviations from this state can arise due to prolonged in vitro culture or exposure to strong plant growth regulators, particularly synthetic auxins such as 2,4-D or high concentrations of cytokinins like TDZ. Cytological tools include chromosome squashes from root tips to assess ploidy, meiotic analysis to detect irregularities in gamete formation, and stomatal measurements (size and chloroplast number), which often correlate with ploidy changes. Although cytological assessments can detect polyploidy or large-scale chromosomal aberrations (Dhooghe et al. 2011), they cannot identify point mutations, small insertions/deletions, or epigenetic alterations. In cannabis, karyotype stability across regenerated lines has been reported, highlighting the utility of these methods as a preliminary layer of evaluation (Adhikary et al. 2021).

12.5.2 Biochemical Markers

Biochemical analyses bridge the gap between phenotype and genotype by assessing whether regenerated plants maintain the same metabolic output as the mother plant. In cannabis, where commercial and medicinal value derive from secondary metabolite composition, these analyses become essential.

Traditional isoenzyme profiling, once used widely in tissue culture studies, provides moderate discrimination by detecting electrophoretic differences among enzyme variants. However, the number of analyzable loci is limited, and isoenzyme expression is influenced by developmental and environmental factors. Because of these constraints, isoenzyme analysis is now largely considered outdated and has been superseded by DNA-based markers in modern fidelity studies (Bairu et al. 2011).

Far more relevant is secondary metabolite profiling, typically performed using high-performance liquid chromatography (HPLC), ultra-high-performance liquid chromatography (UHPLC), or gas chromatography–mass spectrometry (GC-MS) to quantify major cannabinoids (THCA, CBDA, CBGA, etc.) and terpenes (myrcene, limonene, β-caryophyllene, α-pinene). These profiles constitute highly sensitive biochemical fingerprints. Early work by Lata et al. (2016) demonstrated that micropropagated cannabis plants can retain cannabinoid stability comparable to seed-grown controls. More recent studies, such as Ioannidis et al. (2020), reinforce that high-quality in vitro regeneration protocols preserve both cannabinoid and terpene profiles when acclimatization is appropriate, and explant sources are genetically uniform. Although metabolite profiling cannot determine whether observed differences originate from genetic or environmental sources, it remains a functionally essential tool for confirming that micropropagated plants preserve their chemotype.

12.5.3 Molecular Markers

Molecular marker analysis represents the highest standard for confirming genetic fidelity in cannabis, as these methods assess polymorphisms directly at the DNA level and are unaffected by environmental conditions. A broad range of molecular systems has been applied to cannabis genetic studies. These includes random amplified polymorphic DNA (RAPD), restriction fragment length polymorphism (RFLP), amplified fragment length polymorphism (AFLP), inter simple sequence repeat (ISSR), simple sequence repeat (SSR, or microsatellites), short tandem repeat (STR) multiplexes, single nucleotide polymorphisms (SNPs), and PCR allele competitive extension (PACE) assays; each offering different levels of resolution and reproducibility (Kojoma et al. 2002; Alghanim and Almirall 2003; Mendoza et al. 2009; Gao et al. 2014; Henry et al. 2018; Adhikary et al. 2021). Among these, ISSR markers are the most widely applied in tissue-culture systems due to their high reproducibility and ability to detect genome-wide polymorphisms. ISSR analysis has confirmed the clonal fidelity of regenerated cannabis plants in multiple studies, including direct and indirect organogenesis pipelines, synthetic seed storage, and long-term subculturing across more than 30 nodal passages (Lata et al. 2010b; Lata et al. 2012). SSR markers, by contrast, represent the most precise and informative system currently available: they are codominant, highly polymorphic, and locus-specific, making them ideal for detecting even subtle genomic variants. The recent development of cannabis-specific SSR panels (Ioannidis et al. 2022) positions this marker class as the emerging gold standard for modern fidelity testing. RAPD, historically used in early cannabis micropropagation studies (Lata et al. 2016), is now largely deprecated due to its low reproducibility and sensitivity to PCR conditions, but it remains important from a historical perspective. More recently, high-resolution SNP-based approaches, including targeted PACE assays and whole-genome sequencing, have begun to reveal even finer-scale patterns of intra-plant and somaclonal variation, underscoring the need for continued refinement of fidelity assessment tools in cannabis (Adhikary et al. 2021).

Overall, the combined use of at least two complementary molecular systems—most commonly ISSR together with SSR—provides the strongest evidence of genetic stability in regenerated cannabis plants. When integrated with cytological screening and metabolite profiling, these molecular markers form a robust, multidimensional framework capable of ensuring that micropropagated or regenerated material remains genetically uniform and true-to-type.

12.6 Conclusion and Future Perspectives

This chapter has outlined the strategic role of advanced in vitro propagation systems as a robust and scalable platform for the enhanced and standardized production of anti-aging metabolites from *Cannabis sativa*. Central to this biotechnological approach is the precise manipulation of plant growth regulators (PGRs), which act as molecular switches to control morphogenesis. We have detailed how potent cytokinins such as TDZ support high-frequency shoot induction, while more balanced regulators such as mT minimize physiological disorders such as hyperhydricity. Likewise, auxins (particularly IBA) remain indispensable for reliable rhizogenesis, completing an efficient regeneration pipeline capable of producing true-to-type elite genotypes.

A persistent challenge, however, is the pronounced genotype dependence of cannabis. Identical PGR formulations can elicit sharply contrasting responses among cultivars, making universal protocols unrealistic and reinforcing the need for genotype-tailored optimization. Physiological disorders, most notably hyperhydricity, further hinder standardization and often require integrated adjustments in medium composition, gelling strength, osmolarity, and environmental conditions. Equally critical is the assurance of long-term genetic fidelity, as cannabinoid and terpene consistency directly determines industrial and pharmaceutical value. The integration of morphological, cytological, biochemical, and molecular markers increases confidence that regenerated plants remain genetically uniform.

Looking ahead, the future of cannabis biotechnology is shifting toward more integrated and technologically advanced platforms. For large-scale commercialization, bioreactor-based systems such as temporary immersion bioreactors represent the next logical step, offering improved microenvironmental control, automation, and substantial scaling potential. In parallel, genome-editing technologies such as CRISPR-Cas hold exceptional potential for enhancing anti-aging metabolite production, whether by upregulating key biosynthetic enzymes, suppressing competing pathways, or increasing glandular trichome density. Although genome editing was not the primary focus of this chapter, its successful implementation in cannabis critically depends on efficient in vitro regeneration frameworks. Robust micropropagation systems form the biological foundation required for these biotechnologies. Additionally, elicitation strategies and metabolomic-guided selection will likely become central tools for boosting the yield of high-value secondary metabolites within controlled in vitro systems.

In summary, although challenges remain, the convergence of refined micropropagation protocols, rigorous fidelity assurance, and next-generation biotechnologies positions in vitro culture as a cornerstone for the reliable and sustainable production of cannabis-derived anti-aging compounds. These developments will shape the next generation of cosmetic, nutraceutical, and pharmaceutical applications of *Cannabis sativa.*

Acknowledgments Luciano Protti Cosenza, Alexander Peter Aguila Wharton, and Ignacio Scandura are graduate students in the Environment and Health Applied Sciences Doctoral Program (DCAAS) at UNICEN, Argentina. G.B. is an Independent Researcher at CONICET (National Scientific and Technical Research Council, Argentina), M.L.F. is a Researcher of the National University of the Centre of the Province of Buenos Aires (UNICEN, Argentina), A.F. is an Assistant Researcher at CICPBA (Scientific Investigation Commission of the Province of Buenos Aires, Argentina). I.G. is a research fellow at CICPBA. L.P.C., L.M., and A.P.A.W. are research fellows at CONICET.

The authors gratefully acknowledge Natalia Delbón, from the Multidisciplinary Institute of Plant Biology (IMBIV-CONICET-National University of Córdoba (UNC), Córdoba, Argentina), and the Laboratory for Materials Analysis by X-ray Spectrometry (LAMARX, CONICET-UNC, Córdoba, Argentina) for acquiring the microscopy images included in this chapter. The technical assistance and access to specialized equipment provided by the laboratory were essential for obtaining the images presented here.

Competing Interests The authors have no conflicts of interest to declare that are relevant to the content of this chapter.

References

Abiri R, O'Reilly D, Jones AMP (2025) Bottom cooling during culture initiation increases survival and reduces Hyperhydricity in micropropagated cannabis plants. Plants 14(6):886

Abu-Romman SM, Al-Hadid KA, Arabiyyat AR (2015) Kinetin is the Most effective Cytokinin on shoot multiplication from cucumber. J Agric Sci 7(10):159

Adhikary D, Kulkarni M, El-Mezawy A, Mobini S, Elhiti M, Gjuric R, Ray A, Polowick P, Slaski JJ, Jones MP, Bhowmik P (2021) Medical cannabis and industrial hemp tissue culture: present status and future potential. Front Plant Sci 12:627240

Ahidar N, Labhar A, Benamari O, Ahari M, Salhi A, Elyoussfi A, Amhamdi H (2024) Phenolic content and antioxidant activity of *Cannabis sativa* L flowers from the Ketama region in northern Morocco. Ecol Eng Environ Technol 25(1):209–215

Ahsan SM, Injamum-Ul-Hoque M, Das AK, Shaffique S, Hasan M, Kang S-M, Lee I-J, Choi HW (2025) Tuning up in vitro growth and development of *Cannabis sativa*: recent advances in Micropropagational approach. Appl Biosci 4(1):12

Aizpurua-Olaizola O, Navarro P, Vallejo A, Olivares M, Etxebarria N, Usobiaga A (2016) Microencapsulation and storage stability of polyphenols from Vitis vinifera grape wastes. Food Chem 190:614–621

Alghanim HJ, Almirall JR (2003) Development of microsatellite markers in *Cannabis sativa* for DNA typing and genetic relatedness analyses. Anal Bioanal Chem 376(8):1225–1233

Ali HM, Khan T, Khan MA, Ullah N (2022) The multipotent thidiazuron: a mechanistic overview of its roles in callogenesis and other plant cultures in vitro. Biotech Appl Biochem 69(6):2624–2640

Andre CM, Hausman J-F, Guerriero G (2016) *Cannabis sativa*: the Plant of the Thousand and one Molecules. Front Plant Sci 7

Appendino G, Taglialatela-Scafati O (2013) Cannabinoids: chemistry and medicine. In: Ramawat KG, Mérillon J-M (eds) Natural products: Phytochemistry, botany and metabolism of alkaloids, Phenolics and terpenes. Springer Berlin Heidelberg, Berlin, Heidelberg, pp 3415–3435

Aremu AO, Bairu MW, Doležal K, Finnie JF, Van Staden J (2012) Topolins: a panacea to plant tissue culture challenges? Plant Cell Tissue Organ Cult 108(1):1–16

Argueso CT, Kieber JJ (2024) Cytokinin: from autoclaved DNA to two-component signaling. Plant Cell 36(5):1429–1450

Atalay S, Jarocka-Karpowicz I, Skrzydlewska E (2019) Antioxidative and anti-inflammatory properties of Cannabidiol. Antioxidants 9(1):21

Baek S-C, Jeon S-Y, Choi Y-J, Byun B-H, Kim D-H, Yu G-R, Kim H, Lim D-W (2024) Establishment of an In Vitro Micropropagation System for Cannabis sativa 'Cheungsam.'. Horticulturae 10(10):1060

Bairu MW, Aremu AO, Van Staden J (2011) Somaclonal variation in plants: causes and detection methods. Plant Growth Regul 63(2):147–173

Baron-Flores V, Diaz-Ruiz A, Manzanares J, Rios C, Burelo M, Jardon-Guadarrama G, Martínez-Cárdenas MDLÁ, Mata-Bermudez A (2022) Cannabidiol attenuates hypersensitivity and oxidative stress after traumatic spinal cord injury in rats. Neurosci Lett 788:136855

Beleggia R, Menga V, Fulvio F, Fares C, Trono D (2023) Effect of genotype, year, and their interaction on the accumulation of bioactive compounds and the antioxidant activity in industrial hemp (*Cannabis sativa* L.) inflorescences. Int J Mol Sci 24(10):8969

Boivin M (2021) Nabiximols (Sativex®). In: Narouze SN (ed) Cannabinoids and pain. Springer International Publishing, Cham, pp 119–126

Booth JK, Bohlmann J (2019) Terpenes in *Cannabis sativa*—from plant genome to humans. Plant Sci 284:67–72

Booth JK, Page JE, Bohlmann J (2017) Terpene synthases from *Cannabis sativa*. PLoS One 12(3):e0173911

Boucher R, Germain H, Desgagné-Penix I (2025) Exploring the lesser-known bioactive natural products of plant species of the genus *cannabis* L.: alkaloids, Phenolic Compounds, and Their Therapeutic Potential. Plants 14(9):1372

Buirs L, Punja ZK (2025) Endophytes in *Cannabis sativa*: identifying and characterizing microbes with beneficial and detrimental effects on plant health. Plants 14(8):1247

Bukowska B (2024) Current and potential use of biologically active compounds derived from *Cannabis sativa* L. in the treatment of selected diseases. Int J Mol Sci 25(23):12738

Campbell SM, Anderson SL, Brym Z, Pearson BJ (2021) Evaluation of substrate composition and exogenous hormone application on vegetative propagule rooting success of essential oil hemp (*Cannabis sativa* L.). https://doi.org/10.1101/2021.03.15.435449

Cantele C, Bertolino M, Bakro F, Giordano M, Jędryczka M, Cardenia V (2020) Antioxidant effects of hemp (*Cannabis sativa* L.) Inflorescence Extract in Stripped Linseed Oil. Antioxidants 9(11):1131

Caplan D, Dixon M, Zheng Y (2017) Optimal rate of organic fertilizer during the flowering stage for cannabis grown in two coir-based substrates. HortScience 52(12):1796–1803

Casares L, García V, Garrido-Rodríguez M, Millán E, Collado JA, García-Martín A, Peñarando J, Calzado MA, De La Vega L, Muñoz E (2020) Cannabidiol induces antioxidant pathways in keratinocytes by targeting BACH1. Redox Biol 28:101321

Cásedas G, Moliner C, Maggi F, Mazzara E, López V (2022) Evaluation of two different *Cannabis sativa* L. extracts as antioxidant and neuroprotective agents. Front Pharmacol 13:1009868

Chirambo G, Gwedela MNV, Mwambyale T, Chimphepo L, Chikowe I, Jana CE (2024) Cannabis chemotypes and chemovars. In: Cannabis and Khat in Drug Discovery. Elsevier, pp 33–44

Codesido V, Meyer S, Casano S (2020) Influence of media composition and genotype for successful *Cannabis sativa* L. in vitro introduction. Acta Hortic 1285:75–80

Coskun ZM, Bolkent S (2016) Evaluation of Δ9-tetrahydrocannabinol metabolites and oxidative stress in type 2 diabetic rats. Iran J Basic Med Sci 19(2)

Crowley R, Cline K, Hilden D, Beachy M, Hilden D, Beachy M, Carney JK, Foster H, Hoeg A, Hollon M, Jumper C, Mascarenhas SJ, Sagar A, Silberger JR, Slocum J, Tan M, Yousef EA

(2024) Regulatory framework for cannabis: a position paper from the American College of Physicians. Ann Intern Med 177(8):1104–1105
Das P, Srivastav AK (2015) To study the effect of activated charcoal, ascorbic acid and light duration on InVitro micropropagation of Aloe vera L. Int J Innov Res Sci Eng Technol 4(5):3131–3138
Das R, Kretzschmar T, Mieog JC (2024) Importance of media composition and explant type in *Cannabis sativa* tissue culture. Plants 13(18):2544
Dasauni K, Nailwal TK (2025) Efficient micropropagation of Indian Himalayan *Cannabis sativa* L. achieved by using green Sulphur nanoparticles: genetic fidelity analysis using SCoT markers. In Vitro Cell Dev Biol-Plant 61(4):779–788
Dawidowicz AL, Olszowy-Tomczyk M, Typek R (2021) CBG, CBD, Δ9-THC, CBN, CBGA, CBDA and Δ9-THCA as antioxidant agents and their intervention abilities in antioxidant action. Fitoterapia 152:104915
Debnath SC (2009) Characteristics of strawberry plants propagated by *in vitro* bioreactor culture and *ex vitro* propagation method. Eng Life Sci 9(3):239–246
Degrave V, Vega Joubert MB, Ingaramo P, Sedan D, Andrinolo D, D'Alessandro ME, Oliva ME (2023) Effects of full-spectrum cannabis oil with a cannabidiol:tetrahydrocannabinol 2:1 ratio on the mechanisms involved in hepatic steatosis and oxidative stress in rats fed a sucrose-rich diet. Med Cannabis Cannabinoids 6(1):170–183
Deguchi M, Bogush D, Weeden H, Spuhler Z, Potlakayala S, Kondo T, Zhang ZJ, Rudrabhatla S (2020) Establishment and optimization of a hemp (*Cannabis sativa* L.) agroinfiltration system for gene expression and silencing studies. Sci Rep 10(1):3504
Dhooghe E, Van Laere K, Eeckhaut T, Leus L, Van Huylenbroeck J (2011) Mitotic chromosome doubling of plant tissues in vitro. Plant Cell Tissue Organ Cult 104(3):359–373
Elazab, Shaaban (2015) The impact of sucrose concentration on root growth and development in fig (*Ficus carica* L.) in vitro. Assiut J Agric Sci 46(6):67–75
El-Mernissi R, El Menyiy N, Moubachir R, Zouhri A, El-Mernissi Y, Siddique F, Nadeem S, Ibork H, El Barnossi A, Wondmie GF, Bourhia M, Bin Jardan YA, Abboussi O, Hajji L (2024) *Cannabis sativa* L. essential oil: chemical composition, anti-oxidant, anti-microbial properties, and acute toxicity: *In vitro* , *in vivo,* and *in silico* study. Open Chem 22(1):20230214
Ferdous Ona A, Shimasaki K (2023) Effect of citric acid on the organogenesis of *Cymbidium floribundum.* Environ Control Biol 61(4):69–71
Fernández-Ruiz J, Galve-Roperh I, Sagredo O, Guzmán M (2020) Possible therapeutic applications of cannabis in the neuropsychopharmacology field. Eur Neuropsychopharmacol 36:217–234
Fischedick JT, Hazekamp A, Erkelens T, Choi YH, Verpoorte R (2010) Metabolic fingerprinting of *Cannabis sativa* L., cannabinoids and terpenoids for chemotaxonomic and drug standardization purposes. Phytochemistry 71(17–18):2058–2073
Flores-Sanchez IJ, Verpoorte R (2008) Secondary metabolism in cannabis. Phytochem Rev 7(3):615–639
Fordjour E, Manful CF, Sey AA, Javed R, Pham TH, Thomas R, Cheema M (2023) *Cannabis*: a multifaceted plant with endless potentials. Front Pharmacol 14:1200269
Frassinetti S, Moccia E, Caltavuturo L, Gabriele M, Longo V, Bellani L, Giorgi G, Giorgetti L (2018) Nutraceutical potential of hemp (*Cannabis sativa* L.) seeds and sprouts. Food Chem 262:56–66
Frazzini S, Torresani MC, Roda G, Dell'Anno M, Ruffo G, Rossi L (2024) Chemical and functional characterization of the main bioactive molecules contained in hulled *Cannabis sativa* L. seeds for use as functional ingredients. J Agric Food Res 16:101084
Fridia A, Samir CD, Young RY (2014) Effects of calcium gluconate and ascorbic acid on controlling shoot necrosis during micropropagation of primocane-fruiting raspberry (*Rubus idaeus* L.) cultivars. Afr J Biotechnol 13(47):4361–4368
Gabotti D, Locatelli F, Cusano E, Baldoni E, Genga A, Pucci L, Consonni R, Mattana M (2019) Cell suspensions of *Cannabis sativa* (var. Futura): effect of elicitation on metabolite content and antioxidant activity. Molecules 24(22):4056

Galán-Ávila A, García-Fortea E, Prohens J, Herraiz FJ (2020) Development of a direct in vitro plant regeneration protocol from *Cannabis sativa* L. seedling explants: developmental morphology of shoot regeneration and ploidy level of regenerated plants. Front. Plant Sci 11:645

Gantait S, Mitra M (2021) Role of *meta*-topolin on in vitro shoot regeneration: an insight. In: Ahmad N, Strnad M (eds) *Meta*-topolin: a growth regulator for plant biotechnology and agriculture. Springer, Singapore, pp 143–168

Gao C, Xin P, Cheng C, Tang Q, Chen P, Wang C, Zang G, Zhao L (2014) Diversity analysis in *Cannabis sativa* based on large-scale development of expressed sequence tag-derived simple sequence repeat markers. PLoS One 9(10):e110638

Geng F, Moran R, Day M, Halteman W, Zhang D (2016) Increasing in vitro shoot elongation and proliferation of 'G.30' and 'G.41' apple by chilling explants and plant growth regulators. HortScience 51(7):899–904

George EF, Hall MA, de Klerk G (2008) Plant propagation by tissue culture Volume 1: the background

Gul W, Ibrahim EA, Gul SW, Shahzadi I, Radwan MM, Chandra S, Lata H, ElSohly MA (2023) Development and validation of a GC-FID method for the quantitation of 20 different acidic and neutral cannabinoids. Planta Med 89(06):683–696

Gülck T, Møller BL (2020) Phytocannabinoids: origins and biosynthesis. Trends Plant Sci 25(10):985–1004

Hacke ACM, Lima D, De Costa F, Deshmukh K, Li N, Chow AM, Marques JA, Pereira RP, Kerman K (2019) Probing the antioxidant activity of Δ^9 -tetrahydrocannabinol and cannabidiol in *Cannabis sativa* extracts. Analyst 144(16):4952–4961

Hammond CT, Mahlberg PG (1973) Morphology of glandular hairs of cannabis sativa from scanning electron microscopy. Am J Bot 60(6):524–528

Hampson AJ, Grimaldi M, Axelrod J, Wink D (1998) Cannabidiol and (−)Δ^9 -tetrahydrocannabinol are neuroprotective antioxidants. Proc Natl Acad Sci USA 95(14):8268–8273

Hanuš LO, Meyer SM, Muñoz E, Taglialatela-Scafati O, Appendino G (2016) Phytocannabinoids: a unified critical inventory. Nat Prod Rep 33(12):1357–1392

Hazekamp A, Fischedick JT (2012) Cannabis—from cultivar to chemovar: towards a better definition of cannabis potency. Drug Test Anal 4(7–8):660–667

Hazekamp A, Tejkalová K, Papadimitriou S (2016) Cannabis: from cultivar to Chemovar II—a metabolomics approach to cannabis classification. Cannabis Cannabinoid Res 1(1):202–215

Henry P, Hilyard A, Johnson S, Orser C (2018) Predicting chemovar cluster and variety verification in vegetative cannabis accessions using targeted single nucleotide polymorphisms

Hesami M, Jones AMP (2020) Application of artificial intelligence models and optimization algorithms in plant cell and tissue culture. Appl Microbiol Biotechnol 104(22):9449–9485

Hesami M, Pepe M, Monthony AS, Baiton A, Jones AMP (2021) Modeling and optimizing in vitro seed germination of industrial hemp (*Cannabis sativa* L.). Ind Crops Prod 170:113753

Hesami M, Adamek K, Pepe M, Jones AMP (2023a) Effect of explant source on phenotypic changes of in vitro grown cannabis plantlets over multiple subcultures. Biology 12(3):443

Hesami M, Pepe M, Jones AMP (2023b) Morphological characterization of *Cannabis sativa* L. throughout its complete life cycle. Plants 12(20):3646

Hesami M, Pepe M, De Ronne M, Najafabadi MY, Adamek K, Torkamaneh D, Jones AMP (2024) Cannabis leaf arrangement: transcriptome insights into Cannabis sativa phyllotactic regulation. Plant Physiol Rep 29(3):488–498

Hillig KW (2004) A chemotaxonomic analysis of terpenoid variation in cannabis. Biochem Syst Ecol 32(10):875–891

Holmes JE, Lung S, Collyer D, Punja ZK (2021) Variables affecting shoot growth and plantlet recovery in tissue cultures of drug-type *Cannabis sativa* L. Front Plant Sci 12:732344

Hutchison CE, Li J, Argueso C, Gonzalez M, Lee E, Lewis MW, Maxwell BB, Perdue TD, Schaller GE, Alonso JM, Ecker JR, Kieber JJ (2006) The *Arabidopsis* histidine phosphotransfer proteins are redundant positive regulators of cytokinin signaling. Plant Cell 18(11):3073–3087

Ikeuchi M, Favero DS, Sakamoto Y, Iwase A, Coleman D, Rymen B, Sugimoto K (2019) Molecular mechanisms of plant regeneration. Annu Rev Plant Biol 70(1):377–406

Ioannidis K, Dadiotis E, Mitsis V, Melliou E, Magiatis P (2020) Biotechnological approaches on two high CBD and CBG *Cannabis sativa* L. (Cannabaceae) varieties: in vitro regeneration and phytochemical consistency evaluation of micropropagated plants using quantitative 1H-NMR. Molecules 25(24):5928

Ioannidis K, Tomprou I, Mitsis V, Koropouli P (2022) Genetic evaluation of in vitro micropropagated and regenerated plants of *Cannabis sativa* L. using SSR molecular markers. Plants 11(19):2569

Irakli M, Tsaliki E, Kalivas A, Kleisiaris F, Sarrou E, Cook CM (2019) Effect of genotype and growing year on the nutritional, phytochemical, and antioxidant properties of industrial hemp (*Cannabis sativa* L.) seeds. Antioxidants 8(10):491

Isah T (2019) Changes in the biochemical parameters of albino, hyperhydric and normal green leaves of Caladium bicolor cv. "Bleeding hearts" in vitro long-term cultures. J Photochem Photobiol B 191:88–98

Isidore E, Karim H, Ioannou I (2021) Extraction of phenolic compounds and terpenes from *Cannabis sativa* L. by-products: from conventional to intensified processes. Antioxidants 10(6):942

Iuvone T, Esposito G, Esposito R, Santamaria R, Di Rosa M, Izzo AA (2004) Neuroprotective effect of cannabidiol, a non-psychoactive component from *Cannabis sativa*, on β-amyloid-induced toxicity in PC12 cells. J Neurochem 89(1):134–141

Izzo L, Castaldo L, Narváez A, Graziani G, Gaspari A, Rodríguez-Carrasco Y, Ritieni A (2020) Analysis of phenolic compounds in commercial *Cannabis sativa* L. inflorescences using UHPLC-Q-Orbitrap HRMS. Molecules 25(3):631

Jin D, Dai K, Xie Z, Chen J (2020) Secondary metabolites profiled in cannabis inflorescences, leaves, stem barks, and roots for medicinal purposes. Sci Rep 10(1):3309

Johnson GA, Jackson CL, Timoteo A, Sardaru P, Foland MH, Natarajan P, Dhekney SA (2025) Optimizing growth regulator concentrations for *Cannabis sativa* L. micropropagation. Plants 14(16):2586

Judžentienė A, Garjonytė R, Būdienė J (2023) Phytochemical composition and antioxidant activity of various extracts of fibre hemp (*Cannabis sativa* L.) cultivated in Lithuania. Molecules 28(13):4928

Kalinowska M, Płońska A, Trusiak M, Gołębiewska E, Gorlewska-Pietluszenko A (2022) Comparing the extraction methods, chemical composition, phenolic contents and antioxidant activity of edible oils from *Cannabis sativa* and *Silybum marianu* seeds. Sci Rep 12(1):20609

Kassahun Bantte DS, Feyissa T (2015) Effects of polyvinyl Pyrrolidone and activated charcoal to control effect of phenolic oxidation on in vitro culture establishment stage of micropropagation of sugarcane (*Saccharum officinarum* L.). Adv Crop Sci Technol 03(04)

Kastelec D, Troha S, Svetik S, Trafela T, Murovec J (2025) Optimization of micropropagation protocols and assessment of epigenetic changes in tissue cultures of diverse medical cannabis (*Cannabis sativa* L.) genotypes. Ind Crops Prod 229:121015

Khaksar S, Bigdeli M, Samiee A, Shirazi-zand Z (2022) Antioxidant and anti-apoptotic effects of cannabidiol in model of ischemic stroke in rats. Brain Res Bull 180:118–130

Khare S, Singh NB, Singh A, Hussain I, Niharika K, Yadav V, Bano C, Yadav RK, Amist N (2020) Plant secondary metabolites synthesis and their regulations under biotic and abiotic constraints. J Plant Biol 63(3):203–216

Klerk G-J (2002) Rooting of microcuttings: theory and practice. In Vitro Cell Dev Biol Plant 38(5):415–422

Koç E (2021) Effects of meta-topolin on the growth, physiological and biochemical parameters in plant tissue culture. In: Ahmad N, Strnad M (eds) Meta-topolin: a growth regulator for plant biotechnology and agriculture. Springer, Singapore, pp 265–278

Kojoma M, Iida O, Makino Y, Sekita S, Satake M (2002) DNA fingerprinting of *Cannabis sativa* using inter-simple sequence repeat (ISSR) amplification. Planta Med 68(1):60–63

Kopustinskiene DM, Masteikova R, Lazauskas R, Bernatoniene J (2022) *Cannabis sativa* L. bioactive compounds and their protective role in oxidative stress and inflammation. Antioxidants 11(4):660

Kornpointner C, Sainz Martinez A, Marinovic S, Haselmair-Gosch C, Jamnik P, Schröder K, Löfke C, Halbwirth H (2021) Chemical composition and antioxidant potential of *Cannabis sativa* L. roots. Ind Crop Prod 165:113422

Krishna Vrundha CP, Aswathi NV, Thomas TD (2021) The role of meta-topolin in plant morphogenesis in vitro. In: Ahmad N, Strnad M (eds) Meta-topolin: a growth regulator for plant biotechnology and agriculture. Springer, Singapore, pp 93–118

Kumar T, Singh RS, Kumar S, Pal AK (2018) Molecular markers for genetic fidelity assay of tissue cultured crops. Curr J Appl Sci Technol 31(3):1–9

Kurtz LE, Borbas LN, Brand MH, Lubell-Brand JD (2022) Ex vitro rooting of *Cannabis sativa* microcuttings and their performance compared to Retip and stem cuttings. HortScience 57(12):1576–1579

Lapierre É, Monthony AS, Torkamaneh D (2023) Genomics-based taxonomy to clarify cannabis classification. Genome 66(08):202–211

Lata H, Chandra S, Khan I, ElSohly MA (2009) Thidiazuron-induced high-frequency direct shoot organogenesis of *Cannabis sativa* L. In Vitro Cell Dev Biol-Plant 45(1):12–19

Lata H, Chandra S, Khan I, ElSohly M (2010a) High frequency plant regeneration from leaf derived callus of high Δ^9 -tetrahydrocannabinol yielding *Cannabis sativa* L. Planta Med 76(14):1629–1633

Lata H, Chandra S, Techen N, Khan I, ElSohly M (2010b) Assessment of the genetic stability of micropropagated plants of *Cannabis sativa* by ISSR markers. Planta Med 76(01):97–100

Lata H, Chandra S, Mehmedic Z, Khan IA, ElSohly MA (2012) In vitro germplasm conservation of high Δ9-tetrahydrocannabinol yielding elite clones of *Cannabis sativa* L. under slow growth conditions. Acta Physiol Plant 34(2):743–750

Lata H, Chandra S, Techen N, Khan IA, ElSohly MA (2016) In vitro mass propagation of *Cannabis sativa* L.: a protocol refinement using novel aromatic cytokinin meta-topolin and the assessment of eco-physiological, biochemical and genetic fidelity of micropropagated plants. J Appl Res Med Aromat Plants 3(1):18–26

Leyser O (2018) Auxin Signaling. Plant Physiol 176(1):465–479

Liang J, Ji F, Zhou Q, He D (2025) Optimizing LED light intensity and photoperiod to promote growth and rooting of medicinal cannabis in photoautotrophic micropropagation. Biology 14(6):706

Livingston SJ, Quilichini TD, Booth JK, Wong DCJ, Rensing KH, Laflamme-Yonkman J, Castellarin SD, Bohlmann J, Page JE, Samuels AL (2020) Cannabis glandular trichomes alter morphology and metabolite content during flower maturation. Plant J 101(1):37–56

Lowe H, Steele B, Bryant J, Toyang N, Ngwa W (2021) Non-cannabinoid metabolites of *Cannabis sativa* L. with Therapeutic Potential. Plants 10(2):400

Luangpraditkun K, Pimjuk P, Phimnuan P, Wisanwattana W, Wisespongpand C, Waranuch N, Viyoch J (2024) Anti-aging properties of Cannabis sativa leaf extract against UVA irradiation. Cosmetics 11(2):45

Lubell-Brand JD, Kurtz LE, Brand MH (2021) An in vitro–ex vitro micropropagation system for hemp. HortTechnology 31(2):199–207

Mahlberg PG, Kim ES (2004) Accumulation of cannabinoids in glandular trichomes of *cannabis* (Cannabaceae). J Ind Hemp 9(1):15–36

Marsicano G, Moosmann B, Hermann H, Lutz B, Behl C (2002) Neuroprotective properties of cannabinoids against oxidative stress: role of the cannabinoid receptor CB1. J Neurochem 80(3):448–456

McPartland JM (2018) *Cannabis* systematics at the levels of family, genus, and species. Cannabis Cannabinoid Res 3(1):203–212

Mendoza MA, Mills DK, Lata H, Chandra S, ElSohly MA, Almirall JR (2009) Genetic individualization of *Cannabis sativa* by a short tandem repeat multiplex system. Anal Bioanal Chem 393(2):719–726

Mestinsek-Mubi S, Svetik S, Flajsman M, Murovec J (2020) In vitro tissue culture and genetic analysis of two high-CBD medical cannabis (*Cannabis sativa* L.) breeding lines. Genetika 52(3):925–941

Mirzamohammad E, Alirezalu A, Alirezalu K, Norozi A, Ansari A (2021) Improvement of the antioxidant activity, phytochemicals, and cannabinoid compounds of *Cannabis sativa* by salicylic acid elicitor. Food Sci Nut 9(12):6873–6881

Mishchenko S, Laiko I, Tkachenko S, Lavrynenko Y, Marchenko T, Piliarska O (2022) The influence of exogenous growth regulators on the cannabinoid content and the main selection traits of hemp (*Cannabis sativa* L. ssp. Sativa). J Agric Sci 67(3):237–251

Mkpenie VN, Essien EE, Udoh II (2012) Effect of extraction conditions on total polyphenol contents, antioxidant and antimicrobial activities of *Cannabis sativa* L.

Moldzio R, Pacher T, Krewenka C, Kranner B, Novak J, Duvigneau JC, Rausch W-D (2012) Effects of cannabinoids Δ(9)-tetrahydrocannabinol, Δ(9)-tetrahydrocannabinolic acid and cannabidiol in MPP+ affected murine mesencephalic cultures. Phytomedicine 19(8–9):819–824

Monteuuis O, Bon M-C (2000) Influence of auxins and darkness on in vitro rooting of micropropagated shoots from mature and juvenile *Acacia mangium*. Plant Cell Tissue Organ Cult 63(3):173–177

Monthony AS, Bagheri S, Zheng Y, Jones AMP (2021a) Flower power: floral reversion as a viable alternative to nodal micropropagation in *Cannabis sativa*. In Vitro Cell Dev Biol-Plant

Monthony AS, Page SR, Hesami M, Jones AMP (2021b) The past, Present and Future of *Cannabis sativa* Tissue Culture. Plants 10

Mudge EM, Brown PN, Murch SJ (2019) The terroir of cannabis: terpene metabolomics as a tool to understand *Cannabis sativa* selections. Planta Med 85(09/10):781–796

Muhammad F, Liu Y, Wang N, Zhao L, Zhou Y, Yang H, Li H (2022) Neuroprotective effects of cannabidiol on dopaminergic neurodegeneration and α-synuclein accumulation in C. elegans models of Parkinson's disease. Neurotoxicology 93:128–139

Murphy R, Adelberg J (2021) Physical factors increased quantity and quality of micropropagated shoots of *Cannabis sativa* L. in a repeated harvest system with ex vitro rooting. In Vitro Cell Dev Biol-Plant

Muscarà C, Smeriglio A, Trombetta D, Mandalari G, La Camera E, Occhiuto C, Grassi G, Circosta C (2021) Antioxidant and antimicrobial activity of two standardized extracts from a new Chinese accession of non-psychotropic *Cannabis sativa* L. Phytother Res 35(2):1099–1112

Nadeem I, Khan AU, Asghar MN, Ashfaq M, Shahid S, Ahmed D (2012) In vitro Total antioxidant and radical scavenging activities of organic extracts from leaves, stem and inflorescence of *Cannabis sativa* L.

Nafis A, Kasrati A, Jamali CA, Mezrioui N, Setzer W, Abbad A, Hassani L (2019) Antioxidant activity and evidence for synergism of *Cannabis sativa* (L.) essential oil with antimicrobial standards. Ind Crop Prod 137:396–400

Ndlovu SB, Naidoo D, Van Staden J, Gebashe FC (2024) A systematic review of *Cannabis sativa* L. cultivation techniques: a comprehensive overview of tissue culture innovations and growth optimization. Ind Crops Prod 222:119539

Ni B, Liu Y, Dai M, Zhao J, Liang Y, Yang X, Han B, Jiang M (2023) The role of cannabidiol in aging. Biomed Pharmacother 165:115074

Nowak J, Veilleux RE, Turgeon S (2007) Priming for transplant stress resistance in in vitro propagation via plantlet bacterization. Acta Hortic 748:65–75

Nugroho K, Kosmiatin M, Santoso TJ, Sukma D, Purwito A, Husni A, Martasari C (2024) Optimization of kinetin concentrations and medium compositions for citrus shoot multiplication from cotiledonary node. Biotropia 31(1)

Pacifico D, Miselli F, Carboni A, Moschella A, Mandolino G (2008) Time course of cannabinoid accumulation and chemotype development during the growth of *Cannabis sativa* L. Euphytica 160(2):231–240

Pagano C, Savarese B, Coppola L, Navarra G, Avilia G, Laezza C, Bifulco M (2023) Cannabinoids in the modulation of oxidative signaling. Int J Mol Sci 24(3):2513

Page SRG, Monthony AS, Jones AMP (2021) DKW basal salts improve micropropagation and callogenesis compared with MS basal salts in multiple commercial cultivars of *Cannabis sativa*. Botany 99(5):269–279

Palmieri S, Maggio F, Pellegrini M, Ricci A, Serio A, Paparella A, Lo Sterzo C (2021) Effect of the distillation time on the chemical composition, antioxidant potential and antimicrobial activity of essential oils from different *Cannabis sativa* L. Cultivars Mol 26(16):4770

Pardaz J, Ojagh S, Kazamnia HD (2015) Effect of polyvinylpyrrolidone (PVP) on meristem establishment and in-vitro organogenesis of Iranian pear (Pyrus glabra). Int J Agric Biosci 4(5):206–208

Pascual D, Sánchez-Robles EM, García MM, Goicoechea C (2018) Chronic pain and cannabinoids. Great expectations or a christmas carol. Biochem Pharmacol 157:33–42

Pedrosa AM, Caetano TTV, de Andrade CML, Bernardes MM, Alves BC, Belo V, Duarte-Almeida JM, Stein VC (2024) *Cannabis sativa* L. from seized drug material: in vitro germination and establishment. Cannabis Cannabinoid Res 9(3):e924–e932

Pellegrini M, Palmieri S, Ricci A, Serio A, Paparella A, Lo Sterzo C (2021) *In vitro* antioxidant and antimicrobial activity of *Cannabis sativa* L. cv 'Futura 75' essential oil. Nat Prod Res 35(24):6020–6024

Pepe M, Hesami M, Small F, Jones AMP (2021) Comparative analysis of machine learning and evolutionary optimization algorithms for precision micropropagation of *Cannabis sativa*: prediction and validation of in vitro shoot growth and development based on the optimization of light and carbohydrate sources. Front Plant Sci 12:757869

Pepe M, Leonardos ED, Marie TRJG, Kyne ST, Hesami M, Jones AMP, Grodzinski B (2022) A noninvasive gas exchange method to test and model photosynthetic proficiency and growth rates of in vitro plant cultures: preliminary implication for *Cannabis sativa* L. Biology 11(5):729

Pérez-Pérez R, Pinski A, Zaranek M, Beckmann M, Mur LAJ, Nowak K, Rojek-Jelonek M, Kostecka-Gugała A, Petryszak P, Grzebelus E, Betekhtin A (2025) Effect of potent inhibitors of phenylalanine ammonia-lyase and PVP on in vitro morphogenesis of *Cannabis sativa*. BMC Plant Biol 25(1):469

Petrovici AR, Simionescu N, Sandu AI, Paraschiv V, Silion M, Pinteala M (2021) New insights on hemp oil enriched in cannabidiol: decarboxylation, antioxidant properties and in vitro anticancer effect. Antioxidants 10(5):738

Pinto JRS, de Freitas RMO, Praxedes SC (2010) Stimulation of in vitro development of *Cattleya granulosa* by sucrose

Pisanti S, Malfitano AM, Ciaglia E, Lamberti A, Ranieri R, Cuomo G, Abate M, Faggiana G, Proto MC, Fiore D, Laezza C, Bifulco M (2017) Cannabidiol: state of the art and new challenges for therapeutic applications. Pharmacol Ther 175:133–150

Polivanova OB, Bedarev VA (2022) Hyperhydricity in plant tissue culture. Plants 11(23):3313

Poniewozik M, Parzymies M, Szot P (2022) Effect of activated charcoal and ascorbic acid on in vitro morphogenesis and o-dihydroxyphenols content in Paphiopedilum insigne. Hortic Sci 49(1):48–51

Potter D (2009) The propagation, characterisation and optimisation of *Cannabis sativa* L as a phytopharmaceutical

Punja ZK (2021) Emerging diseases of *Cannabis sativa* and sustainable management. Pest Manag Sci 77(9):3857

Punja ZK, Sutton DB, Kim T (2023) Glandular trichome development, morphology, and maturation are influenced by plant age and genotype in high THC-containing cannabis (*Cannabis sativa* L.) inflorescences. J Cannabis Res 5(1):12

Radwan MM, Chandra S, Gul S, ElSohly MA (2021) Cannabinoids, phenolics, terpenes and alkaloids of cannabis. Molecules 26(9):2774

Rico S, Garrido J, Sánchez C, Ferreiro-Vera C, Codesido V, Vidal N (2022) A temporary immersion system to improve *Cannabis sativa* micropropagation. Front Plant Sci 13:895971

Ross SA, ElSohly MA, Sultana GNN, Mehmedic Z, Hossain CF, Chandra S (2005) Flavonoid glycosides and cannabinoids from the pollen of *Cannabis sativa* L. Phytochem Anal 16(1):45–48

Russo EB (2011) Taming THC: potential cannabis synergy and phytocannabinoid-terpenoid entourage effects. Br J Pharmacol 163(7):1344–1364

Salamone S, Waltl L, Pompignan A, Grassi G, Chianese G, Koeberle A, Pollastro F (2022) Phytochemical characterization of *Cannabis sativa* L. chemotype V reveals three new dihydrophenanthrenoids that favorably reprogram lipid mediator biosynthesis in macrophages. Plants 11(16):2130

Schettino L, Prieto M, Benedé JL, Chisvert A, Salvador A (2021) A rapid and sensitive method for the determination of cannabidiol in cosmetic products by liquid chromatography–tandem mass spectrometry. Cosmet 8(2):30

Serventi L, Flores GA, Cusumano G, Barbaro D, Tirillini B, Venanzoni R, Angelini P, Acquaviva A, Di Simone SC, Orlando G, Zengin G, Menghini L, Ferrante C (2023) Comparative investigation of antimicrobial and antioxidant effects of the extracts from the inflorescences and leaves of the *Cannabis sativa* L. cv. Strawberry. Antioxidants 12(2):219

Siano F, Moccia S, Picariello G, Russo GL, Sorrentino G, Di Stasio M, La Cara F, Volpe MG (2018) Comparative study of chemical, biochemical characteristic and ATR-FTIR analysis of seeds, oil and flour of the edible fedora cultivar hemp (*Cannabis sativa* L.). Molecules 24(1):83

Sirikantaramas S, Taura F, Tanaka Y, Ishikawa Y, Morimoto S, Shoyama Y (2005) Tetrahydrocannabinolic acid synthase, the enzyme controlling marijuana psychoactivity, is secreted into the storage cavity of the glandular trichomes. Plant Cell Physiol 46(9):1578–1582

Skoog F, Miller (1957) Chemical regulation of growth and organ formation in plant tissue cultured in vitro, pp 118–131

Skrzypek E, Szechyńska-Hebda M, Dąbrowska G (2007) The influence of phenolic accumulation on callus regeneration abilities of choosen plant species

Ślusarkiewicz-Jarzina A, Ponitka A, Kaczmarek Y (2005) Influence of cultivar, explant source and plant growth regulator on callus induction and plant regeneration of *Cannabis sativa* L. Acta Biol Cracov Ser Bot 47

Small E (2017) Classification of Cannabis sativa L. in relation to agricultural, biotechnological, medical and recreational utilization. In: Chandra S, Lata H, ElSohly MA (eds) *Cannabis sativa* L.—botany and biotechnology. Springer, Cham, pp 1–62

Smeriglio A, Galati EM, Monforte MT, Lanuzza F, D'Angelo V, Circosta C (2016) Polyphenolic compounds and antioxidant activity of cold-pressed seed oil from Finola cultivar of *Cannabis sativa* L. Phytother Res 30(8):1298–1307

Sorokin A, Kovalchuk I (2025) Development of efficient and scalable regeneration tissue culture method for *Cannabis sativa*. Plant Sci 350:112296

Sreelekshmi R, Siril EA (2020) Influence of polyamines on hyperhydricity reversion and its associated mechanism during micropropagation of China pink (*Dianthus chinensis* L.). Physiol Mol Biol Plants 26(10):2035–2045

Stephen C, Zayas VA, Galic A, Bridgen MP (2023) Micropropagation of hemp (*Cannabis sativa* L.). HortScience 58(3):307–316

Subrahmanyeswari T, Gantait S, Sarkar R, Kamble SN, Singh S, Bhattacharyya S (2024) Polyamines- and growth inducers-mediated enhanced mono-phasic in vitro regeneration of sugar leaf plant (*Stevia rebaudiana* Bert.) in liquid medium. S Afr J Bot 173:34–45

Taghinasab M, Jabaji S (2020) Cannabis microbiome and the role of endophytes in modulating the production of secondary metabolites: an overview. Microorganisms 8(3):355

Thomas TD (2008) The role of activated charcoal in plant tissue culture. Biotechnol Adv 26(6):618–631

Velappan Y, Signorelli S, Considine MJ (2017) Cell cycle arrest in plants: what distinguishes quiescence, dormancy and differentiated G1? Ann Bot 120(4):495–509

Wang Z, Arnold JC (2024) Cannabinoids and healthy ageing: the potential for extending healthspan and lifespan in preclinical models with an emphasis on Caenorhabditis elegans. GeroScience 46(6):5643–5661

Wang Z, Zheng P, Xie Y, Chen X, Solowij N, Green K, Chew YL, Huang X-F (2021) Cannabidiol regulates CB1-pSTAT3 signaling for neurite outgrowth, prolongs lifespan, and improves health span in Caenorhabditis elegans of Aβ pathology models. FASEB J 35(5):e21537

Wang F, Jin T, Li H, Long H, Liu Y, Jin S, Lu Y, Peng Y, Liu C, Zhao L, Wang X (2023) Cannabidivarin alleviates α-synuclein aggregation via DAF-16 in *Caenorhabditis elegans*. FASEB J 37(2):e22735

Weijers D, Wagner D (2016) Transcriptional responses to the auxin hormone. Annu Rev Plant Biol 67(1):539–574

Wongchaya P, Bansra T, Mahadtanapuk S (2024) Rapid micro-propagation of disease-free hemp (*Cannabis sativa* L.) to improve industrial production. Trends Sci 22(1):9057

Wróbel T, Dreger M, Wielgus K, Słomski R (2018) The application of plant in vitro cultures in cannabinoid production. Biotechnol Lett 40(3):445–454

Yan X, Tang J, Dos Santos PC, Nurisso A, Simões-Pires CA, Ji M, Lou H, Fan P (2015) Characterization of lignanamides from hemp (*Cannabis sativa* L.) seed and their antioxidant and acetylcholinesterase inhibitory activities. J Agric Food Chem 63(49):10611–10619

Zager JJ, Lange I, Srividya N, Smith A, Lange BM (2019) Gene networks underlying cannabinoid and terpenoid accumulation in cannabis. Plant Physiol 180(4):1877–1897

Zagórska-Dziok M, Bujak T, Ziemlewska A, Nizioł-Łukaszewska Z (2021) Positive effect of *Cannabis sativa* L. herb extracts on skin cells and assessment of cannabinoid-based hydrogels properties. Molecules 26(4):802

Zarei A, Davis B, Feyissa BA, Dinani ET, Simons B (2023) Improvement of mineral nutrition and rooting efficiency of *Cannabis sativa* L. for in vitro large-scale propagation. In Vitro Cell Dev Biol-Plant 59(1):95–105

Zengin G, Menghini L, Di Sotto A, Mancinelli R, Sisto F, Carradori S, Cesa S, Fraschetti C, Filippi A, Angiolella L, Locatelli M, Mannina L, Ingallina C, Puca V, D'Antonio M, Grande R (2018) Chromatographic analyses, in vitro biological activities, and cytotoxicity of *Cannabis sativa* L. essential oil: a multidisciplinary study. Molecules 23(12):3266

Zhang Y, Li H, Jin S, Lu Y, Peng Y, Zhao L, Wang X (2022) Cannabidiol protects against Alzheimer's disease in C. elegans via ROS scavenging activity of its phenolic hydroxyl groups. Eur J Pharmacol 919:174829

Zobayed S (2000) Water control and survival of *Ipomoea batatas* grown photoautotrophically under forced ventilation and photomixotrophically under natural ventilation. Ann Bot 86(3):603–610

Zuardi AW (2008) Cannabidiol: from an inactive cannabinoid to a drug with wide spectrum of action. Braz J Psychiatry 30:271–280

CRISPR/Cas9-Driven Genetic Enhancement of Anti-aging Medicinal Plants

13

Alka Jangra, Garima Sharma, and Vinod Chhokar

Abstract

Medicinal plants continue to attract interest for their role in healthy aging, largely due to their production of antioxidant, anti-inflammatory, and genome-protective secondary metabolites. With the growing availability of high-quality genomic and transcriptomic resources, genome editing, particularly CRISPR/Cas9, offers a precise way to modify the biosynthetic pathways that generate these compounds. Several important species, including *Ocimum sanctum* (Holy Basil), *Withania somnifera* (Ashwagandha), *Aloe vera*, *Centella asiatica* (Gotu kola), *Curcuma longa* (Turmeric), and *Phyllanthus emblica* (Amla), now have well-characterized metabolic genes linked to the synthesis of eugenol, withanolides, acemannan, triterpenoid saponins, curcuminoids, and ascorbate. In Basil and Ashwagandha, pathway reconstruction and enzyme-level studies have identified several rate-limiting steps that could serve as targets for future CRISPR-mediated metabolic enhancement once more reliable transformation systems are established. In species such as *Centella asiatica* and *Aloe vera*, the biochemical characterization of key enzymes including β-amyrin synthase, UDP-glycosyltransferases, mannosyltransferases, and polysaccharide-modifying acetyltransferases is recognized to be a potential target for pathway engineering. Collectively, these advances signal a gradual shift from conventional breeding toward precision genetic approaches aimed at elevating the therapeutic and anti-aging properties of medicinal plants. Despite species-specific recalcitrance and the need to manage off-target effects, CRISPR/Cas9 remains a promising tool for developing next-generation botanicals with enhanced bioactive profiles. This

A. Jangra
Department of Agricultural Biotechnology, College of Biotechnology, Chaudhary Charan Singh Haryana Agricultural University, Hisar, Haryana, India

G. Sharma · V. Chhokar (✉)
Department of Biotechnology, Guru Jambheshwar University of Science and Technology, Hisar, Haryana, India
e-mail: vinodchhokar@gjust.org

S. Gantait et al. (eds.), *Medicinal Plants with Anti-Ageing Properties*,
https://doi.org/10.1007/978-981-95-8064-4_13

chapter provides a comprehensive overview of the current progress, key constrains and omics, and AI-based future prospects of CRISPR/Cas9-driven approaches specifically aimed at enhancing the anti-aging potential of medicinal plants to address the growing global demand for natural therapeutic agents.

Keywords

AI-driven genome editing · Bioactive compounds · Cell repair · Metabolic engineering · Multi-omics

13.1 Introduction

Population aging is a worldwide phenomenon that presents medical, social, and economic challenges. The body accumulates a variety of cellular and molecular damage throughout time, which leads to aging. It is a normal, slow, complicated, and irreversible process that is accompanied by systemic alterations that lower functionality and raise the risk of age-related illnesses that ultimately lead to death (Nunes et al. 2024). The free radical theory of aging proposed by Harman in 1950s, a key concept in the research field of aging, postulates that free radicals generated by internal metabolism and external influences can harm cell structure, impede cell function, and eventually cause cell apoptosis and body aging. As a physiological activity brought on by free radicals, oxidative stress (OS) is a major contributor to the oxidative destruction of cells and tissues as well as the primary cause of aging and several diseases. Numerous studies on aging have demonstrated that OS is crucial to the beginning and progression of aging and associated illnesses. Because persistent OS damages vital cell structures, it not only speeds up the body's aging process but also increases the risk of aging-related disorders (Yang et al. 2024a; Póciniczak et al. 2025). Loss of proteostasis, dysregulated nutrient sensing, an increase in inflammation, oxidative processes, telomere attrition, stem cell exhaustion, mitochondrial dysfunction, cellular senescence, genomic instability, and impaired intercellular communication have all been linked to aging (Nunes et al. 2024). Enzymes including glutathione peroxidase, catalase, and superoxide dismutase (SOD) usually endorse this defense. However, as we become older, our defenses become less effective (Liguori et al. 2018). However, aging is more than just getting older; lifestyle, environment, and genetics all play an integral role. It is essential how our cells communicate with each other through pathways like sirtuins, NF-κB (nuclear factor-kappa-B), and mTOR (mammalian target of rapamycin) (Larocca et al. 2021). Therefore, one of the main goals of anti-aging studies is to find ways to lower OS and inflammation. Medicinal herbs really sparkle in this instance. For thousands of years, they have been used in medical practices like Ayurveda and Traditional Chinese Medicine. These plants comprise compounds that help repair cells and fight oxidation and inflammation (Ding et al. 2025). Ashwagandha (*Withania somnifera*), for instance, contains withanolides which maintain healthy mitochondria and reduce OS indicators, particularly in the brain and liver. Curcumin, major polyphenol found in Turmeric (*Curcuma longa*) is another big one as it is famous for turning down NF-κB-driven inflammation and helping skin and nerve cells bounce back (Chakraborty et al. 2021; Vittal and Vinciguerra 2025).

Apparently, these plant components seem to have been embraced by the cosmetic and supplement businesses. The global market for these natural anti-aging products reached USD 52.44 billion in 2024, and it's set to keep growing (Grand View Research 2024). Yet it is difficult to obtain sufficient amounts of these active compounds as the yield varies greatly amongst plants, and occasionally less than 1% of the beneficial compounds are obtained by weight. Species like *Centella asiatica* and *Phyllanthus emblica* are being overfished due to wild harvesting. In order to generate these molecules, people are searching for more effective and sustainable methods (Rana et al. 2024). Traditional plant breeding is erratic and slow. However, new genetic techniques are altering the landscape.

Entering into the era of CRISPR (Clustered Regularly Interspaced Short Palindromic Repeats)/Cas9 (CRISPR-associated protein 9), this gene-editing tool lets scientists alter plant DNA with high precision. It's fast, accurate, and multiple genes could be targeted at once, which will help to boost the metabolic pathways of interest; maybe by knocking out unhelpful branches or dialing up the machinery that makes the best compounds (Doudna and Charpentier 2014). Results are already rolling in as researchers have used CRISPR to ramp up withanolide production in Ashwagandha by editing genes like *HMGR* and *squalene synthase* (Ahmad et al. 2024). Similarly, CRISPR-based editing of the *GGP* (*VTC2*) gene in *Phyllanthus emblica* increased vitamin C accumulation, suggesting the potential of gene editing in amplifying naturally existing anti-aging compounds (Mahajan et al. 2023). *Centella asiatica* and *Bacopa monnieri* have emerged as candidates for triterpenoid pathway editing, with research focusing on glycosyltransferases and CYP450 monooxygenases, responsible for asiaticoside and bacoside synthesis (Rana et al. 2024). However, working with therapeutic plants is not always straightforward. Each species comes with its own set of transformation hurdles such as species-specific transformation barriers, recalcitrant tissue culture responses, and the long generation cycles of many valuable species and can be tough to grow in tissue culture. Honestly, getting genetic engineering to work in medicinal plants isn't easy. Each species brings its own hurdles; some just don't take up new genes well, others refuse to grow in tissue culture, and a lot of the most valuable plants take ages to mature (Das et al. 2024b). The desire to preserve genetic integrity while improving metabolic routes complicates genetic modification. Furthermore, regulations become more stringent, particularly for plants meant for direct human consumption, requiring demonstration of safety and approval under shifting genome-editing rules (Turnbull et al. 2021).

In light of all of that, this chapter explores how CRISPR/Cas9 can enhance anti-aging chemicals in therapeutic plants. It examines the potential anti-aging plants and their metabolites, species-specific case studies employing CRISPR for metabolic enhancement and discusses the major challenges associated with CRISPR/Cas-mediated metabolic engineering in plants. By combining traditional plant knowledge with cutting-edge genome editing, this chapter paves the way for future advances in plant-based anti-aging products by integration of Omics and AI (Artificial Intelligence).

13.2 Anti-aging Medicinal Plants and Their Bioactive Compounds

Following a time in which natural resources became secondary, herbal medicines have now again become more widely accepted than synthetic alternatives by global consumers. Herbal formulations are an ancient method that is still considered the primary choice for treating adaptive and functional aging-related diseases. It is tempting to hypothesize that herbal remedies could be useful in the field of holistic gerontotherapeutics, or therapeutic management, which is deliberately planned to delay the rapid reduction in performance that occurs regularly after the sixth decade of life. Herbal medicine frequently derives its ability to prevent pathology from folk traditions and millenarian empirical experience, although modern pharmacology appears to examine and validate some of these ideas. Currently, traditional herb medicines have gained special relevance to the point where they are considered prospective candidates for the management and even therapy of chronic and age-related disorders (Kraft and Hobbs 2004; Martins et al. 2016; Bjørklund et al. 2018). Taking such examples into account, it is useful to highlight the findings of multiple research, which show that "anti-aging herbs" have multifunctional effects and can protect the entire body by several methods. Furthermore, additional lines of study have found that those matrices have the potential to prevent and/or treat a variety of age-related neurological illnesses. In reality, the majority of currently available chemical medications are derived, directly or indirectly, from natural molecules, making a significant contribution to drug discovery and development (Chang and So 2008).

According to the World Health Organization (WHO), at least three-quarters of the global population relies on medicinal herbs for health treatment. Actually, outstanding research publications on medicinal plants address the pharmacognosy, chemistry, pharmacology, and clinical therapies of Ayurvedic and Chinese therapeutic systems (Sidhu and Pannu 2010). Current research indicates that the primary causes of inflammatory processes may be dysregulated gene expression in response to age-related OS, as well as the stimulation of redox-sensitive transcription factors. Aging-related up-regulation of inflammatory pathways includes NF-κB, IL-1β, IL-6, TNF-α, cyclooxygenase-2, adhesion molecules, and inducible nitric oxide synthase (Chung et al. 2006; Popoola et al. 2015; Matias et al. 2017). A wide variety of medicinal plants have been recognized for their ability to delay the aging process, owing to the substances they contain that combat OS, chronic inflammation, and cellular degradation.

Withania somnifera (Ashwagandha) is one of the most extensively researched plants in this field. Withanolides have been demonstrated to increase mitochondrial stability, minimize reactive oxygen species, and diminish inflammatory signaling, including NF-κB activity. These advantages are consistent with its long-established Ayurvedic reputation as a Rasayana, or Rejuvena (Basudkar et al. 2024).

Curcuma longa (Turmeric) is another useful example. Curcumin, the main active ingredient, has been demonstrated to help down the aging process and prevent age-related alterations. Its ability to prevent pro-aging proteins like NF-κB and mTOR and change the levels of senescence-related proteins like adenosine 5′-monophosphate-activated protein kinase (AMPK) and sirtuins is what gives it its potential anti-aging properties. Several studies have shown how curcumin can affect the aging process. One study examined the molecular mechanism of curcumin's ability to prolong the lifespan of *Caenorhabditis elegans*, and the findings suggested that this action might have been caused by increased resistance to OS and negative regulation of the AMPK signaling pathway (Nunes et al. 2024).

Phyllanthus emblica (Amla) possesses antioxidant capacity due to its high ascorbic acid content. Numerous researches have investigated the anti-aging potential of *Phyllanthus emblica*. According to the findings, it can increase *Caenorhabditis elegans* cell membrane mobility, resistance to thermal stimulation, and antioxidant damage, as well as Drosophila stress tolerance, weight-bearing swimming time in mice, SOD activity, and MDA (malondialdehyde) levels in mouse serum. It may also extend their lifespan. The research suggests that it has anti-aging qualities and can be used as a dietary intervention to prevent aging-related illnesses. Research indicates that *Phyllanthus emblica* can treat and prevent age-related illnesses (Wu et al. 2025).

Ocimum sanctum (Holy Basil) has traditionally been recognized as a restorative plant, and modern research is gradually reconnecting its phytochemistry to pathways that promote healthy aging. The plant comprises eugenol, ursolic acid, and a number of phenolic antioxidants that shield cells from oxidative and inflammatory damage. These compounds have been shown to inhibit the production of pro-inflammatory mediators, stabilize mitochondrial membranes, and protect neuronal networks, which often degrade with age. Experimental investigations demonstrate that eugenol improves redox balance and protects mitochondrial enzymes under stress, whereas ursolic acid lowers microglial activation and boosts neuronal survival in aging models (Kunkel et al. 2012; Kakkar et al. 2025). *Ocimum sanctum* is a promising natural ingredient for anti-aging skincare and cosmetic applications. It produces rosmarinic acid and phenolics, which exhibit the strongest antioxidant and anti-inflammatory effects. It markedly reduced IL-6 secretion and NF-κB activation and inhibited important skin-aging enzymes such as collagenase, elastase, and hyaluronidase. Rosmarinic acid is the main contributor to these activities (Chaiyana et al. 2019).

Centella asiatica has a slightly different function. The plant's triterpenoid fraction, specifically asiaticoside and madecassoside, promotes collagen turnover and skin healing. Clinical and experimental research suggest that Centella extracts prevent wrinkle formation, block matrix metalloproteinase activity, and improve skin elasticity, therefore their existence in many "pro-collagen" cosmetic product (Hein et al. 2025).

The anti-aging properties of *Aloe vera* are mostly ascribed to acemannan and related polysaccharides. These chemicals promote fibroblast proliferation, aid to retain moisture in skin tissues, and reduce UV-induced oxidative damage. This explains why aloe remains one of the most commonly utilized botanical components in commercial skincare formulas (Surjushe et al. 2008).

Other well-studied plants also have substantial geroprotective properties. *Panax ginseng* and its ginsenosides have been shown to maintain mitochondrial membrane integrity, support telomerase function, and delay senescence markers in vitro and in vivo (Hu et al. 2022). *Ginkgo biloba* contains neuroprotective flavonol glycosides that combat amyloid toxicity and promote synaptic resilience, hence contributing to cognitively healthy aging (Snitz et al. 2009). *Bacopa monnieri*, a long-used cognitive tonic, contains bacosides that reduce neuronal oxidative load and maintain synaptic plasticity in aging animals (Sato et al. 2003). When these medicinal plants are considered together, a common pattern begins to emerge. Most of them help to sustain the body's natural antioxidant defenses, calm persistent low-grade inflammation, and support the stability of mitochondrial activity. Many of these species are also known to help in collagen maintenance, repairing of extracellular matrix, and broader the cellular stability by persuading pathways allied with stress resistance, including Nrf2 and SIRT1. Their rising commercial interest and longstanding use in traditional healthcare systems, reflects not only cultural acquaintance but also a growing scientific appreciation of their biological relevance. With the increasing knowledge of their metabolic pathways, these plants are now being explored more deliberately through pathway engineering and, more recently, CRISPR-driven strategies aimed at improving the production of their key bioactive compounds.

13.3 CRISPR/Cas9 for Enhancement of Anti-aging Traits in Medicinal Plant

13.3.1 The Emergence of CRISPR/Cas9

Targeted genome editing uses a sequence-specific nuclease to cause a DNA double-strand break (DSB) at a target spot, which is then repaired using either the donor-dependent homology-directed repair (HDR) pathway or the erroneous non-homologous end joining (NHEJ) mechanism. Early sequence-specific nucleases that rely on protein-DNA interactions, such as meganuclease, zinc-finger nucleases (ZFNs), and transcriptional activator-like effector nucleases (TALENs), have made significant contributions to plant gene editing. However, the impediment occurs during their synthesis because extensive protein engineering is required. Thus, CRISPR/Cas9 and its other orthologs or alternative CRISPR/Cas systems can introduce DSBs at specified target sites with little effort and expense (Das et al. 2024b). The single-guide RNA (sgRNA), which is made up of a fusion of a trans-activating RNA (tracrRNA) and a CRISPR RNA (crRNA), is a component of the CRISPR/Cas system. It is derived from type II SpCas9 (the most widely used Cas9 protein from *Streptococcus pyogenes*) and is widely used in many species, including plants. The sgRNA-Cas complex attaches to the target DNA sequence, cleaves it, and recognizes a protospacer adjacent motif (PAM) site next to the sgRNA binding site (Jinek et al. 2012).

This shift is particularly significant for medicinal plants because many metabolites along with those having anti-aging properties, occur at low basal levels, are strongly regulated, or respond occasionally to environmental elicitors. In species valued for their anti-aging properties, such as *Withania somnifera*, *Curcuma longa*, *Phyllanthus emblica*, *Ocimum sanctum*, and *Centella asiatica*, CRISPR technology has opened a new avenue to enhance compounds that protect cells from OS, restore mitochondrial function, slow collagen degradation, or maintain neuronal integrity. As shown in Table 13.1, these species produce structurally diverse molecules, including withanolides, curcuminoids, eugenol derivatives, ascorbic acid, asiaticosides, and ginsenosides, but many of their biosynthetic pathways converge on core regulatory steps that can be precisely modulated using CRISPR/Cas9. The increasing availability of high-quality genomes, improved transformation protocols, and functional genomics datasets now allows for systematic interrogation of these pathways. Although the application of CRISPR in medicinal plants is still emerging compared to major crops, the past 5 years have seen significant gains, with multiple species demonstrating stable editing, validated gene knockouts, and measurable changes in metabolite profiles. These early studies provide a foundation for more targeted manipulation of anti-aging pathways.

13.3.2 Linking Anti-aging Metabolites with CRISPR-Editable Targets

Anti-aging activity in medicinal plants is mediated through several overlapping mechanisms, chief among them antioxidant defense, anti-inflammatory activity, neuroprotection, mitochondrial stability, and support of extracellular matrix integrity. Table 13.1 outlines major anti-aging plants, their characteristic compounds, and gene targets that regulate those pathways. CRISPR/Cas9 can intervene at three main levels:

(a) *Upregulation of precursor supply*
Genes in the MVA, MEP, and phenylpropanoid pathways (e.g., DXS, DXR, HMGR, PAL) can be edited to increase metabolic flux into downstream compounds.

(b) *Modification of committed enzymes*
Genes like CURS, DCS, WSCYP450s, EOMT, β-AS, and GGP determine specific metabolite outcomes and can be altered to enhance signature anti-aging compounds.

(c) *Editing of negative regulators.*
Knockout of transcriptional repressors or competing pathways can redirect flux toward desirable metabolites. Through these mechanisms, CRISPR/Cas9 can enhance metabolites such as withanolides, curcuminoids, ascorbic acid, rosmarinic acid, eugenol, asiaticosides, bacosides, and ginsenosides. The case studies below

Table 13.1 Anti-aging medicinal plants, key bioactive compounds, and CRISPR targets

Plant	Major bioactive compounds	Anti-aging justification	Potential CRISPR targets	References
Withania somnifera (Ashwagandha)	Withanolides (withaferin A, withanolide A)	Enhances endogenous antioxidant enzymes (SOD, CAT), reduces cortisol, and prevents telomere shortening, delaying cellular senescence	*HMGR*, *SQS*, *CYP450s*, *WRKY* TFs	Thorat et al. (2022), Ahmad et al. (2024), and Singh et al. (2016)
Curcuma longa (turmeric)	Curcumin, demethoxycurcumin, bis-demethoxycurcumin	Neutralizes free radicals, inhibits NF-κB and mTOR pathways, and enhances collagen synthesis, preventing photo-aging	*DCS*, *CURS1–3*, *PKS* genes	Chakraborty et al. (2021), Al Dayel and El Sherif (2022), and Ghose et al. (2022)
Phyllanthus emblica (Amla)	Ascorbic acid, ellagitannins, gallic acid	Natural vitamin C regenerates collagen, prevents oxidative DNA damage, and restores antioxidant defense in aged cells	*GGP (VTC2)*, *GalDH*, *GalUR*	Mellidou and Kanellis (2017), Mahajan et al. (2023), and Das et al. (2024b)
Centella asiatica	Asiaticoside, madecassoside, flavonoids	Promotes fibroblast proliferation and collagen synthesis; protects skin from UV-B OS	*UGT73*, *CYP716A*, *β-amyrin synthase*	Jo et al. (2017) and Kim et al. (2017)
Ocimum sanctum (holy basil)	Eugenol, ursolic acid, rosmarinic acid	Anti-inflammatory and antioxidant; eugenol inhibits ROS-mediated lipid peroxidation and skin cell aging	*EOMT*, *URS*, *TPS* genes	Singh et al. (2015), Kim et al. (2017), and Banerjee et al. (2023)

(continued)

Table 13.1 (continued)

Plant	Major bioactive compounds	Anti-aging justification	Potential CRISPR targets	References
Aloe vera	Acemannan, anthraquinones (aloin, emodin)	Enhances collagen and elastin fibers, increases skin hydration, and accelerates wound healing in aged tissue	*GMMT (CSLA9), general GT/ UGT involved in polysaccharide modification, aloe poly-pathway genes*	Salinas et al. (2019), Jangra et al. (2023), and Comas-Serra et al. (2024)
Panax ginseng	Ginsenosides (Rb1, Rg1)	Enhances collagen production, activates SIRT1 pathway, and scavenges ROS to prevent cellular aging	*CYP716A47, UGT71A27, DDS*	Yang et al. (2018, 2024b) and Jiang et al. (2024)
Bacopa monnieri	Bacosides A & B	Improves neuronal signaling, reduces OS, and protects against age-related cognitive decline	*SQS, β-amyrin synthase, UGT74*	Vishwakarma (2012) and Jeena et al. (2017)
Ginkgo biloba	Ginkgolides, bilobalide, flavonol glycosides	Prevents oxidative neuronal damage and improves mitochondrial function, delaying brain aging	*GGPPS, CYP716B, DFR*	Wang et al. (2025)

illustrate how these pathways align with Table 13.1 and how genome editing is being applied or is technically feasible.

• ***Withania somnifera*** **(Ashwagandha): Engineering the Withanolide Pathway**
Ashwagandha is widely recognized as an important anti-aging medicinal plant owing to the neuroprotective, anti-inflammatory and stress-mitigating properties of its withanolides. Advances in genome sequencing and pathway reconstruction have greatly improved our understanding of withanolide biosynthesis. Several rate-limiting enzymes such as 1-deoxy-D-xylulose-5-phosphate synthase (DXS), 3-hydroxy-3-methylglutaryl-CoA reductase (HMGR), and multiple cytochrome P450s involved in lactone formation have been functionally characterized through overexpression, silencing, and biochemical studies. Manipulation of specific

glycosyltransferases, including WsGT4 and WsGT6, using virus-induced gene silencing and transient overexpression, has been shown to alter withanolide glycosylation demonstrating that this step of the pathway is responsive to targeted intervention (Singh et al. 2016). The availability of genomic resources and recent proof-of-concept genome-editing attempts in related metabolic pathways underscore the potential of CRISPR/Cas9 for generating heritable modifications affecting withanolide content and glycosylation (Thorat et al. 2022; Ahmad et al. 2024).

- ***Ocimum sanctum* (Holy Basil): Modulating Eugenol, Rosmarinic Acid, and Ursolic Acid**

Holy Basil produces several molecules relevant to healthy aging, including eugenol, ursolic acid, and rosmarinic acid. Because these compounds play central roles in redox regulation, neuroinflammation, and mitochondrial stability, their pathways offer multiple CRISPR entry points. Genes such as EOMT, C4H, 4CL, and PAL govern the formation of aromatic compounds like eugenol and rosmarinic acid (Kim et al. 2017; Banerjee et al. 2023). Editing of orthologous genes in closely related Ocimum species has successfully altered phenylpropanoid profiles, providing a foundation for similar approaches in *Ocimum sanctum*. Ursolic acid biosynthesis involves the MVA pathway genes (HMGR, SQS, β-AS) which have been validated in multiple triterpenoid-rich plants. Given the strong mechanistic links between ursolic acid and anti-aging processes (Kunkel et al. 2012), CRISPR-mediated enhancement is a logical next step.

- ***Curcuma longa* (Turmeric): Curcuminoid Engineering**

Turmeric is globally recognized for its curcuminoids, which exhibit broad anti-aging activity through antioxidant, anti-inflammatory, and signaling modulation effects. The curcuminoid biosynthetic pathway is unusually well characterized, with clear gene-level targets such as CURS1–3, DCS, and various PKSs. Although *Curcuma longa* is recalcitrant in tissue culture, studies in related Zingiberaceae have shown successful protoplast transformation and CRISPR/RNP delivery, establishing a methodological pathway for turmeric editing. The targets outlined in Table 13.1 provide a direct blueprint for enhancing curcumin and demethoxycurcumin levels once efficient editing systems are optimized (Chakraborty et al. 2021; Deepa et al. 2017; Ghose et al. 2022).

- ***Phyllanthus emblica* (Amla): CRISPR Targeting of the Ascorbate Pathway**

Amla's relevance to anti-aging stems from its high ascorbic acid content, which supports collagen synthesis, redox balance, and cellular resilience. The Smirnoff–Wheeler pathway is highly conserved, and GGP has been identified as the primary control point in multiple fruit species. Recent genomic studies on *Phyllanthus emblica* (Mahajan et al. 2023) have mapped the ascorbate biosynthesis genes, making CRISPR editing feasible. Among the enzymes of the L-galactose pathway, GGP, GME, and GalDH offer the most leverage for increasing ascorbate production, as demonstrated in multiple fruit and vegetable crops. Although CRISPR editing of these targets has not yet been implemented in amla, the species benefits from a well-resolved vitamin C biosynthetic map and characterized gene orthologs. Together, these resources create a ready platform for applying genome editing to elevate vitamin C levels beyond natural limits (Mahajan et al. 2023).

- ***Centella asiatica* and *Aloe vera*: Triterpenoid and Polysaccharide Enhancement**

  Centella asiatica is valued for asiaticoside and madecassoside; titrated triterpenoids widely used in anti-wrinkle, wound-healing, and skin-repair formulations. Table 13.1 highlights targets such as β-amyrin synthase, SQS, and key UGTs, all of which have been effectively edited in other triterpenoid-producing plants (Jo et al. 2017; Kim et al. 2017). *Aloe vera*, on the other hand, derives much of its anti-aging value from acemannan and related polysaccharides. While CRISPR research is limited in Aloe species, functional studies show that targeted manipulation of acetyltransferases and mannosyltransferases can alter polysaccharide composition (Salinas et al. 2019; Comas-Serra et al. 2024). These gene families are included in Table 13.1 as forward-looking CRISPR targets.

- **Prospective CRISPR Targets in *Panax ginseng*, *Ginkgo biloba*, and *Bacopa monnieri***

  Even though complete CRISPR workflows are still emerging, these plants have highly studied phytochemical pathways: In *Panax ginseng*, CRISPR has been successfully applied to ginsenoside biosynthesis genes, and edited lines show altered triterpenoid profiles (Yang et al. 2018). In *Bacopa monnieri*, pathway-level analyses have identified candidate genes for bacoside regulation, making β-AS, UGTs, and CYP450s the prime CRISPR targets (Vishwakarma 2012; Jeena et al. 2017). In *Ginkgo biloba*, despite technical challenges, its flavonol biosynthesis pathway is well mapped, and several studies have characterized gene targets relevant to anti-aging flavonoids (Wang et al. 2025). Collectively, these studies demonstrate that even in species lacking robust transformation systems, pathway knowledge is mature enough to define precise CRISPR strategies.

13.4 Bottlenecks of Using CRISPR/Cas for Plant Metabolic Engineering

13.4.1 Off-Target Effect with Cas9

The CRISPR/Cas system has revolutionized genome editing with its precision and efficiency; however, off-target effects, unintended DNA cleavages at genomic regions that share sequence similarity with the target site (Fig. 13.1), remain a substantial technical limitation. The number and position of polymorphisms within the guide RNA (sgRNA), sgRNA structure, GC content, nuclease variants, and delivery methods were all identified as significant determinants of off-target activity through a logistic regression analysis. The optimization of sgRNA design and the increase in the number of mismatches have been demonstrated to significantly reduce off-target interactions (Modrzejewski et al. 2020). Particularly, the inaccurate binding is significantly reduced by mismatches within the seed region (the first 8–12 nucleotides

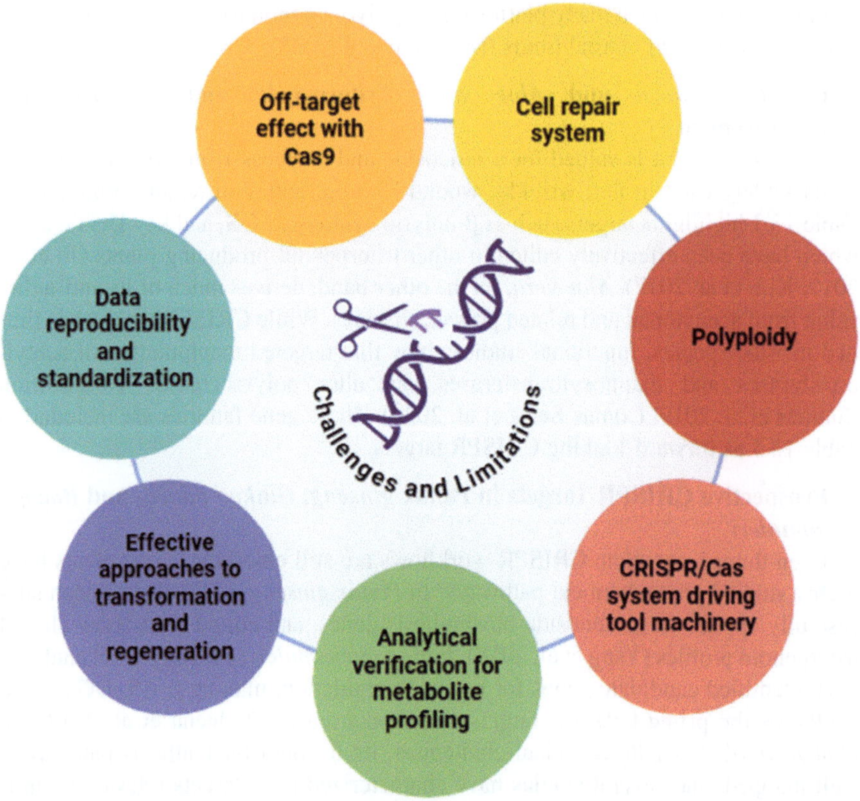

Fig. 13.1 Key factors limiting the efficiency of CRISPR/Cas-mediated metabolic engineering in plants

adjacent to the PAM site), and a single mismatch can reduce off-target frequency by nearly 59% (Das et al. 2024a).

However, double-strand breaks can still be induced by Cas9 binding to unintended loci, leading to chromosomal rearrangements such as deletions, inversions, and translocations (Brunet and Jasin 2018; Yin et al. 2022). Genotoxic stress and potential compromise of genomic integrity in plants may result from an elevated frequency of such events. Furthermore, asymmetric tolerance is demonstrated by mismatches within the sgRNA sequence. The 3′ end of the sequence is notably detrimental to target recognition, while certain 5′ mismatches can also reduce Cas9 activity (Semenova et al. 2011; Fu et al. 2013). Several optimization strategies have been devised to address these challenges, such as the modulation of Cas9 and sgRNA expression levels, the truncation of sgRNA 3′ ends, and the incorporation of additional guanine residues at the 5′ terminus. Although these refinements can enhance the on−/off-target ratio, they may occasionally result in a decrease in on-target editing efficacy (Mipeshwaree Devi et al. 2023).

13.4.2 Cell Repair System

Genome-editing tools typically introduce double-strand breaks (DSBs) in DNA, which are subsequently repaired by the cell's natural repair mechanisms, either HDR or NHEJ. NHEJ repairs breaks by introducing insertions or deletions (indels), often resulting in small sequence alterations, while HDR occurs when homologous DNA templates are available, allowing for precision gene replacement. The CRISPR/Cas system is capable of generating targeted DSBs, which are subsequently repaired through these pathways to achieve specific gene modifications that can modify the expression of metabolic pathways, in contrast to conventional repair mechanisms (Yan et al. 2020). Gene rearrangement, deletion, and replacement have emerged as critical strategies for the engineering of plant genomes to increase the production of secondary metabolites as a result of these processes (Weeks et al. 2016). Nevertheless, the sporadic nature of NHEJ-mediated indels can occasionally lead to frameshift mutations, which can result in unintended gene knockouts; a persistent challenge in precise genome editing (Mipeshwaree Devi et al. 2023).

13.4.3 Polyploidy

Genome editing presents a substantial challenge for plants with polyploid genomes, as they contain lengthy repetitive sequences, high gene copy numbers, and the duplication of genes that may have diverged in function over time. This genetic complexity complicates the process of achieving precise and uniform mutations across all gene copies. Furthermore, gene-editing outcomes are further complicated by dosage effects among paralogous gene copies, particularly in gene knockout trials. In numerous polyploid plant species, the simultaneous editing of multiple alleles of a target gene is necessary to achieve a desired mutant phenotype, a task that is both technically challenging and intricate (Mout et al. 2017; Schaart et al. 2021).

13.4.4 CRISPR Cas System Driving Tool Machinery

The efficiency and reliability of the CRISPR/Cas toolkit in endogenous plant systems must be improved by making substantial improvements to its fundamental components, including promoters, terminators, selectable markers, and transformation protocols. The biosynthesis of valuable secondary metabolites and the overall genome-editing efficiency are frequently restricted by the complex regulatory networks that are inherent to medicinal plants, which frequently impede the successful expression and activity of CRISPR components (Jiang and Doudna 2017; Zhou et al. 2018; Tong et al. 2019).

13.4.5 Analytical Verification for Metabolite Profiling

Metabolomics has emerged as a critical instrument in the field of natural product–based drug discovery, offering a rapid and potent method for the identification of bioactive compounds in plants. Metabolite profiling is frequently conducted using sophisticated analytical platforms, including GC-MS, LC-MS, CE-MS, and NMR. Nevertheless, these methodologies are subject to numerous constraints when utilized to verify metabolic modifications in CRISPR/Cas9-edited plants. The comprehensive detection and quantification of metabolites within plant systems are a challenge due to their extreme chemical complexity and enormous diversity. Additionally, the lack of spectral databases and the presence of overlapping chemical signatures continue to hinder the accurate annotation of metabolites, which is a significant impediment. The differentiation between native metabolites and CRISPR-induced variants is frequently impeded by incomplete reference libraries. Furthermore, the precise correlation between genomic alterations and metabolic shifts is limited by the absence of robust bioinformatic pipelines that can integrate metabolomic datasets with gene-editing outcomes. The necessity for enhanced standardization, database expansion, and data integration tools to ensure the reliable metabolic verification of genome-edited medicinal plants is underscored by these analytical challenges (Salem et al. 2020).

13.4.6 Effective Approaches to Transformation and Regeneration

Effective plant transformation and regeneration systems are essential for enhancing the efficacy of CRISPR/Cas9 across a variety of plant species. Nevertheless, the absence of dependable tissue culture or transformation protocols in numerous species remains a significant obstacle, substantially hindering the success of transformations and the efficacy of editing. In an effort to surmount this obstacle, optimized in vitro culture systems, somatic embryogenesis, and protoplast-based regeneration methods have emerged as indispensable instruments for improving the susceptibility of plant cells to CRISPR/Cas9-mediated modifications. Additionally, the integration of genotype-independent regeneration techniques and transient expression systems can further accelerate the generation of stable, edited lines while reducing somaclonal variation and transformation-induced stress, which are crucial steps toward the full potential of genome editing in medicinal plants (Zhang et al. 2013; Altpeter et al. 2016; Mout et al. 2017).

13.4.7 Data Reproducibility and Standardization

The absence of standardized procedures for the development of effective pharmaceuticals from natural products continues to be a significant obstacle. Additionally, the field of bioinformatics is undergoing rapid evolution as a result of the continual updating or redesigning of bioinformatics tools and databases to accommodate emerging data types. Consequently, bioinformatics software frequently lags behind the development of high-throughput data generation technologies. This divide requires ongoing collaboration among molecular biologists, computer scientists, and mathematicians to develop software and databases that are both efficient and robust. This will thereby improve research on natural products and their potential as therapeutic candidates (Aborode et al. 2022).

13.5 Ethical and Societal Considerations

Recent technological advancements in plant breeding, particularly gene editing and cisgenesis, necessitate a reassessment of current biotechnology regulations. In addition to the legal debates surrounding the definition of genetic modification under the Cartagena Protocol and other national or international laws, broader societal discussions have arisen concerning the environmental impact, ethics, and safety of these technologies. Consequentialism, deontology, and virtue ethics are the three primary philosophical frameworks that are frequently employed in the ethical analyses of such innovations. These frameworks offer unique perspectives on the potential benefits, risks, and moral dilemmas associated with contemporary crop biotechnology (Louwaars and Jochemsen 2021).

13.5.1 Consequentialism

Consequence ethics, also known as consequentialism, stresses the need to judge actions based on their outcomes. This approach focuses on how gene-edited crops affect sustainability, biodiversity, and social equity in plant breeding, with a special focus on CRISPR/Cas9 uses. Gene editing could make the People, Planet, and Profit dimensions better by making crops more stable, more efficient at using nutrients, and more resistant to biotic and abiotic challenges from a sustainability point of view. However, trade-offs can lead to ethical and environmental issues, such as how herbicide-tolerant crops lead to more chemical use (Louwaars 2018). The main goal of biodiversity concerns is to keep genetic variety among and within species. Precision breeding and the introduction of novel traits may enhance crop diversity by gene editing; yet, worries persist over the constriction of genetic pools and the transfer of edited traits to wild relatives. CRISPR has reduced off-target and unexpected impacts, but they still present safety and regulatory issues that require a level of scrutiny similar to that of conventional breeding methods.

Ultimately, issues of social justice emerge in connection with market regulation and intellectual property rights. Even if CRISPR is cheaper and easier to get than classic GMOs, severe biosafety rules and patent restrictions could make it harder for smaller breeders to get involved in new ideas. So, ethical governance frameworks need to find a compromise between long-term sustainability, environmental integrity, and fair access to new technologies (Louwaars and Jochemsen 2021).

13.5.2 Deontology

Deontology emphasizes ethics that are founded on moral obligations and universal principles, with an emphasis on intrinsic values and rights. This encompasses the ethical treatment and integrity of plants in the field of plant biotechnology. The concept of plant rights encompasses the preservation of genetic and phenotypic integrity, the completion of life cycles, and the fulfilment of plants' inherent purposes. Critics contend that genetic modification, particularly through CRISPR/Cas9, may contravene these principles by disrupting natural reproductive barriers or altering intrinsic genetic composition (Bueren and Struik 2005). Genomic integrity is the preservation of the natural DNA sequence in the organic sector, which is in opposition to direct human manipulation, such as transgenesis or gene editing. Phenotypic integrity is concerned with the preservation of plant form and function in accordance with natural expression, which is a cause for concern when plants are engineered for industrial or pharmaceutical purposes. The concept of intrinsic value acknowledges that plants possess value that exceeds their utility to humans. Some frameworks, such as intra-species gene editing or cisgenesis, are considered ethically permissible alternatives under these perspectives, despite the ethical objections that arise when genetic modification crosses species barriers or alters natural evolutionary processes (Louwaars and Jochemsen 2021).

13.5.3 Virtue Ethics

Virtue ethics underscores the importance of moral character and the pursuit of a fulfilling life by emphasizing virtues such as responsibility, care, and integrity. In the field of plant biotechnology, it underscores the breeder's moral obligation to guarantee that their work is beneficial to society and consistent with ethical principles that extend beyond regulatory compliance.

Moral Responsibility Breeders and associated stakeholders are ethically obligated to act with honesty and care, ensuring that technological advancements like CRISPR/Cas9 serve sustainable and socially just purposes. Sectoral initiatives and sustainability frameworks should supplement ethical conduct, which should not be exclusively based on laws. Rather, it should be characterized by internalized virtues.

Transparency Trust is fostered by the disclosure of breeding methods and genetic techniques. Breeders are required to provide consumers and supply chains with information regarding gene-editing methods and their implications, despite the fact that full traceability may be challenging due to global material exchanges and undetectable alterations.

Certification Product labelling and certification systems (e.g., organic, biodynamic, fair trade) can guide consumer choice; however, they encounter challenges as a result of the complexity of contemporary breeding methods. Production costs may be elevated, and consumers may be perplexed by excessive labelling.

Governmental Responsibility Regulators are required to strike a balance between feasibility and public transparency. Mandatory labeling may be impractical due to the fact that gene modifications are frequently undetectable. Consequently, in order to preserve consumer trust and freedom of choice, ethical governance should integrate regulatory supervision, voluntary certification, and transparent communication (Davarnejad 2011; Louwaars and Jochemsen 2021).

13.6 Future Perspectives and Integration with Omics and AI

Pharmacology, biomedicine, and nutrition industries are dramatically influenced by medicinal plants. The accumulation of valuable specialized metabolites can be improved by reengineering their metabolic pathways, particularly by removing competitive or rate-limiting steps. CRISPR/Cas9 technology has emerged as a highly precise and versatile instrument for targeted gene modification, facilitating the development of plant varieties with enhanced metabolite yields and reduced synthesis of undesirable compounds. Refined control over the genome, epigenome, and transcriptional regulation, as well as post-translational modifications, is now possible due to advancements in CRISPR/Cas systems. Improvements in base editing, HDR, and chromosomal rearrangements have expanded its scope. The technology also enables the construction of large-scale mutant libraries and supports gene knockout (CRISPR-KO), repression (CRISPRi), and activation (CRISPRa), thereby expediting the discovery of novel biosynthetic pathways and plant metabolic engineering. Additionally, multiplex genome editing allows for the simultaneous targeting of multiple genes, thereby dodging the limitations of single-gene genetic manipulation. New opportunities for optimizing specialized metabolite biosynthesis in medicinal plants are continually emerging as a result of a more comprehensive understanding of gene regulatory networks and transcription factors. This has the potential to facilitate the production of bioactive compounds that are both sustainable and efficient (Bradbury et al. 2025).

13.6.1 Role of Multi-omics

The swift progression of omics technologies has significantly augmented our capacity to conduct exhaustive biological analyses in medicinal plants. Proteomics (iTRAQ and label-free quantification), transcriptomics (microarray, RNA-seq, scRNA-seq, spRNA-seq), metabolomics (bulk and spatial), and genomics (NGS and long-read sequencing) each have distinctive advantages and disadvantages

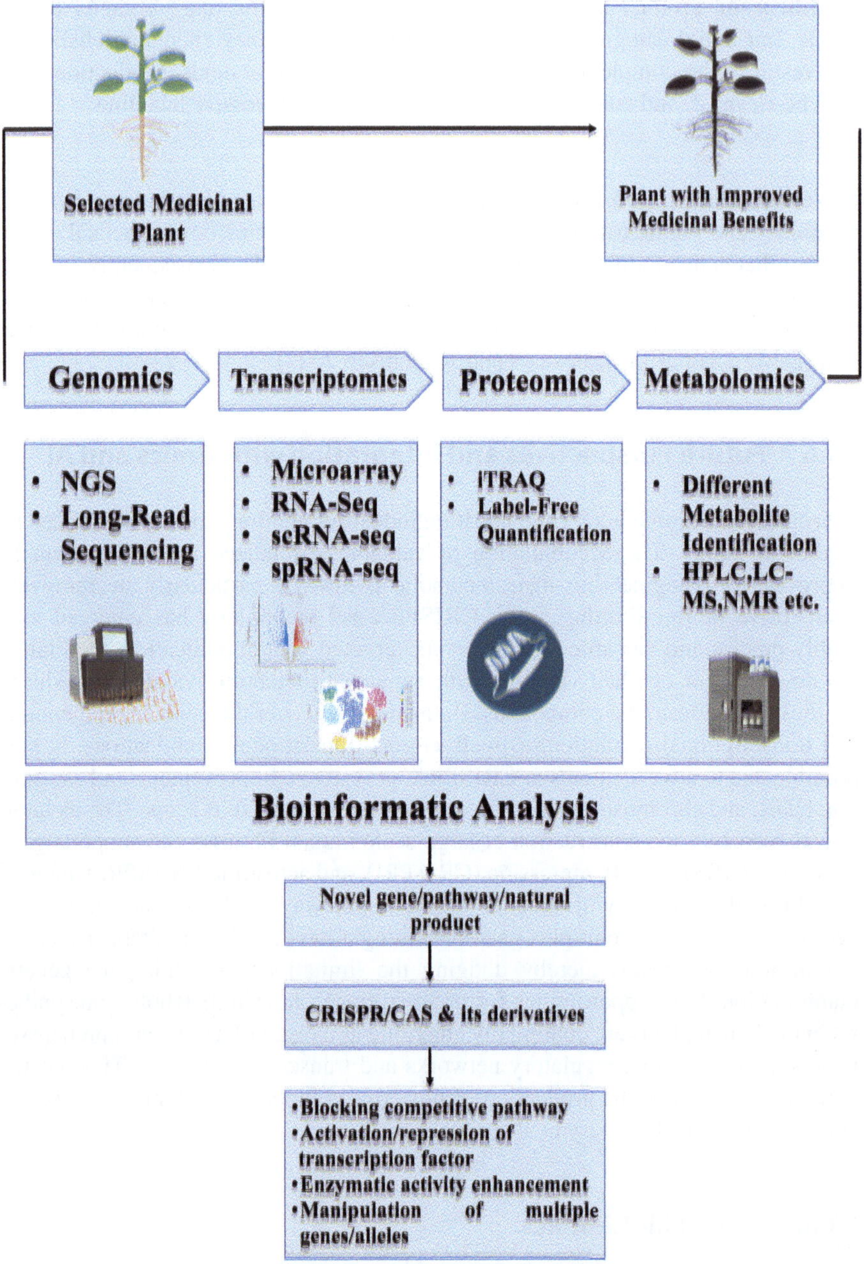

Fig. 13.2 An overview of integrated multi-omics pipelines and CRISPR/Cas9-based genome editing strategies employed in medicinal plants. The primary methodologies of genomics, transcriptomics, proteomics, and metabolomics are illustrated as foundational components of multi-omics research. Following data integration and analysis, these approaches facilitate the discovery of novel secondary metabolites and the identification of key functional genes, gene clusters, and biosynthetic pathways. The combined application of CRISPR/Cas9 and its derivatives with multi-omics tools enables the targeted enhancement of desired metabolites

(Fig. 13.2). Integrating these datasets offers a more comprehensive understanding of complex biological processes, surpassing the limitations of single-omics analyses. Multi-omics integration has transformed medicinal plant research by facilitating the accurate identification of plant species and processed materials, elucidating functional genes that govern crucial agronomic and therapeutic characteristics, clarifying secondary metabolite biosynthetic pathways, and enhancing comprehension of stress response mechanisms that affect the production and quality of bioactive compounds (Yang et al. 2023).

Numerous medicinal species have metabolic gene clusters linked to the manufacture of secondary metabolites, according to recent genome-mining research. To address the rising demand for pharmacologically significant natural products, researchers are progressively adopting multi-omics-guided metabolic engineering approaches (Denoeud et al. 2014). For instance, the CYP719 gene encoding (S)-canadine synthase, a crucial enzyme in the berberine biosynthesis pathway, was discovered in *Coptis chinensis* using an integrated omics study that incorporated high-quality genome assembly, annotation, and metabolomic profiling (Liu et al. 2021). The integrated application of transcriptomics and metabolomics offers complementary perspectives on the correlation between genes and metabolites, thereby enabling the identification of novel biosynthetic pathways and regulatory networks. Ongoing advancements in bioinformatics, computational genome mining, and the decreasing costs of high-throughput omics technologies have further expedited the discovery of pathways for specialized metabolites (Di Filippo et al. 2022). Our knowledge of the intricate molecular interactions controlling the metabolism of medicinal plants is currently being reshaped by integrative analyses that link genetic and metabolic profiles using systems biology frameworks. This is opening the door to precision metabolic engineering and the sustainable production of therapeutically significant compounds (Borah et al. 2024).

13.6.2 Potential of AI-Driven Genome Editing Design

Genetic research and innovative therapeutics have been revolutionized by genome editing tools, especially CRISPR-Cas9. These techniques have been further improved by the incorporation of AI, which has increased their precision, effectiveness, and affordability, particularly when it comes to treating complicated genetic illnesses. By examining genomic context and Cas protein type, AI-driven systems like DeepCRISPR, CRISTA, and DeepHF greatly improve gRNA creation, maximizing editing efficiency, specificity, and reducing off-target impacts. By analyzing patient genomic data to find disease-related mutations, biomarkers, and therapeutic targets, AI enhances personalized and precision medicine beyond gRNA design. In order to determine efficacy, toxicity, and resistance for more specialized therapies, it also aids in drug response prediction. Additionally, through promoter selection, intelligent vector engineering, and targeted delivery to particular cell types, AI helps to optimize the transport and expression of genome editing components, such as Cas proteins, gRNAs, and vectors. By forecasting sgRNA performance for accurate gene regulation without changing DNA sequences, tools such as EpiCas-DL significantly broaden AI's use to epigenetic genome editing. Notwithstanding these developments, there are still issues with lowering operating expenses, enhancing delivery methods, guaranteeing clinical safety, and creating reliable CRISPR databases. To improve the dependability and transparency of AI-guided editing models, future research should concentrate on explainable AI, transfer learning, and uncertainty quantification. A significant step toward safe, accurate, and customized genetic therapeutics, the combination of AI with genome editing will speed up advancements in biomedicine and healthcare for the benefit of human health (Rynjah et al. 2025).

13.7 Conclusion

Plants with long histories in traditional medicine, such as *Withania somnifera, Ocimum sanctum, Curcuma longa, Centella asiatica, Aloe vera, Panax ginseng, Bacopa monnieri, Ginkgo biloba,* and many more, are now being revisited through the lens of modern biology for their possible roles in healthy aging. Their bioactive compounds support many of the processes that weaken over time: they cushion cells against oxidative stress, slow low-grade inflammation, help maintain mitochondrial function, encourage collagen and extracellular-matrix repair, and protect neurons as they grow more vulnerable with age. These effects explain why such plants have remained important across cultures and why interest in them continues to grow. It is now viable to consider enhancing these plants in ways that were previously impractical due to the expansion of genetic resources and a better understanding of metabolic pathways. CRISPR/Cas9 provides a possible prospect where researchers may commence manipulating metabolic flux in potentially healthier directions by focusing on certain enzymes and regulatory genes in the biosynthetic pathways that produce anti-aging compounds such as withanolides, ginsenosides, asiaticosides, curcuminoids, bacosides, etc. However, making the leap from thought to practice still remains difficult. The outcomes of genome editing are made difficult by polyploidy or genomic complexity, and many therapeutic plants have been difficult to regenerate or transform. Progress is additionally slowed by off-target effects, a lack of molecular markers, and the requirement for metabolite verification. Simultaneously, there are ethical and regulatory concerns when modified plants are intended for medical applications. Effectually, the field continues to grow swiftly despite these constraints. The potential to fine-tune the production of age-modulating medications becomes more plausible as tissue-culture techniques become more advance, editing tools get more accurate, and multi-omics datasets develop broader. In order to demonstrate how conventional knowledge, plant biochemistry, and the latest developments in genome editing are starting to converge, this chapter links these threads together. It opens up new revenues to develop therapeutic plants that are both genetically improved to promote healthy aging and scientifically determined.

Acknowledgments The authors acknowledge the assistance of OpenAI for language refinement, with final oversight and revisions made by the authors themselves.

Competing Interests The authors have no relevant financial or non-financial interests to disclose.

References

Aborode AT, Awuah WA, Mikhailova T et al (2022) OMICs technologies for natural compounds-based drug development. Curr Top Med Chem 22:1751–1765

Ahmad Z, Shareen, Ganie IB, Firdaus F, Ramakrishnan M, Shahzad A, Ding Y (2024) Enhancing withanolide production in the Withania species: advances in in vitro culture and synthetic biology approaches. Plants 13:2171

Al Dayel MF, El Sherif F (2022) Spirulina platensis foliar spraying Curcuma longa has improved growth, yield, and curcuminoid biosynthesis gene expression, as well as curcuminoid accumulation. Horticulturae 8:469

Altpeter F, Springer NM, Bartley LE et al (2016) Advancing crop transformation in the era of genome editing. Plant Cell 28:1510–1520

Banerjee J, Samanta S, Ahmed R, Dash SK (2023) Bioactive pentacyclic triterpenes trigger multiple signalling pathways for selective apoptosis leading to anticancer efficacy: recent updates and future perspectives. Curr Protein Pept Sci 24:820–842

Basudkar V, Gujrati G, Ajgaonkar S, Gandhi M, Mehta D, Nair S (2024) Emerging vistas for the nutraceutical Withania somnifera in inflammageing. Pharmaceuticals 17:597

Bjørklund G, Dadar M, Martins N, Chirumbolo S, Goh BH, Smetanina K, Lysiuk R (2018) Brief challenges on medicinal plants: an eye opening look at ageing related disorders. Basic Clin Pharmacol Toxicol 122:539–558

Borah A, Singh S, Chattopadhyay R et al (2024) Integration of CRISPR/Cas9 with multi-omics technologies to engineer secondary metabolite productions in medicinal plant: challenges and prospects. Funct Integr Genomics 24:207

Bradbury A, Clapp O, Biacsi A-S et al (2025) Integrating genome editing with omics, artificial intelligence, and advanced farming technologies to increase crop productivity. Plant Commun 6

Brunet E, Jasin M (2018) Induction of chromosomal translocations with CRISPR-Cas9 and other nucleases: understanding the repair mechanisms that give rise to translocations. Chromosome translocation. Adv Exp Med Biol 1044:15–25

Bueren ETLV, Struik PC (2005) Integrity and rights of plants: ethical notions in organic plant breeding and propagation. J Agric Environ Ethics 18:479–493

Chaiyana W, Anuchapreeda S, Punyoyai C, Neimkhum W, Lee K-H, Lin W-C, Lue S-C, Viernstein H, Mueller M (2019) Ocimum sanctum Linn. As a natural source of skin anti-ageing compounds. Ind Crop Prod 127:217–224

Chakraborty A, Mahajan S, Jaiswal SK, Sharma VK (2021) Genome sequencing of turmeric provides evolutionary insights into its medicinal properties. Commun Biol 4:1193

Chang RC-C, So K-F (2008) Use of anti-ageing herbal medicine, Lycium barbarum, against ageing-associated diseases. What do we know so far? Cell Mol Neurobiol 28:643–652

Chung HY, Sung B, Jung KJ, Zou Y, Yu BP (2006) The molecular inflammatory process in ageing. Antioxid Redox Signal 8:572–581

Comas-Serra F, Miró JL, Umñaa MM, Minjares-Fuentes R, Femenia A, Mota-Ituarte M, Pedroza-Sandoval A (2024) Role of acemannan and pectic polysaccharides in saline-water stress tolerance of Aloe vera (Aloe barbadensis Miller) plant. Int J Biol Macromol 268:131601

Das S, Kwon M, Kim J-Y (2024a) Enhancement of specialized metabolites using CRISPR/Cas gene editing technology in medicinal plants. Front Plant Sci 15:1279738

Das G, Kameswaran S, Ramesh B, Bangeppagari M, Nath R, Das Talukdar A, Shin H-S, Patra JK (2024b) Anti-ageing effect of traditional plant-based food: an overview. Foods 13:3785

Davarnejad L (2011) In the shadow of soft law: the handling of corporate social responsibility disputes under the OECD guidelines for multinational enterprises. J Disp Resol 351–381

Deepa K, Sheeja TE, Rosana OB, Srinivasan V, Krishnamurthy KS, Sasikumar B (2017) Highly conserved sequence of ClPKS11 encodes a novel polyketide synthase involved in curcumin biosynthesis in turmeric (Curcuma longa L.). Ind Crop Prod 97:229–241

Denoeud F, Carretero-Paulet L, Dereeper A et al (2014) The coffee genome provides insight into the convergent evolution of caffeine biosynthesis. Science (1979) 345:1181–1184

Di Filippo M, Pescini D, Galuzzi BG et al (2022) INTEGRATE: model-based multi-omics data integration to characterize multi-level metabolic regulation. PLoS Comput Biol 18:e1009337

Ding F, Yu Y, Zhang Y, Wei S, Han JH, Li Z, Jiang H-B, Ryu D, Park W, Ha K-T (2025) Harnessing nutrients and natural products for sustainable drug development against ageing. Front Pharmacol 16:1579266

Doudna JA, Charpentier E (2014) The new frontier of genome engineering with CRISPR-Cas9. Science (80-) 346:1258096

Fu Y, Foden JA, Khayter C et al (2013) High-frequency off-target mutagenesis induced by CRISPR-Cas nucleases in human cells. Nat Biotechnol 31:822–826

Ghose AK, Abdullah SNA, Md Hatta MA, Megat Wahab PE (2022) DNA free CRISPR/DCAS9 based transcriptional activation system for UGT76G1 gene in *Stevia rebaudiana* Bertoni protoplasts. Plants 11(18):2393

Grand View Research (2024) Anti-ageing products market size, share & trends analysis report, 2024–2030. Grand View Research. Available at: https://www.grandviewresearch.com

Hein ZM, Gopalakrishna PK, Kanuri AK, Thomas W, Hussan F, Naik VR, Shantakumari N, Che Ramli MD, Mohd Moklas MA, Che Mohd Nassir CMN (2025) Centella asiatica: advances in extraction technologies, Phytochemistry, and therapeutic applications. Life 15:1081

Hu Y, He Y, Niu Z, Shen T, Zhang J, Wang X, Hu W, Cho JY (2022) A review of the immunomodulatory activities of polysaccharides isolated from Panax species. J Ginseng Res 46:23–32

Jangra A, Chaturvedi S, Sihag S, Sharma G, Tiwari S, Chhokar V (2023) Identification and functional characterization of a novel aldo-keto reductase from Aloe vera. Planta 258:107

Jeena GS, Fatima S, Tripathi P, Upadhyay S, Shukla RK (2017) Comparative transcriptome analysis of shoot and root tissue of Bacopa monnieri identifies potential genes related to triterpenoid saponin biosynthesis. BMC Genomics 18:490

Jiang F, Doudna JA (2017) CRISPR—Cas9 structures and mechanisms. Annu Rev Biophys 46:505–531

Jiang T, Zhang Y, Zuo G, Luo T, Wang H, Zhang R, Luo Z (2024) Transcription factor PgNAC72 activates DAMMARENEDIOL SYNTHASE expression to promote ginseng saponin biosynthesis. Plant Physiol 195:2952–2969

Jinek M, Chylinski K, Fonfara I, Hauer M, Doudna JA, Charpentier E (2012) A programmable dual-RNA–guided DNA endonuclease in adaptive bacterial immunity. Science (80-) 337:816–821

Jo H-J, Han JY, Hwang H-S, Choi YE (2017) Amyrin synthase (EsBAS) and -amyrin 28-oxidase (CYP716A244) in oleanane-type triterpene saponin biosynthesis in Eleutherococcus senticosus. Phytochemistry 135:53–63

Kakkar A, Singh H, Anand A, Chopra H, Mishra AK (2025) Neuroprotective effects of eugenol in Alzheimer's disease: mitigating oxidative stress, inflammation and amyloid plaques. Curr Pharm Des. https://doi.org/10.2174/0113816128373290250620054118

Kim OT, Jin ML, Lee DY, Jetter R (2017) Characterization of the asiatic acid glucosyltransferase, UGT73AH1, involved in asiaticoside biosynthesis in Centella asiatica (L.) urban. Int J Mol Sci 18:2630

Kraft K, Hobbs C (2004) 20043081041, English, book, Germany, 3-13-126991-X, Stuttgart, Pocket guide to herbal medicine (xi + 491pp.). Georg Thieme Verlag, Leipzig

Kunkel SD, Elmore CJ, Bongers KS, Ebert SM, Fox DK, Dyle MC, Bullard SA, Adams CM (2012) Ursolic acid increases skeletal muscle and brown fat and decreases diet-induced obesity, glucose intolerance and fatty liver disease. PLoS One 7:e39332

Larocca D, Lee J, West MD, Labat I, Sternberg H (2021) No time to age: uncoupling ageing from chronological time. Genes (Basel) 12:611
Liguori I, Russo G, Curcio F, Bulli G, Aran L, Della-Morte D, Gargiulo G, Testa G, Cacciatore F, Bonaduce D (2018) Oxidative stress, ageing, and diseases. Clin Interv Ageing:757–772
Liu Y, Wang B, Shu S et al (2021) Analysis of the Coptis chinensis genome reveals the diversification of protoberberine-type alkaloids. Nat Commun 12:3276
Louwaars NP (2018) Plant breeding and diversity: a troubled relationship? Euphytica 214:114
Louwaars N, Jochemsen H (2021) An ethical and societal analysis for biotechnological methods in plant breeding. Agronomy 11:1183
Mahajan S, Bisht MS, Chakraborty A, Sharma VK (2023) Genome of Phyllanthus emblica: the medicinal plant Amla with super antioxidant properties. Front Plant Sci 14:1210078
Martins N, Barros L, Ferreira ICFR (2016) In vivo antioxidant activity of phenolic compounds: facts and gaps. Trends Food Sci Technol 48:1–12
Matias I, Diniz LP, Buosi A, Neves G, Stipursky J, Gomes FCA (2017) Flavonoid hesperidin induces synapse formation and improves memory performance through the astrocytic TGF-1. Front Ageing Neurosci 9:184
Mellidou I, Kanellis AK (2017) Genetic control of ascorbic acid biosynthesis and recycling in horticultural crops. Front Chem 5:50
Mipeshwaree Devi A, Khedashwori Devi K, Premi Devi P et al (2023) Metabolic engineering of plant secondary metabolites: prospects and its technological challenges. Front Plant Sci 14:1171154
Modrzejewski D, Hartung F, Lehnert H et al (2020) Which factors affect the occurrence of off-target effects caused by the use of CRISPR/Cas: a systematic review in plants. Front Plant Sci 11:574959
Mout R, Ray M, Yesilbag Tonga G et al (2017) Direct cytosolic delivery of CRISPR/Cas9-ribonucleoprotein for efficient gene editing. ACS Nano 11:2452–2458
Nunes YC, Mendes NM, Pereira de Lima E, Chehadi AC, Lamas CB, Haber JFS, dos Santos Bueno M, Araújo AC, Catharin VCS, Detregiachi CRP (2024) Curcumin: a golden approach to healthy ageing: a systematic review of the evidence. Nutrients 16:2721
Póciniczak A, Bukowska-Olech E, Wysocka E (2025) The complexity of oxidative stress in human age-related diseases-a review. Metabolites 15:479
Popoola OK, Marnewick JL, Rautenbach F, Ameer F, Iwuoha EI, Hussein AA (2015) Inhibition of oxidative stress and skin ageing-related enzymes by prenylated chalcones and other flavonoids from Helichrysum teretifolium. Molecules 20:7143–7155
Rana N, Gupta P, Singh H, Nagarajan K (2024) Role of bioactive compounds, novel drug delivery systems, and polyherbal formulations in the management of rheumatoid arthritis. Comb Chem High Throughput Screen 27:353–385
Rynjah D, Sandhanam K, Bhattacharjee B et al (2025) CRISPR/Cas9 gene editing systems for enhancing secondary metabolite biosynthesis via reproductive tissue modification. Discover Plants 2:245
Salem MA, de Souza LP, Serag A et al (2020) Metabolomics in the context of plant natural products research: from sample preparation to metabolite analysis. Metabolites 10(1):37
Salinas P, Salinas C, Contreras RA, Zuiga GE, Dupree P, Cardemil L (2019) Water deficit and abscisic acid treatments increase the expression of a glucomannan mannosyltransferase gene (GMMT) in Aloe vera Burm. F. Phytochemistry 159:90–101
Sato M, Tanaka H, Fujiwara S, Hirata M, Yamaguchi R, Etoh H, Tokuda C (2003) Antibacterial property of isoflavonoids isolated from Erythrina variegata against cariogenic oral bacteria. Phytomedicine 10:427–433
Schaart JG, van de Wiel CCM, Smulders MJM (2021) Genome editing of polyploid crops: prospects, achievements and bottlenecks. Transgenic Res 30:337–351
Semenova E, Jore MM, Datsenko KA et al (2011) Interference by clustered regularly interspaced short palindromic repeat (CRISPR) RNA is governed by a seed sequence. Proc Natl Acad Sci 108:10098–10103

Sidhu K, Pannu K (2010) Indigenous use of medicinal plants for health care. Stud Ethno-Med 4:145–148
Singh P, Kalunke RM, Giri AP (2015) Towards comprehension of complex chemical evolution and diversification of terpene and phenylpropanoid pathways in Ocimum species. RSC Adv 5:106886–106904
Singh G, Tiwari M, Singh SP, Singh S, Trivedi PK, Misra P (2016) Silencing of sterol glycosyltransferases modulates the withanolide biosynthesis and leads to compromised basal immunity of Withania somnifera. Sci Rep 6(1):25562
Snitz BE, O'Meara ES, Carlson MC, Arnold AM, Ives DG, Rapp SR, Saxton J, Lopez OL, Dunn LO, Sink KM (2009) Ginkgo biloba for preventing cognitive decline in older adults: a randomized trial. JAMA 302:2663–2670
Surjushe A, Vasani R, Saple DG (2008) Aloe vera: a short review. Indian J Dermatol 53:163
Thorat SA, Kaniyassery A, Poojari P, Rangel M, Tantry S, Kiran KR, Joshi MB, Rai PS, Botha A-M, Muthusamy A (2022) Differential gene expression and withanolides biosynthesis during in vitro and ex vitro growth of Withania somnifera (L.) Dunal. Front Plant Sci 13:917770
Tong Y, Weber T, Lee SY (2019) CRISPR/Cas-based genome engineering in natural product discovery. Nat Prod Rep 36:1262–1280
Turnbull C, Lillemo M, Hvoslef-Eide TAK (2021) Global regulation of genetically modified crops amid the gene edited crop boom-a review. Front Plant Sci 12:630396
Vishwakarma RK (2012) Isolation, cloning and characterization of terpenoids biosynthesis pathway gene(s) from Bacopa monnieraPh.D. Thesis, CSIR-National Chemical Laboratory, Pune, India
Vittal M, Vinciguerra M (2025) Enhancing healthspan with Ashwagandha (Withania somnifera): a comprehensive review of its multifaceted geroprotective benefits. Biogerontology 26:1–20
Wang C, Zhang W, Liao Y, Ye J, Xu F, Wang Q (2025) Ginkgo biloba flavonoids: analysis of functions, regulatory mechanisms, and extraction. Plant Biol 27:962–974s
Weeks DP, Spalding MH, Yang B (2016) Use of designer nucleases for targeted gene and genome editing in plants. Plant Biotechnol J 14:483–495
Wu N, Pan Y, Liu Q, Shahidi F, Li H-Y, Chen F, Deng Z-Y, Zhang Z-H (2025) Protective benefits and mechanisms of Phyllanthus emblica Linn. on ageing induced by oxidative stress: a system review. Food Med Homol 2:9420029
Yan M-Y, Li S-S, Ding X-Y et al (2020) A CRISPR-assisted nonhomologous end-joining strategy for efficient genome editing in mycobacterium tuberculosis. MBio 11:10–1128
Yang J-L, Hu Z-F, Zhang T-T, Gu A-D, Gong T, Zhu P (2018) Progress on the studies of the key enzymes of ginsenoside biosynthesis. Molecules 23:589
Yang L, Yang Y, Huang L et al (2023) From single-to multi-omics: future research trends in medicinal plants. Brief Bioinform 24:1
Yang J, Luo J, Tian X, Zhao Y, Li Y, Wu X (2024a) Progress in understanding oxidative stress, ageing, and ageing-related diseases. Antioxidants 13:394
Yang Y, Wang J, Han F, Zhang J, Gao M, Zhao Y, Chen Y, Wang Y (2024b) Characterization of UGT71, a major glycosyltransferase family for triterpenoids, flavonoids and phytohormones-biosynthetic in plants. For Res 4:e035
Yin J, Lu R, Xin C et al (2022) Cas9 exo-endonuclease eliminates chromosomal translocations during genome editing. Nat Commun 13:1204
Zhang Y, Zhang F, Li X et al (2013) Transcription activator-like effector nucleases enable efficient plant genome engineering. Plant Physiol 161:20–27
Zhou Z, Tan H, Li Q et al (2018) CRISPR/Cas9-mediated efficient targeted mutagenesis of RAS in Salvia miltiorrhiza. Phytochemistry 148:63–70

Molecular and Multiomics (Genomics, Proteomics, Transcriptomics) Approaches for Enhancing Anti-aging Properties of Plant Secondary Metabolites

14

Iliassou Lah Mouafon, Ebede Guy Roland, Bella Belinga Merveilles Aurore, David R. Katerere, Gaetan Bayiha Ba Njock, and Ndongo Joseph Thierry

Abstract

Molecular and multi-omics approaches to enhance the anti-aging properties of plant secondary metabolites are a highly interdisciplinary pursuit that combines plant chemistry, ethnobotany, biology, biotechnology, pharmacology, and anti-aging research. Although thousands of medicinal plants have yet to be studied, many of them are recognized for their anti-aging properties, mainly due to their anti-inflammatory, antioxidant, and skin-rejuvenating metabolites. However, identifying promising anti-aging compounds is taking time, is scarce, and requires advanced analytical tools due to the complexity of plant metabolomes. Multiomics techniques are increasingly being used and applied to medicinal plants to identify species-specific molecular markers and functional genes underlying key biological traits, to elucidate the regulatory mechanisms of environmental responses and the biosynthetic pathways of bioactive compounds. Thus, the combination of systems biology, artificial intelligence, and biotechnology

I. L. Mouafon · D. R. Katerere
Department of Pharmaceutical Sciences, Tshwane University of Technology, Pretoria, South Africa

E. G. Roland (✉) · B. B. M. Aurore · N. J. Thierry
Department of Chemistry, Higher Training College, University of Yaounde 1, Yaounde, Cameroon
e-mail: guy.ebede@univ-yaounde1.cm

G. B. Ba Njock
Faculty of Science, Department of Chemistry, University of Maroua, Maroua, Cameroon

S. Gantait et al. (eds.), *Medicinal Plants with Anti-Ageing Properties*,
https://doi.org/10.1007/978-981-95-8064-4_14

platforms offers a strategic opportunity for the expansion of innovative, effective, and scalable anti-aging agents. This chapter provides insights into the genetic basis of anti-aging activity, chemistry, and physiology of mode-of-action, as well as cellular mechanisms of anti-aging secondary metabolites. In addition, the chapter presents metabolomics, genomics, proteomics, and transcriptomics approaches that are being used to unravel the mechanism behind understanding the biosynthetic pathways of anti-aging secondary metabolites and their manipulation for genetic improvement of the concerned medicinal plant species.

Keywords

Medicinal plants · Physiology · Omics · Genetic basis · Anti-aging · Mode of action

14.1 Introduction

Aging is a non-reversible and continuing biological process marked by the degenerative changes in tissue and cellular functions, leading to a heightened susceptibility to various age-related diseases. To slow down this process, the use of medicinal plants has proven to be effective as they are rich in secondary metabolites with several pharmacological properties. However, the underlying mechanisms are complex, and research methods encounter several limitations. Natural extracts have been widely employed as a key strategy to slow down and prevent skin aging (Duque et al. 2017). Numerous herbs are known to contain biologically active metabolites, particularly polyphenolic compounds with antioxidant and photoprotective properties. These compounds influence cellular signaling pathways that execute a role in skin health. When incorporated into cosmetic formulations, some secondary metabolites are rapidly absorbed through the skin (Duque et al. 2017). Despite this potential, there is a growing need for an integrated approach to identify and evaluate promising raw materials. Over the past years, the adoption and progress of omics technologies have significantly expanded the focus of aging research, expanding from epigenetic, macrogenomic, and transcriptional aspects to metabolic and protein levels. This chapter delves into the advancements in multiomics approaches like metabolomics, genomics, proteomics, transcriptomics, and specifically in relation to anti-aging compounds derived from plants. The goal is to offer fresh insights and strategies for advancing research methodologies in this area.

14.2 Anti-aging Medicinal Plants and Their Benefits

The use of plants and their products as medicines has been well documented in history for centuries. The natural products extracted and isolated from the various plant parts, such as polyphenols, terpenoids, alkaloids, and flavonoids, have demonstrated admirable medicinal values. These properties included antibacterial, antioxidant, anticancer, anti-inflammatory, and cytotoxic effects (Mouafon and Katerere 2025). The anti-aging properties, which are one of the plant's contributions, are primarily mediated by their anti-inflammatory, skin-rejuvenating, and antioxidant potential (Payne et al. 2025; Chaiyana et al. 2019; Anis et al. 2024). The process of aging and related diseases is complex and characterized by cellular damage due to oxidative stress, enzyme degradation of the skin matrix complex, chronic inflammation, pigmentary disorders, and rejuvenation capacity (Chaiyana et al. 2019; Stavropoulou et al. 2021). Skin, which is one of the organs where the signs of aging are most apparent, is affected by intrinsic and extrinsic determinants (Chaiyana et al. 2019). These include the deterioration of collagen, elastin, and hyaluronic acid, ultimately leading to features like wrinkles, laxity, deterioration of elasticity, melasma, a rough-textured appearance, and freckles (Chaiyana et al. 2019; Boonpisuttinant et al. 2023; Kowalczyk et al. 2025). Therefore, substances capable of mitigating oxidative stress are recognized to slow down the aging-related cognitive and physical decay (Rodrigues et al. 2018).

The use of conventional drugs for anti-aging purposes has been associated with various adverse effects and high costs. For these reasons, natural products that are safe, affordable, and sources of lead compounds, like polyphenolic compounds with their reported antioxidant efficacy, remain a promising source. These molecules mitigate oxidative stress by neutralizing reactive oxygen species (ROS), modulating inflammatory routes, and stabilizing collagen and elastin fibers that accelerate cellular aging (Payne et al. 2025; Anis et al. 2024; Gupta et al. 2023; Verma et al. 2023). In addition, several medicinal plants are reported for their anti-aging potential. *Panax ginseng*, rich in ginsenosides, enhances skin elasticity and skin tone, moisturizes skin, promotes collagen formation, improves blood circulation, and strengthens the body's defenses against tiredness, anxiety, stress and trauma through anti-inflammatory and antioxidant mechanisms (Table 14.1) (Anis et al. 2024; Verma et al. 2023; Raina et al. 2023; Chakraborty et al. 2018). The rhizome of *Curcuma longa* (turmeric) provides curcuminoids that reduce wrinkles, improve hydration and skin flexibility, increase skin elasticity, decrease pigmentation, and inhibit elastase and collagenase enzymes that degrade skin proteins (Anis et al. 2024; Verma et al. 2023; Raina et al. 2023; Chakraborty et al. 2018; Lestari et al. 2023). *Aloe vera* leaves, rich in Aloin A and B, enhance collagen production and are used to treat stretch marks, wrinkling, dark or white areas, and acne (Anis et al. 2024; Verma et al. 2023; Chakraborty et al. 2018).

Further information on anti-aging plants and their benefits has been provided in Table 14.1. These benefits extend beyond skin health; they offer antioxidant defense to neutralize free radicals, inhibit enzymes that degrade skin components, reduce chronic inflammation by modulating cytokines, and enhance skin repair and hydration by stimulating fibroblast activity and hyaluronic acid production. This multifaceted approach not only delays skin aging but also improves overall cellular health, contributing to functional longevity.

Table 14.1 Anti-aging medicinal plants and their benefits

Plant family	Species/popular name	Plant part	Preparation and administration	Key bioactive compounds	Anti-aging effects	References
Acoraceae	*Acorus calamus* L./Shobu	Arial parts	NS	Asarone, eugenol	DNA protection, antioxidant	Raina et al. (2023)
Anacardiaceae	*Sclerocarya birrea* Hochst/Marula	Shoot	Methanol extract	NS	Suppression or blockage of collagenase and elastase enzymes	Kazeem et al. (2024)
Apiaceae	*Centella asiatica* (L.) Urb./Kotukola	Leaves	Water or alcohol extracts in creams and face packs	Asiaticoside, madecassoside, centelloside, and asiatic acid	Increases telomerase activity, promotes skin cell regrowth, collagen synthesis, and wound healing. Promote the fibroblast to proliferate. Firm skin reduces scars	Anis et al. (2024), Raina et al. (2023), and Zamzury and Omar (2019)
Apidae	Honey	Nectar	Used directly or in creams	NS	Prevents skin wrinkles, antioxidant activity	Chakraborty et al. (2018)
Apocynaceae	*Hemidesmus indicus* (L.) R.Br.	Aerial parts	Various extracts	Hemidesmin-1 & 2	Antioxidant, wound healing	Raina et al. (2023)
Araliaceae	*Panax ginseng* C.A.Mey./ginseng	Roots	Water or alcohol extracts in cosmetic formulations	Ginsenosides	Protects from UVB damage, improves blood circulation, moisturizes skin, enhances skin tone, improves learning performance and memory, promotes collagen formation, enhances skin elasticity, and strengthens the body defenses against trauma, rejuvenation, tiredness, stress, and anxiety	Verma et al. (2023), Raina et al. (2023), Anis et al. (2024), and Chakraborty et al. (2018)
Asphodelaceae	*Aloe vera* (L.) Burm.f.	Leaves	Extracted gel applied topically or orally; used in creams or lotion	Aloin A and B	Boost collagen production, UV protection, used to treat stretch marks, wrinkling, dark or white areas, and acne.	Anis et al. (2024), Chakraborty et al. (2018), and Verma et al. (2023)

Asteraceae	*Calendula officinalis* L./Marigold	Flowers	Extracts in ointments and face packs	NS	Anti-inflammatory, wound healing. Soothes skin and promotes healing	Anis et al. (2024)
	Matricaria recutita L./Chamomile	Flowers	Extracts applied topically	NS		Anis et al. (2024)
	Pluchea dioscoridis (L.) DC	Shoot	Essential oil	NS	Strong inhibition of elastase, collagenase, tyrosinase enzymes, and hyaluronidase	(Kazeem et al. (2024)
	Pluchea lanceolata (DC.) C.B.Clarke	Shoot	Extracts	Isorhamnetin, quercetin	Antioxidant, anticlastogenic, and lipid peroxidation inhibition	Raina et al. (2023)
	Silybum marianum (L.) Gaertn/milk thistle	Fruits	Water and glycerol macerations	Silymarin	Possessed hepatoprotective properties	Jabłonowska et al. (2021)
	Sonchus oleraceus L.	Leaves	70% methanolic extract	Chicoric acid, caftaric acid, and chlorogenic acid	It possesses antioxidant properties by significantly suppressing the H2O2 stress-induced premature senescence	Ou et al. (2015)
Bixaceae	*Cochlospermum vitifolium* (Willd.) Spreng./silk-cotton tree	Leaves	Aqueous extract	NS	Potent inhibition of elastase, hyaluronidase, and collagenase	Kazeem et al. (2024)
Boraginaceae	*Alkanna tinctoria* L. Tausch/Dyer's bugloss	Root	Acetone extract	NS	Anti-elastase activity	Kazeem et al. (2024)
Caricaceace	*Carica papaya* Linn./papaya	Fruit pulp	Extracts used in skin creams, topical application, and exfoliating face pack	Vitamin C, vitamin A, and vitamin E	They reduce power and free radical scavenging. Promotes skin cell turnover. Exfoliation, rejuvenation	Anis et al. (2024) and Saini et al. (2016)

(continued)

Table 14.1 (continued)

Plant family	Species/popular name	Plant part	Preparation and administration	Key bioactive compounds	Anti-aging effects	References
Clusiaceae	*Garcinia picrorrhiza* Miq.	Fruit	Ethanol extract	NS	Moderate elastase and weak collagenase inhibition	Kazeem et al. (2024)
	Harungana madagascariensis Lam. ex Poir.	Stem	Methylene chloride extract	NS	Strong elastase inhibition, weak tyrosinase	Kazeem et al. (2024)
Combretaceae	*Terminalia arjuna* (Roxb. ex DC.) Wight & Arn./ Arjuna	Stem bark	Extracts in formulations	Arjunolic acid, arjunic acid, arjungenin	Increases collagen synthesis, moisturizes. Strengthens skin barrier, reduces dryness	Verma et al. (2023) and Chakraborty et al. (2018)
Convolvula-ceae	*Convolvulus prostratus* Forssk.	Arial parts	Extracts	Scopoletin,β-sitosterol	Antioxidant, DNA repair and protection, and enhances telomerase	Raina et al. (2023)
Cordycipita-ceae	*Isaria tenuipes* Peck.	Whole plant	Aqueous extract	NS	Hyaluronidase inhibition, tyrosinase, and elastase	Kazeem et al. (2024)
Fabaceae	*Cassia auriculata* Linn.	Flower	Hydro-methanol extract	NS	Moderate tyrosinase inhibition	Kazeem et al. (2024)
	Glycine max (L.) Merr.	Seed	Ethanolic extract	NS	Strong estrogenic. Treatment of post-menopausal symptoms	Yingngam et al. (2011)
	Glycyrrhiza glabra L./Liquorice.	Roots and rhizomes	Extracts in topical products	Glycyrrhizin or glycyrrhizic acid	Improves DNA damage tolerance, prevents wrinkles, soothes skin, shields the brain from oxidative damage, maintains normal nervous system function, and enhances one's memory	Verma et al. (2023), Chakraborty et al. (2018), and Raina et al. 2023
	Psoralea corylifolia L.	Aerial parts	Extracts	Psoralen, bakuchiol	Antioxidant, anti-inflammatory, and collagen promotion	Raina et al. (2023)

Ginkgoaceae	*Ginkgo biloba* L./ Ginkgo	Leaves	Extracts	Ginkgolides A, B, and C and bilobalide	Antioxidant. Helps with memory, promotes blood flow to the brain, and enhances the brain's capacity for thought	Verma et al. (2023)
Gnetaceae	*Gnetum gnemon* L	NS	Extracts	Trans-resveratrol	Potential as a skin-whitening agent by inhibiting tyrosinase in the melanogenesis process	Lestari et al. (2023)
Hippocastanaceae	*Aesculus hippocastanum* L.// horse chestnut	Seeds	Powder or alcohol extract	Oleic acid	Used to clean oily skin in cosmetics	Verma et al. (2023)
Hypericaceae	*Hypericum origanifolium* Willd	Aerial parts	Ethanol extract	NS	Elastase, collagenase, and hyaluronidase inhibition	Kazeem et al. (2024)
Lauraceae	*Litsea cubeba* (Lour.) Pers./ mountain pepper	Fresh fruits	Essential oil and its cream	NS	Anti-pigmentation, antioxidant, and anti-tyrosinase	Kowalczyk et al. (2025)
Linaceae	*Linum usitatissimum* L.	Seed	Ethanolic extract	NS	Strong estrogenic, antioxidant. Treatment of post-menopausal symptoms	Yingngam et al. (2011)
Lythraceae	*Punica granatum* L./ pomegranate	Fruit peel	Water and ethanolic extracts	Anthocyanin	It has the potential to create a barrier to protect the skin and avoid aging of skin. Possesses antioxidant activity, is weakly estrogenic, and can revitalize mature skin	Yingngam et al. (2011) and Varghese et al. (2024)
Malvaceae	*Hibiscus sabdariffa* L./hibiscus	Calyx, flowers	Water extract in face masks	Anthocyanins, protocatechuic acid	Inhibits collagenase and tyrosinase enzymes, protects collagen fibers, enhances skin texture and elasticity, and antioxidant	Anis et al. (2024), Raina et al. (2023), and Kazeem et al. (2024)

(continued)

Table 14.1 (continued)

Plant family	Species/popular name	Plant part	Preparation and administration	Key bioactive compounds	Anti-aging effects	References
Lamiaceae	*Lavandula angustifolia* Mill./ lavender	Flowers	Oil used in cosmetics	NS	Calming, anti-inflammatory, and stress reduction	Anis et al. (2024)
	Ocimum basilicum L.	Aerial parts	NS	Eugenol, methyl eugenol, linalool	Increases telomere length	Raina et al. (2023)
	Ocimum sanctum Linn./queen of herbs	Aerial part	Ethanolic extract	Rosmarinic acid	Exhibited anti-hyaluronidase, anti-elastase, anti-collagenase, anti-inflammatory, and antioxidant efficacy	Chaiyana et al. (2019)
	Rosmarinus officinalis Linn	Leaves and flowers	Methanol extracts	NS	Inhibited elastase, collagenase, and hyaluronidase enzymes. Possessed antioxidant activity	Gupta et al. (2023)
	Salvia officinalis Linn	Leaves and flowers	Methanol extracts	NS	Inhibited elastase, collagenase, and hyaluronidase enzymes. Possessed antioxidant activity	Gupta et al. (2023)
	Sideritis euboea Heldr.	Aerial parts	Ethyl acetate r residuum	Apigenin 7-O-β-D-glucopyranoside	Anti-hyaluronidase activity	Tomou et al. (2021)
Meliaceae	*Azadirachta indica* A. Juss./ neem	Seed oil or leaves	Extracts or oils applied topically	NS	Antiseptic and antimicrobial properties. Treatment of acne and reduction of skin imperfections	Anis et al. (2024)
	Lansium domesticum Corrêa	Leaves	Extracts	NS	Potential for inhibition of the enzymes elastase and collagenase	Lestari et al. (2023)

Leguminosae	*Butea superba* Roxb. Ex Willd.	Root	Ethanolic extract	NS	Moderate estrogenic. Rejuvenation	Yingngam et al. (2011)
	Pueraria candollei var. mirifica	Tuberous root	Ethanolic extract	NS	Strong estrogenic and antioxidant. Used for the treatment of post-menopausal symptoms, rejuvenation	Yingngam et al. (2011)
	Tadehagi godefroyanum (Kuntze) H.Ohashi	Root	Ethanolic extract	NS	Moderate estrogenic. Restoration menstrual cycle	Yingngam et al. (2011)
	Trigonella foenum-graecum L.	Seed	Ethanolic extract	NS	Weak estrogenic, cytotoxic at high doses. Increase breast milk supply	Yingngam et al. (2011)
Myrtaceae	*Eucalyptus camaldulensis* Dehnh.	Bark	70% ethanolic extract	Rutin	It inhibited apoptosis, senescence, and prevented wrinkle formation. Exhibited elastase, collagenase, and tyrosinase enzymes	Radwan et al. (2020)
	Syzygium nervosum DC./Ma Kiang	Leaves	Methanol extract	NS	Strong tyrosinase inhibition	Kazeem et al. (2024)
Polygonaceae	*Persicaria senegalensis* (Meisn.) Sojak/snakeroot or tapertip smartweed	Leaf and stem	Ethanolic crude extract and its Bifidobacterium bifidum fermented	NS	Enhanced elastase inhibition modulated pro-inflammatory cytokines, and regulated TIMP-1/MMP-9. Clinically induced significant wrinkle reduction	Payne et al. (2025)
	Polygonum maritimum L./sea knotgrass	Roots and leaves	Methanol and dichloromethane extracts	NS	Possessed neuroprotective effects by inhibiting AChE and BuChE. They can also reduce the NO production in LPS. Methanol extracts prevent oxidative stress	Rodrigues et al. (2018)
Nelumbonaceae	*Nelumbo nucifera* Gaertn./sacred lotus	Flower (stamen)	Hydro-ethanol extract	NS	Suppressed tyrosinase and collagenase enzymes	Kazeem et al. (2024)

(continued)

Table 14.1 (continued)

Plant family	Species/popular name	Plant part	Preparation and administration	Key bioactive compounds	Anti-aging effects	References
Oleaceae	*Olea europaea* L./ olive oil	Fruit	Oil used in skin preparations	NS	Hydrating, anti-oxidative. Repairs oxidative damage and nourishes skin	Anis et al. (2024)
Phyllanthaceae	*Phyllanthus emblica* L./Amla	Fruit	Water or ethanol extracts in topical formulations	Vitamin C, emblicanin A, gallic acid	Collagenase and elastase inhibitions, ROS scavenging activities, anti-photoaging, antidepressant, and protective effects	Raina et al. (2023), Chakraborty et al. (2018), Samarakoon et al. (2010), and Kazeem et al. (2024)
Plantaginaceae	*Bacopa monnieri* (L.) Wettst.	Whole plant	Extracts	Bacosides A and B, bacosaponins	Antioxidant, anti-inflammatory. Protects DNA and reduces neuroinflammation	Raina et al. (2023)
Rosaceae	*Malus domestica* (Suckow) Borkh./ apple	Fresh fruit	Extracts	Catechin, Rutin	Antioxidant through the viability and survival of HFL-1 primary human fibroblasts, influencing cellular lifespan	Verma et al. (2023)
	Prunus armeniaca L./apricot	Seed	70% ethanol extract or oil used in skin creams	Amygdalin	Significantly inhibited collagenase and elastase enzymes. Prevent skin aging and to dehydrate, lubricate, and nourish the skin	Ramadan et al. (2020)
	Rosa canina L./ rosehip oil	Seeds and fruit	Oil used topically	NS	Promotes collagen synthesis, protects skin from sun damage, and improves skin tone	Anis et al. (2024)
Rutaceae	*Citrus* × *aurantium* L.	Fruit peels	Water or alcohol extracts in creams and face packs	NS	Encourages collagen production. Reduces age spots and uneven pigmentation	Anis et al. (2024)

Sapindaceae	*Nephelium lappaceum* Linn./ pink rambutan	Peels from ripe and young fruits	Ethnolic extracts	NS	Used for whitening and anti-wrinkle	Boonpisuttinant et al. (2023)
Sapotaceae	*Manilkara zapota* P. Royen/Sapota	Fruit	Ethanol extract	NS	Inhibition of collagenase and elastase	Kazeem et al. (2024)
Simmondsia-ceae	*Simmondsia chinensis* (link) C.K.Schneid./Jojoba	Seeds	Oil is applied topically	NS	Moisturizing, skin barrier protection, prevents moisture loss, and maintains elasticity	Anis et al. (2024)
Solanaceae	*Withania somnifera* (L.) Dunal/ Ashwagandha	Roots and leaves	Extracts	NS	Antioxidant properties and improve cognition	Verma et al. (2023)
Theaceae	*Camellia sinensis (L.)* Kuntze/green tea	Leaves	Extracts in creams and lotions	NS	UV radiation protection and antioxidants by protecting from free radical damage	Anis et al. (2024)
Vitaceae	*Cissus quadrangularis* var. pubescens Dewit	Stem	Ethanolic extract	NS	Moderate estrogenic, growth-promoting. Improvement of bone fracture healing	Yingngam et al. (2011)
Verbenaceae	*Premna odorata* Blanco	Leaves	Essential oil	Palmitic acid	Exhibited high anti-elastase, anti-hyaluronidase, and anti-collagenase activities	Altyar et al. (2020)

(continued)

Table 14.1 (continued)

Plant family	Species/popular name	Plant part	Preparation and administration	Key bioactive compounds	Anti-aging effects	References
Zingiberaceae	*Alpinia galanga* L./ galangal	Leaves	Ethanolic extract	NS	Decrease the senescence phenomena in normal fibroblast cells. It possesses antioxidant properties and increases intracellular ROS	Ahlina et al. (2020)
	Curcuma aeruginosa Roxb.	Rhizome	Ethanolic extract	NS	Moderate estrogenic, cytotoxic at high doses. Restoration of menstrual cycle, post-menopausal symptoms	Yingngam et al. (2011)
	Curcuma comosa Roxb.	Rhizome	Ethanolic extract	NS	Moderate estrogenic. Alleviation of menstrual and post-menopausal symptoms	Yingngam et al. (2011)
	Curcuma heyneana Valeton & Zijp	Rhizome	Water or alcohol extract	Curcuminoid	Slow down aging and/or prevent oxidative stress, and inhibit tyrosinase and collagenase enzymes	Lestari et al. (2023)
	Curcuma longa L./ turmeric	Rhizome	Essential oil or water, or alcohol extracts in creams or masks	Curcumin (diferuloylmeth-ane)	DNA repair enhances the Nrf2 pathway, increasing lifespan. It acts as a superoxide scavenger, reducing wrinkles, improving skin flexibility, and increasing skin elasticity. Additionally, it decreases pigmentation, enhances skin hydration, and inhibits the enzymes elastase and collagenase	Anis et al. (2024), Chakraborty et al. (2018), Zamzury and Omar (2019), Lestari et al. (2023), Verma et al. (2023), and Raina et al. (2023)
	Kaempferia rotunda L./resurrection lily	Rhizome	Extracts	NS	Potent radical-scavenging effects and very strong elastase inhibitor	Lestari et al. (2023)

NS not specified

14.3 Anti-aging Properties

Anti-aging properties refer to actions or treatments aimed at slowing down or potentially reversing the effects of aging, while also promoting an increase in lifespan. Anti-aging pertains to the ability of certain derivatives to delay the aging process and/or mitigate age-related diseases, thereby extending lifespan through the regulation of various physiological processes such as antioxidant activity, immune function, anti-inflammatory responses, telomere maintenance, mitochondrial repair, and inhibition of apoptosis (Tundi et al. 2015; Chuang et al. 2014). The primary goal of modern anti-aging research is to extend the healthy lifespan (healthspan), rather than just prolonging overall lifespan.

14.3.1 Genetic Basis of Anti-aging Activity

Studies using model organisms like yeast (*Saccharomyces cerevisiae*), nematodes (*Caenorhabditis elegans*), flies (*Drosophila*), and rodents have identified key genes that strongly influence lifespan (Rattan 2018). Several genetic pathways play crucial roles in aging regulation, including insulin-like growth factor-1 (IGF-1), mTOR, NF-κB, FOXO3, Nrf2, and sirtuins (SIRT) (Zhang et al. 2019; Kuppusamy and Arumugam 2023). Other modulators such as caloric restriction mimetics, cell regeneration enhancers, autophagy inducers, and epigenetic enzyme inhibitors also affect these pathways (Vaiserman et al. 2017). Reducing insulin/IGF-1 signaling has been shown to extend lifespan and improve health by enhancing stress resistance (Iweala et al. 2024). Under nutrient scarcity, reduced insulin and IGF-1 signaling activate FOXO transcription factors, which stimulate protective proteins like glutathione S-transferase, catalase, superoxide dismutase, and chaperones (Murphy et al. 2003; Murphy 2006; Morris et al. 2015). These proteins protect cells from oxidative stress, repair DNA damage, and maintain protein stability. Notably, certain FOXO gene variants have been linked to increased human longevity, indicating that FOXO activation contributes to anti-aging effects in humans (Flachsbart et al. 2009; Willcox et al. 2008; Liu 2022).

INS/IGF-1 (Insulin/Insulin Growth Factor-1) The insulin/insulin-like growth factor-1 (INS/IGF-1) signaling (IIS) pathway was the first pathway identified to regulate aging and age-related disorders in model organisms such as yeast, flies, nematodes, and mice, with implications for humans (Lee et al. 2015). The function of IGF-1 is evolutionarily conserved, maintaining its role in lifespan regulation across species (Abdellatif et al. 2022). Studies have shown that reduced IIS activity extends lifespan and enhances healthspan by increasing stress resistance (Martel et al. 2019). In humans, gene set and GWAS analyses consistently highlight the IIS pathway as significant for longevity (Deelen et al. 2013; Smulders and Deelen 2024). The pathway includes two main molecular cascades: PI3K/Akt signaling and MAPK signaling (Dixit et al. 2025).

Across species, homologous genes map to the IIS pathway, with AKT, FOXO, IGF, INSR, IRS, PIK3C, RPS6K/S6K, and SHC identified as core longevity-related genes supported by multiple genetic studies (Smulders and Deelen 2024). Among these, FOXO3 has been repeatedly proven as a longevity-associated gene in human populations (Smulders and Deelen 2024). The FOXO transcription factors, particularly FOXO3, are key regulators of lifespan and stress resistance, and activating FOXO3 is viewed as a promising anti-aging strategy. The IIS pathway often acts in conjunction with the TOR pathway, and together they modulate lifespan extension through mechanisms involving the DAF-2:FOXO axis (Smulders and Deelen 2024).

Mammalian Target of Rapamycin (mTOR) The mammalian target of rapamycin (mTOR) signaling pathway is an intensely conserved regulator of longevity, metabolism, and cellular homeostasis across species (Lamming et al. 2012). As a central metabolic hub, mTOR governs cell growth, autophagy, protein synthesis, immune response, and stress adaptation (Lamming et al. 2012).

Belonging to the phosphatidylinositol kinase-related kinase (PIKK) family of serine/threonine kinases, mTOR also shares features with phosphatidyl 3-kinases (PI3Ks) (Arriola Apelo and Lamming 2016). It forms two distinct complexes:

(a) mTORC1 (mTOR Complex 1): the nutrient-sensitive, rapamycin-sensitive complex composed of mTOR, raptor, and GbL, acting as the central element of the TOR signaling network (Arriola Apelo and Lamming 2016).
(b) mTORC2 (mTOR Complex 2): nutrient-independent and only chronically sensitive to rapamycin (Arriola Apelo and Lamming 2016).

The TOR pathway functions as a major regulator of growth and aging in diverse model organisms, comprising yeast (*Saccharomyces cerevisiae*), nematodes (*C. elegans*), and fruit flies (*Drosophila melanogaster*) (Singh et al. 2019; Yousefzadeh et al. 2018; Blagosklonny 2021). Human genetic studies on longevity have identified several key genes within the mTOR pathway EIF4E, PRR5, RPS6KB1 (S6K), and ULK1, whose modulation influences aging processes (Passtoors et al. 2013). These genes act downstream of mTORC1, affecting autophagy and protein synthesis. Furthermore, AMPK activation, such as through calorie restriction, stimulates TSC1–TSC2, leading to mTORC1 inhibition and lifespan extension (Wrighton 2011). Experimental evidence demonstrates that inhibition of mTOR signaling extends lifespan and delays aging across species. Partial inhibition of downstream targets like S6K or protein synthesis has similar effects, for instance, deletion of the Scheme 9 gene (yeast ortholog of mammalian S6K) significantly increases chronological lifespan (Johnson 2019).

FOXO3 Transcription Factor The Forkhead box O-3 (FOXO3) pathway is central to the molecular biology of anti-aging mechanisms and interventions. FOXO3 is part of the forkhead transcription factor family, which also includes FOXO1, FOXO4, and FOXO6, all of which regulate essential cellular processes like apoptosis, cell cycle control, DNA repair, metabolism, and oxidative stress response (Park 2024). Among these, FOXO3 stands out as a major human longevity gene, being one of only two loci, along with APOE, that have been consistently replicated across populations and strongly associated with exceptional human longevity (Flachsbart et al. 2017). The association of FOXO3 variants is particularly robust in long-lived individuals (LLI). Functional studies in model organisms demonstrate the evolutionary conservation of FOXO3's role in lifespan regulation. In *C. elegans*, the ortholog DAF-16 is required for the lifespan extension resulting from reduced insulin/IGF-1 signaling (IIS). Similarly, in *Drosophila melanogaster*, overexpression of dFOXO in the adult fat body significantly extends lifespan. The core function of FOXO3 is to act as a cellular stress and nutrient sensor, triggering protective responses that enhance cellular maintenance, stress resistance, and survival, thereby promoting healthspan and longevity. Supporting its genetic relevance, studies have identified FOXO3 polymorphisms (SNPs) significantly linked with longevity, validated through in silico analyses and in vitro luciferase reporter assays (Flachsbart et al. 2017).

Nuclear Factor Erythroid 2-Related Factor 2 The nuclear factor erythroid 2-related factor 2 (Nrf2) pathway is a key genetic mechanism that protects against aging by regulating the cellular response to oxidative stress (Kensler and Wakabayashi 2010). Under normal conditions, Nrf2 remains bound to its inhibitor Keap1, forming an inactive complex in the cytosol (Sykiotis and Bohmann 2008). This pathway functions as a stress-inducible system known as the Keap1/Nrf2/ARE (antioxidant response element) signaling pathway. Nrf2 plays a principal role in defending cells from xenobiotic and oxidative damage and is regarded as a master redox switch because it controls signaling that activates cytoprotective genes (Kensler and Wakabayashi 2010; Sykiotis and Bohmann 2010; Surh et al. 2008). The Nrf2/ARE pathway induces the expression of several antioxidant enzymes, including HO-1, NADPH, GST, SOD, CAT, NQO1, and GPX, which help maintain cellular redox homeostasis (Sykiotis and Bohmann 2010; Kensler and Wakabayashi 2010).

Sirtuins (SIRT) Pathway Sirtuins are a conserved family of proteins found from bacteria to humans that act as NAD^+-dependent deacetylases or ADP-ribosyltransferases, often referred to as longevity genes (Oberdoerffer and Sinclair 2007; Westphal et al. 2007). The name derives from Sir2, the first sirtuin identified in *Saccharomyces cerevisiae*. In mammals, there are seven sirtuins (SIRT1–SIRT7) located in different cellular compartments: nuclear (SIRT1, SIRT6, SIRT7), cytoplasmic (SIRT2), and mitochondrial (SIRT3, SIRT4, SIRT5) (Iside et al. 2020; Smith et al. 2020). Overexpression of sirtuins has been shown to extend lifespan in yeast, *Drosophila*, and *C. elegans* (Viswanathan et al. 2005; Rogina and Helfand 2004). SIRT1 interacts with AMP-activated protein kinase (AMPK), a major regulator of cellular energy metabolism and survival, enhancing its activity by boosting NAD^+ levels and activating the SIRT1–PGC1α pathway. This interaction promotes autophagy and may extend lifespan (Wanyu 2025). Overall, sirtuins regulate essential cellular processes such as DNA repair, metabolism, cell cycle, and oxidative stress, primarily through deacetylation of histone and non-histone proteins. Their function is tightly linked to cellular energy status, as they depend on NAD^+ availability (Iside et al. 2020).

Chemistry and Physiology of Mode of Action

Chemistry of Anti-aging Agents: Their chemical basis lies in their functional groups and structural features that allow them to interact with biomolecules. Natural products, notably secondary metabolites from plants, are rich sources of bioactive compounds with anti-aging effects (Song and Zhang 2023). These metabolites can be categorized into several groups, such as polyphenols, which include flavonoids, lignans, phenolic acids, tannins and stilbenes (Song and Zhang 2023; Luo et al. 2021), carotenoids (xanthophylls and carotenes) (Maoka 2020), phytosterols (sterols and stanols) (Moreau et al. 2002), and terpenoids (triterpenoids, diterpenoids, tetraterpenoids, hemiterpenoids, monoterpenoids, sesquiterpenoids, polyterpenoids and sesterpenoids (Song and Zhang 2023; Huang et al. 2012) (Fig. 14.1).

Physiological Modes of Action

The anti-aging interventions target the underlying mechanisms that drive the gradual deterioration together with aging, like genomic instability, chronic inflammation, mitochondrial dysfunction, and cellular senescence. The anti-aging effects are highly combined with their ability to modulate these pathways. The regulation of stress response and cellular homeostasis is as follows:

Mitigating Oxidative Stress (OS) and Inflammation: Aging is closely associated with oxidative stress (OS) caused by excessive reactive oxygen species (ROS), which destroy DNA, proteins, mitochondria, and lipids. Anti-aging compounds counteract this via multiple mechanisms:

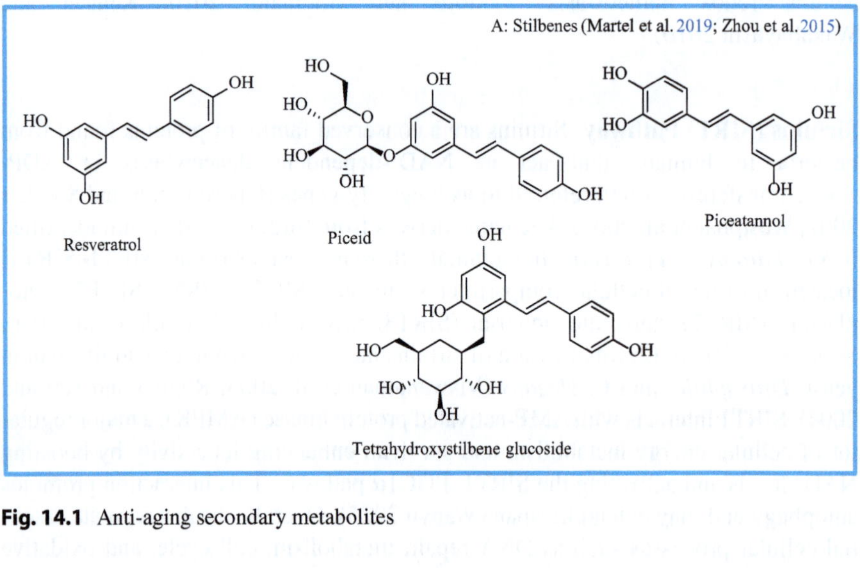

Fig. 14.1 Anti-aging secondary metabolites

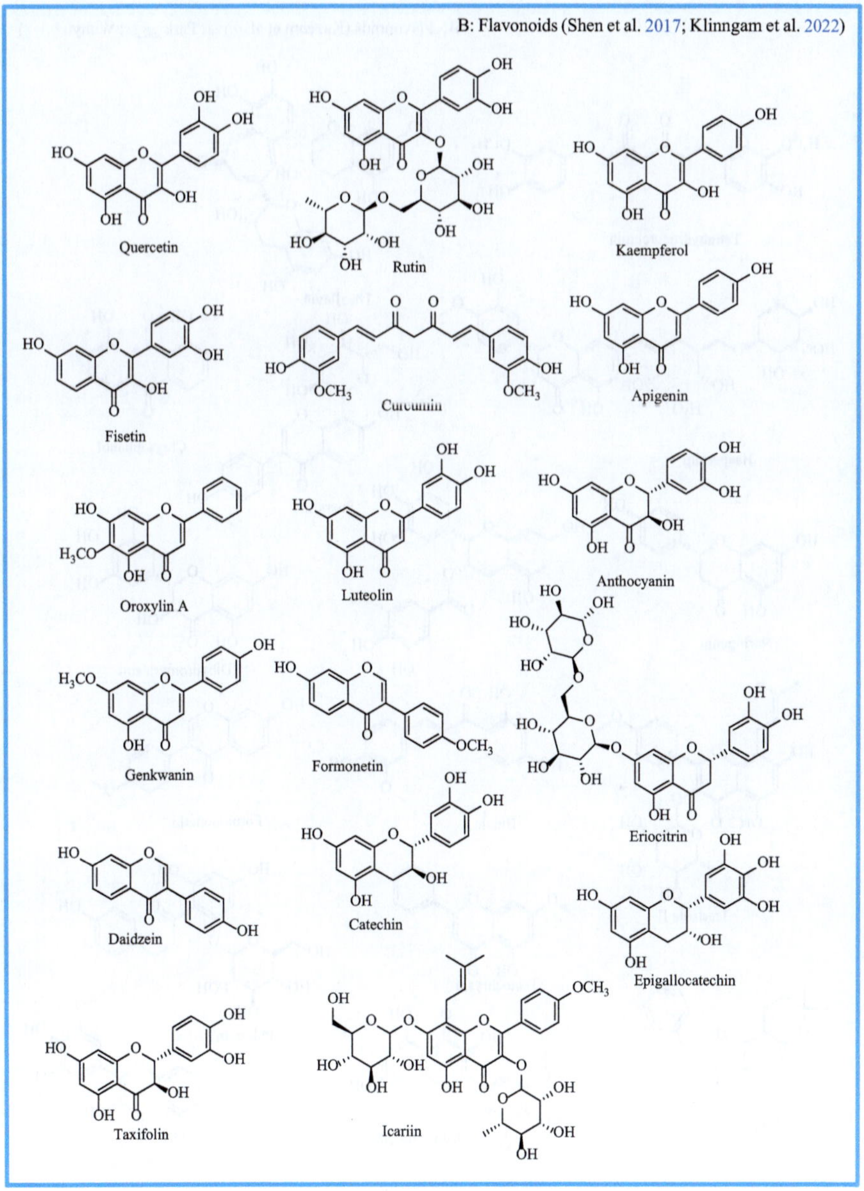

Fig. 14.1 (continued)

B: Flavonoids (Kazeem et al. 2024; Park 2024; Wanyu 2025)

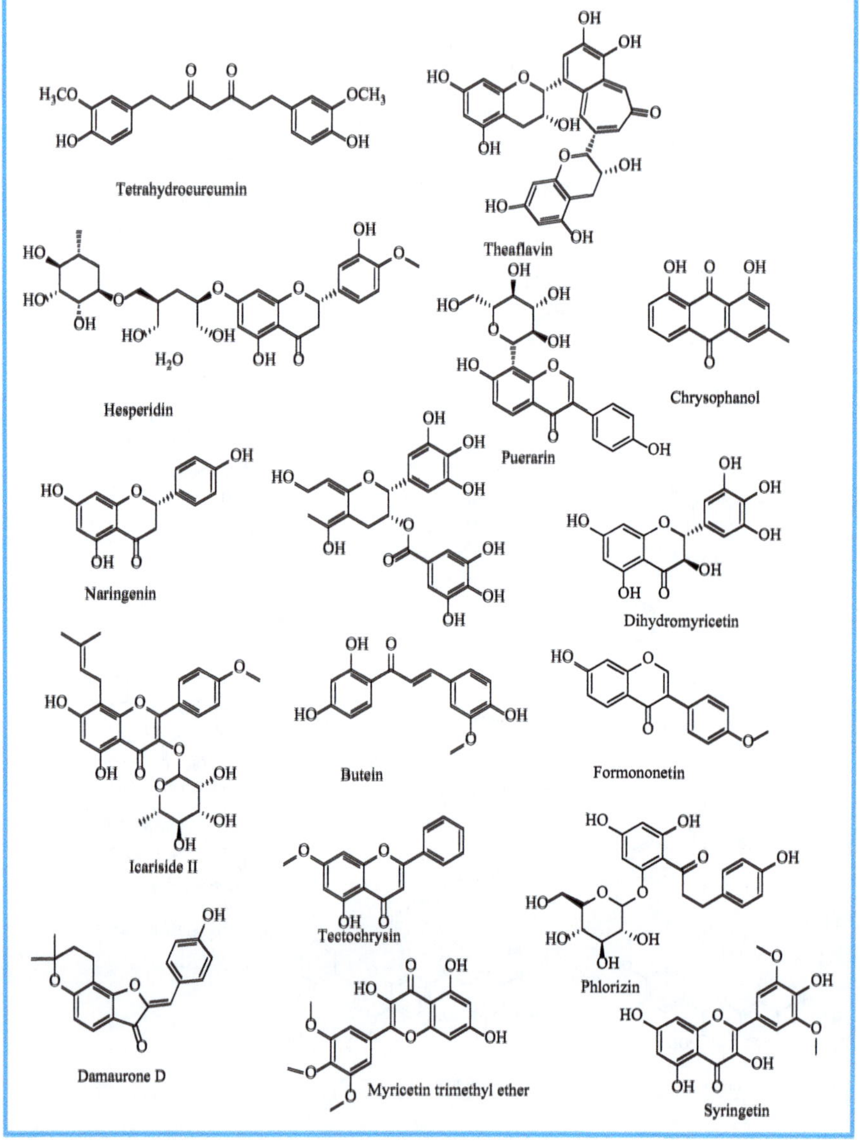

Fig. 14.1 (continued)

C: Lignans (Su et al. 2015, Kazeem et al. 2023)

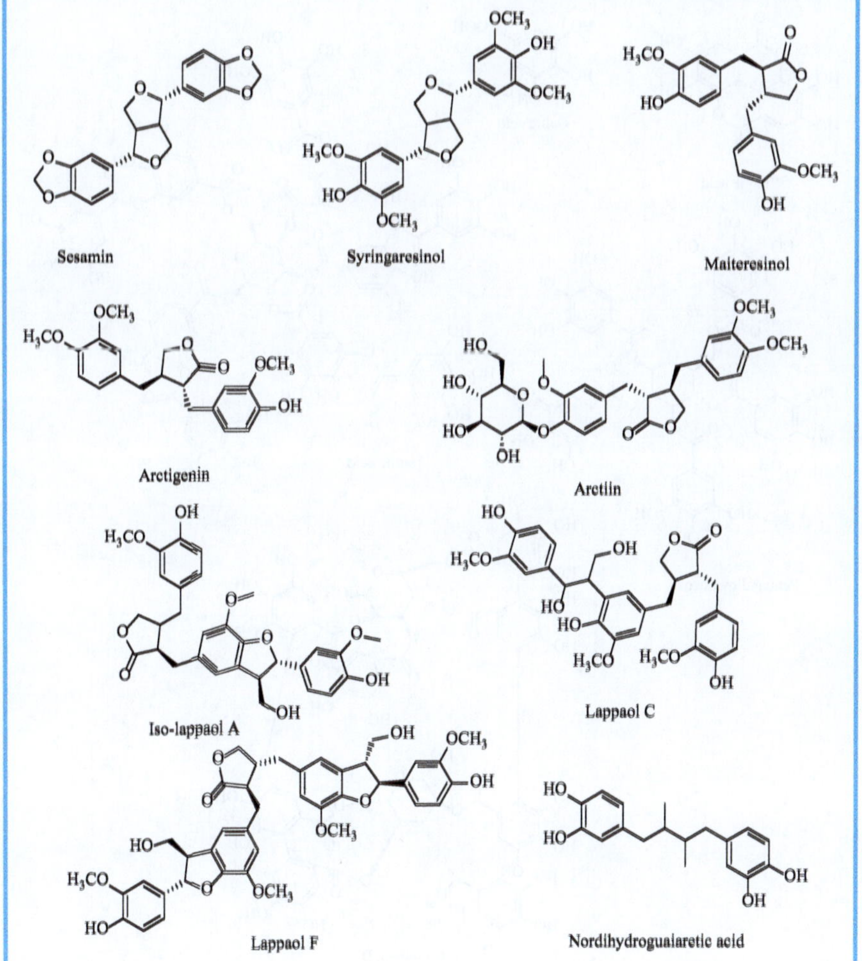

Fig. 14.1 (continued)

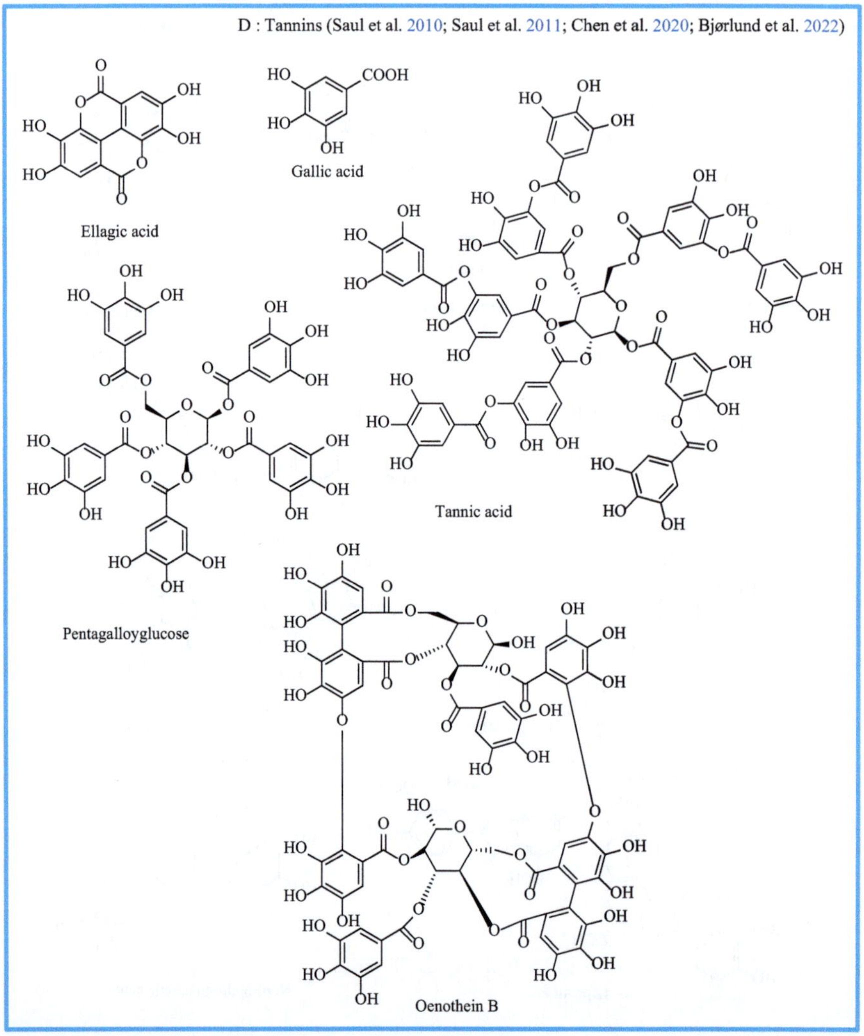

Fig. 14.1 (continued)

E: Carotenoids (Shanaida et al. 2025; Song and Zhang 2023; Shanaida et al. 2025)

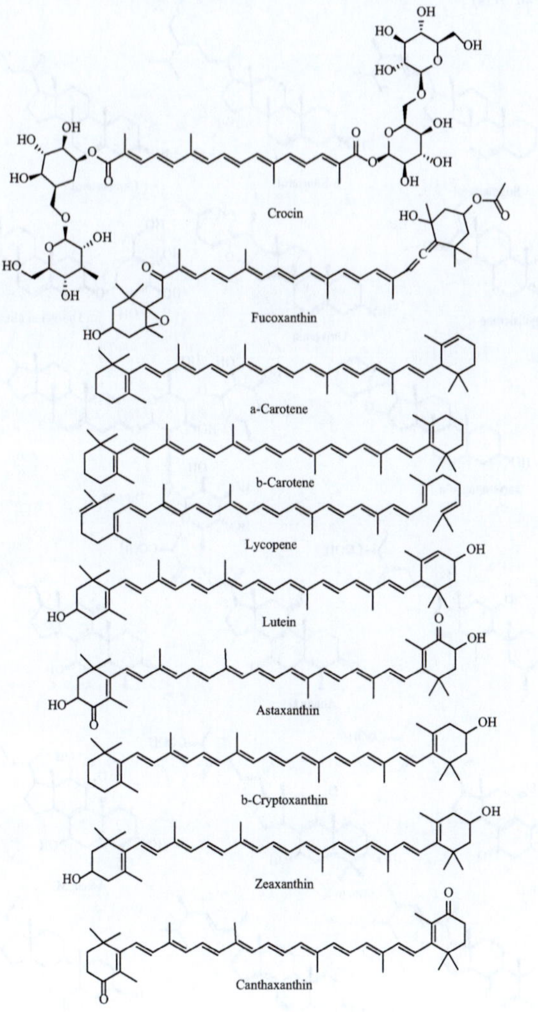

Fig. 14.1 (continued)

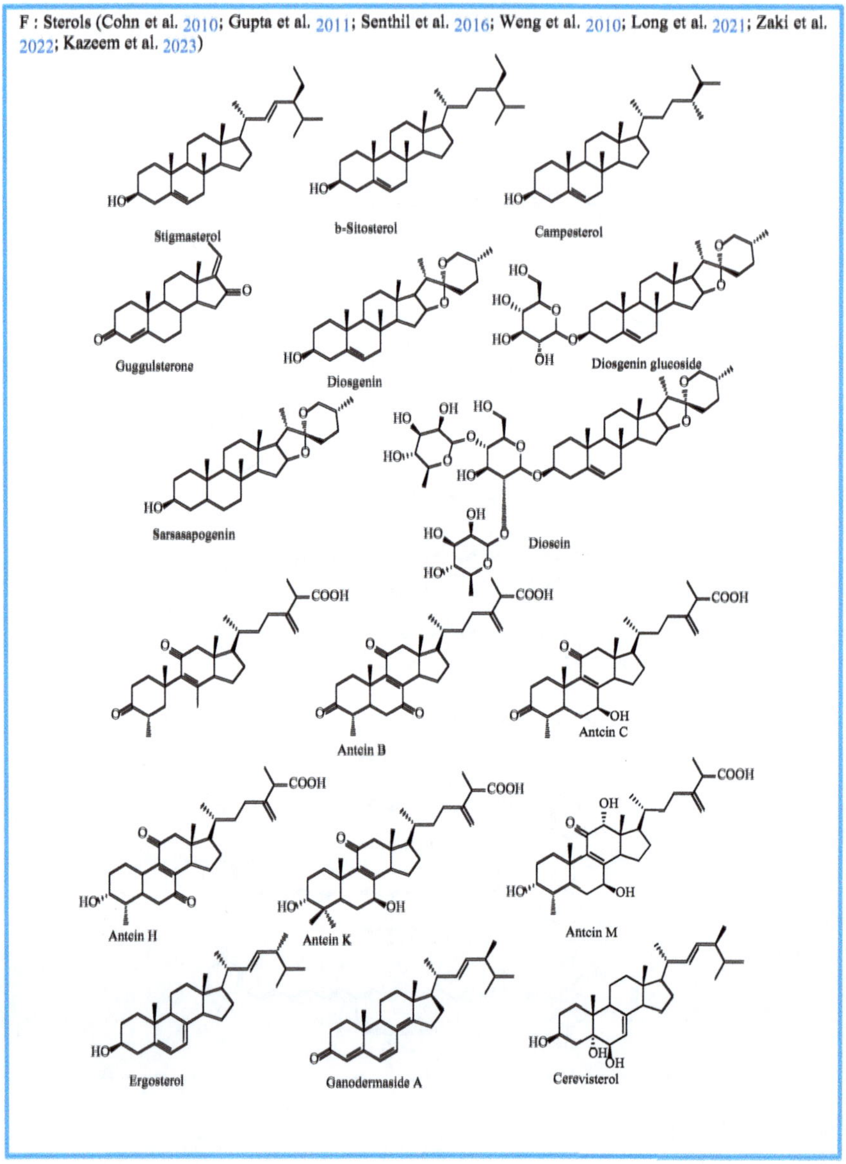

Fig. 14.1 (continued)

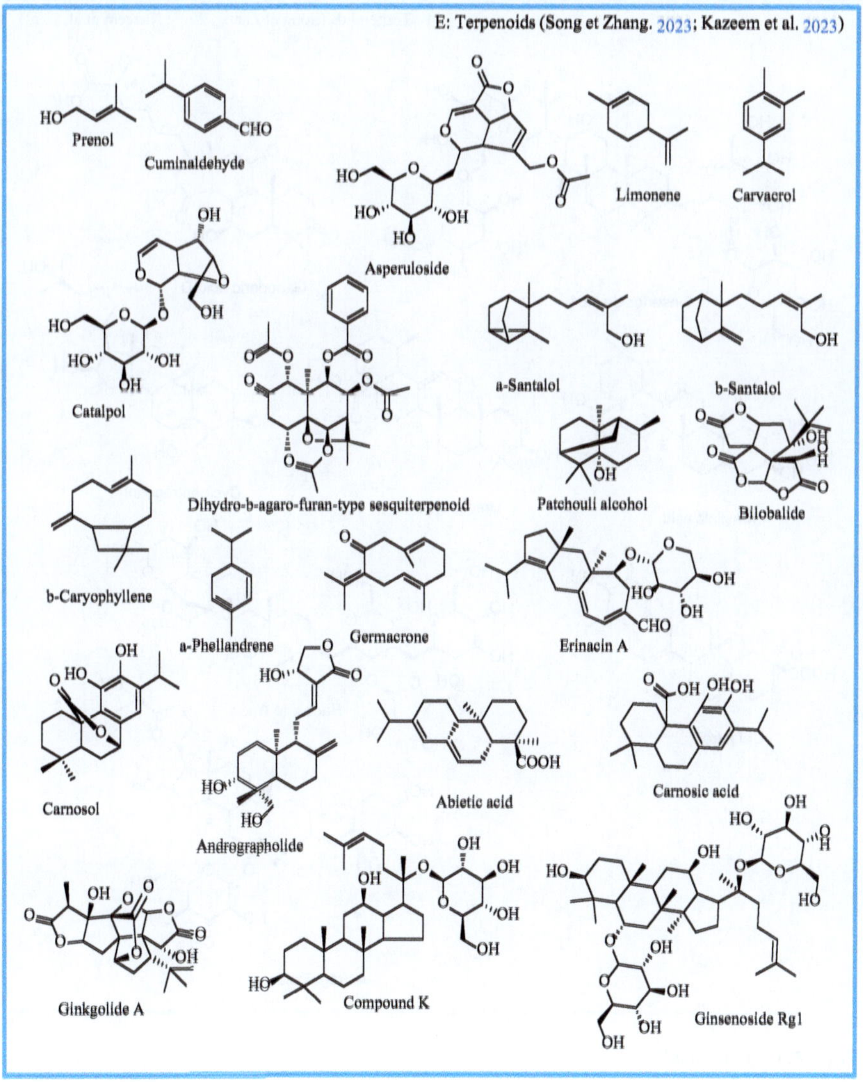

Fig. 14.1 (continued)

F: Terpenoids (Song et Zhang. 2023; Kazeem et al. 2023)

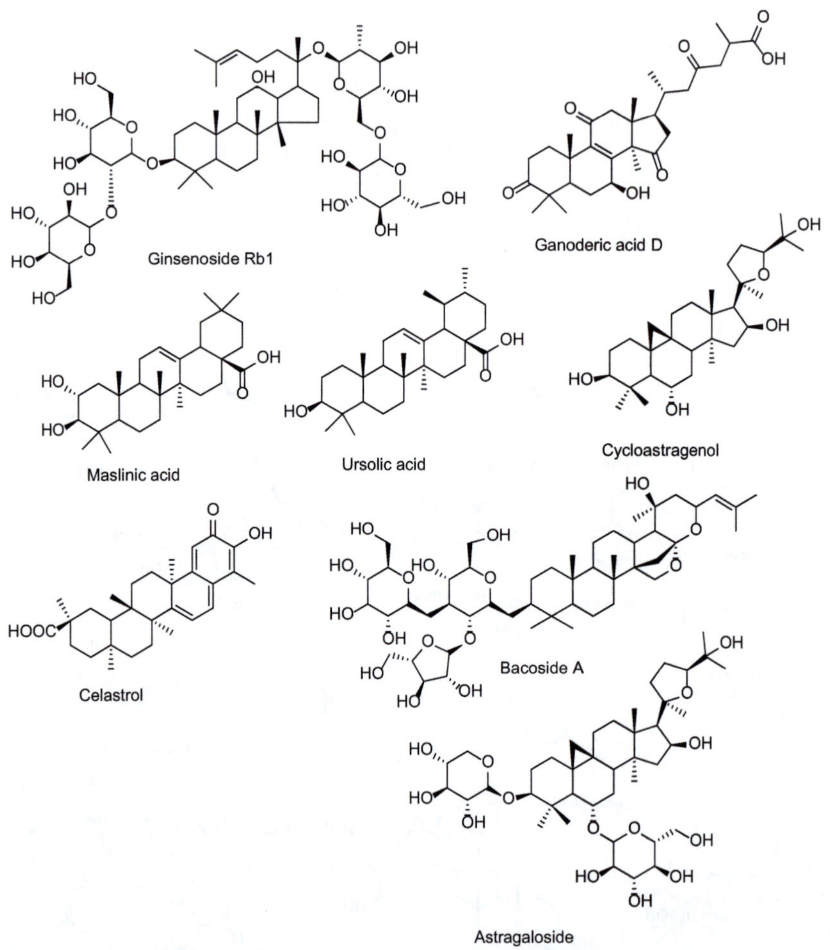

Fig. 14.1 (continued)

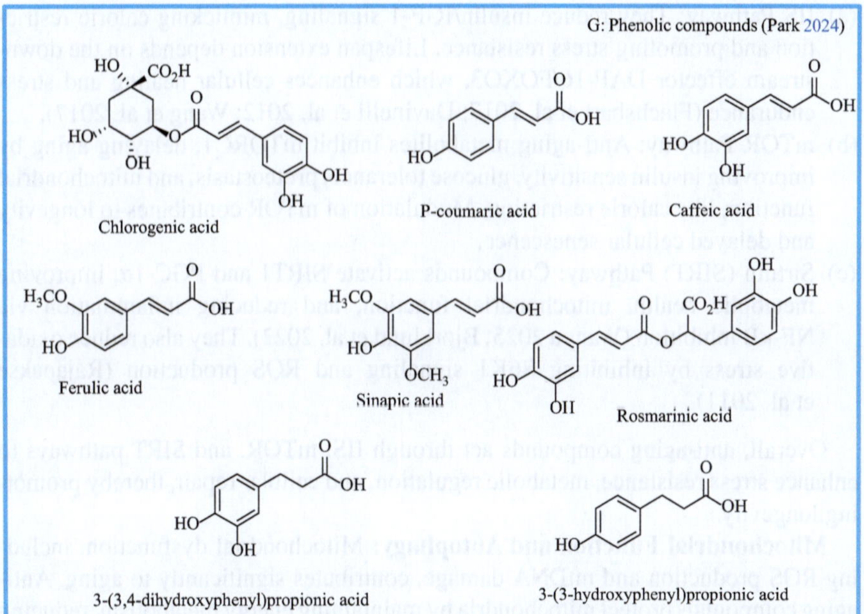

Fig. 14.1 (continued)

(a) ROS scavenging: Certain compounds act directly as antioxidants, enhancing enzyme activity (SOD, GSH) and eliminating excess ROS and H_2O_2 (Wanyu 2025).

(b) Antioxidant enzyme activation: Bioactive compounds like polyphenols, polysaccharides, carotenoids, and terpenoids activate the Keap1/Nrf2/ARE signaling pathway, boosting endogenous antioxidant enzymes (SOD, CAT, GPX) and modulating SIRT1/Nrf2 to reduce apoptosis (Chen et al. 2016; Yang et al. 2021; Wanyu 2025).

(c) Inflammation suppression: Anti-aging compounds inhibit NF-κB activation, reducing chronic low-level inflammation driven by senescent cells and the SASP. Polyphenols specifically block pro-inflammatory pathways, preventing aging-related cellular damage (Izadi et al. 2024).

(d) Integrated pathway modulation: Anti-aging effects involve activating Nrf2/Keap1 or Nrf2/HO-1 for oxidative stress reduction while suppressing NF-κB to maintain immune function and reduce inflammation (Yang et al. 2021; Song et al. 2022a).

This highlights a coordinated approach where anti-inflammatory and antioxidant pathways are modulated simultaneously to delay aging.

Modulation of Nutrient-Sensing Pathways: Anti-aging compounds influence key signaling networks that regulate energy homeostasis, growth, and longevity:

(a) IIS Pathway: They reduce insulin/IGF-1 signaling, mimicking calorie restriction and promoting stress resistance. Lifespan extension depends on the downstream effector DAF-16/FOXO3, which enhances cellular healing and stress endurance (Flachsbart et al. 2017; Davinelli et al. 2012; Wang et al. 2017).
(b) mTOR Pathway: Anti-aging metabolites inhibit mTORC1, delaying aging by improving insulin sensitivity, glucose tolerance, proteostasis, and mitochondrial function, like calorie restriction. Modulation of mTOR contributes to longevity and delayed cellular senescence.
(c) Sirtuin (SIRT) Pathway: Compounds activate SIRT1 and PGC-1α, improving metabolic health, mitochondrial function, and reducing inflammation via NF-κB inhibition (Wanyu 2025; Bjørklund et al. 2022). They also reduce oxidative stress by inhibiting S6K1 signaling and ROS production (Rajapakse et al. 2011).

Overall, anti-aging compounds act through IIS, mTOR, and SIRT pathways to enhance stress resistance, metabolic regulation, and cellular repair, thereby promoting longevity.

Mitochondrial Function and Autophagy: Mitochondrial dysfunction, including ROS production and mtDNA damage, contributes significantly to aging. Anti-aging compounds protect mitochondria by maintaining energy metabolism, reducing ROS generation via inhibition of reverse electron flow through Complex I, and increasing mitochondrial oxygen consumption (Liu 2022). They also induce autophagy and mitophagy, critical for cellular homeostasis, often by inhibiting IGF and mTOR signaling, mimicking the effects of calorie restriction (Liu 2022).

Genomic Stability and Cellular Senescence: For genomic stability, telomeres, composed of TTAGGG repeats, protect chromosome ends from degradation, and their maintenance depends on telomerase, a ribonucleoprotein enzyme that adds telomeric repeats and regulates aging (Blackburn 1991, 2001; Chan and Blackburn 2004). Anti-aging compounds, particularly polysaccharides and terpenoids, act as telomerase activators, extending cellular replicative capacity and improving cell viability (Valenzuela et al. 2009). Cellular growth arrest involves nonreversible cell cycle arrest mediated by p16^INK4a/Rb and p53/p21 pathways in response to stress (Herbig et al. 2004; Campisi and d'Adda di Fagagna 2007). Anti-aging compounds, such as polysaccharides, modulate these pathways and act as senolytics, inducing apoptosis of senescent cells (Kirkland and Tchkonia 2020; Chaib et al. 2022) or senomorphics (mitigating SASP effects; Martel et al. 2020). They also enhance antioxidant defenses via Nrf2-ARE signaling to remove ROS (Mu et al. 2021). This emphasizes that anti-aging strategies target both genomic maintenance and senescence regulation to preserve cellular function and delay aging.

14.3.2 Cellular Mechanisms of Anti-aging Secondary Metabolites

Most of the secondary metabolites (Fig. 14.2) presented exert anti-aging effects and promote lifespan extension through multiple mechanisms, with the suppression of oxidative stress, regulation of age-associated genes, modulation of immune

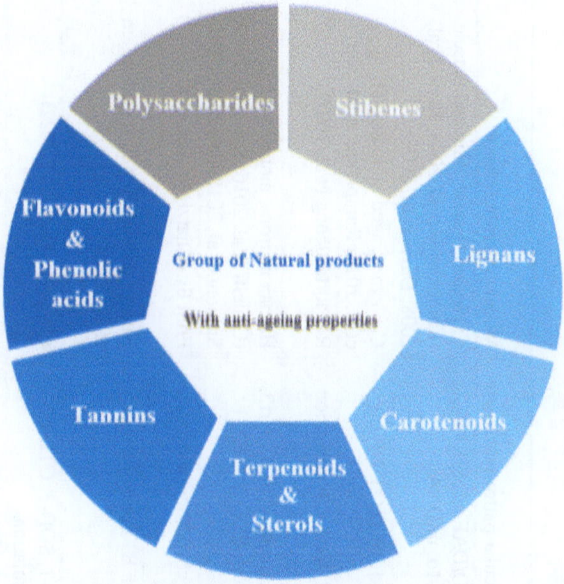

Fig. 14.2 Group of metabolites with anti-aging effects

responses, control of apoptosis and autophagy, inhibition of senescent cell state, and control of mitochondrial function and biogenesis (Table 14.2) (Zhang et al. 2019; Song and Zhang 2023; Iweala et al. 2024; Bjørklund et al. 2022).

14.4 Multiomics Approaches to Enhance the Anti-aging Properties

Therapeutic plants are widely certified for their rich diversity of functional secondary metabolites and have long been used in the treatment of various human diseases. They are also believed to process various biological effects. However, their photoprotective and anti-aging properties remain largely unexplored, particularly through integrative scientific approaches. Over recent times, various omics techniques like proteomics, metabolomics, genomics, and transcriptomics have provided insights into aging processes (Fig. 14.3) (Kudryashova et al. 2020).

14.4.1 Metabolomics

Metabolomics of plants, a vital component of systems biology, targets the in-depth analysis and interpretation of plant metabolites, their composition, variability, and functions. Technological advancements have refined this field into a complex, multi-step process that includes sample harvesting, metabolite extraction, high-capacity experimentation, data processing, and multifactorial statistical assessment. When integrated with other omics approaches, metabolomics offers a holistic view of biological systems, enabling researchers to connect insights from the genomic to

Table 14.2 Summary for some anti-aging compound classes and their cellular mechanisms

Compound class	Specific compound	Anti-aging mechanisms	Molecular pathways/targets	Model system/reference
Flavonoids	Quercetin	Reduces mitochondrial, extends lifespan, increases stress resistance, cellular senescence, and dysfunction	FOXO/DAF-16, SIRT1/PGC-1α, mtDNA	*C. elegans*, mouse brain and muscle (Kampkötter et al. 2008; Kook et al. 2008; Davis et al. 2009)
	Kaempferol	Reduces ROS, protects against oxidative stress, anti-inflammatory	DAF-16/FOXO3, NF-κB	*C. elegans*, aged rat kidney (Kim et al. 2010; Kampkötter et al. 2007)
	Luteolin	Increases insulin sensitivity, improves vascular function	IRS-1/Akt, NO production	Rat aorta (Jeong et al. 2007)
	Curcumin	Scavenges ROS/RNS, reduces exercise-induced oxidative stress, anti-senescence	GSH, catalase, SOD, Nrf2, p16/p21, SOD-1, GPX-1	Human neurons, mouse models (Chilelli et al. 2016; Ren et al. 2019; Simioni et al. 2018; Lee et al. 2020)
	Tiliroside/platanoside	Decreases oxidative stress, improves longevity, and mobility	UPS, chaperone, antioxidant genes	Fruit fly (Chatzigeorgiou et al. 2017)
	Methoxyflavones (from Kaempferia parviflora)	Suppresses cellular senescence and ROS	Lamin B1 upregulation	Human dermal fibroblast (Klinngam et al. 2022)
	Tambulin	Increases longevity, stress tolerance, and reduces ROS	SOD-1, SOD-3, CTL-2 upregulation	*C. elegans* (Liu et al. 2022)
	Epigallocatechin gallate	Lifespan extension, free radical scavenging	SOD-3, HSP-16.2	In vitro/in vivo (Zhang et al. 2019)
	Damaurone A/Tectochrysin	Enhances stress resistance	DAF-16/FOXO3	*C. elegans* (Lu et al. 2020)
Flavonoids	Fisetin	Reduces aging markers, inhibits PI3K/mTOR	PI3K/mTOR	Mice, human adipose tissue (Yousefzadeh et al. 2018; Blagosklonny 2021)
	Hesperidin	Extends lifespan, activates antioxidant genes	SOD, Sir2, represses UTH1	Yeast (Sun et al. 2012)

Stilbenes	Resveratrol	Extends lifespan, reduces inflammation, enhances mitochondrial function	SIRT1/Nrf2, AMPK, Sir2, NF-κB, mTOR, autophagy	*C. elegans*, *Drosophila*, endothelial & fibroblast cells (Bass et al. 2007; Yu and Li 2012; Al-Hassawi and Al-Sammak 2024; Rajapakse et al. 2011)
	Tetrahydroxystilbene glucoside	Improves memory, lifespan, reduces insulin receptor levels	Neural insulin/IGF-1 signaling	SAMP8 mice (Zhou et al. 2015)
Lignans	Sesamin	Lifespan extension, reduces Aβ toxicity	DAF-16, SKN-1 (IGF-1 and p38 MAPK)	*C. elegans* (Keowkase et al. 2018; Yaguchi et al. 2014)
	Matairesinol, arctigenin, arctiin, iso-lappaol A/C/F	Increases stress resistance, extends lifespan	DAF-16, JNK-1	*C. elegans* (Su and Wink 2015)
	Syringaresinol	Delays cellular senescence	FOXO3, SIRT1, insulin/IGF-1	Human cells (Kim et al. 2020)
Tannins	Tannic acid	Lifespan improvement, stress endurance	SEK-1	*C. elegans* (Saul et al. 2010)
	Ellagic acid	Antioxidant, anti-inflammatory, antimicrobial	ROS, TNF-α, GSH	Cells, C. elegans (Saul et al. 2011; Rahimi et al. 2018)
	Pentagalloylglucose, oenothein B	Lifespan extension	Mitochondrial ETC, dietary restriction, IIS	*C. elegans* (Chen et al. 2014, 2020)
Carotenoids	β-Carotene	Reduces senescence markers, ROS, MDA; increases SOD	KAT7-P15 axis	Human cells (Zheng et al. 2022)
	Lycopene	Improves antioxidant enzymes, mitochondrial function	Nrf2/HO-1, NF-κB/TLR4	Hen ovary, granulosa cells, rat brain (Liu et al. 2018; Hu et al. 2013)
	Crocin	Mitigates oxidative stress, enhances mitochondrial function	ROS, neuroinflammation	Rats (Krishnaswamy et al. 2020)
	Astaxanthin	Reduces inflammation, oxidative stress	Nrf2/Keap1, NF-κB	Mice (Al-Amin et al. 2015; Chen et al. 2020)
	β-Cryptoxanthin	Improves mitochondrial function, anti-inflammatory	Nrf2	Human renal cells (Zhang et al. 2023a, b)
	Canthaxanthin	Reduces ROS, MDA, inflammatory factors	SIRT6, SOD, GSH	Human cells (Li et al. 2022)
	Lutein	Protects retinal cells from oxidative stress	Nrf2, HO-1, NQO1, Sirtuins	Human retinal cells (Chae et al. 2018)

(continued)

Table 14.2 (continued)

Compound class	Specific compound	Anti-aging mechanisms	Molecular pathways/targets	Model system/reference
Polysaccharides	*Lycium barbarum polysaccharides* (LBP)	Lifespan extension, stress resistance	MAPK, TOR/S6K, IIS, DAF-16, SOD-3, HSP-16.2	*C. elegans*, *Drosophila* (Wang et al. 2022,Tang et al. 2019)
	Astragalus membranaceus polysaccharides (AMP/APS)	Anti-senescence, telomerase activation, autophagy	p53/p21, p16/pRb, PI3K/AKT/mTOR	Human fibroblasts, hepatocytes (Kim et al. 2019; Song and Zhang 2023)
	Cynomorium songaricum polysaccharide	Telomere protection	Telomerase	D-gal-induced aging mice (Ma et al. 2009)
Polysaccharides	*Cistanche deserticola* polysaccharide	Telomerase activation, immunity, antioxidant	ROS, MDA	D-gal-induced mice (Zhang et al. 2011)
	Chuanminshen violaceum polysaccharides (CVPs)	Free radical scavenging, anti-aging	DPPH, hydroxyl, superoxide anion	D-gal mouse (Fan et al. 2017)
	Ginsenoside residue polysaccharides (GRPs)	Lifespan extension, antioxidant	SOD	*C. elegans* (Sun et al. 2023)
	Rubus suavissmus polysaccharides (STP-60a)	Longevity, stress resistance	Autophagy, IIS, mitochondrial signaling	*C. elegans* (Liu et al. 2022)
	Akebia trifoliate fruit polysaccharides (ATFPs)	Lifespan extension, antioxidant activity	SIRT-2.1, DAF-16, DAF-12	*C. elegans* (Song and Zhang 2023; Zhang et al. 2023a, b)
	Codonopsis pilosula polysaccharides (CPs/SCPs)	Antioxidant, ROS scavenging	Keap1/Nrf2	Various (Yang et al. 2021)

	Red ginseng by-product polysaccharides (RGBPs)	Antioxidant, skin protection	AP-1, TNF-α/IFN-γ signaling	Mice/human skin models (Kim et al. 2019)
	Bletilla striata polysaccharides (BSP)	Lifespan extension, stress resistance	IIS, SOD, CAT	*C. elegans, Drosophila* (Hafeez et al. 2025; Zhang et al. 2015)
Terpenoids	Ursolic acid	Delays aging, improves mitochondrial function	SIRT1, SIRT6, PGC-1β	Cell/mouse models (Bahrami and Bakhtiari 2016)
	Ginsenoside Rg1	Anti-apoptosis, antioxidant, telomerase activation	SIRT3/SOD2, NOX4, NLRP3 inflammasome	Mice (Zhou et al. 2020, 2011; Li et al. 2021)
	Ginsenoside Rb1	Anti-aging, regulates cell cycle and apoptosis	mTOR/p70S6K	Mice (Yu et al. 2020)
	Cycloastragenol	Lifespan extension, T cell proliferation	SKN-1, telomerase	*C. elegans, human T cells* (Chen et al. 2021; Valenzuela et al. 2009)
	Ganoderic acid D	Anti-senescence, telomerase, antioxidant	CaM/CaMKII/Nrf2, P16/P21	Stem cells (Xu et al. 2020; Yuan et al. 2022)
	Carnosol	Healthspan extension, antioxidant	HSF-1, CAT, SOD, GPX	*C. elegans, mice* (Lin et al. 2019)
	Patchouli alcohol	Reduces chondrocyte senescence	Nrf2/HO-1	Mice (Chen et al. 2022)
	Prenol	Stress fight, lifespan prolongation	SKN-1, DAF-16, HSF-1	*C. elegans* (Song and Zhang 2023)
	Carvacrol	Anti-aging, anti-apoptotic	P53, P21, Bax	Aged rats (El-Far et al. 2022)
Sterols	Diosgenin	Lifespan extension, ROS reduction, antioxidant	PI3K/AKT/mTOR inhibition, ubiquitin-proteasome pathway	Male fish (Song et al. 2022a, b)
	Diosgenin glucoside	Reduces dysfunctional autophagy	Rheb-mTOR signaling	Various (Wang et al. 2018)
	Bacoside A	Anti-neuroinflammation	TNF-α, IL-1β	Galactose-induced aging rats (Rastog et al. 2012)
	β-Sitosterol	Prevents sarcopenia, skin aging	FoxO1, hyaluronic acid biosynthesis	Rats/human skin (Hah et al. 2022; Haiyuan et al. 2019)
	Antcin M	Reduces premature senescence	Nrf2, SIRT1	Human dermal fibroblasts (Senthil et al. 2016)

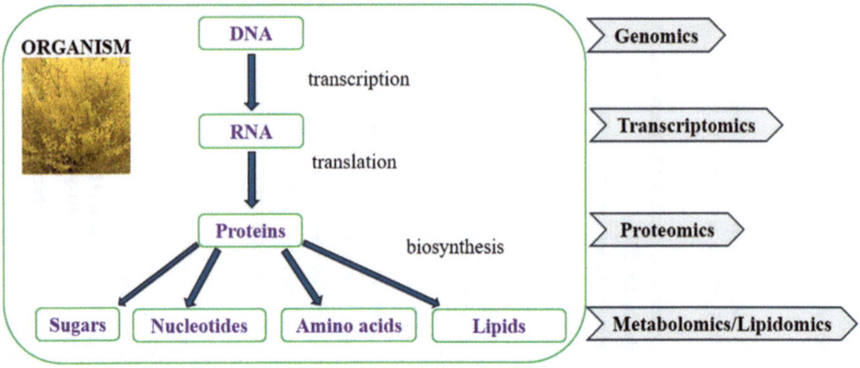

Fig. 14.3 Omics approaches to enhance the anti-aging properties

the metabolic range (Tolani et al. 2021; Hao et al. 2025). In extensive studies of complex plant metabolites with anti-aging properties, researchers commonly employ integrated routes that combine many separation techniques with mass spectrometry (MS). These approaches include gas chromatography mass spectrometry (GC–MS), nuclear magnetic resonance (NMR), capillary-based MS, matrix-assisted laser desorption/ionization (MALDI), liquid chromatography–mass spectrometry (LC–MS), and other MS-driven technologies (Ali et al. 2019; Yu et al. 2023).

14.4.2 Genomics

Genomic analysis is fundamental to plant research, offering deep insights into the genetic structure, evolutionary trajectories, and adaptive strategies of plant species. Plants contain peroxidase enzymes that demonstrate notable anti-aging properties (An et al. 2024).

The advancement of genome sequencing projects is set to significantly accelerate the integration of cutting-edge technologies into medicinal plant research. These developments provide a theoretical foundation for unraveling the biosynthesis and regulatory mechanisms of compounds of bioactivity found in plants, thereby supporting the identification of plant-derived compounds and the investigation of key secondary metabolite pathways. Despite growing interest, current research on the biosynthetic and regulatory pathways of secondary metabolites remains limited, often relying on single-gene approaches. Genomic studies enable the application of multi-omics strategies to explore these complex metabolic networks more comprehensively, offering a roadmap for the large-scale identification of genes involved in these pathways. This progress positively influences research into the biosynthesis and metabolic engineering of secondary metabolites with significant therapeutic potential (An et al. 2024). The emergence of high-throughput sequencing technologies marks a major milestone in genomics, drastically reducing the cost of sequencing and opening new avenues for genomic research (An et al. 2024). These cost-effective technologies make it feasible to sequence the genomes of a broader range of species and decode their genetic blueprints, with next-generation sequencing becoming the preferred tool for whole-genome studies.

Advancements in genomic sequencing technologies have greatly enhanced our understanding of plants, shedding light on their adaptive evolution and medicinal potential. High-throughput platforms like the Illumina HiSeq 4000 have been used to sequence the transcriptomes of yam leaf and rhizome tissues, generating millions of clean reads and thousands of unigenes. These genomic tools support comprehensive analyses of genetic change and underpin the development of molecular markers critical for phylogenetic and population genetic research (Luo et al. 2023; An et al. 2024). Bioinformatics tools are very important for interpreting and processing the large volumes of genomic data produced by sequencing technologies. In plant research, such tools have been employed to analyze transcriptome data, identify differentially expressed genes (DEGs), and confirm gene expression patterns using qRT-PCR. Comparative genomic approaches have also utilized bioinformatics software to align and compare plastome sequences, detect divergent regions, and develop DNA barcodes for species identification (Lu et al. 2023). Additionally, species delimitation has been supported by methods such as Automatic Barcoding Gap Poisson Tree Processes (PTP), Discovery (ABGD), and Generalized Mixed Yule Coalescence (GMYC), which analyze variable loci from plastid genomes to distinguish closely related species (An et al. 2024).

14.4.3 Transcriptomics

Transcriptomics plays a crucial role in profiling the categories and abundance of all RNA molecules, comprising microRNAs (miRNAs) and non-coding RNAs (ncRNAs) tissues or within cells. This approach helps identify functional genomic elements and molecules, thereby deepening our knowledge of disease pathways and organismal processes. Over time, transcriptomic analysis has evolved from microarray-based techniques to advanced high-throughput methods, with RNA sequencing (RNA-seq) emerging as a leading technology due to its accuracy and capability to deliver understanding of gene expression data (Men et al. 2025).

Advances in bioinformatics and biostatistics now make it possible to simultaneously analyze and integrate vast datasets of genes and metabolites, offering novel approaches to uncover key biological pathways involved in aging and to investigate mechanisms relevant to pharmacological research (Ma et al. 2025). The development of high-throughput next-generation sequencing has advanced genomic research, notably with single-cell RNA sequencing (scRNA-seq). This method allows for detailed gene expression profiling and the detection of rare cell types. However, scRNA-seq lacks spatial context, limiting its ability to fully capture cellular interactions and tissue organization. Additionally, cell isolation from preserved samples can introduce artefacts, affecting data reliability.

To solve these limitations, spatial transcriptomics (ST) has emerged as a powerful approach that quantifies RNA expression directly in tissue sections, preserving spatial structure (Tian et al. 2023; Piñeiro et al. 2022). ST is categorized into sequencing-based and imaging-based methods and provides critical insights into how gene expression varies with spatial context, improving our understanding of intercellular communication. Combining scRNA-seq with ST enhances insights into complex biological processes, such as skin aging (Long et al. 2025).

Using single-cell RNA sequencing (scRNA-seq), researchers can detect changes at the single-cell level, uncovering transcriptomic heterogeneity across various keratinocyte subpopulations during aging. Similarly, spatial transcriptomics (ST) provides deeper insights into the molecular and cellular mechanisms underlying human skin aging and regeneration (Long et al. 2025). Significant shifts in gene expression patterns occur as humans age, with transcriptomics, particularly involving mRNA, lncRNA, circRNA, and exosomal RNA, offering key insights. However, due to diverse research approaches, there is no standardized method for gaining a comprehensive view, and full visualization of the transcriptome remains challenging (Roig-Rosello et al. 2024; Peng et al. 2024; Xia et al. 2024). Current research aids in identifying suitable RNA biomarkers for aging-related diagnostics. ABCG1, which regulates lipid balance and immune function, is implicated in vascular protection and longevity. Additionally, a study of over 1000 individuals aged 70–80 found BIRC2, a gene regulating inflammation and cell division, to be the most downregulated with age. Further whole-blood transcriptome analyses revealed that aging correlates positively with 11 genes (e.g., AMZ1, PARP3, PHLDA3) and negatively with 4 genes (e.g., MXRA8, CD248). Five transcriptional biomarkers have also been identified that can differentiate aging profile (Long et al. 2025).

14.4.4 Proteomics

Proteomics is the comprehensive study of all proteins generated by living cells, focusing on their expression, structure, functions, and post-translational modifications. This field includes various areas such as protein identification, analysis of post-translational changes, targeted proteomics, characterization of protein complexes, glycomics, modification genomics, and comparative proteomics (Al-Amrani et al. 2021). Today, proteomics plays a crucial role in exploring molecular drivers of pathology and in the discovery of prospective diagnostic markers and treatment targets. Bioinformatics exhibits a crucial role in understanding protein expression and function under normal and abnormal conditions by enabling functional annotation, interaction analysis, and pathway exploration through large-scale proteomic data (Zhang et al. 2023b). An and collaborators used phosphorylated protein arrays to study human dermal fibroblasts (HDFs) in senescent, quiescent, and proliferative states. They built a Boolean network model to simulate signal transduction in cellular senescence, identifying 3-phosphoinositide-dependent protein kinase 1 as a key regulator capable of reversing senescence to quiescence (Zhang et al. 2024). Further analysis suggested that RPL4, a 60S ribosomal protein, contributed to collagen synthesis and fibroblast structural integrity, mediating CASIN's anti-aging effects (Haydont et al. 2019). The proteome, being more closely linked to the phenotype than the transcriptome or genome, has become a key focus in aging

Table 14.3 Comparison of multiomics approaches to enhance the anti-aging properties

Type of analysis	role	Devices/ technologies	Advantages	Disadvantages
Metabolomics	Metabolite profiling Biomarkers of aging Nutrition & metabolism	Mass spectrometry (MS) GC/LC chromatography NMR	Highly sensitive (MS) Comprehensive metabolite profile Non-destructive NMR	Complex interpretation High cost Requires well-prepared samples
Genomics	Analysis of genetic variations Genes linked to longevity	NGS sequencers qPCR DNA microarrays	Large-scale analysis (NGS) High precision Quantitative data (qPCR)	NGS: massive amounts of data to process Microarrays: less sensitive than NGS
Proteomics	Study of proteins Post-translational modifications Response to cellular stress	Mass spectrometry (MS) LC-MS/MS 2D-PAGE	Detection of thousands of proteins Identification + quantification	Post-translational proteins that are difficult to analyze Expensive instruments
Transcriptomics	Gene expression RNA dynamics Cellular adaptation	RNA-Seq (NGS) qRT-PCR Microarrays ARN	Precise gene expression (RNA-Seq) Highly sensitive (qRT-PCR)	Fragile RNA (rapid degradation) High cost and analysis time (RNA-Seq)

biomarker research. Proteins influence aging directly through signal transduction and inflammatory pathways. Specific peptides, like GDF8 and GDF11, are linked to aging phenotypes and inflammation. Eleven proteins differentially expressed in older adults have potential as biomarkers for aging and related health conditions. Proteomic analyses have identified top proteins associated with unhealthy (e.g., CRTAC1, CD14) and healthy aging (e.g., TGFBI, CLEC3B), offering targets for clinical interventions. Proteins like GDF15, involved in immune aging, stand out due to their consistent increase with age across populations. Despite progress, reliable aging biomarkers remain limited due to challenges like inconsistent sample processing, population diversity, and technical limitations in proteomic analysis (Table 14.3).

14.5 Conclusion

Recent scientific and technological progress has ushered in the "omics era," allowing researchers to analyze aging at multiple molecular levels. Identifying aging biomarkers and therapeutic targets is essential for advancing aging research and geriatric care. Omics approaches spanning genomics, proteomics, transcriptomics, and metabolomics enhance our understanding of aging mechanisms and open new paths for diagnosis and therapy. This work reviews current omics techniques used to identify aging-related biomarkers, proposing the development of an integrated "aging biomarker library" for diagnostic and prognostic use. In geriatrics, identifying new biomarkers and improving biological age measurement methods is crucial. Current research emphasizes the need for more accurate biological age assessments, particularly using DNA methylation-based epigenetic clocks, complemented by transcriptomic aging clocks. These approaches demonstrate that biological age closely reflects the aging process. Transcriptional data, including miRNAs, lncRNAs, and circRNAs, also help uncover novel aging biomarkers. Proteomics is gaining prominence due to its reproducibility and verification ease, though its clinical use is hindered by high costs and limited application. Similarly, metabolomics and microbiomics show promise, but their clinical potential is constrained by the lack of longitudinal studies. A key advancement lies in integrating data across multiple omics fields, genomics, transcriptomics, proteomics, and metabolomics, rather than treating them separately. For example, comparing microbiome-regulated metabolites with blood-based biomarkers in long-lived individuals reveals deeper insights, showing that integrated analysis provides more than the sum of individual methods. Aging clocks and biomarkers, traditionally studied in experimental settings, are gaining traction in clinical research due to their practicality. These tools are enhancing our understanding of human aging and hold promise in areas like drug target identification, where biomarkers can also serve as therapeutic targets. Methylation-based aging clocks are being used to assess anti-aging treatments, and biological age has been linked to outcomes in conditions such as ischemic stroke and brain health. Biomarkers also support drug discovery, improve treatment accuracy, and aid in understanding disease mechanisms through noninvasive methods. Recent advances include miRNA-based therapies for hepatocellular carcinoma. This review emphasizes not only single omics approaches but also highlights the importance of integrating multiple omics to comprehensively analyze aging biomarkers. While multi-omics approaches hold significant promise, they face several limitations. These include high costs, specialized equipment needs, and reliance on skilled personnel for data analysis. Challenges also arise from inconsistent data quality, inaccurate sources, and nonstandard sampling methods. The lack of standardized platforms and bioinformatics tools further complicates large-scale data processing, particularly in metabolomics, where the sheer number of metabolites and individual variability pose major analytical difficulties. To improve the clinical relevance of metabolomics and other omics-based strategies, standardization and cross-disciplinary collaboration are essential. There is also a pressing need for large-scale, longitudinal cohort studies to track biological changes over time. Studying aging in model organisms like fruit flies, nematodes, and mice due to their shorter lifespans and simpler microbiota can help uncover mechanisms relevant to human aging, such as intestinal barrier decline. Given the complexity of aging, anti-aging research and its translation to clinical practice require a comprehensive, multifaceted approach. Cellular aging contributes to tissue damage and age-related diseases. Plant anti-aging research focuses on senolytics and bioflavonoids, along with enhancing the body's natural antioxidant defenses against reactive oxygen species (ROS). These defenses include enzymatic antioxidants like peroxiredoxins, glutathione peroxidase, catalase and superoxide dismutase, as well as nonenzymatic ones such as glutathione, thioredoxin, uric acid, and coenzyme Q. Vitamins (E, A, C), along with minerals like selenium and zinc, also play key antioxidant roles. Flavonoids (e.g., quercetin, resveratrol, curcumin) are potent natural antioxidants that chelate harmful metals, scavenge ROS, and improve overall cellular health. They help regulate blood lipids, extend red blood cell life, and protect against various diseases. Curcumin offers anti-inflammatory and antitumor benefits, while resveratrol and quercetin support antioxidant defense systems. Technological advancements such as AI and meta-analyses are accelerating aging research, particularly in identifying biomarkers. However, before clinical use, these biomarkers require validation in larger populations. Integrating omics data offers a promising path to discover novel biomarkers and therapeutic targets to better understand and combat aging.

Competing Interests The authors have no relevant financial or non-financial interests to disclose.

References

Abdellatif M, Trummer-Herbst V, Heberle AM, Humnig A, Pendl T, Durand S, Cerrato G, Hofer SJ, Islam M, Julia Voglhuber J, Pittol JMR, Kepp O, Hoefler G, Schmidt A, Rainer PP, Scherr D, Lewinski DV, Bisping E, McMullen JR, Diwan A, Eisenberg T, Madeo F, Thedieck K, Kroemer G, Sedej S (2022) Fine-tuning cardiac insulin-like growth factor 1 receptor signaling to promote health and longevity. Circulation 145:1853–1866

Ahlina FN, Nugraheni N, Salsabila IA, Haryanti S, Da'i M, Meiyanto E (2020) Revealing the reversal effect of galangal (*Alpinia galanga* L.) extract against oxidative stress in metastatic breast cancer cells and normal fibroblast cells intended as a co-chemotherapeutic and anti-ageing agent. Asian Pac J Cancer Prev 21:107–117

Al-Amin MM, Akhter S, Hasan AT, Alam T, Nageeb HSM, Saifullah ARM, Shohel M (2015) The antioxidant effect of astaxanthin is higher in young mice than aged: a region-specific study on brain. Metab Brain Dis 30:1237–1246

Al-Amrani S, Al-Jabri Z, Al-Zaabi A, Alshekaili J, Al-Khabori M (2021) Proteomics: concepts and applications in human medicine. World J Biol Chem 12:57–69

Al-Hassawi WW, Al-Sammak MA (2024) Resveratrol as an antiaging drug: a review of articles. Iraqi J Pharm Sci 33:1–13

Ali A, Abouleila Y, Shimizu Y, Hiyama E, Emara S, Mashaghi A, Hankemeier T (2019) Single-cell metabolomics by mass spectrometry: advances, challenges, and future applications. Trac-Trend Anal Chem 120:115436

Altyar AE, Ashour ML, Youssef FS (2020) Premna odorata: seasonal metabolic variation in the essential oil composition of its leaf and verification of its anti-ageing potential via in vitro assays and molecular modelling. Biomolecules 10:879

An SF, Yu YL, Tan ZW, Li CM, Yang Q, Yang HQ, Su XY, Sun Y, Liang HZ (2024) Genomic analysis of yam: understanding its adaptive evolution and medicinal properties. Med Plants Res 14:31–44

Anis A, Tabish A, Das RK, Kohar D, Sharma P, Phogat J (2024) An overview of medicinal plants for the of anti-ageing face packs. World J Pharm Res 13:316–330

Arriola Apelo SI, Lamming DW (2016) Rapamycin: an InhibiTOR of aging emerges from the soil of Easter Island. J Gerontol A Biol Sci Med Sci 71:841–849

Bahrami SA, Bakhtiari N (2016) Ursolic acid regulates aging process through enhancing of metabolic sensor proteins level. Biomed Pharmacother 82:8–14

Bass TM, Weinkove D, Houthoofd K, Gems D, Partridge L (2007) Effects of resveratrol on lifespan in *Drosophila melanogaster* and *Caenorhabditis elegans*. Mech Ageing Dev 128:546–552

Bjørklund G, Shanaida M, Lysiuk R, Butnariu M, Peana M, Sarac I, Strus O, Smetanina K, Chirumbolo S (2022) Natural compounds and products from an anti-aging perspective. Molecules 27:7084

Blackburn EH (1991) Structure and function of telomeres. Nature 350:569–573

Blackburn EH (2001) Switching and signaling at the telomere. Cell 106:661–673

Blagosklonny MV (2021) Anti-aging: senolytics or gerostatics (unconventional view). Oncotarget 12:1821

Boonpisuttinant K, Srisuttee R, Khong HY, Chutoprapat R, Ruksiriwanich W, Udompong S, Chompoo W, Boonbai R, Rattikarl Rakkaew R, Jinapa Sangsee J, Sriprasert K, Malilas W (2023) In vitro anti-ageing activities of ethanolic extracts from Pink rambutan (*Nephelium lappaceum* Linn.) for skin applications. Saudi Pharm J 31:535–546

Campisi J, d'Adda di Fagagna F (2007) Cellular senescence: when bad things happen to good cells. Nat Rev Mol Cell Biol 8:729–740

Chae SY, Park SY, Park G (2018) Lutein protects human retinal pigment epithelial cells from oxidative stress-induced cellular senescence. Mol Med Rep 18:5182–5190

Chaib S, Tchkonia T, Kirkland JL (2022) Cellular senescence and senolytics: the path to the clinic. Nat Med 28:1556–1568

Chaiyana W, Anuchapreeda S, Punyoyai C, Neimkhum W, Kuan-Han Lee KH, Wei-Chao Lin WC, Lue SC, Viernstein H, Mueller M (2019) *Ocimum sanctum* Linn. as a natural source of skin anti-ageing compounds. Ind Crop Prod 127:217–224

Chakraborty A, Sahoo M, Roy SD, Dasgupta RK (2018) Anti-ageing natural herbs: a aystemic review. Indian Res J Pharm Sci 5:1589–1598

Chan SR, Blackburn EH (2004) Telomeres and telomerase. Philos Trans R Soc Lond Ser B Biol Sci 359:109–122

Chatzigeorgiou S, Thai QD, Tchoumtchoua J, Tallas K, Tsakiri EN, Papassideri I, Halabalaki M, Skaltsounis AL, Trougakos IP (2017) Isolation of natural products with anti-ageing activity from the fruits of *Platanus orientalis*. Phytomedicine 33:53–61

Chen Y, Onken B, Chen H, Xiao S, Liu X, Driscoll M, Cao Y, Huang Q (2014) Mechanism of longevity extension of *Caenorhabditis elegans* induced by pentagalloyl glucose isolated from *eucalyptus* leaves. J Agric Food Chem 62:3422–3431

Chen P, He D, Zhang Y, Yang S, Chen L, Wang S, Zou H, Liao Z, Zhang X, Wu M (2016) *Sargassum fusiforme* polysaccharides activate antioxidant defense by promoting Nrf2-dependent cytoprotection and ameliorate stress insult during aging. Food Funct 7:4576–4588

Chen Y, Onken B, Chen H, Zhang X, Driscoll M, Cao Y, Huang Q (2020) Healthy lifespan extension mediated by oenothein B isolated from *Eucalyptus grandis*× *Eucalyptus urophylla* GL9 in *Caenorhabditis elegans*. Food Funct 11:2439–2450

Chen C, Ni Y, Jiang B, Yan S, Xu B, Fan B, Huang H, Chen G (2021) Anti-aging derivatives of cycloastragenol produced by biotransformation. Nat Prod Res 35:2685–2690

Chen M, Wen H, Zhou S, Yan X, Li H (2022) Patchouli alcohol inhibits D-gal induced oxidative stress and ameliorates the quality of aging cartilage via activating the Nrf2/HO-1 pathway in mice. Oxidative Oxid Med Cell Longev 2022:6821170

Chilelli NC, Ragazzi E, Valentini R, Cosma C, Ferraresso S, Lapolla A, Sartore G (2016) Curcumin and *boswellia serrata* modulate the glyco-oxidative status and lipo-oxidation in master athletes. Nutrients 8:745

Chuang SY, Lin CH, Fang JY (2014) Natural compounds and aging: between autophagy and inflammasome. Biomed Res Int 2014:297293

Cohn JS, Kamili A, Wat E, Chung RW, Tandy S (2010) Reduction in intestinal cholesterol absorption by various food components: mechanisms and implications. Atheroscler Suppl 11:45–48

Davinelli S, Willcox DC, Scapagnini G (2012) Extending healthy ageing: nutrient sensitive pathway and centenarian population. Immun Ageing 9:9

Davis JM, Murphy EA, Carmichael MD, Davis B (2009) Quercetin increases brain and muscle mitochondrial biogenesis and exercise tolerance. Am J Physiol Regul Integr Comp Physiol 296:R1071–R1077

Deelen J, Uh HW, Monajemi R, van Heemst D, Thijssen PE, Böhringer S, van den Akker EB, de Craen AJM, Rivadeneira J, Uitterlinden AG, Westendorp RGJ, Goeman JJ, Slagboom PE, Houwing-Duistermaat JJ, Beekman M (2013) Gene set analysis of GWAS data for human longevity highlights the relevance of the insulin/IGF-1 signaling and telomere maintenance pathways. Age 35:235–249

Dixit V, Chaubey KK, Dayal D, Chole PB, Manjunath BT, Bachheti RK, Gupta PC, Bachheti A, Worku LA (2025) Natural products in aging: cellular mechanisms and emerging therapeutics for age-related disorders. J Aging Res 2025:6868732

Duque L, Bravo K, Osorio E (2017) A holistic anti-aging approach applied in selected cultivated medicinal plants: a view of photoprotection of the skin by different mechanisms. Ind Crop Prod 97:431–439

El-Far AH, Mohamed HH, Elsabagh DA, Mohamed SA, Noreldin AE, Al Jaouni SK, Alsenosy AA (2022) Eugenol and carvacrol attenuate brain d-galactose-induced aging-related oxidative alterations in rats. Environ Sci Pollut Res Int 29:47436–47447

Fan J, Feng H, Yu Y, Sun M, Liu Y, Li T, Sun X, Liu S, Sun M (2017) Antioxidant activities of the polysaccharides of *Chuanminshen violaceum*. Carbohydr Polym 157:629–636

Flachsbart F, Caliebe A, Kleindorp R, Blanché H, von Eller-Eberstein H, Nikolaus S, Schreiber S, Nebel A (2009) Association of FOXO3A variation with human longevity confirmed in German centenarians. Proc Natl Acad Sci USA 106:2700–2705

Flachsbart F, Dose J, Gentschew L, Geismann C, Caliebe A, Knecht C, Nygaard M, Badarinarayan N, ElSharawy A, May S, Luzius A, Torres GG, Jentzsch M, Forster M, Häsler R, Pallauf K, Lieb W, Derbois C, Galan P, Drichel D, Arlt A, Till A, Krause-Kyora B, Rimbach G, Blanché H, Deleuze JF, Christiansen L, Christensen K, Nothnagel M, Rosenstiel P, Schreiber S, Franke A, Sebens S, Nebel A (2017) Identification and characterization of two functional variants in the human longevity gene FOXO3. Nat Commun 8:2063

Gupta AK, Savopoulos CG, Ahuja J, Hatzitolios AI (2011) Role of phytosterols in lipid-lowering: current perspectives. QJM 104:301–308

Gupta AK, Yadav R, Sharma S, Sawale J, Kahwa I (2023) Anti-ageing activity and antioxidant activity of methanol extracts of leaves and flowers of *Salvia officinalis* and *Rosmarinus officinalis*. J Young Pharm 15:448–455

Hafeez E, Du D, Ni H, Zhu K, Hu F, Zhou J, Chen D (2025) Regulation and mechanism of Bletilla striata polysaccharide on delaying aging in *Drosophila melanogaster*. Int J Biol Macromol 310:143382

Hah YS, Lee WK, Lee S, Kim EJ, Lee JH, Lee SJ, HoJi Y, Kim GK, Lee HH, Hong SY, Yoo JI (2022) β-Sitosterol attenuates dexamethasone-induced muscle atrophy via regulating FoxO1-dependent signaling in C2C12 cell and mice model. Nutrients 14:2894

Haiyuan YU, Shen X, Liu D, Hong M, Lu Y (2019) The protective effects of β-sitosterol and vermicularin from *Thamnolia vermicularis* (Sw.) Ach. against skin aging *in vitro*. An Acad Bras Cienc 91:e20181088

Hao Y, Zhang Z, Luo E, Yang J, Wang S (2025) Plant metabolomics: applications and challenges in the era of multi-omics big data. aBIOTECH 6:116–132

Haydont V, Bernard BA, Fortunel NO (2019) Age-related evolutions of the dermis: clinical signs, fibroblast and extracellular matrix dynamics. Mech Ageing Dev 177:150–156

Herbig U, Jobling WA, Chen BP, Chen DJ, Sedivy JM (2004) Telomere shortening triggers senescence of human cells through a pathway involving ATM, p53, and p21CIP1, but not p16INK4a. Mol Cell 14:501–513

Hu W, Dai D, Li W (2013) Anti-aging effect of *Blakeslea trispora* powder on adult mice. Biotechnol Lett 35:1309–1315

Huang M, Lu JJ, Huang MQ, Bao JL, Chen XP, Wang YT (2012) Terpenoids: natural products for cancer therapy. Expert Opin Investig Drugs 21:1801–1818

Iside C, Scafuro M, Nebbioso A, Altucci L (2020) SIRT1 activation by natural phytochemicals: an overview. Front Pharmacol 11:1225

Iweala EJ, Adurosakin OE, Innocent U, Omonhinmin CA, Dania OE, Ugbogu EA (2024) Anti-aging potential of bioactive phytoconstituents found in edible medicinal plants: a review. Science 6:36

Izadi M, Sadri N, Abdi A, Zadeh, MMR, Jalaei D, Ghazimoradi MM, Shouri S, Tahmasebi S (2024) Longevity and anti-aging effects of curcumin supplementation. GeroScience 46:2933–2950

Jabłonowska M, Ciganović P, Jablan J, Marguí E, Tomczyk M, Končić MZ (2021) *Silybum marianum* glycerol extraction for the preparation of high-value anti-ageing extracts. Ind Crop Prod 168:113613

Jeong YJ, Choi YJ, Choi JS, Kwon HM, Kang SW, Bae JY, Lee SS, Kang JS, Han SJ, Kang YH (2007) Attenuation of monocyte adhesion and oxidised LDL uptake in luteolin-treated human endothelial cells exposed to oxidised LDL. Br J Nutr 97:447–457

Johnson SC (2019) Nutrient sensing, signaling and ageing: the role of IGF-1 and mTOR in ageing and age-related disease. Biochem Cell Biol Ageing Part I Biomed Sci 90:49–97

Kampkötter A, Gombitang NC, Zurawski RF, Timpel C, Chovolou Y, Wätjen W, Kahl R (2007) Effects of the flavonoids kaempferol and fisetin on thermotolerance, oxidative stress and FoxO transcription factor DAF-16 in the model organism *Caenorhabditis elegans*. Arch Toxicol 81:849–858

Kampkötter A, Timpel C, Zurawski RF, Ruhl S, Chovolou Y, Proksch P, Wätjen W (2008) Increase of stress resistance and lifespan of Caenorhabditis elegans by quercetin. Comp Biochem Physiol Part B Biochem Mol Biol 149:314–323

Kazeem MI, Mellem JJ, Sabiu S (2024) Medicinal foods and plants with antiaging properties: a review of *in vitro* and *in vivo* studies. Food Front 5:24–45

Kensler TW, Wakabayashi N (2010) Nrf2: friend or foe for chemoprevention? Carcinogenesis 31:90–99

Keowkase R, Shoomarom N, Bunargin W, Sitthithaworn W, Weerapreeyakul N (2018) Sesamin and sesamolin reduce amyloid-β toxicity in a transgenic *Caenorhabditis elegans*. Biomed Pharmacother 107:656–664

Kim JM, Lee EK, Kim DH, Yu BP, Chung HY (2010) Kaempferol modulates pro-inflammatory NF-κB activation by suppressing advanced glycation endproducts-induced NADPH oxidase. Age 32:197–208

Kim JE, Jang SG, Lee CH, Lee JY, Park H, Kim JH, Lee S, Kim SH, Park EY, Lee KW, Shin HS (2019) Beneficial effects on skin health using polysaccharides from red ginseng by-product. J Food Biochem 43:e12961

Kim J, Lee SH, Cho M, Lee JY, Choi DH, Lee HY, Choi DH, Lee HY, Cho S, Min KJ, Suh Y (2020) Small molecule from natural phytochemical mimics dietary restriction by modulating FoxO3a and metabolic reprogramming. Adv Biosyst 4:1900248

Kirkland JL, Tchkonia T (2020) Senolytic drugs: from discovery to translation. Intern Med J 288:518–536

Klinngam W, Rungkamoltip P, Thongin S, Joothamongkhon J, Khumkhrong P, Khongkow M, Khongkow M, Namdee K, Tepaamorndech S, Chaikul P, Kanlayavattanakul M, Lourith N, Piboonprai K, Ruktanonchai U, Asawapirom U, Iempridee T (2022) Polymethoxyflavones from Kaempferia parviflora ameliorate skin aging in primary human dermal fibroblasts and ex vivo human skin. Biomed Pharmacother 145:112461

Kook D, Wolf AH, Alice LY, Neubauer AS, Priglinger SG, Kampik A, Welge-Lüssen UC (2008) The protective effect of quercetin against oxidative stress in the human RPE *in vitro*. Invest Ophthalmol Vis Sci 49:1712–1720

Kowalczyk D, Świątek Ł, Zengin G, Sieniawska E, Borowska K (2025) Safety profile of anti-ageing natural cosmetic formulations with Litsea cubeba essential oil. Our Dermatol Online 16:242–248

Krishnaswamy VKD, Alugoju P, Periyasamy L (2020) Effect of short-term oral supplementation of crocin on age-related oxidative stress, cholinergic, and mitochondrial dysfunction in rat cerebral cortex. Life Sci 263:118545

Kudryashova KS, Burka K, Kulaga AY, Vorobyeva NS, Kennedy BK (2020) Aging biomarkers: from functional tests to multi-omics approaches. Proteomics 20:1900408

Kuppusamy UR, Arumugam B (2023) Novel plant bioactives, their antiaging potencies: reality and promises. In: Plant bioactives as natural panacea against age-induced diseases. Elsevier, Amsterdam, pp 359–386

Lamming DW, Ye L, Katajisto P, Goncalves MD, Saitoh M, Stevens DM, Davis JG, Salmon AB, Richardson A, Ahima RS, Guertin DA, Sabatini DM, Baur JA (2012) Rapamycin-induced insulin resistance is mediated by mTORC2 loss and uncoupled from longevity. Science 335:1638–1643

Lee EB, Ahn D, Kim BJ, Lee SY, Seo HW, Cha YS, Jeon H, Eun JS, Cha DS, Kim DK (2015) Genistein from *Vigna angularis* extends lifespan in *Caenorhabditis elegans*. Biomol Ther 23:77

Lee J, Kim YS, Kim E, Kim Y, Kim Y (2020) Curcumin and hesperetin attenuate D-galactose-induced brain senescence *in vitro* and *in vivo*. Nutr Res Pract 14:438–452

Lestari U, Muhaimin M, Chaerunisaa AY, Sujarwo W (2023) Anti-aging potential of plants of the anak dalam tribe, Jambi, Indonesia. Pharmaceuticals 16:1300

Li Y, Zhang D, Li L, Han Y, Dong X, Yang L, Li X, Li W, Li W (2021) Ginsenoside Rg1 ameliorates aging-induced liver fibrosis by inhibiting the NOX4/NLRP3 inflammasome in SAMP8 mice. Mol Med Rep 24:801

Li J, Xian L, Zheng R, Wang Y, Wan X, Liu Y (2022) Canthaxanthin shows anti-liver aging and anti-liver fibrosis effects by down-regulating inflammation and oxidative stress *in vivo* and *in vitro*. Int Immunopharmacol 110:108942

Lin C, Zhang X, Su Z, Xiao J, Lv M, Cao Y, Chen Y (2019) Carnosol improved lifespan and healthspan by promoting antioxidant capacity in *Caenorhabditis elegans*. Oxidative Med Cell Longev 2019:5958043

Liu JK (2022) Antiaging agents: safe interventions to slow aging and healthy life span extension. Nat Prod Bioprospect 12:18

Liu X, Lin X, Zhang S, Guo C, Li J, Mi Y, Zhang C (2018) Lycopene ameliorates oxidative stress in the aging chicken ovary via activation of Nrf2/HO-1 pathway. Aging (Albany NY) 10:2016

Liu M, Li N, Lu X, Shan S, Gao X, Cao Y, Lu W (2022) Sweet tea (*Rubus Suavissmus* S. Lee) polysaccharides promote the longevity of *Caenorhabditis elegans* through autophagy-dependent insulin and mitochondrial pathways. Int J Biol Macromol 207:883–892

Long H, Qiu X, Cao L, Han R (2021) Discovery of the signal pathways and major bioactive compounds responsible for the anti-hypoxia effect of Chinese cordyceps. J Ethnopharmacol 277:114215

Long B, Pan W, Wu S, Nong Q, Li W, Chen S, Guo HW (2025) Advances in the application of multi-omics analysis in skin aging. Front Aging 6:1596050

Lu M, Tan L, Zhou XG, Yang ZL, Zhu Q, Chen JN, Luo HR, Wu GS (2020) Tectochrysin increases stress resistance and extends the lifespan of *Caenorhabditis elegans* via FOXO/DAF-16. Biogerontology 21:669–682

Lu Y, Zhao Y, Hu L, Zhang W, Xie Y, Cheng S, Zheng B, Xia, Q (2023) Exploration of multi-gene DNA Barcode markers to reveal the broad genetic diversity of field ticks (Acari: Ixodidae) in a tropical environment of Hainan Island, China. Cytogenet Genome Res 163:59–73

Luo J, Si H, Jia Z, Liu D (2021) Dietary anti-aging polyphenols and potential mechanisms. Antioxidants 10:283

Luo B, Li D, Hu K, Chen M, Sun X, Lu R, Wei S (2023) Genome survey sequencing of the medical plant *Dioscorea futschauensis*: genomic characteristics estimation and genomic resources development. J Biobaased Mater Bioenergy 17:803–811

Ma LJ, Chen GL, Jia HY, Xie J (2009) Anti-senescence effect of *Cynomorium songaricum* polysaccharide on D-galactose-induced aging mice. Chin J Hosp Pharm 14:1186–1189

Ma SD, Yuan R, Huang MW, Xin QQ, Miao Y, Zhu YZ, Chen KJ, Cong WH (2025) Natural anti-aging herb: role and potential of *Astragalus membranaceus*. Chin J Integr Med 31:743–751

Maoka T (2020) Carotenoids as natural functional pigments. J Nat Med 74:1–16

Martel J, Ojcius DM, Ko YF, Chang CJ, Young JD (2019) Antiaging effects of bioactive molecules isolated from plants and fungi. Med Res Rev 39:1515–1552

Martel J, Ojcius DM, Wu CY, Peng HH, Voisin L, Perfettini JL, Ko YF, Young JD (2020) Emerging use of senolytics and senomorphics against aging and chronic diseases. Med Res Rev 40:2114–2131

Men Y, Wei P, Huang X, Yang L, Cao Z, Wei J, Zhang X, Zhang F, Yang Z (2025) Proteomic and transcriptomic analysis of the action mechanism of spermidine in mitigating the aging of *Allium mongolicum* seeds. Sci Rep 15:17443

Moreau RA, Whitaker BD, Hicks KB (2002) Phytosterols, phytostanols, and their conjugates in foods: structural diversity, quantitative analysis, and health-promoting uses. Prog Lipid Res 41:457–500

Morris BJ, Willcox DC, Donlon TA, Willcox BJ (2015) FOXO3: a major gene for human longevity-a mini-review. Gerontology 61:515–525

Mouafon IL, Katerere DR (2025) Review of the ethnobotany, phytochemistry, pharmacology, and toxicity studies of the genus *Adenia*. Front Pharmacol 16:1581659

Mu S, Yang W, Huang G (2021) Antioxidant activities and mechanisms of polysaccharides. Chem Biol Drug Des 97:628–632

Murphy CT (2006) The search for DAF-16/FOXO transcriptional targets: approaches and discoveries. Exp Gerontol 41:910–921

Murphy CT, McCarroll SA, Bargmann CI, Fraser A, Kamath RS, Ahringer J, Li H, Kenyon C (2003) Genes that act downstream of DAF-16 to influence the lifespan of *Caenorhabditis elegans*. Nature 424:277–283

Oberdoerffer P, Sinclair DA (2007) The role of nuclear architecture in genomic instability and ageing. Nat Rev Mol Cell Biol 8:692–702

Ou ZQ, Rades T, McDowell A (2015) Anti-ageing effects of Sonchus oleraceus L.(pūhā) leaf extracts on H2O2-induced cell senescence. Molecules 20:4548–4564

Park SH (2024) Role of phytochemicals in treatment of aging and cancer: focus on mechanism of FOXO3 activation. Antioxidants 13:1099

Passtoors WM, Beekman M, Deelen J, van der Breggen R, Maier AB, Guigas B, Derhovanessian E, van Heemst D, de Craen AJM, Gunn DA, Pawelec G, Slagboom P (2013) Gene expression analysis of mTOR pathway: association with human longevity. Aging Cell 12:24–31

Payne BD, Gafar M, Govender T, Ashmawy NS, Lall N (2025) The anti-ageing potential of *Persicaria senegalensis* (Meisn.) Sojak and its Bifidobacterium bifidum 11863 fermented counterpart. J Ethnopharmacol 355:120512

Peng X, Zhong Y, Mao R, He F, Cheng Y, Chen M, Zhou L, Xie H, Li J, Zhang Y (2024) Integrated bioinformatics analysis and experimental validation identifies CPE as a potential biomarker and therapeutic target for skin aging. Exp Dermatol 33:e15120

Piñeiro AJ, Houser AE, Ji AL (2022) Research techniques made simple: spatial transcriptomics. J Invest Dermatol 142:993–1001

Radwan RA, El-Sherif YA, Salama MM (2020) A novel biochemical study of anti-ageing potential of Eucalyptus camaldulensis bark waste standardized extract and silver nanoparticles. Colloids Surf B Biointerfaces 191:111004

Rahimi VB, Askari VR, Mousavi SH (2018) Ellagic acid reveals promising anti-aging effects against d-galactose-induced aging on human neuroblastoma cell line, SH-SY5Y: a mechanistic study. Biomed Pharmacother 108:1712–1724

Raina K, Kumari R, Thakur P, Sharma R, Singh R, Thakur A, Anand V, Sharma R, Chaudhary A (2023) Mechanistic role and potential of Ayurvedic herbs as anti-aging therapies. Drug Metab Pers Ther 38:211–226

Rajapakse AG, Yepuri G, Carvas JM, Stein S, Matter CM, Scerri I, Ruffieux J, Montani JP, Ming XF, Yang Z (2011) Hyperactive S6K1 mediates oxidative stress and endothelial dysfunction in aging: inhibition by resveratrol. PLoS One 6:e19237

Ramadan A, Kamel G, Shokry AA, El-Shiekh RA (2020) A potent anti-ageing and immunomodulatory activity of apricot seed standardized extract and its major compound; amygdalin. Trop J Nat Prod Res 4:728–733

Rastog M, Ojha RP, Devi BP, Aggarwal A, Agrawal A, Dubey GP (2012) Amelioration of age associated neuroinflammation on long term bacosides treatment. Neurochem Res 37:869–874

Rattan SI (2018) Biogerontology: research status, challenges and opportunities. Acta Biomed Atenei Parmensis 89:291

Ren L, Zhan P, Wang Q, Wang C, Liu Y, Yu Z, Zhang S (2019) Curcumin upregulates the Nrf2 system by repressing inflammatory signaling-mediated Keap1 expression in insulin-resistant conditions. Biochem Biophys Res Commun 514:691–698

Rodrigues MJ, Slusarczyk S, Pecio Ł, Matkowski A, Salmas RE, Durdagi S, Pereira C, Varela J, Barreira L, Custódio L (2018) In vitro and in silico approaches to appraise *Polygonum maritimum* L. as a source of innovative products with anti-ageing potential. Ind Crop Prod 111:391–399

Rogina B, Helfand SL (2004) Sir2 mediates longevity in the fly through a pathway related to calorie restriction. Proc Natl Acad Sci USA 101:15998–16003

Roig-Rosello E, Dayan G, Bovio S, Manissier P, Errazuriz E, Rousselle P (2024) Dermal stiffness governs the topography of the epidermis and the underlying basement membrane in young and old human skin. Aging Cell 23:e14096

Saini R, Mittal A, Rathi V (2016) Formulation & in-vitro antioxidant analysis of anti-ageing cream of Carica papaya fruit extract. Indian J Drugs 4:8–14

Samarakoon SMS, Chandola HM, Shukla VJ (2010) Pharmacognosy and phytochemical study of Amalakayas Rasayana; an anti ageing ayurvedic formulation. Indian J Ancient Med Yoga 3:181–190

Saul N, Pietsch K, Menzel R, Stürzenbaum SR, Steinberg CE (2010) The longevity effect of tannic acid in *Caenorhabditis elegans*: disposable soma meets hormesis. J Gerontol A Biol Sci Med Sci 65:626–635

Saul N, Pietsch K, Stürzenbaum SR, Menzel R, Steinberg CE (2011) Diversity of polyphenol action in *Caenorhabditis elegans*: between toxicity and longevity. J Nat Prod 74:1713–1720

Senthil KK, Gokila VM, Mau JL, Lin CC, Chu FH, Wei CC, Liao VHC, Wang SY (2016) A steroid like phytochemical Antcin M is an anti-aging reagent that eliminates hyperglycemia-accelerated premature senescence in dermal fibroblasts by direct activation of Nrf2 and SIRT-1. Oncotarget 7:62836

Shanaida M, Mykhailenko O, Lysiuk R, Hudz N, Balwierz R, Shulhai A, Lysiuk R, Hudz N, Shapovalova N, Shanaida V, Balwierz R, Bjørklund G (2025) Carotenoids for antiaging: nutraceutical, pharmaceutical, and cosmeceutical applications. Pharmaceuticals 18:403

Shen CY, Jiang JG, Yang L, Wang DW, Zhu W (2017) Anti-ageing active ingredients from herbs and nutraceuticals used in traditional Chinese medicine: pharmacological mechanisms and implications for drug discovery. Br J Pharmacol 174:1395–1425

Simioni C, Zauli G, Martelli AM, Vitale M, Sacchetti G, Gonelli A, Neri LM (2018) Oxidative stress: role of physical exercise and antioxidant nutraceuticals in adulthood and aging. Oncotarget 9:17181

Singh S, Garg G, Singh AK, Bissoyi A, Rizvi SI (2019) Fisetin, a potential caloric restriction mimetic, attenuates senescence biomarkers in rat erythrocytes. Biochem Cell Biol 97:480–487

Smith HJ, Sharma A, Mair WB (2020) Metabolic communication and healthy aging: where should we focus our energy? Dev Cell 54:196–211

Smulders L, Deelen J (2024) Genetics of human longevity: from variants to genes to pathways. J Intern Med 295:416–435

Song L, Zhang S (2023) Anti-aging activity and modes of action of compounds from natural food sources. Biomolecules 13:1600

Song L, Leng K, Xiao K, Zhang S (2022a) Administration of krill oil extends lifespan of fish *Nothobranchius guentheri* via enhancement of antioxidant system and suppression of NF-κB pathway. Fish Physiol Biochem 48:1057–1073

Song L, Li C, Wu F, Zhang S (2022b) Dietary intake of diosgenin delays aging of male fish *Nothobranchius guentheri* through modulation of multiple pathways that play prominent roles in ROS production. Biogerontology 23:01–213

Stavropoulou MI, Stathopoulou K, Cheilari A, Benaki D, Gardikis K, Chinou I, Aligiannis N (2021) NMR metabolic profiling of Greek propolis samples: comparative evaluation of their phytochemical compositions and investigation of their anti-ageing and antioxidant properties. J Pharm Biomed Anal 194:113814

Su S, Wink M (2015) Natural lignans from *Arctium lappa* as antiaging agents in *Caenorhabditis elegans*. Phytochemistry 117:340–350

Sun K, Xiang L, Ishihara S, Matsuura A, Sakagami Y, Qi J (2012) Anti-aging effects of hesperidin on Saccharomyces cerevisiae via inhibition of reactive oxygen species and UTH1 gene expression. Biosci Biotechnol Biochem 76:640–645

Sun J, Zhong X, Sun D, Xu L, Shi L, Sui J, Liu Y (2023) Anti-aging effects of polysaccharides from ginseng extract residues in *Caenorhabditis elegans*. Int J Biol Macromol 225:1072–1084

Surh YJ, Kundu JK, Na HK (2008) Nrf2 as a master redox switch in turning on the cellular signaling involved in the induction of cytoprotective genes by some chemopreventive phytochemicals. Planta Med 74:1526–1539

Sykiotis GP, Bohmann D (2008) Keapl/Nrf2 signaling regulates oxidative stress tolerance and lifespan in drosophila. Dev Cell 14:76–85

Sykiotis GP, Bohmann D (2010) Stress-activated cap 'n'collar transcription factors in aging and human disease. Sci Signal 3:re3

Tang R, Chen X, Dang T, Deng Y, Zou Z, Liu Q, Gong G, Song S, Ma F, Huang L, Wang Z (2019) *Lycium barbarum* polysaccharides extend the mean lifespan of *Drosophila melanogaster*. Food Funct 10:4231–4241

Tian L, Chen F, Macosko EZ (2023) The expanding vistas of spatial transcriptomics. Nat Biotechnol 41:773–782

Tolani P, Gupta S, Yadav K, Aggarwal S, Yadav AK (2021) Big data, integrative omics and network biology. In: Donev R, Karabencheva-Christova T (eds) Proteomics and systems biology, Advances in Protein Chemistry and Structural Biology, vol 127. Academic Press, Cambridge, pp 127–160

Tomou EM, Papaemmanouil CD, Diamantis DA, Kostagianni AD, Chatzopoulou P, Mavromoustakos T, Tzakos AG, Skaltsa H (2021) Anti-ageing potential of S. euboea Heldr. phenolics. Molecules 26:3151

Tundi R, Loizzo MR, Bonesi M, Menichini F (2015) Potential role of natural compounds against skin aging. Curr Med Chem 22:1515–1538

Vaiserman AM, Koliada AK, Marotta F (2017) Gut microbiota: a player in aging and a target for anti-aging intervention. Ageing Res Rev 35:36–45

Valenzuela HF, Fuller T, Edwards J, Finger D, Molgora B (2009) Cycloastragenol extends T cell proliferation by increasing telomerase activity (90.30). J Immunol 182:90–30

Varghese A, Devassy AM, Aswathy V, Bhagyasree S, Mahesh A (2024) Formulation and evaluation of anti ageing cream from herbal extract. World J Pharm Res 13:943–956

Verma A, Kumar A, Kumar R (2023) Medicinal plants as source of anti-ageing agents: a review. World J Pharm Pharm Sci 12:946–957

Viswanathan M, Kim SK, Berdichevsky A, Guarente L (2005) A role for SIR-2.1 regulation of ER stress response genes in determining *C. elegans* life span. Dev Cell 9:605–615

Wang X, Meng L, Zhao L, Wang Z, Liu H, Liu G, Guan G (2017) Resveratrol ameliorates hyperglycemia-induced renal tubular oxidative stress damage via modulating the SIRT1/FOXO3a pathway. Diabetes Res Clin Pract 126:172–181

Wang L, Du J, Zhao F, Chen Z, Chang J, Qin F, Wang Z, Wang F, Chen N (2018) *Trillium tschonoskii* maxim saponin mitigates D-galactose-induced brain aging of rats through rescuing dysfunctional autophagy mediated by Rheb-mTOR signal pathway. Biomed Pharmacother 98:516–522

Wang Z, Sun Q, Fang J, Wang C, Wang D, Li M (2022) The anti-aging activity of *Lycium barbarum* polysaccharide extracted by yeast fermentation: *in vivo* and *in vitro* studies. Int J Biol Macromol 209:2032–2041

Wanyu L (2025) Advancements in research on the anti-aging effects and mechanisms of flavonoids in natural products. Front Med 12:1637992

Weng Y, Xiang L, Matsuura A, Zhang Y, Huang Q, Qi J (2010) Ganodermasides A and B, two novel anti-aging ergosterols from spores of a medicinal mushroom *Ganoderma lucidum* on yeast via UTH1 gene. Bioorg Med Chem 18:999–1002

Westphal CH, Dipp MA, Guarente L (2007) A therapeutic role for sirtuins in diseases of aging? Trends Biochem Sci 32:55–560

Willcox BJ, Donlon TA, He Q, Chen R, Grove JS, Yano K, Masaki KH, Willcox DC, Rodriguez B, Curb JD (2008) FOXO3A genotype is strongly associated with human longevity. Proc Natl Acad Sci USA 105:13987–13992

Wrighton KH (2011) Staying alive without CRTC-1. Nat Rev Mol Cell Biol 12:206–207

Xia Y, Zhang H, Wu X, Xu Y, Tan Q (2024) Resveratrol activates autophagy and protects from UVA-induced photoaging in human skin fibroblasts and the skin of male mice by regulating the AMPK pathway. Biogerontology 25:649–664

Xu Y, Yuan H, Luo Y, Zhao YJ, Xiao JH (2020) Ganoderic acid D protects human amniotic mesenchymal stem cells against oxidative stress-induced senescence through the PERK/NRF2 signaling pathway. Oxidative Med Cell Longev 2020:8291413

Yaguchi Y, Komura T, Kashima N, Tamura M, Kage-Nakadai E, Saeki S, Terao K, Nishikawa Y (2014) Influence of oral supplementation with sesamin on longevity of *Caenorhabditis elegans* and the host defense. Eur J Nutr 53:1659–1668

Yang S, Sun J, Gu D, Li P, Yao L, Shi D, Guo S, Liu C (2021) Antioxidant activities of sulfated Codonopsis polysaccharides in acute oxidative stress. J Food Biochem 45:e13974

Yingngam B, Supaka N, Rungseevijitprapa W (2011) Estrogen-like activities and cytotoxicity effects of Thai herbal medicines as natural ingredients in anti-ageing. J Med Plants Res 5:6832–6838

Yousefzadeh MJ, Zhu YI, McGowan SJ, Angelini L, Fuhrmann-Stroissnigg H, Xu M, Linga YY, Melosa K, Pirtskhalavab T, Inmanb CL, McGuckian C, Wade EA, Kato JI, Grassi D, Wentworth M, Burd CE, Arriaga EA, Ladiges WL, Tchkonia T, Kirkland JL, Robbins PD, Niedernhofer LJ (2018) Fisetin is a senotherapeutic that extends health and lifespan. EBioMedicine 36:18–28

Yu X, Li G (2012) Effects of resveratrol on longevity, cognitive ability and aging-related histological markers in the annual fish *Nothobranchius guentheri*. Exp Gerontol 47:940–949

Yu S, Xia H, Guo Y, Qian X, Zou X, Yang H, Yin M, Liu H (2020) Ginsenoside Rb1 retards aging process by regulating cell cycle, apoptotic pathway and metabolism of aging mice. J Ethnopharmacol 255:112746

Yu X, Liu Z, Sun X (2023) Single-cell and spatial multi-omics in the plant sciences: technical advances, applications, and perspectives. Plant Commun 4:100508

Yuan H, Xu Y, Luo Y, Zhang JR, Zhu XX, Xia JH (2022) Ganoderic acid D prevents oxidative stress induced senescence by targeting 14-3-3ε to activate CaM/CaMKII/NRF2 signaling pathway in mesenchymal stem cells. Aging Cell 21:e13686

Zaki AH, Zahid MT, Haiying B (2022) Bioactive compounds of the culinary-medicinal mushroom *Leucocalocybe mongolica* (Agaricomycetes): pharmacological and therapeutic applications—a review. Int J Med Mushrooms 24:19–33

Zamzury SN, Omar MR (2019) Healthy ageing in women: anti-ageing properties of common herbs from modern and Islamic perspective. Int J Stud Child Women Elder Disabl 7:17–22

Zhang HQ, Li Y, Song YY (2011) Effect of polysaccharides of *Cistanche deserticola* on immune cells and telomerase activity in aging mice. Chin Pharm J 14:1081–1083

Zhang Y, Lv T, Li M, Xue T, Liu H, Zhang W, Ding X, Zhuang Z (2015) Anti-aging effect of polysaccharide from *Bletilla striata* on nematode *Caenorhabditis elegans*. Pharmacogn Mag 11:449

Zhang L, Yousefzadeh MJ, Suh Y, Niedernhofer LJ, Robbins PD (2019) Signal transduction, ageing and disease. In: Biochemistry and cell biology of ageing: Part II clinical science. Springer, Singapore, pp 227–247

Zhang Z, Gao T, Yan N, Duan Z, Tang Z, Zhou L, Chen T, Feng S, Ding C, Yuan S, Yuan M (2023a) Characterization and anti-aging activity of polysaccharides from *Akebia trifoliata* fruit separated by an aqueous two-phase system. Plant Foods Hum Nutr 78:154–159

Zhang F, Ge W, Huang L, Li D, Liu L, Dong Z, Xu L, Ding X, Zhang C, Sun Y, Jun A, Gao J, Guo T (2023b) A comparative analysis of data analysis tools for data-independent acquisition mass spectrometry. Mol Cell Proteomics 22:100623

Zhang Y, Wang X, Huang J, Zhang X, Bu L, Zhang Y, Liang F, Wu S, Zhang M, Zhang L, Zhang L (2024) CASIN exerts anti-aging effects through RPL4 on the skin of naturally aging mice. Aging Cell 23:e14333

Zheng WV, Xu W, Li Y, Qin J, Zhou T, Li D, Xu Y, Cheng X, Xiong Y, Chen Z (2022) Anti-aging effect of β-carotene through regulating the KAT7-P15 signaling axis, inflammation and oxidative stress process. Cell Mol Biol Lett 27:86

Zhou Y, Jiang R, Yang B, Yao X, Wang P, Liu D, Wang Y (2011) Changes of telomere and telomerase in effect of ginsenoside Rg1 to delay hematopoietic stem cell senescence. China J Chin Mater Med 36:3172–3175

Zhou X, Yang Q, Xie Y, Sun J, Hu J, Qiu P, Cao W, Wang S (2015) Tetrahydroxystilbene glucoside extends mouse life span via upregulating neural klotho and downregulating neural insulin or insulin-like growth factor 1. Neurobiol Aging 36:1462–1470

Zhou Y, Wang YP, He YH, Ding JC (2020) Ginsenoside Rg1 performs anti-aging functions by suppressing mitochondrial pathway-mediated apoptosis and activating sirtuin 3 (SIRT3)/superoxide dismutase 2 (SOD2) pathways in Sca-1+ HSC/HPC cells of an aging rat model. Med Sci Monit 26:e920666–e920661

MIX
Papier aus verantwortungsvollen Quellen
Paper from responsible sources
FSC® C105338
FSC
www.fsc.org

If you have any concerns about our products,
you can contact us on
ProductSafety@springernature.com

In case Publisher is established outside the EU,
the EU authorized representative is:
Springer Nature Customer Service Center GmbH
Europaplatz 3, 69115 Heidelberg, Germany

Printed by Libri Plureos GmbH
in Hamburg, Germany